Dichotomies and Stability in Nonautonomous Linear Systems

Stability and Control: Theory, Methods and Applications

A series of books and monographs on the theory of stability and control

Edited by A.A. Martynyuk
Institute of Mechanics, Kiev, Ukraine

and V. Lakshmikantham
Florida Institute of Technology, USA

Volume 1
Theory of Integro-Differential Equations
V. Lakshmikantham and M. Rama Mohana Rao

Volume 2
Stability Analysis: Nonlinear Mechanics Equations
A.A. Martynyuk

Volume 3
Stability of Motion of Nonautonomous Systems (Method of Limiting Equations)
J. Kato, A.A. Martynyuk and A.A.Shestakov

Volume 4
Control Theory and its Applications
E.O. Roxin

Volume 5
Advances in Nonlinear Dynamics
Edited by S. Sivasundaram and A.A. Martynyuk

Volume 6
Solving Differential Problems by Multistep Initial and Boundary Value Methods
L. Brugnano and D. Trigiante

Volume 7
Dynamics of Machines with Variable Mass
L. Cveticanin

Volume 8
Optimization of Linear Control Systems: Analytical Methods and Computational Algorithms
F.A. Aliev and V.B. Larin

Volume 9
Dynamics and Control
Edited by G. Leitmann, F.E. Udwadia and A.V. Kryazhimskii

Volume 10
Volterra Equations and Applications
Edited by C. Corduneanu and I.W. Sandberg

Volume 11
Nonlinear Problems in Aviation and Aerospace
Edited by S. Sivasundaram

Volume 12
Stabilization of Programmed Motion
E.Ya. Smirnov

Please see the back of this book for other titles in the Stability and Control: Theory, Methods and Applications series.

Dichotomies and Stability in Nonautonomous Linear Systems

Yu. A. Mitropolsky, A. M. Samoilenko
and V. L. Kulik
Institute of Mathematics, Kiev, Ukraine

Taylor & Francis
Taylor & Francis Group
LONDON AND NEW YORK

First published 2003
by Taylor & Francis
11 New Fetter Lane, London EC4P 4EE

Simultaneously published in the USA and Canada
by Taylor & Francis Inc,
29 West 35th Street, New York, NY 10001

Taylor & Francis is an imprint of the Taylor & Francis Group

© 2003 Taylor & Francis

Publisher's Note
This book has been prepared from camera-ready copy provided by the authors

Printed and bound in Great Britain by TJ International Ltd, Padstow, Cornwall

All rights reserved. No part of this book may be reprinted or
reproduced or utilized in any form or by any electronic,
mechanical, or other means, now known or hereafter
invented, including photocopying and recording, or in any
information storage or retrieval system, without permission in
writing from the publishers.

Every effort has been made to ensure that the advice and information in this
book is true and accurate at the time of going to press. However, neither
the publisher nor the authors can accept any legal responsibility or
liability for any errors or omissions that may be made. In the case of
drug administration, any medical procedure or the use of technical equipment
mentioned within this book, you are strongly advised to consult the
manufacturer's guidelines.

British Library Cataloguing in Publication Data
A catalogue record for this book is available from the British Library

Library of Congress Cataloging in Publication Data
A catalog record has been requested

ISBN 0-415-27221-1

Contents

Introduction to the Series   ix

Preface   xi

An Overview of the Book   xiii

1 Exponentially Dichotomous Linear Systems of Differential Equations and Lyapunov Functions of Variable Sign   1

    1.0 Introduction   1

    1.1 Exponential Dichotomy Criterion for Linear Systems in Terms of Quadratic Forms   2

    1.2 Decomposition over the Whole R Axis of Linear Systems of Differential Equations Exponentially Dichotomous on Semiaxes R_+ and R_- 17

    1.3 Degeneracy of the Quadratic Form Possessing a Definite-Sign Derivative along the Solutions of the System (1.1.1)   30

    1.4 Integral Representation of Weakly Regular Systems Bounded on the Whole R Axis   41

    1.5 Complement to the Exponentially Dichotomous of Weakly Regular on R Linear Systems   50

    1.6 Regularity of Linear Systems of the Block-Triangular Form   67

    1.7 Perturbation of the Block-Triangular Form Linear Systems which are Regular and Weakly Regular on the Whole R Axis   76

CONTENTS

1.8	Exponentially Dichotomous Linear Systems with Parameters	86
1.9	Comments and References	92

2 Linear Extension of Dynamical Systems on a Torus — **93**

2.0	Introduction	93
2.1	Necessary Existence Conditions for Invariant Tori	94
2.2	The Green Function. Sufficient Existence Conditions for an Invariant Torus	107
2.3	Existence Conditions for an Exponentially Stable Invariant Torus	111
2.4	Uniqueness Conditions for the Green Function and its Properties	116
2.5	Sufficient Conditions for Exponential Dichotomy of the Invariant Torus	125
2.6	Necessary Conditions for Exponential Dichotomy of the Invariant Torus	132
2.7	Existence Criterion for the Green Function	138
2.8	The Non-unique Green Function and the Properties of the System Implied by its Existence	153
2.9	Invariant Tori of Linear Extensions with Slowly Changing Phase	169
2.10	Preserving the Green Function under Small Perturbations of Linear Expansions on a Torus	177
2.11	On the Smoothness of an Exponentially Stable Invariant Torus	191
2.12	On the Dependence of Green Functions on Parameters	203
2.13	Continuity and Differentiability of the Green Function	212
2.14	Invariant Tori of Linear Extensions with a Degenerate Matrix at the Derivatives	222
2.15	Bounded Invariant Manifolds of Dynamical Systems and their Smoothness	231
2.16	Comments and References	250

3 Splittability of Linear Extensions of Dynamical Systems on a Torus — **253**

3.0	Introduction	253

3.1	Sufficient Conditions for Splittability of Linear Extensions of Dynamical Systems on a Torus	254
3.2	Reversibility of the Theorem on Splittability	263
3.3	On Triangulation and the Relationship of C^l-block Splittability of a Linear System with the Problem on r-frame Complementability up to the Periodic Basis in R^n	271
3.4	Reducing of Linearized Systems to a Diagonal Form	278
3.5	On the Relationship of Exponentially Dichotomous Linear Expansions with the Algebraic System Solvability	287
3.6	Three Block Divisibility of Linear Extensions and Lyapunov Functions of Variable Sign	293
3.7	Algebraic Problems of k-blocked Divisibility of Linear Extensions on a Torus	301
3.8	Comments and References	305

4 Problems of Perturbation Theory of Smooth Invariant Tori of Dynamical Systems — 307

4.0	Introduction	307
4.1	Solution Variations on the Manifold M	308
4.2	Exponential Stability and Dichotomy Conditions for Linear Extensions of Dynamical Systems on a Torus	316
4.3	Roughness Conditions for the Green Function of the Linear Extension of a Dynamical System on a Torus with the Index of Smoothness l	320
4.4	A Theorem of Perturbation Theory of an Invariant Torus of a Dynamical System	329
4.5	Green Function for a Linear Matrix Equation	330
4.6	On the Problem of Structure of Some Regular Linear Extensions of Dynamical Systems on a Torus	335
4.7	Invariant Manifolds of Autonomous Differential Equations and Lyapunov Functions with Alternating Signs	343
4.8	Comments and References	353

References — 355

Index — 367

Introduction to the Series

The problems of modern society are both complex and interdisciplinary. Despite the apparent diversity of problems, tools developed in one context are often adaptable to an entirely different situation. For example, consider the Lyapunov's well known second method. This interesting and fruitful technique has gained increasing significance and has given a decisive impetus for modern development of the stability theory of differential equations. A manifest advantage of this method is that it does not demand the knowledge of solutions and therefore has great power in application. It is now well recognized that the concept of Lyapunov-like functions and the theory of differential and integral inequalities can be utilized to investigate qualitative and quantitative properties of nonlinear dynamic systems. Lyapunov-like functions serve as vehicles to transform the given complicated dynamic systems into a relatively simpler system and therefore it is sufficient to study the properties of this simpler dynamic system. It is also being realized that the same versatile tools can be adapted to discuss entirely different nonlinear systems, and that other tools, such as the variation of parameters and the method of upper and lower solutions provide equally effective methods to deal with problems of a similar nature. Moreover, interesting new ideas have been introduced which would seem to hold great potential.

Control theory, on the other hand, is that branch of application-oriented mathematics that deals with the basic principles underlying the analysis and design of control systems. To control an object implies the influence of its behavior so as to accomplish a desired goal. In order to implement this influence, practitioners build devices that incorporate various mathematical techniques. The study of these devices and their interaction with the object being controlled is the subject of control theory. There have been, roughly speaking, two main lines of work in control theory which are complementary. One is based on the idea that a good model of the object to be controlled is available and that we wish to optimize its behavior, and the other is based on the constraints imposed by uncertainty about the model in which the object operates. The control tool in the latter is the use of feedback in order to correct for deviations from the desired behavior. Mathematically, stability theory, dynamic systems and funtional analysis have had a strong influence on this approach.

Due to the increased interdependency and cooperation among the mathematical sciences across the traditional boundaries, and the accomplishments thus far achieved in the areas of stability and control, there is every reason to believe that many breakthroughs await us, offering existing prospects for these versatile techniques to advance further. It is in this spirit that we see the importance of the 'Stability and Control' series, and we are immensely thankful to Taylor & Francis for their interest and cooperation in publishing this series.

Preface

The investigation of bounded solutions to systems of differential equations involves some important toroidal manifolds. This is due to the fact that many stationary processes in mechanics, chemistry and biology are modelled by bounded solutions of differential equations. Also, establishing general conditions under which the system of differential equations possesses one or more solutions bounded on the whole axis is a typical problem. Moreover, it is of importance that this property be preserved when small changes in the system occur.

This volume deals with the application of Lyapunov functions with alternating signs in the form of quadratic forms to the solution of the following problems:

- the existence of solutions to linear systems of differential equations with variable coefficients bounded in the whole axis;
- establishing the conditions for preserving the invariant tori of a dynamical system under perturbations;
- the separation of normal variables in the linear extensions of dynamical systems on a torus.

It should be noted that the investigations of perturbation theory of the invariant tori of dynamical systems presented in the second chapter required fascinating new problems to be solved in the theory of dichotomous linear systems of differential equations with variable coefficients.

In Chapter 1 a theorem is proved on the transformation on the whole axis of every linear nonautonomous system of differential equations which are exponentially dichotomous only on both semiaxes to a specific block-triangular form. This allowed the necessary and sufficient condition to be established for the inhomogeneous linear system to have at least one solution bounded on the whole axis. The criterion is formulated in terms of alternating sign, degenerate quadratic forms, which have a definite sign derivative by virtue of the conjugated system.

Chapter 2 presents a new idea for the investigation of the weakly regular linear expansions of dynamical systems on a torus which provides a complement to the regular systems. On this basis, new possibilities have been opened up for the intriguing analysis of smoothness and parameter dependence of the invariant tori of weakly regular linear expansions of dynamical systems.

Chapter 3 discusses the approach to qualitative problems of the separation of normal variables in linear extensions of the dynamical systems on a torus. The interest in these problems has been evoked by the extensive investigations the possibility of splitting the linear exponentially dichotomous system with quasiperiodic coefficients by the quasiperiodic change of variables. These investigations revealed that the properties of sign-alternating Lyapunov functions, which provide the means to study the exponential dichotomy of linear expansions of a torus, influence the possibility of splitting normal variables in these extensions to a great extent. There relationships are also discussed in Chapter 3.

Chapter 4 presents new results which were obtained recently by the authors after the publication of the Russian edition of the book, which relate to the problems of the existence of Lyapunov functions with alternating signs, the structure of regular linear extensions on a torus, and the variation of the family of solutions to the dynamical systems starting on a torus.

Direct and converse theorems on the exponential dichotomy on the whole axis of linear systems of differential equations with variable coefficients are proved for the first time in terms of non-degenerate Lyapunov functions with alternating signs in the form of quadratic forms.

The properties of weak regularity on the whole axis of linear systems of differential equations are established by means of sign-alternating degenerate Lyapunov functions. An integral representation is first derived for all invariant tori of weakly regular linear extensions of dynamical systems.

A deep relationship is stated between the properties of sign-definite Lyapunov functions and the ability to split linear extensions of dynamical systems on a torus.

New structures are found for regular linear extensions. Criteria are proved for exponential dichotomy of linear extensions on a torus. New theorems are first deduced on the existence of Lyapunov functions with alternating signs.

The authors would like to express their sincere gratitude to Professors V.M.Millionshchikov (Moscow), V.A.Pliss (St.Peterburg) and A.A.Martynyuk (Kiev) for their valuable discussions of the results presented in this volume and to the professional staff of Taylor & Francis.

Yu. A. Mitropolsky
A. M. Samoilenko
V. L. Kulik

An Overview of the Book

Modern problems of nonlinear dynamics speed the development of new mathematical techniques for the investigation of systems of differential equations. In this framework the analysis of boundedness of solutions to these equations is of topical interest. The main problem which arises is the investigation of conditions for which the invariant tora of the dynamical systems are preserved under perturbations. Many stationary processes observed in mechanics, electronics, authomatics, chemistry, biology, astronomy, geology and economics are characterized by bounded solutions to systems of nonlinear differential equations. In many cases these systems are specifically nonlinear and so a specific approach is required for their investigation. In this regard the linearization of such systems proves to be a method which develops the theory embracing many nonlinear systems of differential equations.

Note that the study of invariant tora of the dynamical systems has involved new problems of the theory of linear systems of differential equations with variable coefficients. In turn the solution of the new problems for such systems has given an impetus for new ideas in the theory of invariant tora.

This monograph deals with the application of Lyapunov functions with variable sign (as quadratic forms) in solution of the following problems of the theory of differential equations:
 – the problem of existence of bounded on the axis solutions to the linear systems of differential equations with variable coefficients;
 – the problem of preserving the invariant tora of the dynamical systems under perturbations;
 – the problem of separation of normal variables in linear expansions of the dynamical systems on torus.

We shall now give a brief description of the monograph.

First, the linear system of differential equations

$$\frac{dx}{dt} = A(t)x, \quad x \in R^n \tag{1}$$

is considered with the ficient matrix $A(t)$ continuous and bounded on $R = (-\infty, +\infty)$. One of the main problems in this case is which conditions the matrix $A(t)$ should satisfy so that the nonhomogeneous system

$$\frac{dx}{dt} = A(t)x + f(t) \tag{2}$$

possess solutions bounded on R for every fixed vector function $f(t)$ continuous and bounded on R?

When only semiaxis $t \geq 0$ is considered this problem is well studied (see, for example, Daletskii and Krein [1], Massera and Schaeffer [1], Majzel [1], etc.). It was established that for the system (2) to possess for $t \geq 0$ a solution bounded on the semiaxis R_+ for every vector function $f(t)$ continuous and bounded on R_+ it is necessary and sufficient that system (2) be exponentially dichotomous on R_+, i.e. roughly speaking, a part of solutions on $+\infty$ tends to zero uniformly exponentially while the other part tends to $-\infty$. This is equivalent to the existence of the quadratic form

$$V = \langle S(t)x, x \rangle = s_{11}(t)x_1^2 + 2s_{12}(t)x_1x_2 + \cdots \tag{3}$$

with the coefficient matrix $S(t)$ continuously differentiable and bounded on R_+. The quadratic form has the derivative of variable sign by virtue of system (1), i.e. the inequality

$$\dot{V} = \langle (\dot{S}(t) + S(t)A(t) + S(t)A^*(t))x, x \rangle \leq -\|x\|^2 \tag{4}$$

is satisfied for all $t \geq 0$ and $x \in R^n$. Note that no conditions are required for degeneracy of the matrix $S(t)$ (this matrix can degenerate for some values of t).

The investigation of solutions to system (2) bounded on the whole axis $R = (-\infty, +\infty)$ by means of Lyapunov functions (3) caused some discussions in mathematical papers.

Some papers claimed that condition (4) for all $t \in R$ is sufficient for system (2) to possess a solution bounded on R. The other paper indicated that non-degeneracy of the matrix $S(t)$

$$\det S(t) \neq 0, \quad \forall t \in R \tag{5}$$

and the even more strong assumption

$$|\det S(t)| \geq \delta = \text{const} > 0 \tag{6}$$

does not yet guarantee the system (2) having solutions bounded on R for every fixed $f(t)$.

AN OVERVIEW OF THE BOOK

In the first chapter it is proved that conditions (4) and (5) with n-dimensional symmetric matrix $S(t)$ continuously differential and bounded ensure exponential dichotomy of system (1) on R. This assertion is equivalent to the fact that system (2) has a unique solution bounded on R for every fixed vector function $f(t)$ continuous and bounded on R. It is also proved that every matrix $S(t)$ satisfying conditions (4) and (5) satisfies estimate (6) as well.

Prior to the investigation of the sets of solutions to system (2) bounded on R the systems (1) exponentially dichotomous simulteneously on semiaxies $R_+ = [0, \infty)$ and $R_- = (-\infty, 0]$ should be studied. In this regard a question arises whether such systems can be simplified by means of the Lyapunov transformation

$$x = L(t)y. \tag{7}$$

It is well known that the Lyapunov transformation can always reduce every system (1) exponentially dichotomous on R to the splitted form

$$\frac{dy_1}{dt} = B^+(t)y_1, \quad \frac{dy_2}{dt} = B^-(t)y_2, \quad y_i \in R^{\Gamma_i}, \quad \Gamma_1 + \Gamma_2 = n, \tag{8}$$

where all nontrivial solutions of system

$$\frac{dy_1}{dt} = B^+(t)y_1$$

exponentially damp to zero on $+\infty$ and increase on $-\infty$ while all nontrivial solutions of system

$$\frac{dy_2}{dt} = B^-(t)y_2$$

exponentially damp to zero on $-\infty$ and increase on $+\infty$.

In the second chapter it is proved that every system (1) exponentially dichotomous simulteneously on the semiaxes R_+ and R_- can always be reduced by the Lyapunov transformation of variables (7) on the whole axis R to the form

$$\begin{aligned}\frac{dy_1}{dt} &= B^+(t)y_1 + B_1(t)\bar{y}, \\ \frac{dy_2}{dt} &= B^-(t)y_2 + B_2(t)\bar{y}, \\ \frac{dy_3}{dt} &= \hat{B}(t)y_3, \\ \frac{dy_4}{dt} &= \check{B}(t)y_4, \quad \bar{y} = (y_3, y_4),\end{aligned} \tag{9}$$

where $y_i \in R^{\Gamma_i}$, $\Gamma_1 + \Gamma_2 + \Gamma_3 + \Gamma_4 = n$. Moreover, systems

$$\frac{dy_3}{dt} = \hat{B}(t)y_3$$

are exponentially damping to zero on $+\infty$ and $-\infty$, and all nontrivial solutions of system
$$\frac{dy_4}{dt} = \check{B}(t)y_4$$
are exponentially increasing on $+\infty$ and $-\infty$, the example of such systems are the equations
$$\frac{dy_3}{dt} = -(\tanh t)y_3, \quad \frac{dy_4}{dt} = (\tanh t)y_4.$$

Thus, we can understand what prevents system (2) from having solutions bounded on R, namely, the presence of system
$$\frac{dy_4}{dt} = \check{B}(t)y_4$$
does. On the other hand the presence of the quadratic form (3) with condition (4) on R excludes the existence of nontrivial solutions of system (1) bounded on R. This allows us to obtain necessary and sufficient existence conditions for more than one solution to nonhomogeneous system (2) bounded on R. Namely, for system (2) to have at least one solution bounded on R for every fixed vector function $f(t)$ continuous and bounded on R, it is necessary and sufficient that the quadratic form

$$\bar{V} = \langle \bar{S}(t)y, y \rangle = \bar{s}_{11}(t)y_1^2 + 2\bar{s}_{12}(t)y_1 y_2 + \cdots \tag{10}$$

exist with the coefficient matrix $\bar{S}(t)$ continuously differentiable and bounded on R. The quadratic form has a derivative of definite sign along solutions of system

$$\frac{dy}{dt} = -A^*(t)y, \quad y \in R^n \tag{11}$$

conjugated with system (1), i.e.

$$\dot{V} = \langle (\dot{\bar{S}}(t) - \bar{S}(t)A^*(t) - \bar{S}(t)A^*(t))y, y \rangle \geq \|x\|^2. \tag{12}$$

Moreover, the matrix $\bar{S}(t)$ can degenerate in some points $t = t_1, t_2, \ldots, t_k$.

It is proved that always $k \leq n$. If the matrix $\bar{S}(t)$ degenerates for $t = t_i$, the system (1) has nontrivial solutions bounded on R and all these solutions are exponentially damping on $+\infty$ and $-\infty$. Besides, the quadratic form (3) with estimate (4) does not exist on the whole R.

In the case when $\det \bar{S}(t) \neq 0$, $\forall t \in R$ both systems (11) and (1) are exponentially dichotomous on R and the matrix $\check{S}(t) = -\gamma \bar{S}^{-1}(t)$, where $\gamma = \text{const} > 0$, can be taken as the matrix $S(t)$ in the quadratic form (3).

It is known that, if system (1) is exponentially dichotomous on R, then all solutions to nonhomogeneous system (2) bounded on R can be represented as

$$x = x^*(t) = \int_{-\infty}^{+\infty} G(t,\tau)f(\tau)\,d\tau, \tag{13}$$

where $G(t,\tau)$ is the Green function of the problem on solutions to system (2) bounded on R. It has the following structure

$$G(t,\tau) = \begin{cases} \Omega_0^t P \Omega_\tau^0, & \tau \leq t, \\ \Omega_0^t (P - I_n) \Omega_\tau^0, & \tau > t, \end{cases} \qquad (14)$$

where Ω_0^t is the matriciant of the linear system (1), $\Omega_0^t\big|_{t=0} = I_n$ is an n-dimensional matrix projected on the subspace E^+ along E^- ($P^2 = P$). Note that due to system (1) being dichotomous on R the Green function (14) is unique and satisfies the estimate

$$\|G(t,\tau)\| \leq K e^{-\gamma |t-\tau|}, \quad K, \gamma = \text{const} > 0. \qquad (15)$$

Obviously, the Green function (14) can be written as

$$G(t,\tau) = \begin{cases} \Omega_\tau^t C(\tau), & \tau \leq t, \\ \Omega_\tau^t [C(\tau) - I_n], & \tau > t, \end{cases} \qquad (16)$$

where

$$C(\tau) = \Omega_0^\tau P \Omega_\tau^0. \qquad (17)$$

Besides, $C^2(\tau) \equiv C(\tau)$.

Assume now that nothing is known about the dichotomy properties of system (1). One only knows that there exists a $n \times n$-dimensional continuous matrix $C(\tau)$ such that the function (14) satisfies estimate (15). Evidently, this is sufficient for the equality (13) to determine the solution to nonhomogeneous system (2) bounded on R. In this regard a question arises about the possible structure of the matrix $C(\tau)$. This problem is investigated in section 1.4.

The analysis of the dependence on parameters of solutions to system (2) bounded on R is expedient in the case when the Green function (14) is unique. If not, the separation of variables by the Lyapunov transformation (7) does not yield the desired results, since the size of the block matrices $\hat{B}$ and $\check{B}$ in system (9) may vary depending on parameters. Therefore a new idea has been proposed for the investigation of systems (16) with estimate (15). The idea was: if the system (2) has at least one solution bounded on R for every fixed vector function $f(t)$, then the extended system

$$\begin{aligned} \frac{dx}{dt} &= A(t)x + f(t), \\ \frac{dy}{dt} &= x - A^*(t)y + g(t) \end{aligned} \qquad (18)$$

always has a unique solution bounded on R for every fixed vector function $(f(t), g(t))$. In other words if system (2) has many different Green functions (16) (without condition (17)), then the extended system (18) has only one Green function.

Now we shall give a brief account of some results in theory of the dynamical system extensions on torus. To this end the system of differential equations

$$\frac{d\varphi}{dt} = a(\varphi), \quad \frac{dx}{dt} = A(\varphi)x, \quad x \in R^n \tag{19}$$

is considered, where the vector function $a(\varphi)$ and the matrix function $A(\varphi)$ are given on m-dimensional torus Γ_m, $\varphi = (\varphi_1, \ldots, \varphi_m) \in \Gamma_m$ and continuous with respect to the totality of variables $\varphi_1, \ldots, \varphi_m$. Aditionally it is assumed on the vector function $a(\varphi)$ that the Cauchy problem

$$\frac{d\varphi}{dt} = a(\varphi), \quad \varphi|_{t=0} = \varphi_0$$

for every fixed $\varphi_0 \in \Gamma_m$ has a unique solution $\varphi_t(\varphi_0)$.

The following generalizations are used: $C^0(\Gamma_m)$ is a space of functions $F(\varphi)$ which are continuous with respect to the totality of variables $(\varphi_1, \ldots, \varphi_m)$ and 2π-periodic with respect to every variable φ_j, $j = \overline{1, m}$, i.e. $F(\varphi)$ are given on m-dimensional torus Γ_m, and $C'(\Gamma_m; a)$ is a subspace of $C^0(\Gamma_m)$ containing functions $F(\varphi)$ such that the superposition $F(\varphi_t(\varphi_0))$ being a function of the variable $t \in R$ is continuously differentiable in t and

$$\dot{F}(\varphi_0) \stackrel{\text{def}}{=} \frac{d}{dt} F(\varphi_t(\varphi_0))\bigg|_{t=0}.$$

We shall claim that system (19) has the Green function $G(\tau, \varphi)$ of the problem on invariant tora, if there exists a $n \times n$-dimensional matrix function $C(\varphi) \in C^0(\Gamma_m)$ such that for the function

$$G_0(\tau, \varphi) = \begin{cases} \Omega_\tau^0 C(\varphi_\tau(\varphi)), & \tau \leq 0, \\ \Omega_\tau^0(\varphi)[C(\varphi_\tau(\varphi)) - I_n], & \tau > 0, \end{cases} \tag{20}$$

the estimate

$$\|G_0(\tau, \varphi)\| \leq Ke^{-\gamma|\tau|} \tag{21}$$

holds true, where $K, \gamma = \text{const} > 0$ and $\Omega_\tau^t(\varphi)$ is the matriciant of the linear system

$$\frac{dx}{dt} = A(\varphi_\tau(\varphi))x, \quad \Omega_\tau^t(\varphi)|_{t=\tau} = I_n. \tag{22}$$

In the case when the Green function (20) with estimate (21) is unique the *system* (19) is called *regular*. If it is known that there exists at least one Green function (20) with estimate (21), the *system* (19) is called *weakly regular*.

The precise wordings of some assertions are as follows.

Theorem 1 *Let there exist a n-dimensional symmetric matrix $S(\varphi) \in C'(\Gamma_m; a)$ satisfying conditions*

$$\det S(\varphi) \neq 0, \quad \forall \varphi \in \Gamma_m, \tag{23}$$

$$\langle [\dot{S}(\varphi) + S(\varphi)A(\varphi) + S(\varphi)A^*(\varphi)]x, x \rangle \geq \|x\|^2, \quad \forall x \in R^n. \tag{24}$$

Then the system (19) is regular. Besides, the matrix function $C(\varphi)$ in the structure of the Green function (20) has the properties

$$C^2(\varphi) \equiv C(\varphi), \quad C(\varphi_\tau(\varphi)) \equiv \Omega_0^t(\varphi)C(\varphi)\Omega_t^0(\varphi) \tag{25}$$

and the constants K and γ in estimate (21) can be represented by the equalities

$$K = (2 + \sqrt{2})(\|A\|_0\|S\|_0)^{3/2}, \quad \gamma = (2\|S\|_0)^{-1},$$

where $\|A\|_0 = \max\limits_{\varphi \in \Gamma_m; \|x\|=1} \|A(\varphi)x\|$.

Theorem 2 *Let system (19) be weakly regular and one of the identities (25) be satisfied. Then*

(1) *the other identity (25) is satisfied;*

(2) *system (19) is regular;*

(3) *there exist symmetric matrices $S(\varphi) \in C'(\Gamma_m; a)$ satisfying condition (24) and all these matrices are non-degenerated.*

Theorem 3 *For system (19) to be weakly regular it is necessary and sufficient that n-dimensional symmetric matrix $\bar{S}(\varphi) \in C'(\Gamma_m; a)$ exist satisfying condition*

$$\langle [\dot{\bar{S}}(\varphi) - A^*(\varphi)\bar{S}(\varphi) - A(\varphi)\bar{S}(\varphi)]y, y \rangle \geq \|y\|^2 \tag{26}$$

for all $y \in R^n$, $\varphi \in \Gamma_m$.

Condition (26) means that the derivative of the quadratic form $\langle \bar{S}(\varphi)y, y \rangle$ along solutions of system

$$\frac{d\varphi}{dt} = a(\varphi), \quad \frac{dy}{dt} = -A^*(\varphi)y$$

is positive definite. Besides, the system $\langle \bar{S}(\varphi)y, y \rangle$ can change its sign when $y \in R^n$ is changed and even the matrix $\bar{S}(\varphi)$ can degenerate for some $\varphi = \varphi_0 \in \Gamma_m$.

Theorem 4 *Let there exist two n-dimensional symmetric matrices $S(\varphi), \bar{S}(\varphi) \in C'(\Gamma_m; a)$ satisfying inequalities (24) and (26). Then $\det S(\varphi) \neq 0$ and system (19) is regular.*

Theorem 5 *Let system (19) be weakly regular. Then the system of equations*

$$\frac{d\varphi}{dt} = a(\varphi), \quad \frac{dx}{dt} = A(\varphi)x, \quad \frac{dy}{dt} = x - A^*(\varphi)y \tag{27}$$

is regular. Besides, the derivative of non-degenerate quadratic form

$$p\langle x, y \rangle + \langle \bar{S}(\varphi)y, y \rangle, \quad p > 0$$

by virtue of system (27) is positive definite for sufficiently large values of the parameter $p > 0$.

Theorem 6 *Let there exist n-dimensional symmetric matrix $\bar{S}(\varphi) \in C'(\Gamma_m; a)$ satisfying inequality (26) and moreover $\det \bar{S}(\varphi_0) = 0$ for some value $\varphi = \varphi_0 \in \Gamma_m$. Then*

(1) *system (19) has an infinite number of different Green functions (20) with estimate (21);*

(2) *system of equations*

$$\frac{d\varphi}{dt} = a(\varphi), \quad \frac{dx}{dt} = A(\varphi)x + f(\varphi)$$

has for every fixed vector function $f(\varphi) \in C^0(\Gamma_m)$ an infinite set of invariant tora

$$x = \int_{-\infty}^{+\infty} G_0(\tau, \varphi) f(\varphi_\tau(\varphi)) \, d\tau + \int_{-\infty}^{+\infty} W(\varphi)(\Omega_0^\tau(\varphi))^* g(\varphi_\tau(\varphi)) \, d\tau,$$

where $g(\varphi) \in C^0(\Gamma_m)$ is an arbitral vector function and $W(\varphi)$ is a n-dimensional symmetric matrix with the properties

$$\|\Omega_0^t(\varphi) W(\varphi)(\Omega_0^\tau(\varphi))^*\| \leq K e^{-\gamma|t-\tau|}, \quad K, = \mathrm{const} > 0,$$

$$W(\varphi_\tau(\varphi)) \equiv \Omega_0^t(\varphi) W(\varphi)(\Omega_0^t(\varphi))^*,$$

$$W(\varphi) \equiv \int_{-\infty}^{+\infty} W(\varphi)(\Omega_0^\tau(\varphi))^* \Omega_0^\tau(\varphi) W(\varphi) \, d\tau;$$

(3) *for the existence of the invariant torus $y = v(\varphi)$ of system*

$$\frac{d\varphi}{dt} = a(\varphi), \quad \frac{dy}{dt} = -A^*(\varphi)y + f(\varphi)$$

it is necessary and sufficient that

$$\int_{-\infty}^{+\infty} \langle f(\varphi_\tau(\varphi)), u(\varphi_\tau(\varphi)) \rangle \, d\tau \equiv 0 \tag{28}$$

for every function $x = u(\varphi)$ determining the invariant torus of system (19). Additionally, the identity (28) is equivalent to

$$\int_{-\infty}^{+\infty} W(\varphi)(\Omega_0^\tau(\varphi))^* f(\varphi_\tau(\varphi)) \, d\tau \equiv 0.$$

This book contains the proofs of these and new results in the theory of stability and dichotomy of linear and co-linear systems. The exercises presented at the end of the book will help students to ensure that the time and efforts spent in studying the book are not in vain.

1 Exponentially Dichotomous Linear Systems of Differential Equations and Lyapunov Functions of Variable Sign

1.0 Introduction

One of the important problems of the theory of nonlinear multifrequency oscillations is to establish the conditions for preserving invariant tori of the dynamical systems under perturbations. This problem is closely connected with problems of the existence and integral representation of solutions, bounded on the whole R axis, to systems of linear differential equations

$$\frac{dx}{dt} = A(t)x + f(t), \quad x \in R^n, \tag{1.0.1}$$

with a matrix $A(t)$ which is continuous and bounded on R, $f(t) \in C^0(R)$. These problems are investigated in this chapter. It is known that the existence of a bounded, for example, on the semiaxis $R_+ = [0, \infty)$ solution of the system (1.0.1) is equivalent for every fixed vector function $f(t)$ bounded on R_+ to the exponential dichotomy (e-dichotomy) on R_+ of the corresponding homogeneous system

$$\frac{dx}{dt} = A(t)x. \tag{1.0.2}$$

On the other hand the property of e-dichotomy on R_+ is equivalent to the existence of the quadratic form (the Lyapunov function)

$$V(t, x) = \langle S(t)x, x \rangle \tag{1.0.3}$$

with a continuously differentiable and bounded on R_+ matrix of the coefficients $S(t)$ which possesses a sign-definite derivative due to the system (1.0.1)

$$\dot{V}(t,x) = \left\langle \left(\frac{d}{dt}S(t) + S(t)A(t) + A^*(t)S(t)\right)x, x \right\rangle \geq \|x\|^2. \tag{1.0.4}$$

Also, no restrictions are required concerning the matrix $S(t)$, except that it is continuously differentiable and bounded on R_+. We are especially interested in the property of e-dichotomy of the system (1.0.2) on the whole R axis, because this very property solely ensures the existence and uniqueness of the solution of the system (1.0.1) bounded on R. Additionally, the condition of non-degeneracy was to be imposed on the quadratic form (1.0.3)

$$\det S(t) \neq 0 \quad \forall t \in R. \tag{1.0.5}$$

The abandoning of the uniqueness condition for the solution of the system (1.0.1) bounded on R yields, by the same token, the abandonment of conditions (1.0.4) and (1.0.5), requiring instead that the derivative of the quadratic form (1.0.3) due to the system which is conjugated with the system (1.0.2)

$$\frac{dx}{dt} = -A^*(t)x$$

is of definite sign

$$\dot{V}(t,x) = \left\langle \left(\frac{d}{dt}S(t) - S(t)A^*(t) - A(t)S(t)\right)x, x \right\rangle \geq \|x\|^2.$$

In addition, the determinant of the matrix $S(t)$ may be zero for some $t = t_i$, $0 \leq i \leq n$.

A more detailed study of the linear system (1.0.2) dichotomous on semiaxes R_+, R_- enables us to obtain an integral representation of all solutions to the system (1.0.1) which are bounded on R. Further investigation of the parameter dependence of the set of solutions of the system (1.0.1) bounded on R results in a new idea to complement systems (1.0.2) weakly regular on R up to the regular ones. This eliminated interest in the investigation of the regularity of systems of block-triangular type differential equations.

1.1 Exponential Dichotomy Criterion for Linear Systems in Terms of Quadratic Forms

Let us consider a linear homogeneous system of differential equations

$$\frac{dx}{dt} = A(t)x, \tag{1.1.1}$$

where $x = (x_1, \ldots, x_n) \in R^n$, $t \in R = (-\infty, \infty)$, $A(t)$ is a continuous $n \times n$-dimensional matrix function bounded on the whole R axis. We employ the following designations.

1. $\Omega_\tau^t(A)$ is a matriciant of the linear system (1.1.1), $\Omega_\tau^\tau(A) = I_n$ is an n-dimensional identity matrix.

2. $\langle x, y \rangle = \sum\limits_{i=1}^{n} x_i y_i$ is an ordinary scalar product in R^n, $\|x\|^2 = \langle x, x \rangle$. Besides, the norm of the $n \times n$-dimensional matrix A is understood as follows: $\|A\| = \max\limits_{\|x\|=1} \|Ax\|$, $\|A\|_0 = \sup\limits_{t \in R} \|A(t)\|$.

3. $C^0(R)$ is a space of functions (vector or matrix) defined for all $t \in R$, continuous and bounded on the whole R axis.

4. $C^1(R)$ is a subspace of $C^0(R)$ consisting of continuously differentiable functions.

5. A^* is a transposed matrix.

Definition 1.1.1 A *system* of the equations (1.1.1) is called *exponentially dichotomous on the whole R axis*, if the phase space R^n can be represented in the form of a direct sum of two subspaces E^+ and E^- such that for all $x^+ \in E^+$ and all $x^- \in E^-$ the estimates

$$\|\Omega_0^t(A)x^+\| \leq K\|\Omega_0^\tau(A)x^+\| \exp(-\gamma(t-\tau)), \quad \tau \leq t, \qquad (1.1.2)$$

$$\|\Omega_0^t(A)x^-\| \leq K\|\Omega_0^\tau(A)x^-\| \exp(\gamma(t-\tau)), \quad t \leq \tau, \qquad (1.1.3)$$

take place, where K and γ are const > 0, $t, \tau \in R$.

Definition 1.1.2 A *system* of the equations (1.1.1) is *exponentially dichotomous on the semiaxes* $R_+ = [0, \infty)$, $R_- = (-\infty, 0]$, if for the positive semiaxis R_+ the space R^n can be presented in the form of a direct sum of two subspaces $\bar{E}^+$ and $\bar{E}^-$: $R^n = \bar{E}^+ \oplus \bar{E}^-$, such that for every $x^+ \in \bar{E}^+$ and every $x^- \in \bar{E}^-$ the estimates (1.1.2) and (1.1.3) are satisfied, provided all $t, \tau \in R_+$, and for the negative semiaxis R_- the other representation is true in general $R^n = \bar{\bar{E}}^+ \oplus \bar{\bar{E}}^-$, for which the estimates (1.1.2) and (1.1.3) take place when $t, \tau \in R_-$.

It is well known (see Daletsky and Krein [1] and Majzel [1]) that the exponential dichotomy of the system (1.1.1) on the semiaxis R_+ (or R_-) is equivalent to the existence of at least one bounded on R_+ (or R_-) solution to the inhomogeneous system of equations

$$\frac{dx}{dt} = A(t)x + f(t), \qquad (1.1.4)$$

for every continuous and bounded on R_+ (R_-) vector-function $f(t)$. This, in its turn, is equivalent to the existence of the quadratic form

$$V(t, x) = \langle S(t)x, x \rangle \qquad (1.1.5)$$

with a continuously differentiable and bounded matrix of the coefficients $S(t)$ having a fixed sign derivative due to the system (1.1.1),

$$\dot{V}(t,x) \leq -\beta\|x\|^2, \quad \beta > 0.$$

Moreover, the quadratic form (1.1.5) can alternate its sign as t and x vary. There arises a question: where is the system of equations (1.1.1) exponentially dichotomous on the whole R axis in the case of the existence of the quadratic form (1.1.5) with a continuously differentiable matrix of the coefficients $S(t)$ bounded on the whole R axis having a fixed sign derivative due to the system (1.1.1)

$$\dot{V}(t,x) = \langle [\dot{S}(t) + S(t)A(t) + A^*(t)S(t)]x, x \rangle \leq -\beta\|x\|^2, \quad (1.1.6)$$

where β is a const > 0, $\dot{S} = dS/dt$?

The investigations made have demonstrated that for the system (1.1.1) to be exponentially dichotomous on R it is sufficient to assume, in addition to the inequality (1.1.6) being valid, that the determinant of the matrix $S(t)$ differs from zero for all $t \in R$. Here the estimate

$$|\det S(t)| \geq \delta - \text{const} > 0, \quad \tau \in R \quad (1.1.7)$$

follows certainly.

We prove the following.

Theorem 1.1.1 *Let the quadratic form (1.1.5) exist with a continuously differentiable and symmetric matrix of coefficients $S(t)$ bounded on the whole R axis, satisfying the condition (1.1.6), and moreover, $\det S(t) \neq 0, \forall t \in R$.*

Then, the system of the equations (1.1.1) is exponentially dichotomous on the whole R axis and the dimension of the subspace E^+ equals to the number of positive eigenvalues of the matrix $S(t)$, and that of the E^- equals the number of negative eigenvalues of $S(t)$. Also, the constants K and γ in the estimates (1.1.2) and (1.1.3) can be specified as follows

$$K = \left(\frac{2\|A\|_0\|S\|_0}{\beta}\right)^{1/2}, \quad \gamma = \frac{\beta}{2\|S\|_0}. \quad (1.1.8)$$

Proof The non-degenerate symmetric matrix $S(t)$ is considered to have r positive eigenvalues and $n - r$ negative ones. We designate

$$M_t^+ = \{x \in R^n \mid \langle S(t)\Omega_0^t(A)x, \Omega_0^t(A)x \rangle \geq 0\}; \quad (1.1.9)$$

$$M_t^- = \{x \in R^n \mid \langle S(t)\Omega_0^t(A)x, \Omega_0^t(A)x \rangle \leq 0\}. \quad (1.1.10)$$

It is obvious that the sets M_t^+ and M_t^- are non-empty for every fixed $t \in R$, since it is presumed that $0 < r < n$. Otherwise, clearly, $(r = 0, r = n)$. We note that the sets M_t^+ diminish as t increases and embed one into another, and the sets M_t^- enlarge, i.e. for $t_1 < t_2$ the inclusions

$$M_{t_2}^+ \subset M_{t_1}^+, \quad M_{t_1}^- \subset M_{t_2}^- \qquad (1.1.11)$$

take place. In fact, inequality (1.1.6) can be represented as

$$\frac{d}{dt}\langle S(t)\Omega_0^t(A)x, \Omega_0^t(A)x\rangle \leq -\beta\|\Omega_0^t(A)x\|^2. \qquad (1.1.12)$$

Integrating both parts of this inequality within the limits from t_1 to t_2 we get the inequality

$$\begin{aligned}\langle S(t_2)\Omega_0^{t_2}(A)x, \Omega_0^{t_2}(A)x\rangle &- \langle S(t_1)\Omega_0^{t_1}(A)x, \Omega_0^{t_1}(A)x\rangle \\ &\leq -\beta\int_{t_1}^{t_2}\|\Omega_0^t(A)x\|^2\,dt \leq -\beta\int_{t_1}^{t_2}\|\Omega_0^t(A)\|^2\,dt\,\|x\|^2.\end{aligned} \qquad (1.1.13)$$

Thus, if a point $x_0 \in R^n$, $x_0 \neq 0$, belongs to the set $M_{t_2}^+$, i.e.

$$\langle S(t_2)\Omega_0^{t_2}(A)x_0, \Omega_0^{t_2}(A)x_0\rangle \geq 0,$$

then the inequality (1.1.13) yields

$$\langle S(t_1)\Omega_0^{t_1}(A)x_0, \Omega_0^{t_1}(A)x_0\rangle > 0,$$

i.e. $x_0 \in M_{t_1}^+$.

Let us show that the set

$$M^+ = \bigcap_{t=0}^{\infty} M_t^+ \qquad (1.1.14)$$

is non-empty and represents an r-dimensional linear subspace of the space R^n. With this in mind we reduce the quadratic form $\langle S(0)x, x\rangle$ by the orthogonal change of variables

$$x = Qy \qquad (1.1.15)$$

to the canonical form

$$\langle Q^*S(0)Qy, y\rangle = \lambda_1 y_1^2 + \cdots + \lambda_r y_r^2 - \lambda_{r+1}y_{r+1}^2 - \cdots - \lambda_n y_n^2, \qquad (1.1.16)$$

where $\lambda_1, \ldots, \lambda_r, -\lambda_{r+1}, \ldots, -\lambda_n$ are eigenvalues of the matrix $S(0)$, $\lambda_i > 0$, $i = 1, \ldots, n$. We fix the non-zero r-dimensional vector $C = (c_1, \ldots, c_r)$ and consider a closed sphere in the space of variables $y_{r+1}, \ldots, y_n$ which is determined by the inequality $\lambda_{r+1}y_{r+1}^2 + \cdots + \lambda_n y_n^2 \leq \lambda_1 c_1^2 + \cdots + \lambda_r c_r^2$ which is representable in the form of the system

$$\langle Q^* S(0) Q y, y \rangle \geq 0, \quad [I_r, 0] y = C. \tag{1.1.17}$$

Turning back to the variables x one can state by the formula (1.1.15) that the system

$$\langle S(0) x, x \rangle \geq 0, \quad [I_r, 0] Q^* x = C \tag{1.1.18}$$

determines the closed sphere $\mathcal{N}_0$ located in the $(n-r)$-dimensional hyperplane $[I_r, 0] Q^* x = 0$. Now we consider for every fixed value $t > 0$ the set $\mathcal{N}_t$ consisting of the values $x \in R^n$ for which the conditions

$$\langle S(t) \Omega_0^t(A) x, \Omega_0^t(A) x \rangle \geq 0, \quad [I_r, 0] Q^* x = C \tag{1.1.19}$$

are satisfied simultaneously. Here we note that the orthogonal matrix Q is fixed and the same as in (1.1.15) and (1.1.18).

Let us show that for every fixed value of t the sets $\mathcal{N}_t$ are closed spheres located in the $(n-r)$-dimensional hyperplane $[I_r, 0] Q^* x = C$. We fix a value $t = t_0 > 0$ and make sure that there exist non-zero points $\bar{y} \in R^n$ satisfying the conditions

$$\langle \tilde{S}(t_0) \bar{y}, \bar{y} \rangle \geq 0, \quad [I_r, 0] \bar{y} = C \tag{1.1.20}$$

where

$$\tilde{S}(t_0) = Q^* (\Omega_0^{t_0}(A))^* S(t_0) \Omega_0^{t_0}(A) Q. \tag{1.1.21}$$

We reduce the quadratic form $\langle \tilde{S}(t_0) y, y \rangle$ by the orthogonal change of variables $y = P(t_0) z$ to the canonical form $\langle P^*(t_0) \tilde{S}(t_0) P(t_0) z, z \rangle = \lambda_1(t_0) z_1^2 + \cdots + \lambda_r(t_0) z_r^2 - \lambda_{r+1}(t_0) z_{r+1}^2 - \cdots - \lambda_n(t_0) z_n^2$ and consider the inequality

$$\lambda_1(t_0) z_1^2 + \cdots + \lambda_r(t_0) z_r^2 - \lambda_{r+1}(t_0) z_{r+1}^2 - \cdots - \lambda_n(t_0) z_n^2 \geq 0, \tag{1.1.22}$$

which seems to define a conic set $\mathfrak{N}(t_0)$ in the space of variables $z_1, \ldots, z_r$. This set $\mathfrak{N}(t_0)$ contains r-dimensional linear subspaces from R^n. In particular, one of these subspaces has the basis

$$e = [e_1, \ldots, e_r] = \begin{pmatrix} 1 & 0 & \cdots & 0 \\ 0 & 1 & \cdots & 0 \\ \vdots & \vdots & \ddots & \vdots \\ 0 & 0 & \cdots & 1 \\ \vdots & \vdots & \ddots & \vdots \\ 0 & 0 & \cdots & 0 \end{pmatrix} = \begin{pmatrix} I_r \\ 0 \end{pmatrix}. \tag{1.1.23}$$

One can obtain the basis of other subspaces belonging to the set $\mathfrak{N}(t_0)$ by slightly changing the last $n - r$ coordinates in (1.1.23). Returning to the variables y we claim that the conic set $\mathfrak{M}(t_0) = \{y \mid \langle \tilde{S}(t_0)y, y \rangle \geq 0\}$ contains r-dimensional linear subspaces, one of which, E^r, has the basis

$$U = [u_1, \ldots, u_r] = P(t_0)e = \begin{pmatrix} U_1 \\ U_2 \end{pmatrix}. \tag{1.1.24}$$

Here U_1 is the $r \times r$-dimensional matrix.

The inclusion of the sets $M_{t_0}^+ \subset M_0^+$ yields the inclusion $\mathfrak{M}(t_0) \subset \mathfrak{M}(0)$. Therefore

$$E^r \subset \mathfrak{M}(0). \tag{1.1.25}$$

Hence, the non-degeneracy of the matrix U_1. In fact, if $\det U_1 = 0$, then a non-trivial linear combination of columns of this matrix is zero as well, i.e.

$$\sum_{i=1}^r c_i u_i^{(1)} = 0, \quad \sum_{i=1}^r c_i^2 \neq 0.$$

So the first r coordinates of the vector $y = \sum_{i=1}^r c_i u_i$ equal zero, and due to (1.1.16) $y \notin \mathfrak{M}(0) = \{y \mid \langle Q^* S(0) Qy, y \rangle \geq 0\}$, which contradicts the inclusion (1.1.25).

We find a common point belonging to E^r and hyperplane $[I_r, 0]y = C$. This point is searched for in the form of the linear combination

$$\bar{y} = \sum_{i=1}^r \alpha_i u_i = U\alpha,$$

where $\alpha = \text{col}\{\alpha_1, \ldots, \alpha_r\}$ is an unknown vector. Substituting by (1.1.25) into (1.1.20) we get $[I_r, 0]U\alpha = C$. By this the vector of coefficients $\alpha = U_1^{-1}C$ is determined uniquely. Thus, the vector $\bar{y} = UU_1^{-1}C$ satisfies two conditions (1.1.20). Since the space $E^r \subset \mathfrak{M}(t_0)$ can also be taken different, then there are many vectors $\bar{y}$ satisfying the conditions (1.1.20). This demonstrates that the set $\mathcal{N}_t$ is non-empty for every fixed $t = t_0 > 0$. Moreover, we show $\mathcal{N}_t$ to be a closed sphere located on the $(n-r)$-dimensional hyperplane $[I_r, 0]Q^*x = C$. We have, by virtue of inequality (1.1.13) for $t_2 = t_0$, $t_1 = 0$, in view of the representation (1.1.16) and designation (1.1.21)

$$\langle \tilde{S}(t_0)y, y \rangle \leq \lambda_1 y_1^2 + \cdots + \lambda_r y_r^2 - \lambda_{r+1} y_{r+1}^2 - \cdots - \lambda_n y_n^2. \tag{1.1.26}$$

Hence, we have the negative definiteness of the $(n-r)$-dimensional block $\tilde{S}_{22}(t_0)$ of the matrix

$$\tilde{S}(t_0) = \begin{pmatrix} \tilde{S}_{11}(t_0) & \tilde{S}_{12}(t_0) \\ \tilde{S}_{21}(t_0) & \tilde{S}_{22}(t_0) \end{pmatrix}.$$

Let the change of variables $y^{(2)} = T(t_0)v^{(2)}$ ($y^{(2)} = (y_{r+1}, \ldots, y_n)$, $v^{(2)} = (v_{r+1}, \ldots, v_n)$) reduce the quadratic form $\langle \tilde{S}_{22}(t_0)y^{(2)}, y^{(2)} \rangle$ to the canonical form

$$\langle \tilde{S}_{22}(t_0)T(t_0)v^{(2)}, T(t_0)v^{(2)} \rangle = -\tilde{\lambda}_{r+1}v_{r+1}^2 - \cdots - \tilde{\lambda}_n v_n^2 = \langle \tilde{\Lambda}v^{(2)}, v^{(2)} \rangle. \tag{1.1.27}$$

Then we make the change of variables $y = \operatorname{diag}\{I_r, T(t_0)\}v$, $v = (v^{(1)}, v^{(2)})$ in the inequality $\langle \tilde{S}(t_0)y, y \rangle \geq 0$, and substituting by $v^{(1)} = C$ get

$$\langle \tilde{S}_{11}(t_0)C, C \rangle + \langle \tilde{S}_{12}(t_0)T(t_0)v^{(2)}, C \rangle + \langle \tilde{S}_{21}(t_0)C, T(t_0)v^{(2)} \rangle \\ + \langle \tilde{\Lambda}v^{(2)}, v^{(2)} \rangle \geq 0. \tag{1.1.28}$$

Clearly, by selecting the perfect square, the inequality (1.1.28) can be represented as

$$\tilde{\lambda}_{r+1}(v_{r+1} - \beta_1)^2 + \cdots + \tilde{\lambda}_n(v_n - \beta_{n-r})^2 \leq f(C). \tag{1.1.29}$$

The fact that the set $\mathcal{N}_{t_0}$ contains at least two non-zero points from R^n yields $f(C) > 0$. Inequality (1.1.29) shows that $\mathcal{N}_{t_0}$ is a closed sphere. Thus, the cross-section $\bigcap_{t=0}^{\infty} \mathcal{N}_t \subset M^+$ has at least one non-zero point $\bar{x}(C)$. Having changed the r-dimensional vector C we get r linearly independent points belonging to M^+. We designate these points by $x^+_{(1)}, x^+_{(2)}, \ldots, x^+_{(r)}$ and show that the corresponding solutions of system (1.1.1) $\Omega_0^t(A)x^+_{(i)}$, $i = 1, \ldots, r$, satisfy the estimate (1.1.2). Since $x^+_{(i)} \in M^+$, then

$$\langle S(t)\Omega_0^t(A)x^+_{(i)}, \Omega_0^t(A)x^+_{(i)} \rangle > 0, \quad i = 1, \ldots, r, \tag{1.1.30}$$

for all $t \in R$. We designate

$$V_{\pm\varepsilon}(t, x) = \langle S(t)x, x \rangle \pm \varepsilon \|x\|^2, \quad \varepsilon \geq 0. \tag{1.1.31}$$

Since the inequality (1.1.6) is satisfied for constant positive value of β, then for $\varepsilon > 0$ which is small enough the quadratic form (1.1.31) derivative due to system (1.1.1) is also negative definite

$$\dot{V}_{\pm\varepsilon}(t, x) \leq -\beta\|x\|^2 \pm 2\varepsilon\langle A(t)x, x \rangle \leq -(\beta - 2\|A\|_0\varepsilon)\|x\|^2, \tag{1.1.32}$$

provided only that

$$0 < \varepsilon < \frac{\beta}{2\|A\|_0}. \tag{1.1.33}$$

Estimate (1.1.32) and inequalities (1.1.30) and (1.1.33) allow us to establish

$$\frac{d}{dt} V_\varepsilon(t, \Omega_0^t(A)x^+_{(i)}) \leq -(\beta - 2\|A\|_0\varepsilon) \|\Omega_0^t(A)x^+_{(i)}\|^2 \\ \leq -\frac{\beta - 2\|A\|_0\varepsilon}{\|S\|_0 + \varepsilon} V_\varepsilon(t, \Omega_0^t(A)x^+_{(i)}), \tag{1.1.34}$$

$$V_\varepsilon(t, \Omega_0^t(A)x^+_{(i)}) \leq V_\varepsilon(\tau, \Omega_0^\tau(A)x^+_{(i)}) \exp\{-\gamma_1(t - \tau)\}, \quad \tau \leq t,$$

where
$$\gamma_1 = \frac{\beta - 2\|A\|_0 \varepsilon}{\|S\|_0 + \varepsilon} \geq 0. \tag{1.1.35}$$

Hence the following estimate follows immediately

$$\begin{aligned}
\|\Omega_0^t(A)x_{(i)}^+\| &= \{\varepsilon^{-1}[V_\varepsilon(t, \Omega_0^t(A)x_{(i)}^+) - V_0(t, \Omega_0^t(A)x_{(i)}^+)]\}^{1/2} \\
&\leq [\varepsilon^{-1} V_\varepsilon(t, \Omega_0^t(A)x_{(i)}^+)]^{1/2} \\
&\leq [\varepsilon^{-1} V_\varepsilon(\tau, \Omega_0^\tau(A)x_{(i)}^+)]^{1/2} \exp\left\{-\frac{\gamma_1}{2}(t-\tau)\right\} \\
&\leq [\varepsilon^{-1}(\|S\|_0 + \varepsilon)]^{1/2} \|\Omega_0^\tau(A)x_{(i)}^+\| \exp\left\{-\frac{\gamma_1}{2}(t-\tau)\right\},
\end{aligned} \tag{1.1.36}$$

for all $\tau \leq t$, $t, \tau \in R$. If the point $x^* \in R^n$ does not belong to the set M^+, then starting with a value $t = t^*$, $x^* \notin M_t^+$ $\forall t \geq t^*$, i.e. $V_0(t, \Omega_0^t(A)x^*) < 0$, $t^* \leq t$. Therefore, the inequality (1.1.12) implies

$$\frac{d}{dt}[-V_0(t, \Omega_0^t(A)x^*)] \geq \frac{\beta}{\|S\|_0}[-V_0(t, \Omega_0^t(A)x^*)]. \tag{1.1.37}$$

Integrating this inequality from t^* to t yields

$$\begin{aligned}
&[-V_0(t^*, \Omega_0^{t^*}(A)x^*)] \exp\left\{\frac{\beta}{\|S\|_0}(t-t^*)\right\} \\
&\leq -V_0(t, \Omega_0^t(A)x^*) \leq \|S\|_0 \|\Omega_0^t(A)x^*\|^2, \quad t^* \leq t.
\end{aligned} \tag{1.1.38}$$

The estimate (1.1.38) indicates the exponential increase of the norm of the corresponding solution $\Omega_0^t(A)x^*$, $x^* \notin M^+$. Thus, any linear combination of the points $x_{(i)}^+ \in M^+$ also belongs to M^+, since the linear combination of the corresponding solutions $\Omega_0^t(A)x_{(i)}^+$, $i = 1, \ldots, r$, is a solution to the system (1.1.1) decreasing over $+\infty$.

Let us verify that M^+ is the r-dimensional linear subspace as well. To this end, it is sufficient to show that no point $\tilde{x} \in M^+$ exists which could not be represented as a linear combination of the points $x_{(i)}^+$, $i = 1, \ldots, r$. Let such a point $\tilde{x} \in M^+$ exist. Then a linear combination

$$\sum_{i=1}^r c_i x_{(i)}^+ + c_{r+1} \tilde{x} = \bar{x}$$

is not contained in M_0^+, since the set M_0^+ can not embrace $(r+1)$-dimensional subspaces. Then, due to the inclusion $M^+ \subset M_0$ the point $\bar{x} \notin M^+$, which means that M^+ is the r-dimensional linear subspace from R^n which is designated by E^+.

So as to refine estimate (1.1.36) we note that for every non-zero point $x^+ \in E^+$ the corresponding solution of the system (1.1.1) $\Omega_0^t(A)x^+$ satisfies, for all $t \in R$, the inequality
$$V_{-\varepsilon}(t, \Omega_0^t(A)x^+) > 0, \tag{1.1.39}$$
where ε is an arbitrary fixed number satisfying the inequality (1.1.33).

Actually, if the inequality (1.1.39) is violated for some $t = t^*$, i.e.
$$V_{-\varepsilon}(t^*, \Omega_0^{t^*}(A)x^+) \leq 0,$$
then this results in an increase over $+\infty$ of the norm of the solution $x = \Omega_0^t(A)x^+$:
$$[-V_{-\varepsilon}(\tau, \Omega_0^\tau(A)x^+)] \exp\left\{\frac{\beta - \varepsilon\|A\|_0}{\|S\|_0 + \varepsilon}(t - \tau)\right\}$$
$$\leq (\|S\|_0 + \varepsilon)\|\Omega_0^t(A)x^+\|^2, \quad t^* \leq \tau \leq t,$$
which contradicts $x^+ \in E^+$. If the inequality (1.1.39) is satisfied then
$$\|\Omega_0^t(A)x^+\| = \{\varepsilon^{-1}[V_0(t, \Omega_0^t(A)x^+) - V_{-\varepsilon}(t, \Omega_0^t(A)x^+)]\}^{1/2}$$
$$\leq \{\varepsilon^{-1}V_0(t, \Omega_0^t(A)x^+)\}^{1/2}$$
$$\leq \{\varepsilon^{-1}V_0(\tau, \Omega_0^\tau(A)x^+)\}^{1/2} \exp\left\{-\frac{\beta}{2\|S\|_0}(t - \tau)\right\} \tag{1.1.40}$$
$$\leq \{\varepsilon^{-1}\|S\|_0\}^{1/2}\|\Omega_0^\tau(A)x^+\| \exp\left\{-\frac{\beta}{2\|S\|_0}(t - \tau)\right\}, \quad \tau \leq t.$$

Thus, the constant γ in the estimate (1.1.2) can be represented as $\gamma = \beta/2\|S\|_0$. The inequality (1.1.40) implies
$$\sup_{\tau \leq t} \|\Omega_0^t(A)x^+\| \|\Omega_0^\tau(A)x^+\|^{-1} \exp\left\{-\frac{\beta}{2\|S\|_0}(t - \tau)\right\}$$
$$= K(x^+) \leq \{\varepsilon^{-1}\|S\|_0\}^{1/2},$$
where ε can take values arbitrarily close to the value $\frac{\beta}{2\|A\|_0}$. Therefore,
$$K(x^+) \leq \sqrt{\frac{2\|A\|_0\|S\|_0}{\beta}}.$$

We establish in the same manner that the set $M^- = \bigcap_{t=0}^\infty M_t^-$ is the $(n-r)$-dimensional linear subspace of the space R^n, $M^- = E^-$. For every non-zero point the inequality
$$V_\varepsilon(t, \Omega_0^t(A)x^-) < 0 \tag{1.1.41}$$

is satisfied when all $t \in R$ and every fixed $\varepsilon \in \left[0, \frac{\beta}{2\|A\|_0}\right)$. In view of inequality (1.1.12) and the boundedness of quadratic form $-V_0(t,x)$: $-V_0(t,x) \leq \|S\|_0 \|x\|^2$ we get

$$\frac{d}{dt}[-V_0(t, \Omega_0^t(A)x^-)] \geq \beta \|\Omega_0^t(A)x^-\|^2 \geq \frac{\beta}{\|S\|_0}[-V_0(t, \Omega_0^t(A)x^-)],$$

$$-V_0(t, \Omega_0^t(A)x^-) \geq -V_0(\tau, \Omega_0^\tau(A)x^-) \exp\left\{\frac{\beta}{\|S\|_0}(t-\tau)\right\}, \quad \tau \leq t,$$

$$-V_0(\tau, \Omega_0^\tau(A)x^-) \leq -V_0(t, \Omega_0^t(A)x^-) \exp\left\{-\frac{\beta}{\|S\|_0}(t-\tau)\right\}, \quad \tau \leq t, \qquad (1.1.42)$$

$$-V_0(\tau, \Omega_0^\tau(A)x^-) \leq \|S\|_0 \|\Omega_0^t(A)x^-\|^2 \exp\left\{-\frac{\beta}{\|S\|_0}(t-\tau)\right\}, \quad \tau \leq t.$$

Now we have in view of estimates (1.1.41) and (1.1.42)

$$\|\Omega_0^\tau(A)x^-\| = \{\varepsilon^{-1}[V_\varepsilon(\tau, \Omega_0^\tau(A)x^-) - V_0(\tau, \Omega_0^\tau(A)x^-)]\}^{1/2}$$

$$\leq \{\varepsilon^{-1}[-V_0(\tau, \Omega_0^\tau(A)x^-)]\}^{1/2}$$

$$\leq \{\varepsilon^{-1}\|S\|_0\}^{1/2} \|\Omega_0^t(A)x^-\| \exp\left\{-\frac{\beta}{2\|S\|_0}(t-\tau)\right\}, \quad \tau \leq t.$$

Changing the designations $t \to \tau$ and $\tau \to t$ we find

$$\|\Omega_0^t(A)x^-\| \leq \{\varepsilon^{-1}\|S\|_0\}^{1/2} \|\Omega_0^\tau(A)x^-\| \exp\left\{\frac{\beta}{2\|S\|_0}(t-\tau)\right\}, \quad t \leq \tau. \quad (1.1.43)$$

The estimate (1.1.43) is just the estimate (1.1.3) and $\gamma = \frac{\beta}{2\|S\|_0}$ and the value $(\varepsilon^{-1}\|S\|_0)^{1/2}$ can be taken arbitrarily close to $\left\{\frac{2\|A\|_0 \|S\|_0}{\beta}\right\}^{1/2}$. Therefore

$$K = \left\{\frac{2\|A\|_0 \|S\|_0}{\beta}\right\}^{1/2}.$$

This completes the proof of the Theorem 1.1.1.

Let us prove its converse.

Theorem 1.1.2 *Let the system of the equations (1.1.1) be exponentially dichotomous on the whole R axis. Then there exist $n \times n$- dimensional symmetric matrix functions $S(t) \in C^1(R)$ satisfying the condition (1.1.6). Each of these functions is non-degenerate for all $t \in R$. Moreover, for each of the functions the estimate (1.1.7) is valid with a constant $\delta > 0$ independent of $t \in R$.*

Proof The existence of one of such matrix function $S(t)$ becomes obvious immediately by the following representation

$$S(t) = \int_t^\infty [\Omega_0^\sigma(A)C\Omega_t^0(A)]^*[\Omega_0^\sigma(A)C\Omega_t^0(A)]\,d\sigma \qquad (1.1.44)$$
$$- \int_{-\infty}^t [\Omega_0^\sigma(A)(C-I_n)\Omega_t^0(A)]^*[\Omega_0^\sigma(A)(C-I_n)\Omega_t^0(A)]\,d\sigma,$$

where C is a constant matrix of projection on the subspace E^+ along E^-. Immediately computing the derivative $\dot{V}(t,x)$ of the quadratic form $V(t,x) = \langle S(t)x, x\rangle$ due to the system (1.1.1) we verify the validity of the inequality $\dot{V}(t,x) \leq -\frac{1}{2}\|x\|^2$ $\forall x \in R^n$.

Now we show that each of the matrix functions $S(t) \equiv S^*(t) \in C^1(R)$ satisfying the condition (1.1.6) is non-degenerate for all $t \in R$. To this end we transform the system of equations (1.1.1) by a change of Lyapunov variables $x = L(t)y$ to the decomposed form

$$\frac{dy_1}{dt} = B^+(t)y_1, \quad \frac{dy_2}{dt} = B^-(t)y_2, \qquad (1.1.45)$$

corresponding to the representation $R^n = E^+ \oplus E^-$, $y_1 = (y_1, \ldots, y_r)$, $y_2 = (y_{r+1}, \ldots, y_n)$. Since $L(t)$ is a Lyapunov matrix, then for the corresponding matriciants $\Omega_\tau^t(B^+)$, $\Omega_\tau^t(B^-)$ the estimates similar to (1.1.2) and (1.1.3)

$$\|\Omega_\tau^t(B^+)\| \leq K_1 \exp\{-\gamma(t-\tau)\}, \quad \tau \leq t, \qquad (1.1.46)$$
$$\|\Omega_\tau^t(B^-)\| \leq K_1 \exp\{\gamma(t-\tau)\}, \quad t \leq \tau \qquad (1.1.47)$$

take place, where $K_1, \gamma = \text{const} > 0$. Let us verify the estimate

$$\frac{d}{dt}\langle T(t)\Omega_\tau^t(B)y, \Omega_\tau^t(B)y\rangle \leq -\gamma_1\|\Omega_\tau^t(B)y\|^2, \qquad (1.1.48)$$

where $\gamma_1 = \text{const} > 0$, $\Omega_\tau^t(B) = \text{diag}\{\Omega_\tau^t(B^+), \Omega_\tau^t(B^-)\}$, $T(t) = L^*(t)S(t)L(t)$. To this end we represent (1.1.6) as

$$\langle L^*(t)[\dot{S}(t) + S(t)A(t) + A^*(t)S(t)]L(t)y, y\rangle \leq -\beta\|L(t)y\|^2.$$

Hence

$$\langle[L^*(t)S(t)L(t) + \dot{L}^*(t)S(t)L(t) + L^*(t)S(t)\dot{L}(t) - \dot{L}^*(t)S(t)L(t) - L^*(t)S(t)\dot{L}(t)$$
$$+ L^*(t)S(t)L(t)L^{-1}(t)A(t)L(t) + L^*(t)A^*(t)(L^*(t))^{-1}L^*(t)S(t)L(t)]y, y\rangle$$
$$\leq -\beta\|L(t)y\|^2 \leq -\beta\|L^{-1}\|_0^{-2}\|y\|^2.$$

It is clear that

$$\langle [\dot{T}(t) + T(t)B(t) + B^*(t)T(t)]y, y\rangle \leq -\gamma_1 \|y\|^2,$$

i.e. the estimate (1.1.48) is satisfied. Integrating both parts of the inequality (1.1.48) from τ to t yields

$$\langle T(t)\Omega_\tau^t(B)y, \Omega_\tau^t(B)y\rangle - \langle T(\tau)y, y\rangle \leq -\gamma_1 \int_\tau^t \|\Omega_\tau^\sigma(B)y\|^2\, d\sigma. \qquad (1.1.49)$$

The block form of the matrix $T(t)$ is

$$T(t) = \begin{pmatrix} T_{11}(t) & T_{12}(t) \\ T_{21}(t) & T_{22}(t) \end{pmatrix},$$

where $T_{11}(t)$ and $T_{22}(t)$ are quadratic matrices of the same dimensions as $B_1(t)$ and $B_2(t)$. Setting $y = \operatorname{col}\{y_1, 0\}$, $\|y_1\| \neq 0$, in the estimate (1.1.49),

$$\langle T_{11}(t)\Omega_\tau^t(B^+)y_1, \Omega_\tau^t(B^+)y_1\rangle - \langle T_{11}(\tau)y_1, y_1\rangle \leq -\gamma_1 \int_\tau^t \|\Omega_\tau^\sigma(B^+)y_1\|^2\, d\sigma,$$

and passing to the limit $t \to \infty$ we get

$$\langle T_{11}(\tau)y_1, y_1\rangle \geq \gamma_1 \int_\tau^\infty \|\Omega_\tau^\sigma(B^+)y_1\|^2\, d\sigma > 0. \qquad (1.1.50)$$

Now, setting $y = \operatorname{col}\{0, y_2\}$, $\|y_2\| \neq 0$, in (1.1.49) and making the change of variables $y_2 = \Omega_\tau^t(B^-)z_2$ we find

$$\langle T_{22}(t)z_2, z_2\rangle \leq \langle T_{22}(t)\Omega_t^\tau(B^-)z_2, \Omega_t^\tau(B^-)z_2\rangle - \gamma_1 \int_\tau^t \|\Omega_t^\sigma(B^-)z_2\|^2\, d\sigma.$$

Letting $\tau \to -\infty$ and redesignating $z_2 \to y_2$ we obtain

$$\langle T_{22}(\tau)y_2, y_2\rangle \leq -\gamma_1 \int_{-\infty}^t \|\Omega_t^\sigma(B^-)y_2\|^2\, d\sigma < 0. \qquad (1.1.51)$$

Let us now show that the estimates (1.1.50) and (1.1.51) imply non-degeneracy of the matrix $T(t)$ for any side blocks $T_{12}^* = T_{21}$. We fix a value $t = t_0 \in R$ and note that the quadratic form $\langle T_{11}(t_0)y_1, y_1\rangle$ can always be reduced to the algebraic sum

of squares. Therefore, there exist non-degenerate matrices Q_1 and Q_2 satisfying the conditions
$$Q_1^* T_{11}(t_0) Q_1 = I_r, \quad Q_2^* T_{22}(t_0) Q_2 = -I_{n-r}.$$

We consider the matrix
$$M = \begin{pmatrix} Q_1^* & 0 \\ 0 & Q_{22}^* \end{pmatrix} \begin{pmatrix} T_{11}(t) & T_{12}(t) \\ T_{21}(t) & T_{22}(t) \end{pmatrix} \begin{pmatrix} Q_1 & 0 \\ 0 & Q_2 \end{pmatrix} = \begin{pmatrix} I_r & M_{12} \\ M_{12}^* & -I_{n-r} \end{pmatrix}. \quad (1.1.52)$$

Let us verify that the matrix
$$M^2 = \begin{pmatrix} I_r + M_{12} M_{12}^* & 0 \\ 0 & I_{n-r} + M_{12}^* M_{12} \end{pmatrix} \quad (1.1.53)$$

is positive definite. Actually, considering the quadratic form corresponding to the matrix (1.1.53) we get
$$\langle M^2 y, y \rangle = \langle (I_r + M_{12} M_{12}^*) y_1, y_1 \rangle + \langle (I_{n-r} + M_{12}^* M_{12}) y_2, y_2 \rangle$$
$$= \|y_1\|^2 + \|M_{12}^* y_1\|^2 + \|y_2\|^2 + \|M_{12} y_2\|^2 \geq \|y_1\|^2 + \|y_2\|^2.$$

Since the matrix (1.1.53) is positive definite, then $\det M \neq 0$, and therefore
$$\det T(t_0) \neq 0, \quad \det S(t_0) = \det [L^{-1}(t_0)^* T(t_0) L^{-1}(t_0)] \neq 0.$$

Now we show that the determinant of each symmetrical matrix $S(t) \in C^1(R)$ satisfying the condition (1.1.6) not only differs from zero for all $t \in R$, but also satisfies the condition (1.1.7), where the positive constant δ does not depend on $t \in R$. Assume the existence of the symmetric matrix $S(t) \in C^1(R)$ satisfying the condition (1.1.6) and not (1.1.7). Hence, it follows immediately that for every $\varepsilon > 0$ the determinant of at least one of the matrices
$$S(t) + \varepsilon I_n, \quad S(t) - \varepsilon I_n \quad (1.1.54)$$

vanishes for some $t = t(\varepsilon)$. On the other hand, both matrices (1.1.54) satisfy the condition (1.1.6) for $0 < \varepsilon < \beta/2\|A\|_0$. That is why they must be non-degenerate for small $\varepsilon > 0$ when all $t \in R$. The contradiction obtained proves that condition (1.1.7) is satisfied.

This completes the proof of Theorem 1.1.2.

It is known (see Daletsky and Krein [1] and Pliss [2]) that in the case of exponential dichotomy of system (1.1.1) on the whole R axis the inhomogeneous system of equations (1.1.4) has for every fixed vector-function $f(t) \in C^0(R)$ a unique solution $x(t)$ bounded on the whole R axis, that can be represented as
$$x = x(t) = \int_{-\infty}^{\infty} G(t, \tau) f(\tau) \, d\tau. \quad (1.1.55)$$

Here $G(t, \tau)$ is a Green function of the problem on solutions bounded on R, which has the following structure

$$G(t, \tau) = \begin{cases} \Omega_0^t(A) C \Omega_\tau^0(A), & \tau \le t, \\ \Omega_0^t(A) (C - I_n) \Omega_\tau^0(A), & \tau > t, \end{cases} \quad (1.1.56)$$

where C is a constant projection matrix on the subspace E^+ along E^-, $C^2 = C$. In addition, for the Green function (1.1.56) the estimate

$$\|G(t, \tau)\| \le K_0 \exp\{-\gamma |t - \tau|\} \quad (1.1.57)$$

is satisfied for all $t, \tau \in R$, where K_0 and γ are positive constants independent of $t, \tau \in R$. Clearly, under these conditions the Green function (1.1.56) is unique. In fact, if such another function $\tilde{G}(t, \tau)$ is assumed to exist, then the solution of the system (1.1.4) bounded on R can be represented as

$$x = x(t) = \int_{-\infty}^{\infty} \tilde{G}(t, \tau) f(\tau) \, d\tau. \quad (1.1.58)$$

Since the inhomogeneous system (1.1.4) has only one solution bounded on R, the inequalities (1.1.55) and (1.1.58) imply

$$\int_{-\infty}^{\infty} [G(t, \tau) - \tilde{G}(t, \tau)] f(\tau) \, d\tau = 0,$$

for every vector-function $f(t) \in C^0(R)$. Therefore $G(t, \tau) \equiv \tilde{G}(t, \tau)$.

The representation of K_0 and γ immediately via the matrix functions $S(t)$ and $A(t)$ is of great importance. Let us now prove the following result.

Theorem 1.1.3 *Let the conditions of Theorem 1.1.1 be satisfied. Then the constants K_0 and γ in estimate (1.1.57) of the Green function (1.1.56) can be represented by the inequality*

$$K_0 = \left(2 + \sqrt{2}\right) \left(\frac{\|A\|_0 \|S\|_0}{\beta}\right)^{3/2}, \quad \gamma = \frac{\beta}{2\|S\|_0}. \quad (1.1.59)$$

Proof We focus our attention on inequality (1.1.39) which can be represented as

$$\langle [S(t) - \varepsilon I_n] \Omega_0^t(A) C z, \Omega_0^t(A) C z \rangle \ge 0, \quad (1.1.60)$$

where C is the projection matrix on the subspace E^+ along E^-, i.e. the same as in the Green function (1.1.56), ε is a fixed number satisfying the inequality (1.1.33),

$z \in R^n$. Since $(I_n - C)$ is the projection matrix on the subspace E^-, the inequality (1.1.41) becomes

$$\langle [S(t) + \varepsilon I_n]\Omega_0^t(A)(I_n - C)z, \ \Omega_0^t(A)(I_n - C)z \rangle \geq 0. \tag{1.1.61}$$

We introduce the designations $C(t) = \Omega_0^t(A)C\Omega_t^0(A)$ and $u = \Omega_0^t(A)z$ in the inequalities (1.1.60) and (1.1.61) and represent the former as

$$\langle S(t)C(t)u, \ C(t)u \rangle - \varepsilon \|C(t)u\|^2 \geq 0, \tag{1.1.62}$$

$$\langle S(t)[I_n - C(t)]u, \ [I_n - C(t)]u \rangle + \varepsilon \|[I_n - C(t)]u\|^2 \leq 0. \tag{1.1.63}$$

Considering the difference of inequalities (1.1.62) and (1.1.63) we get

$$\varepsilon\{\|C(t)\|^2 + \|[I_n - C(t)]u\|^2\} - 2\langle S(t)C(t)u, u \rangle + \langle S(t)u, u \rangle \leq 0. \tag{1.1.64}$$

The parameter ε takes positive values $0 < \varepsilon < \beta/2\|A\|_0$. In view of this we separate perfect squares in (1.1.64)

$$\left\|C(t)u - \frac{1}{2\varepsilon}S(t)u\right\|^2 + \left\|[I_n - C(t)]u + \frac{1}{2\varepsilon}S(t)u\right\|^2 \leq \frac{1}{2\varepsilon^2}\|S(t)u\|^2. \tag{1.1.65}$$

Then we find the norm estimates of vectors $C(t)u$ and $[I_n - C(t)]u$

$$\|C(t)u\| = \left\|C(t)u - \frac{1}{2\varepsilon}S(t)u + \frac{1}{2\varepsilon}S(t)u\right\| \leq \left\|C(t)u - \frac{1}{2\varepsilon}S(t)u\right\|$$

$$+ \frac{1}{2\varepsilon}\|S(t)u\| \leq \frac{1}{\sqrt{2}\,\varepsilon}\|S\|_0\|u\| + \frac{1}{2\varepsilon}\|S\|_0\|u\|,$$

$$\|[I_n - C(t)]u\| = \left\|[I_n - C(t)]u + \frac{1}{2\varepsilon}S(t)u - \frac{1}{2\varepsilon}S(t)u\right\|$$

$$\leq \left\|[I_n - C(t)]u + \frac{1}{2\varepsilon}S(t)u\right\| + \frac{1}{2\varepsilon}\|S(t)u\|$$

$$\leq \frac{1}{\sqrt{2}\,\varepsilon}\|S\|_0\|u\| + \frac{1}{2\varepsilon}\|S\|_0\|u\|.$$

Thus the variable projection matrices $C(t)$ and $I_n - C(t)$ are bounded

$$\max\{\|C(t)\|_0, \ \|[I_n - C(t)]\|_0\} \leq \left(\frac{1}{\sqrt{2}} + \frac{1}{2}\right)\frac{\|S\|_0}{\varepsilon}. \tag{1.1.66}$$

Since the parameter ε can take values arbitrarily close to $\beta/2\|A\|_0$, the inequality (1.1.66) can be represented as

$$\max\{\|C(t)\|_0, \ \|[I_n - C(t)]\|_0\} \leq (1 + \sqrt{2})\frac{\|A\|_0\|S\|_0}{\beta}. \tag{1.1.67}$$

Taking into account (1.1.2), (1.1.3) and (1.1.66) and equality (1.1.8) we estimate the Green function (1.1.56). We have for $\tau \leq t$

$$\|\Omega_0^t(A) C \Omega_\tau^0(A)\| \leq K \|\Omega_0^t(A) C \Omega_\tau^0(A)\| \exp\{-\gamma(t-\tau)\}$$

$$\leq \left(\frac{2\|A\|_0 \|S\|_0}{\beta}\right)^{1/2} (1+\sqrt{2}) \frac{\|A\|_0 \|S\|_0}{\beta} \exp\left(-\frac{\beta}{2\|S\|_0}(t-\tau)\right)$$

$$= (2+\sqrt{2}) \left(\frac{\|A\|_0 \|S\|_0}{\beta}\right)^{3/2} \exp\left(-\frac{\beta}{2\|S\|_0}(t-\tau)\right).$$

In the case $\tau > t$ we arrive at a similar estimate with the same constants K_0 and γ. This completes the proof of Theorem 1.1.3.

1.2 Decomposition over the Whole R Axis of Linear Systems of Differential Equations Exponentially Dichotomous on Semiaxes R_+ and R_-

The main results of this section is as follows.

Theorem 1.2.1 *Let the system of equations (1.1.1) be exponentially dichotomous on semiaxes R_+ and R_-. Then there exists the Lyapunov transformation of variables*

$$x = L(t)y, \quad t \in R, \tag{1.2.1}$$

reducing the system (1.1.1) to the block-triangular form

$$\begin{aligned}
\frac{dy_1}{dt} &= B^+(t)y_1 + B_1(t)\tilde{y}, \\
\frac{dy_2}{dt} &= B^-(t)y_2 + B_2(t)\tilde{y}, \\
\frac{dy_3}{dt} &= \hat{B}(t)y_3, \\
\frac{dy_4}{dt} &= \check{B}(t)y_4, \quad \tilde{y} = (y_3, y_4).
\end{aligned} \tag{1.2.2}$$

Moreover, for the matricants $\Omega_\tau^t(B^\pm)$, $\Omega_\tau^t(\hat{B})$, $\Omega_\tau^t(\check{B})$ of the corresponding system of equations

$$\frac{dy_i}{dt} = B^\pm(t)y_i, \quad i = 1, 2,$$

$$\frac{dy_3}{dt} = \hat{B}(t)y_3, \quad \frac{dy_4}{dt} = \check{B}(t)y_4$$

the estimates

$$\|\Omega_\tau^t(B^+)\| \le K \exp\{-\gamma(t-\tau)\}, \quad \tau \le t, \quad t, \tau \in R,$$

$$\|\Omega_\tau^t(B^-)\| \le K \exp\{\gamma(t-\tau)\}, \quad t \le \tau,$$

$$\|\Omega_\tau^t(\hat{B})\| \le K \begin{cases} \exp\{-\gamma(t-\tau)\}, & 0 \le \tau \le t, \\ \exp\{\gamma(t-\tau)\}, & 0 \le t \le \tau, \end{cases} \quad (1.2.3)$$

$$\|\Omega_\tau^t(\check{B})\| \le K \begin{cases} \exp\{\gamma(t-\tau)\}, & 0 \le t \le \tau, \\ \exp\{-\gamma(t-\tau)\}, & \tau \le t \le 0, \end{cases}$$

$$K, \gamma = \text{const} > 0$$

hold, where $B_i(t)$, $i = 1, 2$, are some rectangular matrix function.

Proof Turning back to the definition of exponential dichotomy of the system (1.1.1) on the semiaxes R_+ and R_- cited in Section 1.1 we designate $\hat{E} = \bar{E}^+ \cap \bar{\bar{E}}^-$. Obviously, the subspace $\hat{E}$ consists of the points $\hat{x} \in R^n$ for which the corresponding solution of the system (1.1.1) $\Omega_0^t(A)\hat{x}$ damps exponentially on $\pm\infty$. Besides, it is possible that $\hat{E} = \{0\}$, or $\hat{E} = \bar{E}^+$. We designate by E^+ some fixed space complementing $\hat{E}$ up to $\bar{E}^+$. In the case when $\hat{E} = \{0\}$, $E^+ = \bar{E}^+$ and if $\hat{E} = \hat{E}^+$, $E^+ = \{0\}$. Since there is an arbitrariness in the choice of the subspace $\bar{\bar{E}}^+$, one can consider $E^+ \subseteq \bar{\bar{E}}^+$. Similarly, E^- is a fixed subspace complementing $\hat{E}$ up to $\bar{\bar{E}}^-$, $E^- \subseteq \bar{\bar{E}}^-$. We designate by $\check{E}$ a fixed subspace R^n complementing the subspace $E^+ \oplus E^- \oplus \hat{E}$ up to R^n: $R^n = E^+ \oplus E^- \oplus \hat{E} \oplus \check{E}$, $\dim E^\pm = r^\pm$, $\dim \check{E} = \check{r}$, $\dim \hat{E} = \hat{r}$, $r^+ + r^- + \check{r} + \hat{r} = n$. Thus, we get the representation of the space R^n in the form of a direct sum of four subspaces, each of which possesses the following properties

$$\text{if} \quad x^+ \in E^+, \quad \text{then} \quad \lim_{t \to +\infty} \|\Omega_0^t(A)x^+\| = 0, \quad \lim_{t \to -\infty} \|\Omega_0^t(A)x^+\| = \infty;$$

$$\text{if} \quad x^- \in E^-, \quad \text{then} \quad \lim_{t \to -\infty} \|\Omega_0^t x^-\| = 0, \quad \lim_{t \to +\infty} \|\Omega_0^t x^-\| = \infty;$$

$$\text{if} \quad \hat{x} \in \hat{E}, \quad \text{then} \quad \lim_{|t| \to \infty} \|\Omega_0^t(A)\hat{x}\| = 0, \quad \text{and}$$

$$\text{if} \quad \check{x} \in \check{E}, \quad \text{then} \quad \lim_{|t| \to \infty} \|\Omega_0^t \check{x}\| = \infty.$$

Let E be a subspace of R^n, then we designate by $E(t)$ the subspace consisting of the points $\{\Omega_0^t(A)y\}$, where y runs through the whole subspace E.

We take r^+ arbitrary linear independent solutions to the system of equations (1.1.1) $x_+^{(1)}(t), \ldots, x_+^{(r^+)}(t)$ such that $x_+^{(i)}(0) \in E^+$. Clearly, $x_+^{(i)}(t) \in E^+(t)$, $i = 1, \ldots, r^+$, for all $t \in R$. The matrix function composed of the columns $x_+^{(i)}(t)$, $i = 1, \ldots, r^+$, is designated by $X^+(t)$. Similarly $X^-(t)$ denotes the matrix function made up of r^- linear independent solutions to the system (1.1.1)

$x_-^{(1)}(t), \ldots, x_-^{(r^-)}(t)$ with the initial points $x_-^{(i)}(0) \in E^-$, $x_-^{(i)}(t) \in E^-(t)$, $t \in R$, $i = 1, \ldots, r^+$. Applying the Schmidt orthogonality process to the vectors $x_+^{(i)}(t)$, $i = 1, \ldots, r^+$, we get r^+ linear independent orthonormed vectors $\{e_+^{(i)}\}_{i=1}^{r^+}$. Let us represent this in the matrix form

$$Q^+(t) = \left(e_+^{(1)}(t), \ldots, e_+^{(r^+)}(t)\right) = X^+(t)T(t), \qquad (1.2.4)$$

where $T(t)$ is an $r^+ \times r^-$–dimensional triangular non-degenerate matrix function, $\langle e_+^{(i)}(t), e_+^{(j)}(t)\rangle = \delta_{ij}$, $\delta_{ij} = 0$ if $i \neq j$, $\delta_{ii} = 1$. Similarly by orthogonalizing the vectors $x_-^{(i)}(t)$, we get $Q^-(t) = \left(e_-^{(1)}(t), \ldots, e_-^{(r^-)}(t)\right) = X^-(t)\bar{T}(t)$. We designate by $\tilde{Q}(t)$ the matrix function composed of $n - r^+ - r^-$ mutually orthogonal unit columns, and

$$\tilde{Q}^*(t)Q^{\pm}(t) \equiv 0. \qquad (1.2.5)$$

Since the inclusions $E^+(t) \subseteq \bar{E}^+(t)$, $E^+(t) \subseteq \bar{\bar{E}}^+(t)$, $E^-(t) \subseteq \bar{E}^-(t)$, $E^-(t) \in \bar{\bar{E}}^-(t)$ take place for all $t \in R$ and mutual inclinations of subspaces $\bar{E}^+(t)$ and $\bar{E}^-(t)$ on the semiaxis R_+ and that of subspaces $\bar{\bar{E}}^+(t)$ and $\bar{\bar{E}}^-(t)$ on R are larger than zero (see Daletsky and Krein [1])

$$\begin{aligned} \operatorname{Sn}(\bar{E}^+(t), \bar{E}^-(t)) &\geq \gamma > 0, \quad t \in R_+; \\ \operatorname{Sn}(\bar{\bar{E}}^+(t), \bar{\bar{E}}^-(t)) &\geq \gamma, \quad t \in R_-, \end{aligned} \qquad (1.2.6)$$

where $\operatorname{Sn}(E_1, E_2) = \inf_{x_i \in E_i} \left\| \|x_1\|^{-1} x_1 + \|x_2\|^{-1} x_2 \right\|$, then

$$\operatorname{Sn}(E^+(t), E^-(t)) \geq \gamma = \text{const} > 0, \qquad (1.2.7)$$

for all $t \in R$. Equality (1.2.5) and inequality (1.2.7) allow a conclusion on the non-degeneracy of the matrix

$$Q(t) = [Q^+(t), Q^-(t), \tilde{Q}(t)], \qquad (1.2.8)$$

for all $t \in R$, and moreover

$$|\det Q(t)| \geq \delta = \text{const} > 0, \qquad (1.2.9)$$

for all $t \in R$. Actually, let (1.2.9) not be satisfied, then a sequence of values of $t = t_n \xrightarrow[n \to \infty]{} \infty$ is found, for which $\det Q(t_n) \xrightarrow[n \to \infty]{} \infty$. By virtue of the set of matrices $\{Q(t_n)\}_{n=1}^{\infty}$ being compact, a subsequence $t_{n_k} \xrightarrow[n \to \infty]{} \infty$ can be taken so that $\lim_{k \to \infty} Q(t_{n_k}) = \bar{Q}$ exists. On the one hand the inequality $\det \bar{Q} = 0$ holds true. On the other hand we have

$$\lim_{k \to \infty} \operatorname{Sn}(E^+(t_{n_k}), E^-(t_{n_k})) \geq \gamma, \quad \lim_{k \to \infty} \tilde{Q}^*(t_{n_k}) Q^{\pm}(t_{n_k}) = 0.$$

Moreover, the matrices $\bar{Q}$ are the identity ones relative to the norm, which proves the non-degeneracy of the matrix $\bar{Q}$. The contradiction obtained demonstrates that the estimate (1.2.9) is satisfied.

Now we show that the derivative $\frac{d}{dt}Q(t)$ of the matrix function (1.2.8) is bounded on the whole R axis. To this end we represent (1.2.4) as

$$Q^+(t)T^{-1}(t) = X^+(t), \qquad (1.2.10)$$

and differentiate both its parts by t:

$$\frac{d}{dt}(Q^+(t))T^{-1}(t) + Q^+(t)\frac{d}{dt}(T^{-1}(t)) = A(t)X^+(t).$$

Hence it follows that

$$\frac{d}{dt}Q^+(t) = A(t)Q^+(t) - Q^+(t)\frac{d}{dt}(T^{-1}(t))T(t). \qquad (1.2.11)$$

The equality obtained shows that it is sufficient to prove boundedness of the matrix function $\frac{d}{dt}(T^{-1}(t))T(t) = M$ on R. In view of the columns of matrix $Q^+(t)$ being orthonormed we find from (1.2.10)

$$(T^{-1}(t))^*T^{-1}(t) = (X^+(t))^*X^+(t). \qquad (1.2.12)$$

Differentiating both parts of this equality we obtain

$$\frac{d}{dt}(T^{-1}(t))^*T^{-1}(t) + (T^{-1}(t))^*\frac{d}{dt}(T^{-1}(t)) = (X^+(t))^*A^*(t)X^+(t)$$
$$+ (X^+(t))^*A(t)X^+(t).$$

Multiplying the obtained equality by T on the right and by T^* on the left yields

$$T^*(t)\frac{d}{dt}(T^{-1}(t))^* = \frac{d}{dt}(T^{-1}(t))T(t) = (Q^+(t))^*[A^*(t) + A(t)]Q^+(t),$$

or

$$M^*(t) + M(t) = (Q^+(t))^*[A^*(t) + A(t)]Q^+(t). \qquad (1.2.13)$$

As the right-side part of the equality is bounded for all $t \in R$, the matrix $M^*(t) + M(t)$ is bounded on R as well. Since the matrix M, as well as T is of triangular form, the boundedness of the sum $M^*(t) + M(t)$ on R implies the boundedness on R of the matrix $M(t)$. The boundedness on the whole R axis of the derivative $\frac{d}{dt}Q^-(t)$ is proved in the same manner.

Let us now prove that the matrix $\tilde{Q}(t)$ satisfying the condition (1.2.5) and consisting of orthonormed columns can be taken so that the derivative $\frac{d}{dt}\tilde{Q}(t)$

is bounded on the whole R axis. With this in mind we return to the vector-functions $x_+^{(i)}(t) \in E^+(t)$, $x_-^{(j)}(t) \in E^-(t)$, and having orthogonalized the whole set $x_+^{(1)}, \ldots, x_+^{(r^+)}(t)$, $x_-^{(1)}(t), \ldots, x_-^{(r^-)}(t)$, obtain $u_1(t), \ldots, u_{r^++r^-}(t)$, $\langle u_i(t), u_j(t) \rangle \equiv \delta_{ij}$, $\delta_{ii} = 1$, $\delta_{ij} = 0$, $i \neq j$. Clearly, any linear combination of the vectors $u_j(t)$, $j = 1, \ldots, r^+ + r^-$, does not lead out of the space $E^+(t) \oplus E^-(t)$ and the matrix

$$P(t) = \sum_{i=1}^{r^++r^-} u_i(t) u_i^*(t) \tag{1.2.14}$$

is the matrix of the orthogonal projection on the subspace $E^+(t) \oplus E^-(t)$, $P^2(t) \equiv P(t) \equiv P^*$. By virtue of the above-proved result the derivative $\frac{d}{dt} u_i(t)$ is bounded on the whole R axis. Therefore the derivative of the projection matrix (1.2.14) is also bounded on R.

Further (see Bylov et al. [1] and Daletsky and Krein [1]) we consider the system of differential equations which is exponentially dichotomous on the whole R axis:

$$\frac{dz}{dt} = \left[\frac{d}{dt} P(t) + I_n\right][2P(t) - I_n]z = \bar{A}(t)z. \tag{1.2.15}$$

One of the solution subspaces of this system coincides with $E^+(t) \oplus E^-(t) = \tilde{E}(t)$ and the other one, $\bar{E}(t)$ is orthogonal to $\tilde{E}(t)$. Now, we take $n - r^+ - r^-$ linearly independent solutions of the system (1.2.15) $z^{(i)}(t)$, $i = 1, \ldots, (n - r^+ - r^-)$, in the subspace $\bar{E}(t)$ which are then orthonormed, and compose the matrix $\tilde{Q}(t)$ of the obtained vectors. All columns of the matrix $\tilde{Q}(t)$ are obviously orthogonal to all columns of the matrices $Q^+(t)$ and $Q^-(t)$. Since the matrix function $\bar{A}(t)$ is bounded on the whole R axis, then it follows from the above mentioned facts that the derivative $\frac{d}{dt} \tilde{Q}(t)$ is bounded on R. Thus, the matrix (1.2.8) is a Lyapunov matrix on the whole R axis, i.e. it is bounded on R together with its derivative $\frac{d}{dt} Q(t)$ and satisfies condition (1.2.9).

In the system (1.1.1.) we make the change of variables

$$x = Q(t)y. \tag{1.2.16}$$

Since the matrix $Q(t)$ can be represented in the form of the product

$$[X^+(t), X^-(t), \tilde{Q}(t)] \operatorname{diag}\{T(t), \bar{T}(t), I_{n-r^+-r^-}\},$$

the change of variables (1.2.16) is made in two stages. First, we set

$$x = [X^+(t), X^-(t), \tilde{Q}(t)]z,$$

and then

$$z = \operatorname{diag}\{T(t), \bar{T}(t), I_{n-r^+-r^-}\}y.$$

Also, the vectors z and y are expressed ($z = \mathrm{col}\{z_1, z_1, \tilde{z}\}$, $y = \mathrm{col}\{y_1, y_2, \tilde{y}\}$) so that the dimensions of the coordinates z_1, y_1, z_2, y_2 correspond to the dimensions of the matrices X^+, T, X^-, $\bar{T}$. As a result of the first stage of the change of variables the system of equations (1.1.1) becomes

$$\frac{dz_1}{dt} = B_{13}(t)\tilde{z}, \quad \frac{dz_2}{dt} = B_{23}(t)\tilde{z}, \quad \frac{d\tilde{z}}{dt} = B_{33}(t)\tilde{z},$$

and the second stage yields

$$\frac{dy_1}{dt} = B^+(t)y_1 + \bar{B}_{13}(t)\tilde{y},$$
$$\frac{dy_2}{dt} = B^-(t)y_2 + \bar{B}_{23}(t)\tilde{y}, \quad (1.2.17)$$
$$\frac{d\tilde{y}}{dt} = B_3(t)\tilde{y},$$

where

$$B^+(t) = -T^{-1}(t)\frac{d}{dt}T(t), \quad B^-(t) = -\bar{T}^{-1}(t)\frac{d}{dt}\bar{T}(t),$$
$$\bar{B}_{13}(t) = -T^{-1}(t)B_{13}(t), \quad \bar{B}_{23}(t) = \bar{T}^{-1}(t)B_{23}(t).$$

Since (1.2.16) is the *Lyapunov transformation*, the coefficients of the system of equations (1.2.17) are bounded on the R axis and all exponential dichotomy conditions are preserved on the semiaxes of the system (1.1.1) for the system (1.2.17) as well.

Let us now verify whether the equations of the subspaces E^+ and E^- for system (1.2.17) are of the form $\tilde{y} = 0$, $y_2 = 0$, and $\tilde{y} = 0$, $y_1 = 0$ respectively. Let $y_1(t)$ be a solution of system

$$\frac{dy_1}{dt} = B^+(t)y_1,$$

then the function $x(t) = Q^+(t)y_1(t)$ is the solution of the system (1.1.1). Since the initial point of this solution $x(0) = Q^+(0)y_1(0) \in E^+$ and the equality $(Q^+(t))^*x(t) = y_1(t)$ occur, then the behaviour of the solution $y_1(t)$ on $\pm\infty$ is the same as that of $x(t)$.

Let us now show that the system of equations

$$\frac{d\tilde{y}}{dt} = B_{33}(t)\tilde{y} \quad (1.2.18)$$

is also exponentially dichotomous on the semiaxes R_+ and R_-. Actually, for example, considering for $t \in R_+$ the inhomogeneous system

$$\frac{d\tilde{y}}{dt} = B_{33}(t)\tilde{y} + f(t),$$

we make sure that for every vector-function $f(t) \in C^0(R_+)$ this system has a solution bounded on R_+, because the whole system (1.2.17) is exponentially dichotomous on R. This is sufficient for the system (1.2.18) to be exponentially dichotomous on R_+ (see Majzel [1] and Massera and Schaeffer [1]).

Note that the system of the equations (1.2.18) has no solutions which are vanishing on $+\infty$ and increasing on $-\infty$, or vice versa decreasing on $-\infty$ and increasing on $+\infty$. In fact, if such a solution $\tilde{y}^*(t)$ exists, $\|\tilde{y}^*(t)\| \underset{t \to +\infty}{\longrightarrow} 0$ and $\|\tilde{y}^*(t)\| \underset{t \to -\infty}{\longrightarrow} \infty$, the subspace E^+ has the dimensions $r^+ + 1$, which is impossible.

Let us show the system (1.2.18) to have a $\hat{r}$-dimensional subspace of the solutions vanishing as $|t| \to \infty$. To this end we consider $\hat{r}$ linearly independent solutions of the system (1.2.17), damping at $\pm\infty$. Then $y^{(i)}(t) = \text{col}\{y_1^{(i)}(t), y_2^{(i)}(t), \tilde{y}^{(i)}(t)\}$, $i = 1, \ldots, \hat{r}$. Note that the system of $\hat{r}$ vectors $\{y^{(i)}\}_{i=1}^{\hat{r}}$ is also linearly independent for all $t \in R$. If not, then for some $t = t^*$ the equality

$$\tilde{y}(t^*) = \sum_{i=1}^{\hat{r}} c_i \tilde{y}^{(i)}(t^*) = 0$$

takes place, provided that $\sum_{i=1}^{\hat{r}} c_i^2 \neq 0$. Now, on one hand the solution $y(t) = \sum_{i=1}^{\hat{r}} c_i \tilde{y}^{(i)}(t^*)$ of the system (1.2.17) vanishes as $|t| \to \infty$ and, on the other hand, since $\tilde{y}(t^*) = 0 \to \tilde{y}(t) \equiv 0$ the system of equations (1.2.17) possesses no such nontrivial solutions. Thus, the system (1.2.18) has an $\hat{r}$-dimensional subspace $\hat{E}(t)$ of solutions bounded on the whole R axis (every solution decreases exponentially to zero as $|t| \to \infty$).

We take the $\check{r}$-dimensional subspace $\check{E}(t)$ of the solutions to the system of equations (1.2.18) increasing on $\pm\infty$ so that $\check{E}(t) \oplus \hat{E}(t) = R^{\tilde{r}}$. Similarly to the above, by virtue of separability of the subspaces $\hat{E}(t)$ and $\check{E}(t)$ on the whole R axis $(\text{Sn}(\hat{E}(t), \check{E}(t)) \geq \delta > 0 \; \forall t \in R)$ the unitary tracking hedrons ($\hat{r}$-hedron and $\check{r}$-hedron respectively) can be taken as the subspaces $\hat{E}(t)$ and $\check{E}(t)$ so that the square matrix $\Lambda(t)$ composed of them is a Lyapunov matrix. Now, making the change of variables $y = \Lambda(t) \text{col}\{y_3, y_4\}$ in the system of equations (1.2.17) we arrive at equations (1.2.2). Theorem 1.2.1 is proved.

Assume the system (1.1.1) is exponentially dichotomous on the semiaxes R_+ and R_-. Then by virtue of the system (1.1.1) for $t \in R_+$ there exist the quadratic forms $V^+(t, x)$ for the right semiaxis R^+ having a derivative $\dot{V}^+(t, x)$ of definite sign, and the quadratic forms $V^-(t, x)$ for the left semiaxis R_- having derivative $\dot{V}^-(t, x)$ of definite sign. There arises a question: whether a quadratic form $V(t, x)$ exists on the whole R axis under these conditions, which has a derivative of definite sign due to the system (1.1.1)? The investigations which have been made have shown that such a quadratic form does not exist every time. The basic result is as follows.

Theorem 1.2.2 *In order that the quadratic form* $V(t,x) = \langle S(t)x, x\rangle$, $S^*(t) \equiv S(t) \in C^1(R)$, *exists, which satisfies estimate (1.1.6) for all* $t \in R$, *it is necessary and sufficient that the system of equations (1.1.1) be exponentially dichotomous on the semiaxes* R_+ *and* R_- *and have no non-trivial solution bounded on* R.

Proof *Necessity.* The proof of the exponential dichotomy of system (1.1.1) on the axes R_+ and R_- is cited in Majzel [1] and Massera and Schaeffer [1]. That is why we only show that the system (1.1.1) has no non-trivial solutions bounded on the whole R axis. Let such a solution $x = x^*(t)$ exist, then for all $t \in R$

$$V(t, x^*(t)) > 0. \tag{1.2.19}$$

Actually, if this inequality is violated for some $t = t_1$, $V(t_1, x^*(t_1)) \leq 0$ then by virtue of the monotonicity $V(t_2, x^*(t_2)) < 0$, $t_1 < t_2$, and for all $t \geq t_2$ the estimate

$$\|S\|_0 \|x^*(t)\|^2 \geq -V(t, x^*(t)) \geq -V(t_2, x^*(t_2)) \exp\left\{\frac{\beta}{\|S\|_0}(t - t_2)\right\}$$

is valid, which contradicts the boundedness of the solution $x = x^*(t)$ on the whole R axis. Therefore, the inequality (1.2.19) must be satisfied for all $t \in R$. On the other hand, this inequality provides the estimate

$$0 < V(t, x^*(t)) \leq V(\tau, x^*(\tau)) \exp\left\{-\frac{\beta}{\|S\|_0}(t - \tau)\right\}.$$

Hence it follows that $\lim_{\tau \to -\infty} V(\tau, x^*(\tau)) = +\infty$, which contradicts the estimate $V(\tau, x^*(\tau)) \leq \|S\|_0 \|x^*(\tau)\|^2$ and the boundedness on R of the solution $x = x^*(t)$. Thus, we obtain that the system (1.1.1) has no non-trivial solutions bounded on the whole R axis.

Sufficiency. We transform the system (1.1.1) to the form (1.2.2) by the change of Lyapunov variables (1.2.1). Since the system (1.1.1) has no non-trivial solutions bounded on R, then neither has the system (1.2.2). Therefore, in the decomposed system (1.2.2) the subsystem

$$\frac{dy_3}{dt} = \hat{B}(t) y_3$$

is absent and, moreover, $\tilde{y} = y_4$. Without loss of generality, one can consider the norms of matrices $B_1(t)$ and $B_3(t)$ to be sufficiently small. This can be obtained by the change of variables $y_4 = \varepsilon \bar{y}_4$ and by taking the fixed parameter $\varepsilon > 0$ sufficiently small.

Consider the quadratic form

$$V(t, y_1, y_2, y_4) = \langle v^+(t) y_1, y_1 \rangle + \langle v^-(t) y_2, y_2 \rangle + \langle \tilde{v}(t) y_4, y_4 \rangle, \tag{1.2.20}$$

where

$$v^+(t) = \int_t^\infty [\Omega_t^\tau(B^+)]^* \Omega_t^\tau(B^+)\, d\tau, \quad v^-(t) = -\int_{-\infty}^t [\Omega_t^\tau(B^-)]^* \Omega_t^\tau(B^-)\, d\tau,$$

$$\check{v}(t) = \int_t^\infty [\Omega_t^\tau(\check{B})]^* \Omega_t^\tau(\check{B})\, d\tau. \tag{1.2.21}$$

We make sure that all that the matrix functions (1.2.21) belong to the space $C^1(R)$. Its continuous differentiability is obvious:

$$\frac{d}{dt} v^+(t) = -I_{r^+} - [B^+(t)]^* v^+(t) - v^+(t) B^+(t),$$

$$\frac{d}{dt} v^-(t) = -I_{r^-} - [B^-(t)]^* v^-(t) - v^-(t) B^-(t), \tag{1.2.22}$$

$$\frac{d}{dt} \check{v}(t) = -I_{\check{r}} - [\check{B}(t)]^* \check{v}(t) - \check{v}(t) \check{B}(t).$$

The estimates (1.2.3) imply boundedness of the functions (1.2.21) on the whole R axis

$$\|v^+(t)\| \leq K^2 \int_t^\infty \exp\{-2\gamma(\tau - t)\}\, d\tau = \frac{K^2}{2\gamma},$$

$$\|v^-(t)\| \leq K^2 \int_{-\infty}^t \exp\{2\gamma(t - \tau)\}\, d\tau = \frac{K^2}{2\gamma}.$$

For all $t < 0$

$$\|\check{v}(t)\| \leq K^2 \int_t^0 \exp\{-2\gamma(\tau - t)\}\, d\tau = \frac{K^2}{2\gamma}(1 - \exp\{2\gamma t\}) \leq \frac{K^2}{2\gamma}.$$

If $t > 0$, then

$$\|\check{v}(t)\| \leq K^2 \int_0^t \exp\{2\gamma(\tau - t)\}\, d\tau = \frac{K^2}{2\gamma}(1 - \exp\{-2\gamma t\}) \leq \frac{K^2}{2\gamma}.$$

The representation of derivatives (1.2.22) provides that the derivative of the quadratic form (1.2.20) due to the system

$$\frac{dy_1}{dt} = B^+(t) y_1, \quad \frac{dy_2}{dt} = B^-(t) y_2, \quad \frac{dy_4}{dt} = \check{B}(t) y_4 \tag{1.2.23}$$

is of definite sign

$$\dot{V}(t, y_1, y_2, y_4) \leq -(\|y_1\|^2 + \|y_2\|^2 + \|y_4\|^2).$$

Since we presume the norm of the matrix functions B_1 and B_2 to be sufficiently small, the derivative of the quadratic form (1.2.20) due to the perturbed system

$$\frac{dy_1}{dt} = B^+(t)y_1 + B_1(t)y_4,$$

$$\frac{dy_2}{dt} = B^-(t)y_2 + B_2(t)y_4, \qquad (1.2.24)$$

$$\frac{dy_4}{dt} = \check{B}(t)y_4$$

is also negative definite. Actually, computing the derivative $\dot{V}(t, y_1, y_2, y_4)$ in view of the system (1.2.24) yields

$$\begin{aligned}\dot{V}(t, y_1, y_2, y_4) &= -\|y_1\|^2 - \|y_2\|^2 - \|y_4\|^2 + \langle v^+(t)B_1(t)y_4, y_1\rangle \\ &+ \langle v^+(t)y_1, B_1(t)y_4\rangle + \langle v^-(t)B_2(t)y_4, y_2\rangle + \langle v^-(t)y_2, B_2(t)y_4\rangle \\ &\leq -\|y_1\|^2 - \|y_2\|^2 - \|y_4\|^2 + 2\|v^+\|_0 \|B_1\|_0 \|y_4\| \|y_1\| \\ &+ 2\|v^-\|_0 \|B_2\|_0 \|y_4\| \|y_2\|.\end{aligned} \qquad (1.2.25)$$

Using the inequality $2ab \leq a^2/\sigma^2 + \sigma^2 b^2$, valid for any value of the parameter $\sigma \neq 0$ and any $a, b \in R$ we represent the estimate (1.2.25) as

$$\dot{V}(t, y_1, y_2, y_4) \leq -\left(1 - \frac{\|v^+\|_0^2 \|B_1\|_0^2}{\sigma_1^2}\right)\|y_1\|^2 \qquad (1.2.26)$$
$$-\left(1 - \frac{\|v^-\|_0^2 \|B_2\|_0^2}{\sigma_2^2}\right)\|y_2\|^2 - (1 - \sigma_1^2 - \sigma_2^2)\|y_4\|^2.$$

Hence it is clear that for the derivative $\dot{V}(t, y_1, y_2, y_4)$ to be negative definite, it is sufficient to take the values of parameters σ_1 and σ_2 such that the inequalities

$$\sigma_1^2 - \|v^+\|_0^2 \|B_1\|_0^2 > 0, \quad \sigma_2^2 - \|v^-\|_0^2 \|B_2\|_0^2 > 0, \quad 1 - \sigma_1^2 - \sigma_2^2 > 0 \qquad (1.2.27)$$

be satisfied simultaneously. Adding the first and second inequalities and in view of the third one, we get

$$\|v^+\|_0^2 \|B_1\|_0^2 + \|v^-\|_0^2 \|B_2\|_0^2 < 1. \qquad (1.2.28)$$

Obviously, if the inequality (1.2.28) holds, the values of parameters σ_1 and σ_2 can always be found such that each of the inequalities (1.2.27) is satisfied. Thus, if the matrix-functions B_1 and B_2 satisfy the inequality (1.2.28), i.e. its norm is sufficiently small, then the derivative of the quadratic form (1.2.20) due to the system (1.2.24) is negative definite.

Passing back to the variables x: $x = L(t)\operatorname{col}\{y_1, y_2, y_4\}$ in the quadratic form (1.2.20) we get the quadratic form $V(t,x)$ satisfying the estimate (1.1.6) for all $t \in R$, $x \in R^n$. This completes the proof of the Theorem 1.2.2.

A complement to the proofs of the Theorem 1.2.1 and 1.2.2 results in the following assertion.

Theorem 1.2.3 *Let there exist the symmetric matrix function $S(t) \in C^0(R)$ satisfying the condition*

$$\langle [\dot{S}(t) - S(t)A^*(t) - A(t)S(t)]x,\, x \rangle \leq -\beta \|x\|^2 \qquad (1.2.29)$$

for all $t \in R$, $x \in R^n$, where β is a const > 0. Then there is a change of Lyapunov variables $x = L(t)y$ reducing the system of the equations (1.1.1) to the block-triangular form

$$\frac{dy_1}{dt} = \Phi^+(t)y_1, \qquad \frac{dy_2}{dt} = \Phi^-(t)y_2,$$
$$\frac{dy_3}{dt} = \Phi_1(t)y_1 + \Phi_2(t)y_2 + \hat{\Phi}_3(t)y_3, \qquad (1.2.30)$$

where $\Phi_1(t)$ and $\Phi_2(t)$ are some rectangular matrix functions belonging to the space $C^0(R)$, and the quadratic matrix functions $\Phi^\pm(t)$ and $\hat{\Phi}(t)$ are such that the corresponding matriciants satisfy the estimates

$$\|\Omega_\tau^t(\Phi^+)\| \leq K \exp\{-\gamma(t-\tau)\}, \quad \tau \leq t,$$
$$\|\Omega_\tau^t(\Phi^-)\| \leq K \exp\{\gamma(t-\tau)\}, \quad t \leq \tau, \qquad (1.2.31)$$
$$\|\Omega_\tau^t(\hat{\Phi})\| \leq K \begin{cases} \exp\{-\gamma(t-\tau)\}, & 0 \leq \tau \leq t, \\ \exp\{\gamma(t-\tau)\}, & t \leq \tau < 0, \end{cases}$$

where K and γ are const > 0.

Proof Inequality (1.2.29) means negative definiteness of the derivative $\dot{V}$ of the quadratic form $V = \langle S(t)x, x \rangle$ computed along the solutions of the linear system of differential equations conjugated with (1.1.1)

$$\frac{dx}{dt} = -A^*(t)x. \qquad (1.2.32)$$

Therefore, by Theorem 1.2.2 the system (1.2.32) is exponentially dichotomous on the semiaxes R_+ and R_- and possesses no non-trivial solutions bounded on the whole R axis. There is a change of Lyapunov variables $x = \bar{L}(t)y$ reducing the system (1.2.32) to (1.2.24), i.e.

$$-\bar{L}^{-1}(t)A^*(t)\bar{L}(t) - \bar{L}^{-1}(t)\dot{\bar{L}}(t) = \begin{pmatrix} B^+(t) & 0 & B_1(t) \\ 0 & B^-(t) & B_2(t) \\ 0 & 0 & \check{B}(t) \end{pmatrix} = B(t). \quad (1.2.33)$$

In view of the identity

$$\left(\frac{d}{dt}\bar{L}(t)\right)^* (\bar{L}^{-1}(t))^* \equiv -\bar{L}^*(t) \frac{d}{dt} (\bar{L}^{-1}(t))^*$$

we get from (1.2.32) that

$$\bar{L}^*(t) A(t) (\bar{L}^{-1}(t))^* - \bar{L}^*(t) \frac{d}{dt}(\bar{L}^{-1}(t))^* = -B^*(t). \quad (1.2.34)$$

Hence, it becomes clear that if in the initial system (1.1.1) the change of variables $x = (\bar{L}^{-1}(t))^* y$ is made, one gets the system conjugated with the system (1.2.24)

$$\frac{dy_1}{dt} = -(B^+(t))^* y_1, \quad \frac{dy_2}{dt} = -(B^-(t))^* y_2,$$
$$\frac{dy_3}{dt} = -B_1^*(t) y_1 - B_2^*(t) y_2 - (\check{B}(t))^* y_3, \quad (1.2.35)$$

(here $y_4 \to y_3$). Now let us show that if for the matricant $\Omega_\tau^t(B^+)$ the first of the estimates (1.2.3) is satisfied, then for the matricant $\Omega_\tau^t(-(B^+)^*) = \Omega_\tau^t(\Phi^-)$ the second of the estimates (1.2.31) is also satisfied. In fact, since the identity $\Omega_\tau^t(-B^*) \equiv (\Omega_t^\tau(B))^*$ occurs, the first of the estimates (1.2.3) leads to

$$\|\Omega_\tau^t(\Phi^-)\| = \|\Omega_\tau^t(-(B^+)^*)\| = \|\Omega_t^\tau(B^+)\| \le K \exp\{-\gamma(\tau - t)\}$$
$$= K \exp\{\gamma(t - \tau)\}, \quad t \le \tau, t, \tau \in R.$$

Thus, having designated $\Phi^+(t) = -(B^-(t))^*$, $\Phi^-(t) = -(B^+(t))^*$, $\hat{\Phi}(t) = -(\check{B}(t))^*$ and changing $y_1 \to y_2$, $y_2 \to y_1$ we pass from the system (1.2.35) to (1.2.30) with the estimates of the matricants (1.2.31). This completes the proof of Theorem 1.2.3.

Then a question arises: whether the transformation of system (1.1.1) to the block-diagonal form (1.2.2) is better than the possible transformation of this system to the triangular form (the known Perron theorem)? To answer this question let us consider two simple examples

$$\frac{dx_1}{dt} = x_1, \quad \frac{dx_2}{dt} = x_1 - x_2; \quad (1.2.36)$$

$$\frac{dx_1}{dt} = -(\tanh t) x_1, \quad \frac{dx_2}{dt} = x_1 + (\tanh t) x_2. \quad (1.2.37)$$

Both systems are exponentially dichotomous on the whole R axis. In addition, the corresponding diagonal system for system (1.2.36)

$$\frac{dx_1}{dt} = x_1, \quad \frac{dx_2}{dt} = -x_2$$

is also exponentially dichotomous on R, and the corresponding diagonal system for the system (1.2.37)

$$\frac{dx_1}{dt} = -(\tanh t)x_1, \quad \frac{dx_2}{dt} = (\tanh t)x_2$$

is no longer exponentially dichotomous on R.

Thus, if the block-diagonal system

$$\frac{d}{dt} \begin{pmatrix} x^{(1)} \\ x^{(2)} \\ \vdots \\ x^{(k)} \end{pmatrix} = \begin{pmatrix} A_1(t) & 0 & \cdots & 0 \\ 0 & A_2(t) & \cdots & 0 \\ \vdots & \vdots & \ddots & \vdots \\ 0 & 0 & \cdots & A_k(t) \end{pmatrix} \begin{pmatrix} x^{(1)} \\ x^{(2)} \\ \vdots \\ x^{(k)} \end{pmatrix} \quad (1.2.38)$$

is exponentially dichotomous on the whole R axis, the corresponding system of block-triangular form

$$\frac{d}{dt} \begin{pmatrix} x^{(1)} \\ x^{(2)} \\ \vdots \\ x^{(k)} \end{pmatrix} = \begin{pmatrix} A_1(t) & 0 & \cdots & 0 \\ B_{21}(t) & A_2(t) & \cdots & 0 \\ \vdots & \vdots & \ddots & \vdots \\ B_{k1}(t) & B_{k2}(t) & \cdots & A_k(t) \end{pmatrix} \begin{pmatrix} x^{(1)} \\ x^{(2)} \\ \vdots \\ x^{(k)} \end{pmatrix} \quad (1.2.39)$$

is also exponentially dichotomous on R for any fixed matrix functions $B_{ij}(t) \in C^0(R)$. However, the converse assertion is not true. The system of equations (1.2.39) can be exponentially dichotomous on R for some matrix functions A_i and B_j, although the corresponding system (1.2.38) may not be exponentially dichotomous on the whole R axis (it is exponentially dichotomous only on the semiaxes R_+ and R_-). It is known (see Daletsky and Krein [1]) that the system (1.1.1) which is exponentially dichotomous on R can always be decomposed by the transformation of Lyapunov variables into two subsystems

$$\frac{dy_1}{dt} = \Phi^+(t)y_1, \quad \frac{dy_2}{dt} = \Phi^-(t)y_2, \quad (1.2.40)$$

corresponding to the exponential dichotomy of the system (1.1.1). Therefore for the system (1.1.1) which is exponentially dichotomous on the semiaxes R_+ and R_- a problem of its transformation was posed such that the diagonal matrix functions determine exponentially dichotomous behaviour of all solutions on the R axis.

Remark 1.2.1 If in the conditions of Theorem 1.2.3 it is additionally required that $\det S(t) \neq 0 \ \forall t \in R$, then in the system (1.2.30) the last subsystem is absent, i.e. system (1.1.1) is transformed to the decomposed form (1.2.40).

1.3 Degeneracy of the Quadratic Form Possessing a Definite-Sign Derivative along the Solutions of the System (1.1.1)

The fact that inequality (1.1.6) is satisfied for some matrix function $S(t) \in C^1(R)$ does not ensure that $\det S(t) \neq 0$ for all $t \in R$. The determinant of the matrix $S(t)$ may vanish for some values of t. The question is: how many such values of t exist and how they are located on the numerical R axis?

Theorem 1.3.1 *Let there exist a symmetric matrix function $S(t) \in C^1(R)$, satisfying the estimate (1.1.6). Then, the determinant of the matrix $S(t)$ may vanish at not more than n isolated points $t_1, \ldots, t_k$, $k \leq n$ (n is the dimension of the system (1.1.1)).*

Proof We assume that there exist $n+1$ different values of t: $t_1, \ldots, t_{n+1}$ ($t_j < t_{j+1}$, $j = 1, \ldots, n$) for which $\det S(t_i) = 0$. We take and fix n-dimensional numerical vectors $\eta^{(i)}$, $\|\eta^{(i)}\| = 1$, $i = 1, \ldots, n+1$, satisfying the systems of algebraic equations

$$S(t_i)\eta = 0, \quad i = 1, \ldots, n+1. \tag{1.3.1}$$

For the sake of simplicity let us consider the case when rank $S(t_i) = n - 1$ for all $i = 1, \ldots, n-1$. We designate by $x^{(i)}(t)$ the solution of the system (1.1.1) with the initial condition $x^{(i)}(t)|_{t=t_i} = \eta^{(i)}$, $i = 1, \ldots, n+1$. Note that the solutions $x^{(1)}(t)$, $x^{(2)}(t)$ cannot be linearly dependent. Actually, if $cx^{(2)}(t) \equiv x^{(1)}(t)$, then the function $m_1(t) = \langle S(t)x^{(1)}(t), x^{(1)}(t)\rangle = c^2 \langle S(t)x^{(2)}(t), x^{(2)}(t)\rangle$ takes zero values $m_1(t_1) = m_1(t_2) = 0$, $t_1 < t_2$, and this is impossible due to the strict monotonicity of the function $m_1(t)$.

We discuss now the function

$$m_2(t) = \langle S(t)[x^{(1)}(t) + x^{(2)}(t)], [x^{(1)}(t) + x^{(2)}(t)]\rangle. \tag{1.3.2}$$

Let us show that the inequalities

$$m_2(t_1) > 0, \quad m_2(t_2) < 0 \tag{1.3.3}$$

are satisfied at two different points $t = t_1$ and $t = t_2$. The computation of the value of function $m_2(t)$ at the point $t = t_1$ and in view of the fact that the vector $\eta^{(1)}$ satisfies the system of algebraic equations (1.3.1) for $i = 1$ yields

$$m_2(t_1) = \langle S(t_1)[\eta^{(1)} + x^{(2)}(t_1)], [\eta^{(1)} + x^{(2)}(t_1)]\rangle = \langle S(t_1)x^{(2)}(t_1), x^{(2)}(t_1)\rangle.$$

Because the function $m(t) = \langle S(t)x^{(2)}(t), x^{(2)}(t)\rangle$ is strictly monotonous decreasing on the whole R axis and $m(t_2) = 0$, it takes positive values at $t = t_1 < t_2$, and hence $m_2(t_1) = m(t_1) > 0$. Now we compute the value of the function (1.3.2) at point $t = t_2$: $m_2(t_2) = \langle S(t_2)x^{(1)}(t_2), x^{(1)}(t_2)\rangle = m_1(t_1)$. Since the function $m_1(t)$ is strictly monotonous decreasing and $m_1(t_1) = 0$, $t_1 < t_2$, then $m_1(t_2) < 0$. The validity of the inequalities (1.3.3) ensures the existence of the point $t = t^* \in (t_1, t_2)$, at which the continuous function (1.3.2) takes the value zero, $m_2(t^*) = 0$. This allows us to state linear independence of solutions $x^{(i)}(t)$, $i = 1, 2, 3$. In fact if these solutions are linearly dependent, i.e. $x^{(3)}(t) \equiv c_1 x^{(1)}(t) + c_2 x^{(2)}(t)$, then the strictly monotonous decreasing function $\langle S(t)x^{(3)}(t), x^{(3)}(t)\rangle$ takes zero values at two different points $t = t_3$ and $t = t^* \in (t_1, t_2)$, which is impossible.

Now we consider the function

$$m_3(t) = \langle S(t)[x^{(1)}(t) + x^{(2)}(t) + x^{(3)}(t)], [x^{(1)}(t) + x^{(2)}(t) + x^{(3)}(t)]\rangle \quad (1.3.4)$$

and show that $m_3(t_1) > 0$ and $m_3(t_3) < 0$. Taking into account the strict monotonicity of the functions (1.3.2) and (1.3.4) and the fact that $m_2(t^*) = 0$, $t^* \in (t_1, t_2)$, we get

$$m_3(t_1) > m_3(t^*) = \langle S(t^*)x^{(3)}(t^*), x^{(3)}(t^*)\rangle > \langle S(t_3)x^{(3)}(t_3), x^{(3)}(t_3)\rangle = 0.$$

Similarly $m_3(t_3) = m_2(t_3) < m_2(t^*) = 0$. Thus there exists a point $t = t^{**} \in (t_1, t_3)$, where $m_3(t^{**}) = 0$. This again allows us to conclude that the solution $x^{(4)}(t)$ cannot be represented as the linear combination of solutions $x^{(i)}(t)$, $i = 1, 2, 3$.

Proceeding in a similar manner results in a contradiction for $n + 1$, since the linear system of n equations (1.1.1) can have n linearly independent solutions, and no more. This completes the proof of Theorem 1.3.1.

Remark 1.3.1 The proof of the Theorem 1.3.1 shows that for the number of a values of $t = t_1, \ldots, t_k$ for which $\det S(t_i) = 0$ a more refined estimate can be found taking into account the rank of matrices $S(t_i)$, namely

$$\sum_{i=1}^{k} [n - \mathrm{rank}\ S(t_i)] \leq n.$$

Remark 1.3.2 The condition of the Theorem 1.3.1 can be weakened by the requirement that the inequality

$$\langle [\dot{S}(t) + S(t)A(t) + A^*(t)S(t)]x, x\rangle < 0 \quad \forall x \in R^n, \quad x \neq 0 \quad (1.3.5)$$

be satisfied for the symmetric matrix $S(t)$ instead of (1.1.6).

Now let us discuss the question: what matrix function $S(t) \in C^1(R)$ does the matrix $A(t)$ exist for, such that the inequality (1.1.6) takes place? The answer is provided by the following result.

DICHOTOMIES AND STABILITY

Theorem 1.3.2 *Let the symmetric matrix function $S(t) \in C^1(R^n)$ satisfy the following conditions*

(1) $\langle \dot{S}(t)x, x \rangle \leq m\|x\|^2$, $x \in R^n$, where m is a const > 0 $\forall t \in R$;

(2) *the determinant of the matrix $S(t)$ vanishes at a finite number of isolated points $t = t_1, \ldots, t_k$, $t_1 < t_2 < \cdots < t_k$;*

(3) *for all $x \in R^n$ the estimate*

$$\langle \dot{S}(t_i)P_i x, P_i x \rangle \leq -\bar{\beta}\|P_i x\|^2, \quad i = 1, \ldots, k, \tag{1.3.6}$$

is valid, where $\bar{\beta}$ is const > 0, $x \in R^n$, P_i is a matrix of the orthogonal projection on the subspace of solutions to the algebraic system of equations $S(t_i)x = 0$, $P_i^2 = P_i = P_i^*$;

(4) *for every $\varepsilon > 0$ there exists a $\delta(\varepsilon) > 0$, such that*

$$\|S(t)x\|^2 \geq \delta(\varepsilon)\|x\|^2 \quad \text{for all} \quad t \in (-\infty, t_1 - \varepsilon] \cup [t_k + \varepsilon, \infty).$$

Then there exists a matrix function $A(t) \in C^0(R)$ such that the estimate (1.1.6) is satisfied, and the matrix $-\lambda S(t)$, in particular, can be taken for $A(t)$, where the parameter λ is sufficiently large.

Proof Let us first show that for λ being large enough the estimate

$$\dot{V} = \langle [\dot{S}(t_i) - \lambda S^2(t_i)]x, x \rangle \leq -\beta_1 \|x\|^2 \tag{1.3.7}$$

takes place for all $x \in R^n$, where β_1 is const > 0, $i = 1, \ldots, k$. In fact, we designate by $\bar{m}$ a positive number which satisfies the inequality

$$\|S(t_i)x\|^2 \geq \bar{m} \|(I_n - P_i)x\|^2, \quad i = 1, \ldots, k. \tag{1.3.8}$$

The following number can be taken to be such a number

$$\bar{m} = \min_{1 \leq i \leq k} \{m_i\}, \tag{1.3.9}$$

where $m_i = \min\{\lambda_1^{-2}, \lambda_2^{-2}, \ldots, \lambda_{p_i}^{-2}\}$, $\lambda_1, \lambda_2, \ldots, \lambda_{p_i}$ are non-zero eigenvalues of the matrix $S(t_i)$. If all eigenvalues of some matrix $S(t_0)$ are zero, i.e. $S(t_0) = 0$, then $m_{i_0} > 0$ in (1.3.9) can be arbitrary.

We consider the following transformation

$$\dot{V} = \langle [\dot{S}(t_i) - \lambda S^2(t_i)]x, x \rangle = \langle \dot{S}(t_i)P_i x, P_i x \rangle + 2\langle \dot{S}(t_i)P_i x, (I_n - P_i)x \rangle$$
$$+ \langle \dot{S}(t_i)(I_n - P_i)x, (I_n - P_i)x \rangle - \lambda \|S(t_i)x\|^2 \leq -\bar{\beta}\|P_i x\|^2 \tag{1.3.10}$$
$$+ 2m\|P_i x\| \|(I_n - P_i)x\| + m\|(I_n - P_i)x\|^2 - \lambda \bar{m}\|(I_n - P_i)x\|^2.$$

Introducing the non-zero parameter σ in the same way as for inequality (1.1.26) we proceed with the inequality (1.3.10) as follows

$$\dot{V} \leq -\left(\bar{\beta} - \frac{m^2}{\sigma^2}\right)\|P_i x\|^2 - (\lambda\bar{m} - m - \sigma^2)\|(I_n - P_i)x\|^2.$$

Hence, it becomes clear that as the inequalities

$$\gamma_1 = \bar{\beta} - \frac{m^2}{\sigma^2} > 0, \quad \gamma_2 = \lambda\bar{m} - m - \sigma^2 > 0, \tag{1.3.11}$$

are satisfied, the estimate (1.3.7) is satisfied too, where $\beta_1 = \frac{1}{2}\min\{\gamma_1, \gamma_2\}$. Having analysed (1.3.11) we make sure that provided

$$\lambda > m/\bar{m}\left(1 + m/\bar{\beta}\right), \tag{1.3.12}$$

there always exist the value of parameter σ, for which inequalities (1.3.11) hold true simultaneously.

Since the matrix functions $S(t)$ and $\dot{S}(t)$ depend continuously on t, the estimate (1.3.7) also takes place for all t from some neighbourhoods of the points $t = t_i$, i.e. there exists an $\varepsilon > 0$ which is sufficiently small, such that for all $t \in \bigcup_{i=1}^{k}[t_i-\varepsilon, t_i+\varepsilon]$ the inequality

$$\langle[\dot{S}(t) - \lambda S^2(t)]x, x\rangle \leq -\beta_1^*\|x\|^2, \quad x \in R^n \tag{1.3.13}$$

holds. We represent the whole R axis in the form of the unification of two sets M_1 and M_2, where $M_1 = \bigcup_{i=1}^{k}[t_i - \varepsilon, t_i + \varepsilon]$, $M_2 = R \setminus M_1$. With regard to the third and fourth condition of Theorem 1.3.2 we have for all $t \in M_2$

$$\langle[\dot{S}(t) - \lambda S^2(t)]x, x\rangle \leq m\|x\|^2 - \lambda\|S(t)x\|^2 \leq -(\lambda\bar{\delta}(\varepsilon) - m)\|x\|^2,$$

where

$$\bar{\delta}(\varepsilon) = \min\{\delta(\varepsilon), \tilde{\delta}(\varepsilon)\}, \quad \tilde{\delta}(\varepsilon) = \inf_{t \in M_1, \|x\|=1}\|S(t)x\|^2.$$

Thus provided that the parameter λ is large enough, the inequality (1.3.13) is satisfied for all $t \in R$. Theorem 1.3.2 is proved.

Remark 1.3.3 In the case when $\det S(t)$ does not vanish at any point of the R axis, we assume that the estimate $|\det S(t)| \geq \delta = \text{const} > 0$ in Theorem 1.3.2 is satisfied for all $t \in R$ instead of conditions (2)–(4).

Theorem 1.3.3 *In order that the inhomogeneous system of the equations (1.1.4) have at least one solution bounded on the whole R axis for every fixed vector-function $f(t) \in C^0(R)$, it is necessary and sufficient that an $n \times n$-dimensional symmetric matrix function $S(t) \in C^1(R)$ exists satisfying the condition (1.2.29).*

Proof Let the system (1.1.4) possess at least one solution bounded on R for every vector-function $f(t) \in C^0(R)$. Then the system (1.1.1) is exponentially dichotomous on the semiaxes R_+ and R_- (see Daletsky and Krein [1] and Pliss [1]). Therefore, it can be reduced to the form (1.2.2) by means of the Lyapunov variable transformations (1.2.1). We show that in system (1.2.2) the last subsystem

$$\frac{dy_4}{dt} = \check{B}(t)y_4$$

is absent, all non-trivial solutions of which increase exponentially over $\pm\infty$. In fact, if this subsystem is present, the inhomogeneous system

$$\frac{dy_4}{dt} = \check{B}(t)y_4 + \varphi(t) \qquad (1.3.14)$$

does not have a solution bounded on R for every vector-function $\varphi(t) \in C^0(R)$. The general solution of the system (3.1.14) is

$$y_4(t) = \Omega_0^t(\check{B})\left[y_4(0) + \int_0^t \Omega_\sigma^t(\check{B})\varphi(\sigma)\,d\sigma\right]. \qquad (1.3.15)$$

In view of the last estimate in (1.2.3) the initial point of the solution bounded on the semiaxis R_+ of the system (1.3.4) is

$$y_4^+(0) = -\int_0^\infty \Omega_\sigma^0(\check{B})\varphi(\sigma)\,d\sigma. \qquad (1.3.16)$$

Similarly, we find the initial point of the unique solution bounded on R_- from the general solution (1.3.15)

$$y_4^-(0) = \int_{-\infty}^0 \Omega_\sigma^0(\check{B})\varphi(\sigma)\,d\sigma. \qquad (1.3.17)$$

Hence we conclude that if the system (1.3.4) has a solution bounded on R, then $y_4^+(0) = y_4^-(0)$, and we consequently get from (1.3.16) and (1.3.17)

$$\int_{-\infty}^\infty \Omega_\sigma^0(\check{B})\varphi(\sigma)\,d\sigma = 0. \qquad (1.3.18)$$

Obviously, the condition (1.3.18) does not hold for every vector-function $\varphi(t) \in C^0(R)$. Therefore, the inhomogeneous system (1.3.14) does not have a solution bounded on R for every vector-function $\varphi(t) \in C^0(R)$. Thus, in the decomposed system (1.2.2) the last subsystem is absent, and hence, the system conjugated with (1.2.2) is

$$\frac{dy_1}{dt} = -(B^+(t))^* y_1, \quad \frac{dy_2}{dt} = -(B^-(t))^* y_2,$$

$$\frac{dy_3}{dt} = -B_1^*(t) y_1 - B_2^*(t) y_2 - (\check{B}(t))^* y_3.$$

Therefore, the quadratic form is constructed for it in a form similar to (1.2.20) and (1.2.21).

Assume that there exists an n-dimensional symmetric matrix $S(t) \in C^1(R)$ satisfying the condition (1.2.29). Then, by Theorem 1.2.3 there is a substitution of Lyapunov variables $x = L(t)y$ transforming the inhomogeneous system (1.1.4) to the form

$$\frac{dy_1}{dt} = \Phi^+(t) y_1 + \varphi_1(t), \quad \frac{dy_2}{dt} = \Phi^-(t) y_2 + \varphi_2(t), \qquad (1.3.19)$$

$$\frac{dy_3}{dt} = \Phi_1(t) y_1 + \Phi_2(t) y_2 + \hat{\Phi}(t) y_3 + \varphi_3(t),$$

where $\text{col}\{\varphi_1(t), \varphi_2(t), \varphi_3(t)\} = L^{-1}(t) f(t)$. As $L(t)$ is the Lyapunov matrix, the existence of the solutions of the system (1.1.4) bounded on R is equivalent to that of solutions of the system (1.3.19) bounded on R which exist by the estimates (1.2.31) and are

$$y_1(t) = \int_{-\infty}^{t} \Omega_\sigma^t(\Phi^+) \varphi_1(\sigma) \, d\sigma, \quad y_2(t) = -\int_{t}^{\infty} \Omega_\sigma^t(\Phi^-) \varphi_2(\sigma) \, d\sigma,$$

$$y_3(t) = \Omega_0^t(\hat{\Phi}) y_3(0) + \int_0^t \Omega_\sigma^t(\hat{\Phi}) [\varphi_3(\sigma) + \Phi_1(\sigma) y_1(\sigma) + \Phi_2(\sigma) y_2(\sigma)] \, d\sigma.$$

Also, the initial point $y_3(0) \in R^{\hat{n}}$ can be taken in an arbitrary way. This completes the proof of Theorem 1.3.3.

Remark 1.3.4 Condition (1.3.18) is necessary and sufficient for the system of the equations (1.3.14) to have a solution bounded on R.

Actually, if this condition is satisfied, we take one of the values (1.3.16) or (1.3.17) as the initial point $y_4(0)$ in the general solution (1.3.15). Taking, for example, (1.3.16) and substituting it into (1.3.15) we find $y_4 = -\int_t^\infty \Omega_\sigma^t(\check{B}) \varphi(\sigma) \, d\sigma$. Estimating $y_4(t)$ for $t \geq 0$ yields

$$\|y_4(t)\| \leq K \|\varphi\|_0 \int_t^\infty \exp\{\gamma(t-\sigma)\} \, d\sigma = K \|\varphi\|_0 \gamma^{-1}.$$

If $t < 0$ is considered, then in view of (1.3.18) we have $y_4^-(0) = y^+(0)$, and hence,

$$y_4 = \int_{-\infty}^{t} \Omega_\sigma^t(\check{B})\varphi(\sigma)\,d\sigma$$

and $\|y_4(t)\| \leq K\|\varphi\|_0 \gamma^{-1}$.

The presence of the quadratic form with the coefficients, bounded on R and continuously differentiable, which satisfies inequality (1.1.6) excludes non-trivial solutions of the system (1.1.1) bounded on R, though its degeneracy is possible. This allows us to formulate the condition of exponential dichotomy on R for the system (1.1.1) by means of two quadratic forms.

Theorem 1.3.4 *In order that the system of equations (1.1.1) be exponentially dichotomous on the whole R axis it is necessary and sufficient that two symmetric matrices $S_1(t), S_2(t) \in C^1(R)$ exist satisfying the conditions*

$$\langle [\dot{S}_1(t) + S_1(t)A(t) + A^*(t)S_1(t)]x, x \rangle \leq -\gamma_1 \|x\|^2; \quad (1.3.20)$$

$$\langle [\dot{S}_2(t) - S_2(t)A^*(t) - A(t)S_2(t)]x, x \rangle \leq -\gamma_2 \|x\|^2, \quad \gamma_i > 0. \quad (1.3.21)$$

Then the matrices $S_1(t)$ and $S_2(t)$ are non-degenerate for all $t \in R$.

Remark 1.3.5 If the n-dimensional symmetric matrix $S_1(t)$ is non-degenerate for all $t \in R$ and satisfies condition (1.3.20), then the matrix $S_2(t) = -S_1^{-1}(t)$ can be taken for the matrix $S_2(t)$ satisfying condition (1.3.21).

In fact, making the substitution of variables $x = S_1^{-1}(t)y$ in (1.3.20) we get

$$\langle [S_1^{-1}(t)\dot{S}_1(t)S_1^{-1}(t) + A(t)S_1^{-1}(t) + S_1^{-1}(t)A^*(t)]y, y \rangle$$
$$\leq -\gamma_1 \|S_1^{-1}(t)y\|^2 \leq -\gamma_1 \|S_1\|_0^{-2}\|y\|^2.$$

In view of the identity

$$\frac{d}{dt}[-S_1^{-1}(t)] \equiv S_1^{-1}(t)\dot{S}_1(t)S_1^{-1}(t), \quad \forall\, t \in R,$$

and redesignating $y \to x$ and $\gamma_2 = \gamma_1 \|S_1\|_0^{-2}$ we arrive at the inequality (1.3.21) where $S_1 = -S_1^{-1}$.

In the conditions of the Theorem 1.3.3 the definition of the dimensions of the subspace $\hat{E}(t)$ of the solutions to (1.1.1) bounded on the whole R axis is of interest. We assume that the matrix $S(t) \equiv S^*(t) \in C^1(R)$ exists satisfying the condition (1.2.29). Let $t_1, \ldots, t_k$ be the points where $\det S = 0$. We designate by $n^+(+\infty)$ the number of positive eigenvalues of the matrix $S(t)$ for sufficiently large t: $t > \max_i \{t_i\}$, $n^+(-\infty)$ is a number of positive eigenvalues of the matrix $S(t)$

for $t < \min_i \{t_i\}$. Similarly, $n^-(+\infty)$ is the number of negative eigenvalues of the matrix $S(t)$ for $t > \max_i \{t_i\}$. The proof of Theorem 1.1.1 demonstrated that if, for example, on the interval $[T, +\infty)$ the matrix $S(t)$ is non-degenerate, then the dimensions of the subspace $E^+(t)$ of the solutions to (1.1.1) which damp going to $+\infty$ equals the number of positive eigenvalues of the matrix $S(t)$. Thus, if the matrices $B^\pm$ and $\check{B}$ in (1.2.33) have dimensions $r_*^\pm \times r_*^\pm$ and $\check{r}_* \times \check{r}_*$ correspondingly, then $n^+(+\infty) = r_*^+$, $n^-(+\infty) = r_*^- + \check{r}_*$, $n^+(-\infty) = r_*^+ + \check{r}_*$, $n^-(-\infty) = r_*^-$. Therefore $\check{r}_* = n^+(-\infty) - n^+(+\infty) = n^-(+\infty) - n^-(-\infty)$. When passing to the conjugated system (1.2.30), $\check{r}_* = \hat{r}$, because all solutions increasing on both sides become damping ones. Thus the following result holds.

Theorem 1.3.5 *Let an n-dimensional symmetric matrix $S(t) \in C^1(t)$ exist, satisfying the inequality (1.2.29). Then the dimensions of the subspace $\hat{E}(t)$ of the solutions to (1.1.1) bounded on the whole R axis is represented by the formula*

$$\dim \hat{E}(t) = n^-(+\infty) - n^-(-\infty) = n^+(-\infty) - n^+(+\infty). \qquad (1.3.22)$$

Further we need the following definition.

Definition 1.3.1 *The system of equations (1.1.1) is called weakly regular on the whole R axis, if for every vector-function $f(t) \in C^0(R)$ the inhomogeneous system (1.1.4) has at least one solution bounded on R, and regular on R, if this system has precisely one solution bounded on R for every fixed function $f(t)$.*

Obviously, the regularity of the system (1.1.1) on R is equivalent to its exponential dichotomy on the whole R axis.

Remark 1.3.6 Not every system (1.1.1) exponentially dichotomous on the semiaxes R_+ and R_- is weakly regular on the R axis (see Persidsky [1]). For it to be weakly regular on R it is necessary to require additionally that the system (1.2.32) conjugated with it possesses no non-trivial solutions bounded on R.

Obviously there exist systems (1.1.1) exponentially dichotomous on the semiaxes R_+ and R_- such that not only these systems are not weakly regular on R, but the system (1.2.32) conjugated with them is not also weakly regular on R. In addition, it becomes clear that one can make either (1.1.1) or (1.2.32) conjugated with it, by an arbitrarily small perturbation on one of the systems, weakly regular on R. To formulate a result obtained in this direction we introduce the designations:

M is a set of matrix-functions $A(t) \in C^0(R)$ such that the system (1.1.1) is weakly regular on the whole R axis;

M^* is a set of matrix functions $A(t) \in C^0(R)$ such that the conjugated system (1.2.32) is weakly regular on R;

D is a set of matrices $A(t) \in C^0(R)$ such that the system (1.1.1) and evidently also system (1.2.32) are exponentially dichotomous on the semiaxes R_+ and R_-.

Theorem 1.3.6 *For every matrix function $A(t) \in D$ and every number $\varepsilon > 0$ there exists a matrix function $A_\varepsilon(t) \in M \cup M^*$ such that $\|A_\varepsilon(t) - A(t)\|_0 \leq \varepsilon$.*

Proof We reduce the system (1.1.1) to the form of (1.2.2) under the conditions of Theorem 1.2.2. Note that the condition (1.3.18) is necessary and sufficient for the system (1.2.14) to have a solution bounded on the whole R axis.

Let the number of coordinates y_3 not be smaller than that of y_4. Then, we show that the system of equations

$$\frac{dy_3}{dt} = \hat{B}(t)y_3, \quad \frac{dy_4}{dt} = \check{B}(t)y_4, \tag{1.3.23}$$

can be made by an arbitrarily small perturbation weakly regular on R. Without loss of generality, we assume that the matrices $\hat{B}(t)$ and $\check{B}(t)$ have a triangular form. In addition, the matrix $\hat{B}(t)$ is represented as

$$\hat{B}(t) = \begin{pmatrix} \hat{B}_{11}(t) & 0 \\ \hat{B}_{21}(t) & \hat{B}_{22}(t) \end{pmatrix}.$$

Here the dimensions of the block $\hat{B}_{11}(t)$ are the same as those of the matrix $\check{B}(t)$. Designating $y_3 = (y_{31}, y_{32})$ let us show that the system of equations

$$\frac{dy_{31}}{dt} = \hat{B}_{11}(t)y_{31}, \quad \frac{dy_{32}}{dt} = \hat{B}_{21}(t)y_{31} + \hat{B}_{22}(t)y_{32},$$

$$\frac{dy_4}{dt} = \varepsilon y_{31} + \check{B}(t)y_4, \tag{1.3.24}$$

is weakly regular on the whole R axis for every $\varepsilon \neq 0$. To this end we make sure that there is a solution which is bounded on R of the inhomogeneous system of equations

$$\frac{dy_{31}}{dt} = \hat{B}_{11}(t)y_{31} + f_1(t), \quad \frac{dy_{32}}{dt} = \hat{B}_{21}(t)y_{31} + \hat{B}_{22}(t)y_{32} + f_2(t),$$

$$\frac{dy_4}{dt} = \varepsilon y_{31} + \check{B}(t)y_4 + f_3(t). \tag{1.3.25}$$

Substituting by the general solution of the first subsystem from (1.3.25)

$$y_{31}(t) = \Omega_0^t(\hat{B}_{11})\left[y_{31}(0) + \int_0^t \Omega_\sigma^0(\hat{B}_{11})f_1(\sigma)\,d\sigma\right],$$

into the third subsystem yields

$$\frac{dy_4}{dt} = \check{B}(t)y_4 + f_3(t) + \varepsilon\Omega_0^t(\hat{B}_{11})y_{31}(0) + \varepsilon \int_0^t \Omega_\sigma^t(\hat{B}_{11})f_1(\sigma)\,d\sigma. \tag{1.3.26}$$

For system (1.3.26) the existence condition for a solution to (1.3.18) which is bounded on R is

$$\varepsilon \int_{-\infty}^{\infty} \Omega_\sigma^0(\check{B})\Omega_0^\sigma(\hat{B}_{11})\, d\sigma\, y_{31}(0) + \int_{-\infty}^{\infty} \Omega_\sigma^0(\check{B})\varphi_\varepsilon(\sigma)\, d\sigma = 0, \quad (1.3.27)$$

where

$$\varphi_\varepsilon(\sigma) = f_3(\sigma) + \varepsilon \int_0^\sigma \Omega_\tau^\sigma(\hat{B}_{11}) f_1(\tau)\, d\tau \in C^0(R).$$

Clearly, for the algebraic system (1.3.2) to be solvable relative to $y_{31}(0)$ for all $\varphi_\varepsilon(t) \in C^0(R)$ it is necessary and sufficient that

$$\det \int_{-\infty}^{\infty} \Omega_\sigma^0(\check{B})\Omega_0^\sigma(\hat{B}_{11})\, d\sigma \neq 0. \quad (1.3.28)$$

Taking into account the triangular form of the matrices $\check{B}$ and $\hat{B}_{11}$

$$\check{B}(t) = \begin{pmatrix} \check{b}_{11}(t) & 0 & \cdots & 0 \\ \check{b}_{21}(t) & \check{b}_{22}(t) & \cdots & 0 \\ \vdots & \vdots & \ddots & \vdots \\ \check{b}_{k1}(t) & \check{b}_{k2}(t) & \cdots & \check{b}_{kk}(t) \end{pmatrix},$$

$$\hat{B}_{11}(t) = \begin{pmatrix} \hat{b}_{11}(t) & 0 & \cdots & 0 \\ \bar{b}_{21}(t) & \hat{b}_{22}(t) & \cdots & 0 \\ \vdots & \vdots & \ddots & \vdots \\ \bar{b}_{k1}(t) & \bar{b}_{k2}(t) & \cdots & \hat{b}_{kk}(t) \end{pmatrix},$$

one gets

$$\Omega_\sigma^0(\check{B})\Omega_0^\sigma(\hat{B}_{11})$$

$$= \begin{pmatrix} \exp\left\{\int_\sigma^0 \check{b}_{11}(t)\, dt\right\} & 0 & \cdots & 0 \\ \omega_{21}(\sigma) & \exp\left\{\int_\sigma^0 \check{b}_{22}(t)\, dt\right\} & \cdots & 0 \\ \vdots & \vdots & \ddots & \vdots \\ \omega_{k1}(\sigma) & \omega_{k2}(\sigma) & \cdots & \exp\left\{\int_\sigma^0 \check{b}_{kk}(t)\, dt\right\} \end{pmatrix}$$

$$\times \begin{pmatrix} \exp\left\{\int_0^\sigma \hat{b}_{11}(t)\,dt\right\} & 0 & \cdots & 0 \\ \bar{\omega}_{21}(\sigma) & \exp\left\{\int_0^\sigma \hat{b}_{22}(t)\,dt\right\} & \cdots & 0 \\ \vdots & \vdots & \ddots & \vdots \\ \bar{\omega}_{k1}(\sigma) & \bar{\omega}_{k2}(\sigma) & \cdots & \exp\left\{\int_0^\sigma \hat{b}_{kk}(t)\,dt\right\} \end{pmatrix}$$

$$= \begin{pmatrix} e^{\left\{\int_0^\sigma [\hat{b}_{11}-\check{b}_{11}(t)]\,dt\right\}} & 0 & \cdots & 0 \\ W_{21}(\sigma) & e^{\left\{\int_0^\sigma [\hat{b}_{22}-\check{b}_{22}(t)]\,dt\right\}} & \cdots & 0 \\ \vdots & \vdots & \ddots & \vdots \\ W_{k1}(\sigma) & W_{k2}(\sigma) & \cdots & e^{\left\{\int_0^\sigma [\hat{b}_{kk}-\check{b}_{kk}(t)]\,dt\right\}} \end{pmatrix}.$$

Hence it becomes clear that the correlation

$$\det \int_{-\infty}^{\infty} \Omega_\sigma^0(\check{B})\Omega_0^\sigma(\hat{B}_{11})\,d\sigma = \prod_{i=1}^{k} \int_{-\infty}^{\infty} \exp\left\{\int_0^\sigma [\hat{b}_{ii}(t) - \check{b}_{ii}(t)]\,dt\right\} d\sigma > 0$$

ensures the condition (1.3.28) being satisfied and so the solvability relative to $y_{31}(0)$ of the algebraic system (1.3.27). Thus, in the case when the dimension of y_4 does not exceed that of y_3 the system of equations (1.2.2) which is exponentially dichotomous on the semiaxes R_+ and R_- can be made weakly regular on R by an arbitrarily small perturbation, replacing, in addition, the last two systems in (1.2.2) by the system (1.3.24).

In the case when the dimension of y_4 is larger than that of y_3, we consider the system conjugated with system (1.2.2). Further we proceed as above. Since (1.2.1) is the transformation of Lyapunov variables on the whole R axis, then all conclusions made for system (1.2.2) remain valid for the initial system (1.1.1). This completes the proof of Theorem 1.3.6.

The theorem cited can be represented in terms of quadratic forms as follows.

Theorem 1.3.7 *For sufficiently large $t: t \geq t_0 > 0$ let there exists an $n \times n$-dimensional symmetric differentiable matrix function $S_1(t) \in C^1(R_+)$ satisfying the condition*

$$\langle [\dot{S}_1(t) - S_1(t)A^*(t) - A(t)S_1(t)]x,\, x \rangle \leq -\|x\|^2, \quad t \geq t_0,$$

and for all $t \leq -t_0 < 0$ there exist a similar matrix function

$$\langle [\dot{S}_2(t) - S_2(t)A^*(t) - A(t)S_2(t)]x,\, x \rangle \leq -\|x\|^2, \quad t \leq t_0.$$

Then, for every $\varepsilon > 0$ there exists an $n \times n$-dimensional matrix function $B(t) \in C^0(R)$ such that $\|B(t)\| \le \varepsilon$ and continuously differentiable and bounded on R symmetric matrix function $S(t)$ satisfying one of the conditions:

(1) if $n_2^-(-\infty) < n_1^-(+\infty)$, then

$$\langle [\dot{S}(t) - S(t)(A^*(t) + B^*(t)) - (A(t) + B(t))S(t)]x, x \rangle \le -\|x\|^2; \qquad (1.3.29)$$

(2) if $n_2^-(-\infty) > n_1^-(+\infty)$, then

$$\langle [\dot{S}(t) + S(t)(A(t) + B(t)) + (A^*(t) + B^*(t))S(t)]x, x \rangle \le -\|x\|^2; \qquad (1.3.30)$$

(3) if $n_1^-(+\infty) = n_2^-(-\infty)$, then there exist non-degenerate matrix functions $S(t) = S_i(t) \in C^1(R)$, $i = 1, 2$, satisfying the condition (1.3.29) for $S = S_1$ and condition (1.3.30) for $S = S_2$, i.e. the system of the equations

$$\frac{dx}{dt} = [A(t) + B(t)]x$$

is exponentially dichotomous on the whole R axis.

Corollary 1.3.1 *For the matrix function $A(t)$ let there exist finite limits $\lim_{t \to +\infty} A(t) = A_+$ and $\lim_{t \to -\infty} A(t) = A_-$. Then for every $\varepsilon > 0$ there exists a matrix function $B(t) \in C^0(R)$, continuous on R, with the norm $\|B(t)\| \le \varepsilon$ such that one of the systems*

$$\frac{dx}{dt} = [A(t) + B(t)]x, \qquad \frac{dx}{dt} = -[A^*(t) + B^*(t)]x$$

is weakly regular on R.

1.4 Integral Representation of Weakly Regular Systems Bounded on the Whole R Axis

Let the system (1.1.1) be exponentially dichotomous on R. We consider the structure of the Green function (1.1.56) which can be represented as

$$G(t, \tau) = \begin{cases} \Omega_\tau^t(A)C(\tau), & \tau \le t, \\ \Omega_\tau^t(A)[C(\tau) - I_n], & t < \tau, \end{cases} \qquad (1.4.1)$$

where the matrix function
$$C(\tau) = \Omega_0^\tau(A) C \Omega_\tau^0(A). \tag{1.4.2}$$

Since $C^2 = C$, then
$$C^2(\tau) \equiv C(\tau). \tag{1.4.3}$$

Now we assume that it is unknown whether the system (1.1.1) is exponentially dichotomous or not. It is only known that there exists an $n \times n$-dimensional function $C(\tau) \in C^0(R)$ such that the function (1.4.1) satisfies the estimate (1.1.57). The questions are: are the identities (1.1.42) and (1.1.43) satisfied necessarily, and what can be said about the system (1.1.1) in this regard? We note first that the presence of the matrix function $C(\tau) \in C^0(R)$ ensures the existence of a solution to the inhomogeneous system (1.1.4) which is bounded on R for every vector-function $f(t) \in C^0(R)$ which can be represented by (1.1.55) (other solutions which are bounded on R may also exist, but this is unknown so far). Moreover, there exists a symmetric matrix function $S(t) \in C^1(R)$ satisfying the condition (1.2.29). It can be represented as

$$S(t) = \int_t^{+\infty} \Omega_\tau^t(A)(C(\tau) - I_n)[\Omega_\tau^t(A)(C(\tau) - I_n)]^* d\tau \tag{1.4.4}$$
$$- \int_{-\infty}^{t} \Omega_\tau^t(A) C(\tau) [\Omega_\tau^t(A) C(\tau)]^* d\tau.$$

Thus, under the conditions of Theorems 1.2.3 and 1.3.3 the exponential behaviour of solutions to the system (1.1.1) becomes entirely clear.

Definition 1.4.1 We state that the system of equations (1.1.1) *has a Green function* of the problem of solutions (1.4.1) bounded on R, provided that a continuous matrix function $C(\tau)$ exists such that the function (1.4.1) satisfies the estimate (1.1.57).

Theorem 1.4.1 *Every system (1.1.1) which is weakly regular on R has a Green function of the problem on solutions bounded on R.*

Proof Let the matriciant $\Omega_\tau^t(\Phi)$ for the system (1.2.30) be of the form

$$\Omega_\tau^t(\Phi) = \begin{pmatrix} \omega_1(t,\tau) & 0 & 0 \\ 0 & \omega_2(t,\tau) & 0 \\ \omega_3(t,\tau) & \omega_4(t,\tau) & \omega_5(t,\tau) \end{pmatrix}, \tag{1.4.5}$$

where

$$w_1(t,\tau) = \Omega_\tau^t(\Phi^+), \quad w_2(t,\tau) = \Omega_\tau^t(\Phi^-), \quad w_5(t,\tau) = \Omega_\tau^t(\hat{\Phi}),$$

$$w_3(t,\tau) = \int_\tau^t w_5(t,\sigma)\Phi_1(\sigma)w_1(\sigma,\tau)\,d\sigma,$$

$$w_4(t,\tau) = \int_\tau^t w_5(t,\sigma)\Phi_2(\sigma)w_2(\sigma,\tau)\,d\sigma.$$

Let us first make clear what structure the matrix $\bar{C}(\tau)$ should have in order that the estimates

$$\begin{aligned}\|\Omega_\tau^t(\Phi)\bar{C}(\tau)\| &\leq K_1 \exp\{-\gamma_1(t-\tau)\}, \quad \tau \leq t, \\ \|\Omega_\tau^t(A)[\bar{C}(\tau) - I_n]\| &\leq K_1 \exp\{\gamma_1(t-\tau)\}, \quad t \leq \tau,\end{aligned} \quad (1.4.6)$$

be satisfied, where $K_1, \gamma_1 = \text{const} > 0$. Assuming that the estimates (1.4.6) are satisfied, we represent the matrix $\bar{C}(\tau)$ in the block form

$$\begin{pmatrix} C_{11}(\tau) & C_{12}(\tau) & C_{13}(\tau) \\ C_{21}(\tau) & C_{22}(\tau) & C_{23}(\tau) \\ C_{31}(\tau) & C_{32}(\tau) & C_{33}(\tau) \end{pmatrix}, \quad (1.4.5a)$$

correspondingly to the representation of the matriciant (1.4.5) and consider one of the estimates

$$\|\Omega_\tau^t(\Phi)C_{12}(\tau)\| \leq K_1 \exp\{-\gamma_1|t-\tau|\},$$

which follows from (1.4.6). Hence we find

$$\begin{aligned}\|C_{12}(\tau)\| &= \|\Omega_t^\tau(\Phi^+)\Omega_\tau^t(\Phi^+)C_{12}(\tau)\| \leq K_1 \|\Omega_t^\tau(\Phi^+)\| \exp\{-\gamma|t-\tau|\} \\ &\leq K_1 K \exp\{-\gamma(\tau-t)\}\exp\{-\gamma_1(\tau-t)\}, \quad t \leq \tau.\end{aligned}$$

Passing to the limit of $t \to -\infty$ in the last inequality we get $C_{12}(\tau) \equiv 0$. Similarly we establish that $C_{13}(\tau) \equiv 0$, $C_{21}(\tau) \equiv 0$, $C_{23}(\tau) \equiv 0$ for all $\tau \in R$. Now we show that $C_{11}(\tau) \equiv I_{r^+}$. In fact, if at some point $\tau = \tau_0$ $C_{11}(\tau_0) \neq I_{r^+}$ then a $x_0 \in R^{r^+}$ exists such that $[C_{11}(\tau_0) - I_{r^+}]x_0 \neq 0$. Therefore,

$$y_1^*(t) = \Omega_{\tau_0}^t(\Phi^+)[C_{11}(\tau_0) - I_{r^+}]x_0$$

is a non-trivial solution of the first subsystem from (1.2.30). On the one hand, by virtue of estimates (1.2.31) $\lim_{t \to +\infty} y_1^*(t) = 0$; on the other hand, due to the second estimate (1.4.6) $\lim_{t\to -\infty} y_1^*(t) = 0$, and the first subsystem (1.2.30) has no such nontrivial solutions. So, there is one possibility left: $C_{11}(\tau) \equiv I_{r+}$. Proceeding in the same way we get to know that $C_{22}(\tau) \equiv 0$ for all $\tau \in R$.

Let us clarify what sort the functions $C_{3i}(\tau)$, $i = 1, 2, 3$ should be. Estimates (1.4.6) imply

$$\|\omega_3(t,\tau) + \omega_5(t,\tau)C_{31}(\tau)\| \leq K_1 \exp\{-\gamma_1(t-\tau)\}, \quad \tau \leq t,$$
$$\|\omega_5(t,\tau)C_{31}(\tau)\| \leq K_1 \exp\{\gamma_1(t-\tau)\}, \quad t \leq \tau. \tag{1.4.7}$$

We show that the matrix function $C_{31}(\tau)$ approximates exponentially to some function bounded on R_- as $\tau \to -\infty$, and tends to zero as $\tau \to +\infty$. In view of the estimates (1.4.7) and (1.2.31) we get for $\tau \in R_-$

$$\|\omega_5(\tau,0)\omega_3(0,\tau) + C_{31}(\tau)\| \leq \|\omega_5(\tau,0)\| \|\omega_3(0,\tau) + \omega_5(0,\tau)C_{31}(\tau)\|$$
$$\leq KK_1 \exp\{(\gamma+\gamma_1)\tau\}, \quad \tau \leq 0,$$

and if $\tau \in R_+$, then

$$\|C_{31}(\tau)\| \leq \|\omega_5(\tau,0)\| \|\omega_5(0,\tau)C_{31}(\tau)\| \leq KK_1 \exp\{-(\gamma+\gamma_1)\tau\}, \quad \tau > 0.$$

Now we make sure that the function $F_1(\tau) = -\omega_5(\tau,0)\omega_3(0,\tau)$ is bounded on the semiaxis R_-. Estimating it for $\tau < 0$ yields

$$\|F_1(\tau)\| = \left\| \Omega_0^\tau(\hat{\Phi}) \int_\tau^0 \Omega_\sigma^0(\hat{\Phi}) \Phi_1(\sigma) \Omega_\tau^\sigma(\Phi^+) \, d\sigma \right\|$$

$$\leq K_2 \int_\tau^0 \|\Omega_\sigma^\tau(\hat{\Phi})\| \|\Omega_\tau^\sigma(\Phi^+)\| \, d\sigma \leq K_2 K^2 \int_\tau^0 \exp\{2\gamma(\tau-\sigma)\} \, d\sigma \leq K_2 K^2 (2\gamma)^{-1}.$$

Similarly we establish the exponential vanishing of the matrix $C_{32}(\tau)$ as $\tau \to -\infty$ and the approximation of it to some function as $\tau \to +\infty$ which is bounded on R_+.

We investigate the behaviour of the matrix function $C_{33}(\tau)$ over infinity. Estimates (1.4.6) immediately imply

$$\|\omega_5(t,\tau)C_{33}(\tau)\| \leq K_1 \exp\{-\gamma_1(t-\tau)\}, \quad \tau \leq t,$$
$$\|\omega_5(t,\tau)[C_{33}(\tau) - I_{\hat{r}}]\| \leq K_1 \exp\{\gamma_1(t-\tau)\}, \quad t \leq \tau.$$

Hence

$$\|C_{33}(\tau)\| \leq \|\omega_5^{-1}(0,\tau)\| \|\omega_5(0,\tau)C_{33}(\tau)\| \leq K_1 K \exp\{(\gamma+\gamma_1)\tau\}, \quad \tau < 0,$$
$$\|C_{33}(\tau) - I_{\hat{r}}\| \leq \|\omega_5^{-1}(0,\tau)\| \|\omega_5(0,\tau)[C_{33}(\tau) - I_{\hat{r}}]\|$$
$$\leq K_1 K \exp\{-(\gamma+\gamma_1)\tau\}, \quad \tau > 0.$$

Thus, for the estimates (1.4.6) to be satisfied, it is necessary that the matrix (1.4.5a) is of the form

$$\bar{C}(\tau) = \begin{pmatrix} I_{r^+} & 0 & 0 \\ 0 & 0 & 0 \\ C_{31}(\tau) & C_{32}(\tau) & C_{33}(\tau) \end{pmatrix}, \qquad (1.4.8)$$

with the matrix functions $C_{3i}(\tau)$, $i = 1, 2, 3$, satisfying the estimates

$$\begin{aligned} \|C_{31}(\tau) - F_1(\tau)\| &\leq K_2 \exp\{\alpha\tau\}, & \tau < 0; \\ \|C_{31}(\tau)\| &\leq K_2 \exp\{-\alpha\tau\}, & \tau > 0; \\ \|C_{32}(\tau)\| &\leq K_2 \exp\{\alpha\tau\}, & \tau < 0; \\ \|C_{32}(\tau) - F_2(\tau)\| &\leq K_2 \exp\{-\alpha\tau\}, & \tau \geq 0; \\ \|C_{33}(\tau)\| &\leq K_2 \exp\{\alpha\tau\}, & \tau < 0; \\ \|C_{33}(\tau) - I_{\hat{r}}\| &\leq K_2 \exp\{-\alpha\tau\}, & \tau \geq 0, \end{aligned} \qquad (1.4.9)$$

where K_2 and α are const > 0,

$$\begin{aligned} F_1(\tau) &= -\int_\tau^0 \omega_5(\tau,\sigma)\Phi_1(\sigma)\omega_1(\sigma,\tau)\,d\sigma, \\ F_2(\tau) &= -\int_0^\tau \omega_5(\tau,\sigma)\Phi_2(\sigma)\omega_2(\sigma,\tau)\,d\sigma. \end{aligned} \qquad (1.4.10)$$

We show that for any matrix functions $C_{3i}(\tau)$, $i = 1, 2, 3$, which satisfy the estimates (1.4.9) for sufficiently large fixed $\alpha > 0$ the main inequalities (1.4.6) are satisfied, where the matrix $C(\tau)$ is of the form of (1.4.8).

We consider several cases in this regard.

Case 1. $0 \leq \tau \leq t$. Then

$$\|\omega_3(t,\tau) + \omega_5(t,\tau)C_{31}(\tau)\| \leq \|\omega_3(t,\tau)\| + \|\omega_5(t,\tau)\|\,\|C_{31}(\tau)\|$$

$$\leq \bar{K}\int_\tau^t \|\Omega_\sigma^t(\hat{\Phi})\|\,\|\Omega_\tau^\sigma(\Phi^+)\|\,d\sigma + KK_2 \exp\{-\gamma(t-\tau) - \alpha\tau\}$$

$$\leq \bar{K}K^2 \int_\tau^t \exp\{-\gamma(t-\sigma) - \gamma(\sigma-\tau)\}\,d\sigma + KK_2 \exp\{-\gamma(t-\tau)\}$$

$$\leq K^* \exp\{-\gamma^*(t-\tau)\}, \quad 0 < \gamma^* < \gamma.$$

Case 2. $\tau \leq 0 \leq t$. Then

$$\|\omega_3(t,\tau) + \omega_5(t,\tau)C_{31}(\tau)\| = \left\| \int_\tau^0 \Omega_\sigma^t(\hat{\Phi})\Phi_1(\sigma)\Omega_\tau^\sigma(\Phi^+)\, d\sigma \right.$$

$$\left. + \int_0^t \Omega_\sigma^t(\hat{\Phi})\Phi_1(\sigma)\Omega_\tau^\sigma(\Phi^+)\, d\sigma + \Omega_\tau^t(\hat{\Phi})C_{31}(\tau) \right\|$$

$$\leq \|\Omega_\tau^t(\hat{\Phi})\| \left\| \int_\tau^0 \Omega_\sigma^\tau(\hat{\Phi})\Phi_1(\sigma)\Omega_\tau^\sigma(\Phi^+)\, d\sigma + C_{31}(\tau) \right\| + \bar{K}\int_0^t \|\Omega_\sigma^t(\Phi)\|\,\|\Omega_\tau^\sigma(\Phi^+)\|\, d\sigma$$

$$\leq K\left[\exp\{-\gamma t - m\tau + \alpha\tau\} + \int_0^t \exp\{-\gamma(t-\sigma) - \gamma(\sigma-\tau)\}\, d\sigma\right]$$

$$\leq K^* \exp\{-\gamma^*(t-\tau)\}.$$

Case 3. $\tau \leq t \leq 0$. Then

$$\|\omega_3(t,\tau) + \omega_5(t,\tau)C_{31}(\tau)\| = \left\| \int_\tau^0 \Omega_\sigma^t(\hat{\Phi})\Phi_1(\sigma)\Omega_\tau^\sigma(\Phi^+)\, d\tau + \Omega_\tau^t(\hat{\Phi})C_{31}(\tau) \right.$$

$$\left. - \int_t^0 \Omega_\sigma^t(\hat{\Phi})\Phi_1(\sigma)\Omega_\tau^\sigma(\Phi^+)\, d\sigma \right\| \leq \|\Omega_\tau^t(\hat{\Phi})\|\,\| - F_1(\tau) + C_{31}(\tau)\|$$

$$+ \bar{K}\int_t^0 \|\Omega_\sigma^t(\hat{\Phi})\|\,\|\Omega_\tau^\sigma(\Phi^+)\|\, d\sigma \leq K\left[\exp\{-\gamma(t-\tau) + \alpha\tau\}\right.$$

$$\left. + \int_t^0 \exp\{\gamma(t-\sigma) + \gamma(\sigma-\tau)\}\, d\sigma \right] \leq K^* \exp\{-\gamma^*(t-\tau)\}.$$

Case 4. $0 < t \leq \tau$. Then

$$\|\Omega_\tau^t(\hat{\Phi})C_{31}(\tau)\| \leq \|\Omega_0^t(\hat{\Phi})\|\,\|\Omega_\tau^0(\hat{\Phi})\|\,\|C_{31}(\tau)\| \leq K^* \exp\{-\gamma^*(t-\tau)\}.$$

In view of the boundedness on R_- of the matrix function $C_{31}(\tau)$ we get the following case

Case 5. $t \leq \tau \leq 0$. Then

$$\|\Omega_\tau^t(\hat{\Phi})C_{31}(\tau)\| \leq K^* \exp\{\gamma(t-\tau)\}.$$

Similarly the estimates are found using the matrix functions $C_{32}(\tau)$ and $C_{33}(\tau)$. Now we recall that the passage from (1.1.1) to the decomposed system (1.2.30) was made by means of the change of Lyapunov variables $x = L(t)y$. Therefore, if in the structure of the Green function (1.4.1) we set $C(\tau) = L(\tau)\bar{C}(\tau)L(\tau)^{-1}$, where $\bar{C}(\tau)$ is of the form of (1.4.8), then by virtue of (1.4.6) the estimate (1.1.57) is satisfied. This completes the proof of the Theorem 1.4.1.

The proof of the theorem yields

Theorem 1.4.2 *Let there exist a matrix function $S(t) \equiv S^*(t) \in C^1(R)$ satisfying the inequality (1.2.29) and for some value $t = t_0$ det $S(t_0) = 0$. Then the system of equations (1.1.1) has many Green functions (1.4.1), for which estimate (1.1.57) is satisfied. Besides,*

(1) *neither of the matrices $C(\tau)$ is projecting, i.e. $C^2(\tau) \not\equiv C(\tau)$;*
(2) *neither of the matrices $C(\tau)$ satisfies (1.4.2), where*

$$C = C(0) = \text{const};$$

(3) *there exists a fixed Lyapunov matrix $L(\tau) \in C^1(R)$ such that for every matrix $C(\tau)$ the matrix*

$$L^{-1}(\tau)C(\tau)L(\tau) = \bar{C}(\tau)$$

is of the form of (1.4.8).

In the representation (1.4.8) the matrix functions $C_{3i}(\tau)$ can be changed arbitrarily on the finite interval of τ change. Although it is of importance that these functions behave in a certain way (i.e. the estimates (1.4.9) are satisfied) on $\pm\infty$. There arises a question in this regard: whether the equality (1.1.55) can determine all solutions of the inhomogeneous system (1.1.4) bounded on R by changing the matrix functions $C_{3i}(\tau)$, $i = 1, 2, 3$, appropriately in the structure of the Green function (1.4.1)? We prove the following.

Theorem 1.4.3 *Let all conditions of the Theorem 1.4.2 be satisfied. Then all solutions of the inhomogeneous system of equations (1.1.4) bounded on R for every fixed vector-function $f(t) \in C^0(R)$, $f(t) \not\equiv 0$, can be represented by the equality (1.1.55) by the appropriate choice of the matrix function $C(\tau)$ in (1.4.1).*

Proof We transform the system (1.1.4) in the conditions of Theorem 1.2.3 to the form (1.3.19). Evidently, it is enough to make sure that every solution of the inhomogeneous system (1.3.19) bounded on R can be represented in the form

$$y(t) = \int_{-\infty}^{t} \Omega_\tau^t(\Phi)\bar{C}(\tau)\varphi(\tau)\,d\tau + \int_{t}^{\infty} \Omega_\tau^t(\Phi)[\bar{C}(\tau) - I_n]\varphi(\tau)\,d\tau, \qquad (1.4.11)$$

under an appropriate choice of the matrix function $\bar{C}(\tau)$ of the form (1.4.8). Let $y = \bar{y}(t)$ be a fixed solution of the system (1.3.19) bounded on R, and $y = y(t)$ be a solution which can be expressed in the form of (1.4.11). Then the difference $\bar{y}(t) - y(t)$ is a solution of the homogeneous system (1.2.30) bounded on R. Let us show that every solution $y = \tilde{y}(t)$ bounded on R of the system (1.2.30) can be represented as

$$\tilde{y}(t) = \begin{pmatrix} \tilde{y}_1(t) \\ \tilde{y}_2(t) \\ \tilde{y}_3(t) \end{pmatrix}$$

$$= \int\limits_{-\infty}^{+\infty} \begin{pmatrix} \omega_1(t,\tau) & 0 & 0 \\ 0 & \omega_2(t,\tau) & 0 \\ \omega_3(t,\tau) & \omega_4(t,\tau) & \omega_5(t,\tau) \end{pmatrix} \begin{pmatrix} 0 \\ 0 \\ \sum\limits_{i=1}^{3} \hat{C}_{3i}(\tau)\varphi_i(\tau) \end{pmatrix} d\tau, \quad (1.4.12)$$

where $\hat{C}_{3i}(\tau)$ are continuously differentiable finite matrix functions (they are equal to zero outside of some finite interval of τ variation) and (1.4.2) implies

$$\tilde{y}_1(t) \equiv 0, \quad \tilde{y}_2(t) \equiv 0, \quad \tilde{y}_3(t) \equiv \int\limits_{-\infty}^{\infty} \omega_5(t,\tau) \sum_{i=1}^{3} \hat{C}_{3i}(\tau)\varphi_i(\tau)\, d\tau.$$

Let us prove that for any numerical vector $\tilde{y}_3(0)$ a function $\hat{C}_{3i}(\tau)$ can be found so that the equality

$$\tilde{y}_3(0) = \int\limits_{-\infty}^{\infty} \omega_5(t,\tau) \sum_{i=1}^{3} \hat{C}_{3i}(\tau)\varphi_i(\tau)\, d\tau \quad (1.4.13)$$

is valid.

Let at some point $\tau = \tau_0 \in R$, for example, the inequality $\|\varphi_2(\tau_0)\| > 0$ holds. Then putting in (1.4.13) $\hat{C}_{31}(\tau) \equiv 0$, $\hat{C}_{33}(\tau) \equiv 0$ we get

$$\tilde{y}_3(0) = \int\limits_{-\infty}^{\infty} \omega_5(0,\tau)\hat{C}_{32}(\tau)\varphi_2(\tau)\, d\tau. \quad (1.4.14)$$

We designate the matrix function $\omega_5(0,\tau)\hat{C}_{32}(\tau)$ by $U(\tau)$. Then the right-hand side of (1.4.14) becomes

$$\int\limits_{-\infty}^{\infty} U(\tau)\varphi_2(\tau)\, d\tau = \int\limits_{-\infty}^{\infty} \begin{pmatrix} u_{11}(\tau) & \cdots & u_{1r}(\tau) \\ \vdots & \ddots & \vdots \\ u_{k1}(\tau) & \cdots & u_{kr}(\tau) \end{pmatrix} \begin{pmatrix} \varphi_{21}(\tau) \\ \vdots \\ \varphi_{2r}(\tau) \end{pmatrix} d\tau. \quad (1.4.15)$$

One can consider, without loss of generality, that $\varphi_{21}(\tau_0) \neq 0$ in (1.4.15). Now all columns, but the first one, of the matrix $U(\tau)$ are set equal to zero, and the first column elements are taken in the form of finite scalar functions $u_{i1}(\tau)$, $i = 1, \ldots, k$, such that

$$\int_{-\infty}^{\infty} u_{i1}(\tau)\varphi_{21}(\tau) \, d\tau \neq 0.$$

This can always be done, since $\varphi_{21}(\tau_0) \neq 0$. Consequently, the functions $u_{i1}(\tau)$ can be taken so that

$$\int_{-\infty}^{\infty} u_{i1}(\tau)\varphi_{21}(\tau) \, d\tau = \tilde{y}_{3i}(0)$$

for every prescribed number $\tilde{y}_{3i}(0)$. Since $\hat{C}_{32}(\tau) = \omega_5(\tau, 0)U(\tau)$, a finite matrix function $\hat{C}_{32}(\tau)$ can always be taken so that the equality (1.4.14) is satisfied for every prescribed numerical vector $\tilde{y}_3(0)$. This completes the proof of the Theorem 1.4.3.

Remark 1.4.1 In the definition of the Green function $G(t, \tau)$ of the problem on solutions to the system (1.1.1) which is bounded on R it is possible to require that instead of the estimate (1.1.57) a weaker condition

$$\int_{-\infty}^{\infty} \|G(t,\tau)\| \, d\tau \leq K = \text{const} < \infty, \tag{1.4.16}$$

is satisfied, which ensures the existence of at least one solution to the inhomogeneous system (1.1.4) which is bounded on R for every vector-function $f(t) \in C^0(R)$. In this case, generally speaking, the estimate (1.1.57) cannot be satisfied at all, though another Green function $G(t, \tau)$ is always found for which this estimate is valid. This is implied by the following. The condition (1.4.16) ensures the existence of solutions to the inhomogeneous system (1.1.4) bounded on R for every vector-function $f(t) \in C^0(R)$. Therefore, the system (1.1.1) is exponentially dichotomous on the semiaxes R_+ and R_-, and hence, it can be reduced to the form of (1.2.30) by means of the transformation of the Lyapunov variables, and consequently, the Green function exists for it which satisfies estimate (1.1.57).

Simple examples (in particular, the equation

$$\frac{dx}{dt} = -(\tanh t)x$$

can be analysed) show that under condition (1.4.16) the matrix function $C(\tau)$ in the structure of the Green function (1.4.1) can be taken to be unbounded on R only in the case when the system (1.1.1) has solutions which are non-trivial and bounded on R.

1.5 Complement to the Exponentially Dichotomous of Weakly Regular on R Linear Systems

We focus our attention on example (1.2.37). The first equation

$$\frac{dx}{dt} = -(\tanh t)x$$

is weakly regular on R, because all solutions to the inhomogeneous equation

$$\frac{dx}{dt} = -(\tanh t)x + f(t), \quad f(t) \in C^0(R),$$

are bounded on R, and the extended system (1.2.37) is already exponentially dichotomous on R. Therefore, a question arises: whether the system (1.1.1) which is regular on R can be extended so that the block-triangular system is regular, i.e. exponentially dichotomous on R? It turns out that this can always be done in the following manner

$$\frac{dx}{dt} = A(t)x, \quad \frac{dy}{dt} = -x - A^*(t)y. \qquad (1.5.1)$$

Moreover, the system

$$\frac{dx}{dt} = A(t)x, \quad \frac{dy}{dt} = px - A^*(t)y, \qquad (1.5.2)$$

can be viewed, which is exponentially dichotomous on the whole R axis for every fixed value $p \neq 0$, $p \in R$.

Theorem 1.5.1 *Let there exist an n-dimensional symmetric matrix function $S(t) \in C^1(R)$ satisfying the inequality (1.2.29).*

Then, for every fixed definite-sign symmetric matrix $B(t) \in C^0(R)$, $|\langle B(t)x, x\rangle| \geq \beta_0 \|x\|^2$, $\beta_0 > 0$, the system of differential equations of the block-triangular form

$$\frac{dx}{dt} = A(t)x, \quad \frac{dy}{dt} = B(t)x - A^*(t)y \qquad (1.5.3)$$

is exponentially dichotomous on R, and $\dim E^{\pm} = n$.

Proof Assume that the matrix $B(t)$ is positive definite and $\langle B(t)x, x\rangle \geq \beta_0 \|x\|^2$, $\beta_0 = \text{const} > 0$. We compute the derivative of the quadratic form

$$V(t, x, y) = -\lambda \langle x, y\rangle + \langle S(t)y, y\rangle \qquad (1.5.4)$$

(λ is a fixed parameter) along the solutions of the system (1.5.3):

$$\dot{V}(t,x,y) = -\lambda\langle A(t)x, y\rangle - \lambda\langle x, B(t)x - A^*(t)y\rangle + \langle \dot{S}(t)y, y\rangle$$
$$+ 2\langle S(t)y, B(t)x - A^*(t)y\rangle$$
$$\leq -\lambda\langle B(t)x, x\rangle - \beta\|y\|^2 + +2\langle S(t)B(t)x, y\rangle$$
$$\leq -\lambda\beta_0\|x\|^2 - \beta\|y\|^2 + 2\|S\|_0\|B\|_0\|x\|\,\|y\|.$$

We incorporate the inequality $2ab \leq a^2\sigma^{-2} + b^2\sigma^2$ with the introduced parameter $\sigma \neq 0$ and obtain, estimating the derivative $\dot{V}(t,x,y)$,

$$\dot{V}(t,x,y) \leq -(\lambda\beta_0 - \|S\|_0^2\|B\|_0^2\sigma^{-2})\|x\|^2 - (\beta - \sigma^2)\|y\|^2. \tag{1.5.5}$$

It is clear then that when the inequalities

$$\lambda\beta_0 - \|S\|_0^2\|B\|_0^2\sigma^{-2} > 0, \quad \beta - \sigma^2 > 0 \tag{1.5.6}$$

are satisfied, the derivative $\dot{V}(t,x,y)$ is negative definite. For the parameter $\sigma \neq 0$ to exist which satisfies two inequalities (1.5.6), it is necessary and sufficient that the parameter λ satisfies the inequality

$$\lambda > \|S\|_0^2\|B\|_0^2\beta^{-1}\beta_0^{-1}. \tag{1.5.7}$$

Besides, we note that the quadratic form (1.5.4) is non-degenerate for all $t \in R$:

$$\det \bar{S}_\lambda(t) \neq 0, \quad \bar{S}_\lambda(t) = \begin{pmatrix} 0 & -\frac{\lambda}{2}I_n \\ -\frac{\lambda}{2}I_n & S(t) \end{pmatrix},$$

and hence the system of the equations (1.5.3) is exponentially dichotomous on the whole R axis.

Now we show the symmetric matrix $\bar{S}_\lambda(t)$ to have n positive eigenvalues and n negative ones independently of the matrix $S(t)$. Let $TST^{-1} = D$ be a diagonal matrix, then

$$\det[\bar{S}_\lambda(t) - \sigma I_{2n}] = \det T^{-1}\begin{pmatrix} -\sigma I_n & -\frac{\lambda}{2}I_n \\ 0 & S - \sigma I_n + \frac{\lambda^2}{4\sigma}I_n \end{pmatrix} T$$
$$= \det\left[\sigma^2 I_n - \sigma S - \frac{\lambda^2}{4}I_n\right] = \det T^{-1}\left[\sigma^2 I_n - \sigma D - \frac{\lambda^2}{4}I_n\right]T$$
$$= \det\left[\sigma^2 I_n - \sigma D - \frac{\lambda^2}{4}I_n\right].$$

From here it is obvious that the characteristic equation $\det[\bar{S}_\lambda(t) - \sigma I_{2n}] = 0$ has n positive and n negative roots and hence, $\dim E^- = \dim E^+ = n$.

The Theorem 1.5.1 is proved.

When the system (1.5.3) is considered a question arises: for what matrix functions $B(t) \in C^0(R)$ (not necessaryly symmetric and of definite sign) is the system of the equations (1.5.3) exponentially dichotomous on R? First, let us solve this problem by a concrete example

$$\frac{dx}{dt} = -(\tanh t)x, \quad \frac{dy}{dt} = b(t)x + (\tanh t)y. \tag{1.5.8}$$

Clearly, for the system (1.5.8) to be exponentially dichotomous on R a function $b(t) \in C^0(R)$ should be taken so that the inhomogeneous system

$$\frac{dx}{dt} = -(\tanh t)x + f_1(t), \quad \frac{dy}{dt} = b(t)x + (\tanh t)y + f_2(t), \tag{1.5.9}$$

possesses a unique solution which is bounded on R, for every fixed function $f_i(t) \in C^0(R)$, $i = 1, 2$. Substituting the general solution of the first equation (1.5.9)

$$x(t) = \cosh^{-1} t \left[x(0) + \int_0^t f_1(\tau) \cosh \tau \, d\tau \right]$$

into the second one yields

$$\frac{dy}{dt} = (\tanh t)y + b(t) \cosh^{-1} t \, x(0) + \varphi(t), \tag{1.5.10}$$

where

$$\varphi(t) = f_2(t) + b(t) \cosh^{-1} t \int_0^t f_1(\tau) \cosh \tau \, d\tau \in C^0(R).$$

Condition (1.3.18) for the inhomogeneity in (1.5.10) is

$$\int_{-\infty}^{\infty} b(\sigma) \cosh^{-2} \sigma \, d\sigma \, x(0) + \int_{-\infty}^{\infty} \varphi(\sigma) \cosh^{-1} \sigma \, d\sigma = 0.$$

Thus, for the system of equations (1.5.8) to be exponentially dichotomous on the whole R axis it is necessary and sufficient that for the function $b(t) \in C^0(R)$ the following condition

$$\int_{-\infty}^{\infty} b(\sigma) \cosh^{-2} \sigma \, d\sigma \neq 0 \tag{1.5.11}$$

is satisfied.

Theorem 1.5.2 *Let the system (1.1.1) be weakly regular on the whole R axis. Then, for the exponential dichotomy on R of the system (1.5.3) it is necessary and sufficient that the equality*

$$\operatorname{rank} \int_{-\infty}^{\infty} P^*[\Omega_0^\sigma(A)]^* B(\sigma) \Omega_0^\sigma(A) P \, d\sigma = \hat{r} \tag{1.5.12}$$

be satisfied, where P are the projection matrices on the subspace $\hat{E}(0)$ of the initial points for $t = 0$ of solutions to the system (1.1.1) bounded on the whole R axis, $\hat{r} = \dim \hat{E}(0)$.

If, in addition, equality (1.5.12) is satisfied for one of the matrices P, it holds true for all such matrices P ($P^2 = P$) as well.

Proof In the conditions of Theorem 1.2.3 and 1.3.3 the system of equations (1.5.3) is reduced by the change of Lyapunov variables

$$x = L(t)z, \quad y = [L^{-1}(t)]^* v, \tag{1.5.13}$$

to the form

$$\frac{dz_1}{dt} = \Phi^+(t)z_1, \quad \frac{dz_2}{dt} = \Phi^-(t)z_2,$$

$$\frac{dz_3}{dt} = \Phi_1(t)z_1 + \Phi_2(t)z_2 + \hat{\Phi}(t)z_3,$$

$$\frac{dv_1}{dt} = \sum_{i=1}^{3} D_{1i}(t)z_i - [\Phi^+(t)]^* v_1 - [\Phi_1(t)]^* v_3,$$

$$\frac{dv_2}{dt} = \sum_{i=1}^{3} D_{2i}(t)z_i - [\Phi^-(t)]^* v_2 - [\Phi_2(t)]^* v_3, \tag{1.5.14}$$

$$\frac{dv_3}{dt} = \sum_{i=1}^{3} D_{3i}(t)z_i - [\hat{\Phi}(t)]^* v_3,$$

where

$$\{D_{ij}(t)\}_{i,j=\overline{1,3}} = D(t) = L^*(t)B(t)L(t). \tag{1.5.15}$$

Moreover, for the corresponding matricient $\Omega_\tau^t(\Phi^\pm)$ and $\hat\Omega_\tau^t(\Phi)$ the estimates (1.2.31) hold. The solutions of the system (1.5.4) are represented as

$$z_1(t) = \Omega_0^t(\Phi^+)z_1(0), \quad z_2(t) = \Omega_0^t(\Phi^-)z_2(0),$$

$$z_3(t) = \Omega_0^t(\hat\Phi)\left\{z_3(0) + \int_0^t \Omega_\sigma^0(\hat\Phi)[\Phi_1(\sigma)\Omega_0^\sigma(\Phi^+)z_1(0)\right.$$

$$\left. + \Phi_2(\sigma)\Omega_0^\sigma(\Phi^-)z_2(0)]\,d\sigma\right\}, \qquad (1.5.16)$$

$$v_1(t) = [\Omega_\tau^0(\Phi^+)]^*\left\{v_1(0) + \int_0^t [\Omega_0^\sigma(\Phi^+)]^*\left[\sum_{i=1}^3 D_{1i}(\sigma)z_i(\sigma) - \Phi_1^*(\sigma)v_3(\sigma)\right]d\sigma\right\},$$

$$v_2(t) = [\Omega_t^0(\Phi^-)]^*\left\{v_2(0) + \int_0^t [\Omega_0^\sigma(\Phi^-)]^*\left[\sum_{i=1}^3 D_{2i}(\sigma)z_i(\sigma) - \Phi_2^*(\sigma)v_3(\sigma)\right]d\sigma\right\},$$

$$v_3(t) = [\Omega_t^0(\hat\Phi)]^*\left\{v_3(0) + \int_0^t [\Omega_0^\sigma(\hat\Phi)]^*\left[\sum_{i=1}^3 D_{3i}(\sigma)z_i(\sigma)\right]d\sigma\right\}.$$

Assume that the system of equations (1.5.3) is exponentially dichotomous on R. Clearly, this is equivalent to the exponential dichotomy on R of the system (1.5.14). Let the initial point $(z_1(0), z_2(0), z_3(0), v_1(0), v_2(0), v_3(0))$ belong to the subspace E^+. Then from (1.5.16) we get, in view of (1.2.31)

$$z_2(0) = 0, \quad v_1(0) + d_1 z_1(0) + d_3 z_3(0) + d_4 v_3(0) = 0,$$
$$v_3(0) + k_1 z_1(0) + k_3 z_3(0) = 0, \qquad (1.5.17)$$

where d_i and k_i are some constant matrices,

$$k_3 = \int_0^\infty [\Omega_0^\sigma(\hat\Phi)]^* D_{33}(\sigma)\Omega_0^\sigma(\Phi)\,d\sigma. \qquad (1.5.18)$$

Equalities (1.5.17) can be viewed as the equations with respect to the variables $z_i(0)$ and $v_i(0)$ of some subspace $\tilde E^+ \subseteq E^+$. Similarly, we get the equations for the subspace $\tilde E^- \subseteq E^-$ from (1.5.16):

$$z_1(0) = 0, \quad v_2(0) + \bar d_2 z_2(0) + \bar d_3 z_3(0) + \bar d_4 v_3(0) = 0,$$
$$v_3(0) + \bar k_2 z_2(0) + \bar k_3 z_3(0) = 0, \qquad (1.5.19)$$

where $\bar d_i$ and $\bar k_i$ are constant matrices,

$$\bar k_3 = -\int_{-\infty}^0 [\Omega_0^\sigma(\hat\Phi)]^* D_{33}(\sigma)\Omega_0^\sigma(\hat\Phi)\,d\sigma. \qquad (1.5.20)$$

Since the ranks of matrices in (1.5.17) and (1.5.19) coincide

$$\operatorname{rank}\begin{pmatrix} 0 & I_{r-} & 0 & 0 & 0 & 0 \\ d_1 & 0 & d_3 & I_{r+} & 0 & d_4 \\ k_1 & 0 & k_3 & 0 & 0 & I_{\hat{r}} \end{pmatrix} = \operatorname{rank}\begin{pmatrix} I_{r+} & 0 & 0 & 0 & 0 & 0 \\ 0 & \bar{d}_2 & \bar{d}_3 & 0 & I_{r-} & \bar{d}_4 \\ 0 & \bar{k}_2 & \bar{k}_3 & 0 & 0 & I_{\hat{r}} \end{pmatrix} = n,$$

then $\tilde{E}^+ = E^+$, $\tilde{E}^- = E^-$. The fact that n-dimensional subspaces E^+ and E^- complement each other up to the complete space R^{2n} and intersect only at zero demonstrates the non-degeneracy of the main determinant of the system of algebraic equations (1.5.17) and (1.5.19)

$$\det\begin{pmatrix} 0 & I_{r-} & 0 & 0 & 0 & 0 \\ d_1 & 0 & d_3 & I_{r+} & 0 & d_4 \\ k_1 & 0 & k_3 & 0 & 0 & I_{\hat{r}} \\ I_{r+} & 0 & 0 & 0 & 0 & 0 \\ 0 & \bar{d}_2 & \bar{d}_3 & 0 & I_{r-} & \bar{d}_4 \\ 0 & \bar{k}_2 & \bar{k}_3 & 0 & 0 & I_{\hat{r}} \end{pmatrix} = (\pm)\det\begin{pmatrix} d_3 & I_{r+} & 0 & d_4 \\ k_3 & 0 & 0 & I_{\hat{r}} \\ \bar{d}_3 & 0 & I_{r-} & \bar{d}_4 \\ \bar{k}_3 & 0 & 0 & I_{\hat{r}} \end{pmatrix}$$

$$= (\pm)\det\begin{pmatrix} d_3 & I_{r+} & 0 & d_4 \\ k_3 & 0 & 0 & I_{\hat{r}} \\ \bar{d}_3 & 0 & I_{r-} & \bar{d}_4 \\ \bar{k}_3 - k_3 & 0 & 0 & 0 \end{pmatrix} = (\pm)\det(\bar{k}_3 - k_3) \neq 0.$$

Thus, the following condition is necessary for the exponential dichotomy on R of the systems (1.5.3) and (1.5.14)

$$\det(\bar{k}_3 - k_3) = \det \int_{-\infty}^{\infty} [\Omega_0^\sigma(\hat{\Phi})]^* D_{33}(\sigma)\Omega_0^\sigma(\hat{\Phi})\, d\sigma \neq 0. \tag{1.5.21}$$

Let us show this condition to be sufficient as well.

As the system (1.5.14) is of the block-triangular form, it seems sufficient to verify the existence of a unique solution to the inhomogeneous system which is bounded on R

$$\begin{aligned} \frac{dz_3}{dt} &= \hat{\Phi}(t)z_3 + f_1(t), \\ \frac{dv_3}{dt} &= D_{33}(t)z - [\hat{\Phi}(t)]^* v_3 + f_2(t), \end{aligned} \tag{1.5.22}$$

for every vector-function $f_i(t) \in C^0(R)$. Substituting the general solution to the first subsystem (1.5.22)

$$z_3(t) = \Omega_0^t(\hat{\Phi})\left[z_3(0) + \int_0^t \Omega_\sigma^0(\hat{\Phi}) f_1(\sigma)\, d\sigma\right]$$

into the second subsystem, we get

$$\frac{dv_3}{dt} = -[\hat{\Phi}(t)]^* v_3 + D_{33}(t)\Omega_0^t(\hat{\Phi})z_3(0) + \psi(t), \qquad (1.5.23)$$

where

$$\psi(t) = f_2(t) + D_{33}(t)\int_0^t \Omega_\sigma^t(\hat{\Phi}) f_1(\sigma)\, d\sigma.$$

The inhomogeneous system (1.5.23) is similar to the system (1.3.14). Therefore, for the solution to this system which is bounded on R to exist, it is necessary and sufficient that the equality (1.3.18) is satisfied, which in this case takes the form

$$\int_{-\infty}^{\infty} [\Omega_0^\sigma(\hat{\Phi})]^* D_{33}(\sigma)\Omega_0^\sigma(\hat{\Phi})\, d\sigma\, z_3(0) + \int_{-\infty}^{\infty} [\Omega_0^\sigma(\hat{\Phi})]^* \psi(\sigma)\, d\sigma = 0. \qquad (1.5.24)$$

The above equality is considered as the algebraic system relative to $z_3(0)$. Obviously, under the condition (1.5.21) this system has a unique solution $z_3(0)$, and consequently, the system of differential equations (1.5.22) has a unique solution bounded on R as well.

Now we show the conditions (1.5.21) and (1.5.12) to be equivalent. To this end, we first make sure that (1.5.21) can be replaced by

$$\operatorname{rank} \int_{-\infty}^{\infty} \bar{P}^* [\Omega_0^\sigma(\Phi)]^* D(\sigma)\Omega_0^\sigma(\Phi)\bar{P}\, d\sigma = \hat{r}, \qquad (1.5.25)$$

where

$$\bar{P} = \begin{pmatrix} 0 & 0 & 0 \\ 0 & 0 & 0 \\ P_1 & P_2 & P_3 \end{pmatrix},$$

the matriciant $\Omega_0^\sigma(\Phi)$ is of the form (1.4.5), $t = \sigma$, $\tau = 0$, P_1 and P_2 are arbitrary constant matrices, the dimensions of which coincide with those of the matrices Φ_1 and Φ_2. By means of the immediate computation we verify that

$$\int_{-\infty}^{\infty} \bar{P}^* [\Omega_0^\sigma(\Phi)]^* D(\sigma)\Omega_0^\sigma(\Phi)\bar{P}\, d\sigma = \begin{pmatrix} P_1^* k P_1 & P_1^* k P_2 & P_1^* k \\ P_2^* k P_1 & P_2^* k P_2 & P_2^* k \\ k P_1 & k P_2 & k \end{pmatrix}, \qquad (1.5.26)$$

where $k = k_3 - \bar{k}_3$. This shows that if (1.5.25) is satisfied for some projection matrix $\bar{P}$, then it is satisfied for every other such projection matrix $\bar{P}$, because the rank of the matrix on the right-hand side of (1.5.26) coincides with the dimensions of the non-degenerate matrix $\bar{P}$. We transform the left-hand side of (1.5.25) as follows

$$\operatorname{rank} \int_{-\infty}^{\infty} \bar{P}^* L^*(0)[\Omega_0^\sigma(A)]^*[L^{-1}(\sigma)]^* L^*(\sigma) B(\sigma) L(\sigma) L^{-1}(\sigma) \Omega_0^\sigma(A) L(0) \bar{P} \, d\sigma$$

$$= \operatorname{rank} \int_{-\infty}^{\infty} L^*(0) P^* [\Omega_0^\sigma(A)]^* B(\sigma) \Omega_0^\sigma(A) P L(0) \, d\sigma$$

$$= \operatorname{rank} \int_{-\infty}^{\infty} P^* [\Omega_0^\sigma(A)]^* B(\sigma) \Omega_0^\sigma(A) P \, d\sigma,$$

where $P = L(0) \bar{P} L^{-1}(0)$. This completes the proof of Theorem 1.5.2.

Corollary 1.5.1 *If the system (1.1.1) is weakly regular on R and all solutions of it are bounded on R, the necessary and sufficient condition for the system (1.5.3) to be exponentially dichotomous on R is*

$$\det \int_{-\infty}^{\infty} [\Omega_0^\sigma(A)]^* B(\sigma) \Omega_0^\sigma(A) \, d\sigma \neq 0. \tag{1.5.27}$$

In particular, in the scalar case the inequality (1.5.11) is equivalent to (1.5.27).

Remark 1.5.1 In the case when the matrix $\bar{B}(t) = \frac{1}{2}[B(t) + B^*(t)]$ is positive (or negative) for every fixed $t \in R$, the condition (1.5.12) is satisfied.

Actually, if the matrix $\bar{B}(t)$ is positive definite for every $t \in R$, the matrix $\bar{D}(t) = \frac{1}{2}[D(t) + D^*(t)]$ is also positive definite, where $D(t)$ is determined by (1.5.15). Therefore, for the diagonal block $D_{33}(t)$ of the matrix $D(t)$ the property $\langle D_{33}(t) z_3, z_3 \rangle \geq d(t) \|z_3\|^2$, $d(t) > 0$ is also valid. Hence, it follows that

$$\left\langle \int_{-\infty}^{\infty} [\Omega_0^\sigma(\hat{\Phi})]^* D_{33}(\sigma) \Omega_0^\sigma(\hat{\Phi}) \, d\sigma \, z_3, \, z_3 \right\rangle = \int_{-\infty}^{\infty} \langle D_{33}(\sigma) \Omega_0^\sigma(\hat{\Phi}) z_3, \, \Omega_0^\sigma(\hat{\Phi}) z_3 \rangle \, d\sigma$$

$$\geq \int_{-\infty}^{\infty} d(\sigma) \|\Omega_0^\sigma(\hat{\Phi})\|^{-2} d\sigma \|z_3\|^2 \geq \varepsilon_0 \|z_3\|^2, \quad \varepsilon_0 = \mathrm{const} > 0.$$

Thus, the positive definiteness of the matrix $k = k_3 - \bar{k}_3$ yields $\det k \neq 0$ and consequently, the equality (1.5.12) is satisfied.

We assume system (1.5.3) to be exponentially dichotomous on the whole R axis. Then the inhomogeneous system

$$\frac{dx}{dt} = A(t) + f_1(t), \quad \frac{dy}{dt} = B(t)x - A^*(t)y + f_2(t), \tag{1.5.28}$$

has a unique solution which is bounded on R for every fixed vector function $f_i \in C^0(R)$, and this solution can be represented via the Green function $\bar{G}(t,\tau)$

$$\begin{pmatrix} x \\ y \end{pmatrix} = \begin{pmatrix} x(t) \\ y(t) \end{pmatrix} = \int_{-\infty}^{\infty} \bar{G}(t,\tau) \begin{pmatrix} f_1(t) \\ f_2(t) \end{pmatrix} d\tau. \tag{1.5.29}$$

The Green function of the problem on solutions to the system (1.5.3) bounded on R is unique and possesses the same structure as the function (1.1.56)

$$\bar{G}(t,\tau) = \begin{cases} \begin{pmatrix} \Omega_0^t(A) & 0 \\ \omega(t,0) & [\Omega_t^0(A)]^* \end{pmatrix} \begin{pmatrix} C_{11} & C_{12} \\ C_{21} & C_{22} \end{pmatrix} \\ \times \begin{pmatrix} \Omega_\tau^0(A) & 0 \\ \omega(0,\tau) & [\Omega_0^\tau(A)]^* \end{pmatrix}, & \tau \leq t, \\ \begin{pmatrix} \Omega_0^t(A) & 0 \\ \omega(t,0) & [\Omega_t^0(A)]^* \end{pmatrix} \begin{pmatrix} C_{11} - I_n & C_{12} \\ C_{21} & C_{22} - I_n \end{pmatrix} \\ \times \begin{pmatrix} \Omega_\tau^0(A) & 0 \\ \omega(0,\tau) & [\Omega_0^\tau(A)]^* \end{pmatrix}, & \tau > t, \end{cases} \tag{1.5.30}$$

where

$$\omega(t,\tau) = \int_\tau^t [\Omega_t^\sigma(A)]^* B(\sigma) \Omega_\tau^\sigma(A) \, d\sigma, \tag{1.5.31}$$

$\{C_{ij}\}_{i,j=1,2} = C$ is a constant projection matrix; $C^2 = C$. Furthermore, the estimate

$$\|\bar{G}(t,\tau)\| \leq \bar{K}_0 \exp\{-\bar{\gamma}|t-\tau|\}, \quad \bar{K}_0, \bar{\gamma} = \text{const} > 0,$$

is valid. Evidently, the Green function (1.5.30) can be represented as

$$\bar{G}(t,\tau) = \begin{cases} \begin{pmatrix} \Omega_\tau^t(A) & 0 \\ \omega(t,\tau) & [\Omega_t^\tau(A)]^* \end{pmatrix} \begin{pmatrix} C_{11}(\tau) & C_{12}(\tau) \\ C_{21}(\tau) & C_{22}(\tau) \end{pmatrix}, & \tau \leq t, \\ \begin{pmatrix} \Omega_\tau^t(A) & 0 \\ \omega(t,\tau) & [\Omega_t^\tau(A)]^* \end{pmatrix} \begin{pmatrix} C_{11}(\tau) - I_n & C_{12}(\tau) \\ C_{21}(\tau) & C_{22}(\tau) - I_n \end{pmatrix}, & \tau > t, \end{cases}$$

where

$$\begin{pmatrix} C_{11}(\tau) & C_{12}(\tau) \\ C_{21}(\tau) & C_{22}(\tau) \end{pmatrix} \equiv \begin{pmatrix} \Omega_0^\tau(A) & 0 \\ \omega(\tau,0) & [\Omega_\tau^0(A)]^* \end{pmatrix} \begin{pmatrix} C_{11} & C_{12} \\ C_{21} & C_{22} \end{pmatrix} \\ \times \begin{pmatrix} \Omega_\tau^0(A) & 0 \\ \omega(0,\tau) & [\Omega_0^\tau(A)]^* \end{pmatrix} \tag{1.5.32}$$

Hence, in particular it follows that

$$\Omega_\tau^0(A)C_{11}(\tau)\Omega_0^\tau(A) \equiv C_{11} + C_{12}\int_\tau^0 [\Omega_0^\sigma(A)]^*B(\sigma)\Omega_0^\sigma(A)\,d\sigma, \tag{1.5.33}$$

$$\Omega_\tau^0(A)C_{12}(\tau)[\Omega_\tau^0(A)]^* \equiv C_{12}.$$

Setting $f_1(t) \equiv 0$ in the equality (1.2.29), we get for the first n coordinates of x

$$x = x(t) = \int_{-\infty}^\infty \Omega_0^t(A)C_{12}[\Omega_0^\tau(A)]^*f_2(\tau)\,d\tau. \tag{1.5.34}$$

Moreover, the estimate

$$\|\Omega_0^t(A)C_{12}[\Omega_0^\tau(A)]^*\| \leq \bar{K}_0\exp\{-\bar{\gamma}|t-\tau|\}$$

is valid, which follows from (1.5.30).

Proposition 1.5.1 *The equality (1.5.34) determines every solution of the system (1.1.1) bounded on R under the appropriate choice of the vector-function $f_2(t) \in C^0(R)$.*

In fact, let $x = x_0(t)$ be the solution of the system (1.1.1) bounded on R. Then the inhomogeneous system of equations

$$\frac{dx}{dt} = A(t)x, \qquad \frac{dy}{dt} = B(t)x - A^*(t)y - B(t)x_0(t)$$

must possess a unique solution bounded on R (due to the regularity of the system (1.5.3) on R) $x = x^0(t)$, $y = y^0(t)$. This solution is determined by (1.5.29), where $f_1(\tau) \equiv 0$, $f_2(\tau) \equiv -B(\tau)x_0(\tau)$. However, on the other hand $x = x_0(t)$, $y = 0$ is the solution of this system bounded on R. Hence, because of the uniqueness of $x_0(t) \equiv x^0(t)$, $y^0(t) \equiv 0$ and in view of (1.5.34) we get

$$x_0(t) \equiv -\int_{-\infty}^\infty \Omega_0^t(A)C_{12}[\Omega_0^\tau(A)]^*B(\tau)x_0(\tau)\,d\tau. \tag{1.5.35}$$

Proposition 1.5.2 *The matrix C_{12} satisfies the equation*

$$C_{12} = -\int_{-\infty}^\infty C_{12}[\Omega_0^\tau(A)]^*B(\tau)\Omega_0^\tau(A)C_{12}\,d\tau. \tag{1.5.36}$$

In fact, due to estimate (1.5.35) $x = \Omega_0^t(A)C_{12}\eta$ is a solution to the system (1.1.1) bounded on R for every fixed $\eta \in R^n$. That is why equation (1.5.35) must also hold for it

$$\Omega_0^t(A)C_{12}\eta \equiv -\int_{-\infty}^{\infty} \Omega_0^t(A)C_{12}[\Omega_0^\tau(A)]^* B(\tau)\Omega_0^\tau(A)C_{12}\eta \, d\tau. \qquad (1.5.37)$$

In view of the non-degeneracy of the matrix $\Omega_0^t(A)$ and the arbitrary choice of $\eta \in R^n$ we get the equation (1.5.36) from (1.5.37).

Remark 1.5.2 The operator $\mathfrak{M}$ acting on the vector-function $f(t)$ from the space $C^0(R)$ and determined by

$$\mathfrak{M}f = -\int_{-\infty}^{\infty} \Omega_0^t(A)C_{12}[\Omega_0^\tau(A)]^* B(\tau)f(\tau) \, d\tau \qquad (1.5.38)$$

is projecting, $\mathfrak{M}^2 = \mathfrak{M}$.

Theorem 1.5.3 *Under conditions of the Theorem 1.4.2 the inhomogeneous system of equations*

$$\frac{dy}{dt} = -A^*(t)y + \varphi(t) \qquad (1.5.39)$$

does not possess a solution which is bounded on R for every vector-function $\varphi(t) \in C^0(R)$. For such a solution to exist, it is necessary and sufficient that the equality

$$\int_{-\infty}^{\infty} \langle x_0(t), \varphi(t)\rangle \, dt = 0 \qquad (1.5.40)$$

is satisfied for every non-trivial solution $x = x_0(t)$ of the system (1.1.10) which is bounded on R.

Proof Let the system (1.5.39) have a solution $y = y_0(t)$ which is bounded on R for some vector-function $\varphi(t) \in C^0(R)$. Then the extended system

$$\frac{dx}{dt} = A(t)x, \qquad \frac{dy}{dt} = -x - A^*(t)y + \varphi(t) \qquad (1.5.39a)$$

has a solution $x = 0$, $y = y_0(t)$ bounded on R which is unique. Due to (1.5.34)

$$0 = \int_{-\infty}^{\infty} C_{12}[\Omega_0^\tau(A)]^* \varphi(\tau) \, d\tau = \int_{-\infty}^{\infty} [\Omega_0^\tau(A)C_{12}]^* \varphi(\tau) \, d\tau. \qquad (1.5.41)$$

Consequently, for every numerical vector $\eta \in R^n$

$$0 = \int_{-\infty}^{\infty} [\Omega_0^\tau(A)C_{12}\eta]^* \varphi(\tau)\, d\tau = \int_{-\infty}^{\infty} x_0^*(\tau)\varphi(\tau)\, d\tau = \int_{-\infty}^{\infty} \langle x_0(\tau), \varphi(\tau)\rangle\, d\tau. \quad (1.5.42)$$

Since the equality $x = \Omega_0^t(A)C_{12}\eta$ determines every solution to the system (1.1.1) bounded on R, then (1.5.42) immediately implies (1.5.40) for every solution $x = x_0(t)$ bounded on R of this system.

Now let the equality (1.5.40) be satisfied for some vector-function $\varphi(t) \in C^0(R)$ for every solution $x = x_0(t)$ of the system (1.1.1) bounded on R. From this it follows that (1.5.42) holds for all $\eta \in R^n$ and hence, the validity of (1.5.41). Therefore, system (1.5.39a) has a solution bounded on R of the form $x = 0$, $y = \bar{y}(t)$, where $y = \bar{y}(t)$ is the solution of the system (1.5.39).

Remark 1.5.3 In the case when all solutions of the system (1.1.1) are exponentially damping at $\pm \infty$ the equality (1.5.41) becomes

$$0 = \int_{-\infty}^{\infty} [\Omega_0^\tau(A)]^* \varphi(\tau)\, d\tau = \int_{-\infty}^{\infty} \Omega_\tau^0(-A^*)\varphi(\tau)\, d\tau,$$

i.e. the previously established condition (1.3.18) is satisfied.

Now we return to the system of the equations (1.5.1) which is exponentially dichotomous on R and for $B(t) \equiv -I_n$ represent equality (1.5.36) in the form

$$C_{12} = -\int_{-\infty}^{\infty} C_{12}[\Omega_0^\tau(A)]^* \Omega_0^\tau(A) C_{12}\, d\tau. \quad (1.5.43)$$

There arises a question: how many n-dimensional symmetric constant matrices C_{12} exist satisfying the equation (1.5.43)?

Lemma 1.5.1 *Let there exist n-dimensional symmetric constant matrices W and $\bar{W}$ satisfying the equations*

$$W = \int_{-\infty}^{\infty} W[\Omega_0^\tau(A)]^* \Omega_0^\tau(A) W\, d\tau, \quad (1.5.43a)$$

$$\bar{W} = \int_{-\infty}^{\infty} \bar{W}[\Omega_0^\tau(A)]^* \Omega_0^\tau(A) \bar{W}\, d\tau, \quad (1.5.43b)$$

and every column of the matrix W is represented in the form of a linear combination of columns of the matrix $\bar{W}$, and vice versa, every column of the matrix $\bar{W}$ can be represented in the form of a linear combination of columns of the matrix W. Then $W = \bar{W}$.

Proof Under the conditions of the Lemma 1.5.1 there exist matrices D and $\bar{D}$ such that
$$W = \bar{W}D, \quad \bar{W} = W\bar{D}. \tag{1.5.44}$$

Multiplying (1.5.43a) on the right by the matrix $\bar{D}$ one gets
$$\bar{W} = \int_{-\infty}^{\infty} W[\Omega_0^\tau(A)]^*\Omega_0^\tau(A)\bar{W}\,d\tau. \tag{1.5.45}$$

Subtracting (1.5.45) from (1.5.43a) yields
$$W - \bar{W} = \int_{-\infty}^{\infty} W[\Omega_0^\tau(A)]^*\Omega_0^\tau(A)[W - \bar{W}]\,d\tau. \tag{1.5.46}$$

Now we subtract (1.5.45) from (1.5.43)
$$0 = \int_{-\infty}^{\infty} (\bar{W} - W)[\Omega_0^\tau(A)]^*\Omega_0^\tau(A)\bar{W}\,d\tau$$

and multiply on the right by the matrix D. In view of (1.5.44) we get
$$0 = \int_{-\infty}^{\infty} (\bar{W} - W)[\Omega_0^\tau(A)]^*\Omega_0^\tau(A)W\,d\tau. \tag{1.5.47}$$

Thus, with regard to the matrices W and $\bar{W}$ being symmetric, and comparing both parts of (1.5.46) and (1.5.47) we conclude that $W = \bar{W}$.

Note that by means of the symmetric matrix C_{12} found in (1.5.43) it is possible to define all solutions to the system (1.1.1) bounded on R
$$x = \Omega_0^t(A)C_{12}\eta, \quad \eta \in R^n. \tag{1.5.48}$$

So, if another such matrix $\bar{C}_{12}$ exists, they are correlated by equalities similar to (1.5.44). Summing up the above we come to the following result.

Theorem 1.5.4 *Let the system of the equations (1.1.1) be weakly regular on R, then there exists a unique n-dimensional symmetric matrix C_{12} satisfying the equality (1.5.43) and such that the equality (1.5.48) specifies all solutions to the system (1.1.1) bounded on R.*

Remark 1.5.4 If for the symmetric matrix C_{12} the equality (1.5.48) is not required to determine all solutions to the system (1.1.1) bounded on R, then it is possible that there exists more than one matrix C_{12} satisfying (1.5.43).

To determine the structure of all symmetric matrices W satisfying (1.5.43) the weakly regular on R system (1.1.1) is considered to have the split form (1.2.30). Moreover, estimates (1.2.31) are satisfied. Since the product $\Omega_0^t(\Phi)W$ is to be bounded on R, and the matriciant $\Omega_0^t(\Phi)$ has the structure of (1.4.5), it is necessary that the matrix W has the form $W = \text{diag}\{0,0,W_{33}\}$ (also, the dimensions of block W_{33} coincides with those of the matrix $\hat{\Phi}$). Since the matriciant $\Omega_0^t(\hat{\Phi})$ is exponentially damping on $\pm\infty$, we get from (1.5.43)

$$W_{33} = W_{33} \int_{-\infty}^{\infty} [\Omega_0^\tau(\hat{\Phi})]^* \Omega_0^\tau(\hat{\Phi})\, d\tau\, W_{33}, \qquad (1.5.49)$$

i.e. the constant matrix W_{33} can be taken out of the integral sign. Note that the matrix

$$M = \int_{-\infty}^{\infty} [\Omega_0^\tau(\hat{\Phi})]^* \Omega_0^\tau(\hat{\Phi})\, d\tau$$

is symmetric and positive definite

$$\langle M\eta, \eta\rangle = \int_{-\infty}^{\infty} \|\Omega_0^\tau(\hat{\Phi})\eta\|^2\, d\tau \geq \int_{-\infty}^{\infty} \|\Omega_0^\tau(\hat{\Phi})\|^{-2}\, d\tau\, \|\eta\|^2, \quad \eta \in R^{\hat{r}}.$$

Hence it follows, in particular, that $\det M \neq 0$.

Now we treat equality (1.5.49) as the equation relative to the matrix $X = W_{33}$

$$X = XMX. \qquad (1.5.49a)$$

The general solution of equation (1.5.49a) can be obtained as follows. We multiply the equality (1.5.49) on the right (on the left) by the matrix M and designate $P = XM$, $P^2 = P$. Thus, the general solution of equation (1.5.49a) is $X = PM^{-1}$, where P is an arbitrary projection matrix, $P^2 = P$. However, all symmetric matrices X are also to be obtained which satisfy the equation (1.5.49). Therefore, we multiply equality (1.5.49a) on the left and on the right by the matrix $M^{1/2}$: $M^{1/2}XM^{1/2} = M^{1/2}XM^{1/2}M^{1/2}XM^{1/2}$ and designate $P = M^{1/2}XM^{1/2}$. Hence we get a general solution of (1.5.49a) in terms of the symmetric matrices

$$X = M^{-1/2}PM^{-1/2}, \qquad (1.5.50)$$

where P is an arbitrary matrix of the orthogonal projection, $P^2 = P = P^*$. Thus, in the case of the system (1.1.1) which is weakly regular on R of the split form (1.2.30) all the matrices $W = W^*$ satisfying the equality (1.5.43) are representable in the form

$$W = \text{diag}\{0, 0, M^{-1/2}PM^{-1/2}\}. \qquad (1.5.51)$$

Also, P is an orthogonal projection matrix. In particular, it becomes clear that the additional requirement of the matrix W, rank $W = \hat{r}$, where $\hat{r}$ is the dimension of the subspace of damping on $\pm\infty$ solutions of the system (1.1.1) results in its uniqueness $W = \mathrm{diag}\,\{0, 0, M^{-1}\}$, because we set $P = I_{\hat{r}}$ in (1.5.51).

Now we return to the system (1.5.3) and note that the $2n$-dimensional matrix

$$\mathcal{P}(t) = \begin{pmatrix} A(t) & 0 \\ B(t) & -A^*(t) \end{pmatrix} \qquad (1.5.52)$$

possesses, in the case of $B^*(t) \equiv B(t)$, the following property

$$\mathcal{P}(t) \equiv J\mathcal{P}^*(t)J, \qquad (1.5.53)$$

where

$$J = \begin{pmatrix} 0 & I_n \\ -I_n & 0 \end{pmatrix}. \qquad (1.5.54)$$

Clearly, the identity (1.5.53) is also satisfied for the matrix of a more general form than (1.5.52)

$$\mathcal{P}(t) = \begin{pmatrix} A(t) & D(t) \\ B(t) & -A^*(t) \end{pmatrix}, \qquad (1.5.52a)$$

provided that $B^*(t) \equiv B(t)$ and $D^*(t) \equiv D(t)$.

Lemma 1.5.2 *Let the two $2n$-dimensional linear systems*

$$\frac{dx}{dt} = A_1(t)x, \qquad \frac{dx}{dt} = A_2(t)x, \qquad x \in R^{2n}, \qquad (1.5.55)$$

be weakly regular on R and the $2n \times 2n$-dimensional matrices $A_i(t) \in C^0(R)$, $i = 1, 2$, be correlated by

$$A_2(t) = JA_1^*J, \qquad (1.5.56)$$

where J is a matrix of the form of (1.5.54). Then both systems (1.5.55) are regular on R (exponentially dichotomous on R) and its unique Green functions $G_i(t, \tau)$, $i = 1, 2$, of the problem on solutions bounded on R are

$$G_i(t, \tau) = \begin{cases} \Omega_0^t(A_i)C_i\Omega_\tau^0(A_i), & \tau \leq t, \\ \Omega_0^t(A_i)[C_i - I_{2n}]\Omega_\tau^0(A_i), & \tau > t, \end{cases} \qquad (1.5.57)$$

satisfy the identity

$$G_2(t, \tau) \equiv JG_1^*(\tau, t)J, \qquad (1.5.58)$$

and the projection matrices C_i are correlated by

$$C_2 = JC_1^*J + I_{2n}. \qquad (1.5.59)$$

Proof The weak regularity on R of each of the systems (1.5.55) is equivalent to the existence of $2n$-dimensional symmetric matrices $S_i(t) \in C^1(R)$, $i = 1,2$, satisfying the condition

$$\langle [\dot{S}_i(t) - S_i(t)A_i^*(t) - A_i(t)S_i(t)]x, x\rangle \leq -\|x\|^2. \tag{1.5.60}$$

Making the change of variables $x \to Jx$, in view of $J^* = -J$ we get

$$\langle J[\dot{S}_i(t) - S_i(t)A_i^*(t) - A_i(t)S_i(t)]Jx, x\rangle \geq \|x\|^2.$$

Hence

$$\langle [J\dot{S}_i(t)J + JS_i(t)JJA_i^*(t)J + JA_i(t)JJS_i(t)J]x, x\rangle \geq \|x\|^2.$$

Taking into account (1.5.56) we arrive at

$$\langle [\dot{\bar{S}}_i(t) + \bar{S}_i(t)A_j(t) + A_j^*(t)\bar{S}_i(t)]x, x\rangle \geq \|x\|^2, \quad i \neq j. \tag{1.5.61}$$

Setting $i = 1$ in (1.5.60) and in (1.5.61) $i = 2$, $j = 1$ we note that there exist two quadratic forms $\langle S_1(t)x, x\rangle$ and $-\langle \bar{S}_2(t)x, x\rangle$ that possess negative definite derivatives along the solutions of mutually conjugated linear systems

$$\frac{dx}{dt} = -A_1^*(t)x \quad \text{and} \quad \frac{dx}{dt} = A_1(t)x.$$

From here it follows (see Theorem 1.3.4) that $\det S_1(t) \neq 0$, $\det \bar{S}_2(t) \neq 0$ $\forall t \in R$ and the system

$$\frac{dx}{dt} = A_1(t)x$$

is exponentially dichotomous on the whole R axis. Similarly, setting $i = 2$ in (1.5.60), and $i = 1$, $j = 2$ in (1.5.61) we conclude that the system

$$\frac{dx}{dt} = A_2(t)x$$

is also exponentially dichotomous on R_-.

Consider the Green function $G_1(t,\tau)$ and compose the following expression for it

$$JG_1^*(\tau,t)J = \begin{cases} J[\Omega_0^\tau(A_1)C_1\Omega_t^0(A_1)]^*J, & t \leq \tau, \\ J[\Omega_0^\tau(A_1)[C_1 - I_{2n}]\Omega_t^0(A_1)]^*J, & t > \tau. \end{cases} \tag{1.5.62}$$

Further, it follows, for example, for $t \leq \tau$

$$J[\Omega_0^\tau(A_1)C_1\Omega_t^0(A_1)]^*J = J[\Omega_t^0(A_1)]^*C_1^*[\Omega_0^\tau(A_1)]^*J$$
$$= \{J[\Omega_t^0(A_1)]^*J\}\{JC_1^*J\}\{J[\Omega_0^\tau(A_1)]^*J\}. \tag{1.5.63}$$

Obviously, due to (1.5.56) the expression $-J[\Omega_t^0(A_1)]^*J$ is the matriciant of the linear system
$$\frac{dx}{dt} = A_2(t)x.$$
Therefore,
$$-J[\Omega_t^0(A_1)]^*J \equiv \Omega_0^t(A_2). \tag{1.5.64}$$

Now let us verify that the matrix
$$C = JC_1^*J + I_{2n} \tag{1.5.65}$$
is a projection matrix. Actually
$$C^2 = (JC_1^*J + I_{2n})(JC_1^*J + I_{2n}) = -JC_1^*J + JC_1^*J + JC_1^*J + I_{2n} = JC_1^*J + I_{2n} = C.$$

Thus, in view of (1.5.63), (1.5.64) and (1.5.65), the expression (1.5.62) becomes
$$JG_1^*(\tau,t)J = \begin{cases} \Omega_0^t(A_2)C\Omega_\tau^0(A_2), & \tau \leq t, \\ \Omega_0^t(A_2)(C - I_{2n})\Omega_\tau^0(A_2), & \tau > t. \end{cases} \tag{1.5.62a}$$

We compare (1.5.62a) with the Green function $G_2(t,\tau)$. Since the estimate $\|JG_1^*(\tau,t)J\| \leq Ke^{\gamma|t-\tau|}$ is valid and C is the projection matrix, then by virtue of the uniqueness of the Green function $G_2(t,\tau)$ the identity (1.5.58) is valid, and the matrix C is equal to C_2, i.e. (1.5.59) is satisfied.

Remark 1.5.5 If the weak regularity is assumed in the conditions of the Lemma 1.5.2 for only one of the systems (1.5.55), the first one for example, then by virtue of equality (1.5.56) the weak regularity on R is only implied for the conjugated system
$$\frac{dx}{dt} = -A_2^*(t)x.$$

The proved Lemma 1.5.2 easily yields the following assertion.

Theorem 1.5.5 *Let the linear system of differential equations*
$$\frac{dz}{dt} = \mathcal{P}(t)z, \quad z \in R^{2n}, \tag{1.5.66}$$

be weakly regular on R and the $2n$-dimensional matrix function $\mathcal{P}(t) \in C^0(R)$ satisfies (1.5.53). Then the system (1.5.66) is exponentially dichotomous on R, and the Green function $G(t,\tau)$ satisfies the identity
$$G(t,\tau) \equiv JG^*(\tau,t)J. \tag{1.5.67}$$

Remark 1.5.6 The identity (1.5.67) provides the following equalities for $n \times n$-dimensional blocks C_{ij} of the projection matrix $C = \{C_{ij}\}_{i,j=1,2}$

$$C_{12} = C_{12}^*, \quad C_{21} = C_{21}^*, \quad C_{11} + C_{22}^* = I_n. \tag{1.5.68}$$

In fact, in the case when $C_2 = C_1 = C$, where C is a $2n \times 2n$-dimensional projection matrix, the equality (1.5.59) becomes

$$\begin{pmatrix} C_{11} & C_{12} \\ C_{21} & C_{22} \end{pmatrix} = \begin{pmatrix} 0 & I_n \\ -I_n & 0 \end{pmatrix} \begin{pmatrix} C_{11}^* & C_{21}^* \\ C_{12}^* & C_{22}^* \end{pmatrix} \begin{pmatrix} 0 & I_n \\ -I_n & 0 \end{pmatrix} + \begin{pmatrix} I_n & 0 \\ 0 & I_n \end{pmatrix}.$$

With further operating as usual with the matrices, we easily arrive at the equalities (1.5.68).

1.6 Regularity of Linear Systems of the Block-Triangular Form

With regard to the proposed extensions (1.5.1)–(1.5.3) of weakly linear systems (1.1.1) which are regular on R it is of interest to discuss the systems of block-triangular form

$$\frac{dx}{dt} = A_1(t)x, \quad \frac{dy}{dt} = B(t)x + A_2(t)y, \tag{1.6.1}$$

where $A_i(t), B(t) \in C^0(R)$, $x \in R^n$, $y \in R^m$. Additionally, the existence of two symmetric matrix functions $S_i(t) \in C^1(R)$, $i = 1, 2$, is to be assumed which satisfy the condition

$$\begin{aligned} \langle [\dot S_1(t) - S_1(t)A_1^*(t) - A_1(t)S_1(t)]x, x \rangle &\leq -\|x\|^2, \\ \langle [\dot S_2(t) + S_2(t)A_2(t) + A_2^*(t)S_2(t)]y, y \rangle &\leq -\|y\|^2. \end{aligned} \tag{1.6.2}$$

Clearly, under these conditions the non-degeneracy of the matrix $S_2(t)$ provides the weak regularity on R of the system (1.6.1) for all $t \in R$ for every matrix function $B(t) \in C^0(R)$. Therefore the case of the degeneracy of both matrices $S_i(t)$, $i = 1, 2$, for some, probably different, values of $t = t_i, \bar t_j$ is of interest: $\det S_1(t_i) = 0$, $i = 1, \ldots, k$, $1 \leq k \leq n$; $\det S_2(\bar t_j) = 0$, $j = 1, \ldots, k$, $1 \leq k \leq m$. Such values of t can only be isolated and the number of them should not be larger than the dimensions of the system (1.6.1).

The question is: what additional properties must the matrix functions $S_i(t)$, $i = 1, 2$, possess for the system (1.6.1) to be regular or weakly regular on R for some matrix function $B(t) \in C^0(R)$?

To answer this question we designate the number of positive eigenvalues of the matrix

$$S(t) = \text{diag}\{S_1(t), -S_2(t)\} \tag{1.6.3}$$

by $n^+(+\infty)$ for sufficiently large values of t: $t > \max\{t_i, \bar t_j\}$, and by $n^+(-\infty)$ the number of positive eigenvalues of the matrix (1.6.3) for $t < \min\{t_i, \bar t_j\}$.

Theorem 1.6.1 *Let there exist symmetric matrix functions $S_1(t)$, $S_2(t) \in C^1(R)$ bounded on R which satisfy the conditions (1.6.2). Then for some matrix function $B(t) \in C^0(R)$ the system of equations (1.6.1) is*

(1) *regular on R if and only if*

$$n^+(+\infty) = n^+(-\infty); \qquad (1.6.4)$$

(2) *weakly regular on R if and only if*

$$n^+(+\infty) \leq n^+(-\infty). \qquad (1.6.5)$$

Moreover, the dimensions of the subspace $\hat{E}$ of solutions to the system (1.6.1) bounded on R is determined by $\dim \hat{E} = n^+(-\infty) - n^+(+\infty)$.

Proof We investigate the existence conditions for the solutions which are bounded on the whole R axis to the inhomogeneous system of equations

$$\frac{dx}{dt} = A_1(t)x + f(t), \quad \frac{dy}{dt} = B(t)x + A_2(t)y + \bar{f}(t), \qquad (1.6.6)$$

for any fixed vector-functions $f(t)$, $\bar{f}(t) \in C^0(R)$. We note that if both matrix functions $S_1(t)$ and $S_2(t)$ are not degenerate for all $t \in R$, then the systems

$$\frac{dx}{dt} = A_1(t)x$$

and

$$\frac{dy}{dt} = A_2(t)y$$

are exponentially dichotomous on the whole R axis, and hence, the system (1.6.6) has a unique solution bounded on R. Moreover, it seems that the condition (1.6.4) is satisfied. In the conditions of Theorem 1.2.3 both systems of equations

$$\frac{dx}{dt} = A_1(t)x$$

and

$$\frac{dy}{dt} = -A_2^*(t)y$$

can be transformed to a form similar to (1.2.30). Therefore, system (1.6.6) is considered in the form

$$\frac{dx_1}{dt} = \Phi^+(t)x_1 + f_1(t),$$

$$\frac{dx_2}{dt} = \Phi^-(t)x_2 + f_2(t),$$

$$\frac{dx_3}{dt} = \Phi_1(t)x_1 + \Phi_2(t)x_2 + \hat{\Phi}(t)x_3 + f_3(t),$$

$$\frac{dy_1}{dt} = \sum_{i=1}^{3} B_{1i}(t)x_i + Y^+(t)y_1 + Y_1(t)y_3 + f_4(t), \qquad (1.6.6a)$$

$$\frac{dy_2}{dt} = \sum_{i=1}^{3} B_{2i}(t)x_i + Y^-(t)y_2 + Y_2(t)y_3 + f_5(t),$$

$$\frac{dy_3}{dt} = \sum_{i=1}^{3} B_{3i}(t)x_i + \check{Y}(t)y_3 + f_6(t),$$

where $f_i(t)$, $i = 1, \ldots, 6$, are arbitrary fixed vector-functions from the space $C^0(R)$, $\hat{\Phi}$ is a $\hat{r} \times \hat{r}$-dimensional and $\check{Y}$ is an $\check{r} \times \check{r}$-dimensional matrices.

The first two subsystems (1.6.6) have unique solutions bounded on R

$$x_1 = x_1^0(t) = \int_{-\infty}^{t} \Omega_\tau^t(\Phi^+) f_1(\tau) \, d\tau, \qquad x_2 = x_2^0(t) = -\int_{t}^{\infty} \Omega_\tau^t(\Phi^-) f_2(\tau) \, d\tau.$$

We substitute the solutions into the third subsystem (1.6.6a). Then its general solution is

$$x_3(t) = \Omega_0^t(\hat{\Phi})\eta + x_3^0(t), \qquad (1.6.7)$$

where η is an arbitrary constant vector from $R^{\hat{r}}$,

$$x_3^0(t) = \int_0^t \Omega_\tau^t(\hat{\Phi})[\Phi_1(\tau)x_1^0(\tau) + \Phi_2(\tau)x_2^0(\tau) + f_3(\tau)] \, d\tau.$$

Now substituting (1.6.7) into the last subsystem (1.6.6a) we get

$$\frac{dy_3}{dt} = \check{Y}(t)y_3 + \sum_{i=1}^{3} B_{3i}(t)x_i^0(t) + B_{33}(t)\Omega_0^t(\hat{\Phi})\eta + f_6(t). \qquad (1.6.8)$$

We recall that for the matriciant $\Omega_\tau^t(\check{Y})$ the estimate

$$\|\Omega_\tau^t(\check{Y})\| \leq K \begin{cases} \exp\{\gamma(t-\tau)\}, & 0 \leq t \leq \tau, \\ \exp\{-\gamma(t-\tau)\}, & \tau \leq t < 0 \end{cases}$$

is valid. Therefore, the inhomogeneous system

$$\frac{dy_3}{dt} = \check{Y}(t)y_3 + \varphi(t)$$

does not have a solution bounded on R for every vector-function $\varphi(t) \in C^0(R)$. In order that such a solution exists, it is necessary and sufficient that the equality

$$\int_{-\infty}^{\infty} \Omega_\sigma^0(\check{Y})\varphi(\sigma)\, d\sigma = 0 \qquad (1.6.9)$$

is satisfied (see (1.3.18)). Returning to the system of equations (1.6.8) we represent the condition (1.6.9) for the vector-function

$$\varphi(t) = \sum_{i=1}^{3} B_{3i}(t)x_i^0(t) + B_{33}(t)\Omega_0^t(\hat{\Phi})\eta + f_6(t),$$

in the form

$$\int_{-\infty}^{\infty} \Omega_\tau^0(\check{Y})B_{33}(\tau)\Omega_0^\tau(\hat{\Phi})\, d\tau\, \eta = \lambda, \qquad (1.6.10)$$

where

$$\lambda = -\int_{-\infty}^{\infty} \Omega_\tau^0(\check{Y}) \left[\sum_{i=1}^{3} B_{3i}(\tau)x_i^0(\tau) + f_6(\tau)\right] d\tau. \qquad (1.6.11)$$

Then we show that for every numerical vector $\lambda \in R\check{r}$ the vector-function $f_6(t) \in C^0(R)$ exists satisfying (1.6.11). Let us make sure that the following form can be taken for such a vector-function

$$f_6(t) = -\sum_{i=1}^{3} B_{3i}(t)x_i^0(t) - [\Omega_t^0(\check{Y})]^* N^{-1}\lambda, \qquad (1.6.12)$$

where the matrix N is determined by the equality

$$N = \int_{-\infty}^{\infty} \Omega_\tau^0(\check{Y})[\Omega_\tau^0(\check{Y})]^*\, d\tau. \qquad (1.6.13)$$

Since the matrix N is symmetric and positive definite

$$\langle Nx, x \rangle = \int_{-\infty}^{\infty} \|[\Omega_\tau^0(\check{Y})]^* x\|^2\, d\tau \geq \int_{-\infty}^{\infty} \|\Omega_0^\tau(\check{Y})\|^{-2}\, d\tau\, \|x\|^2,$$

then $\det N \neq 0$, and therefore, there exists an inverse matrix N^{-1}. Substituting equation (1.6.12) into the right-hand side of (1.6.11) and taking into account (1.6.13) we get

$$-\int_{-\infty}^{\infty} \Omega_\tau^0(\check{Y})[-\Omega_\tau^0(Y)]^* N^{-1}\lambda\, d\tau = NN^{-1}\lambda = \lambda.$$

Thus, if the vector-function $f_6(t)$ runs through the whole space $C^0(R)$, the numerical vector λ takes all values from $R^{\check{r}}$. So, for the system of the equations (1.6.10) to have a solution relative to η for any $\lambda \in R^{\check{r}}$ it is necessary and sufficient that the equality

$$\operatorname{rank}\left(\int_{-\infty}^{\infty} \Omega_\tau^0(\check{Y})B_{33}(\tau)\Omega_0^\tau(\hat{\Phi})\, d\tau\right) = \check{r}. \tag{1.6.14}$$

be satisfied. The matrix

$$M = \int_{-\infty}^{\infty} \Omega_\tau^0(\check{Y})B_{33}(\tau)\Omega_0^\tau(\hat{\Phi})\, d\tau \tag{1.6.14a}$$

consists of $\hat{r}$ columns and $\check{r}$ rows. Condition (1.6.14) can be satisfied only when

$$\hat{r} \geq \check{r}. \tag{1.6.15}$$

Furthermore, if $\hat{r} = \check{r}$, then (1.6.14) means that $\det M \neq 0$ and the equation (1.6.10) $M\eta = \lambda$ has a unique solution for every fixed $\lambda \in R^{\check{r}}$ i.e. the system of equations (1.6.1) is regular on R. If $\hat{r} > \check{r}$, condition (1.6.14) provides the existence of a $\hat{r}-\check{r}$-dimensional subspace of solutions to the system (1.6.10) for every $\lambda \in R^{\check{r}}$.

Now we represent $\hat{r}$ and $\check{r}$ via the eigenvalues of the matrices $S_1(t)$ and $S_2(t)$. We designate the number of positive eigenvalues of the matrix $S_i(t)$ for sufficiently large t: $t \to +\infty$ ($t \to -\infty$) by $n_i^+(+\infty)$ ($n_i^+(-\infty)$). Similarly $n_i^-(\pm\infty)$ is the number of negative eigenvalues of the matrix $S_i(t)$ as $t \to \pm\infty$. Clearly (see Theorem 1.3.5), the equation $\check{r} = n_2^+(-\infty) - n_2^+(+\infty)$, $\hat{r} = n_1^+(-\infty) - n_1^+(+\infty)$ are valid. Then the inequality (1.6.15) becomes

$$n_1^+(-\infty) - n_1^+(+\infty) \geq n_2^+(-\infty) - n_2^+(+\infty).$$

Replacing $n_2^+(-\infty) = m - n_2^-(-\infty)$ and $n_2^+(+\infty) = m - n_2^-(+\infty)$ we get

$$n_1^+(-\infty) + n_2^-(-\infty) \geq n_2^-(+\infty) + n_1^+(+\infty). \tag{1.6.16}$$

Returning to the matrix (1.6.3) we note that

$$n^+(+\infty) = n_1^+(+\infty) + n_2^-(+\infty),$$
$$n^+(-\infty) = n_1^+(-\infty) + n_2^-(-\infty). \qquad (1.6.17)$$

Thus, inequality (1.6.16) is identical to (1.6.5). If we replace all the inequalities in the above arguments by the equalities $\check{r} = \hat{r}$ we finally arrive at (1.6.4).

Finally, it remains to be shown that when inequality (1.6.15) is satisfied the matrix function $B_{33}(t) \in C^0(R)$ is always found, for which condition (1.6.14) holds. Evidently, one can assume the matricants $\Omega_\tau^0(\check{Y})$ and $\Omega_0^\tau(\hat{\Phi})$ are of triangular form (Perron theorem)

$$\Omega_\tau^0(\check{Y}) = \begin{pmatrix} \exp\left\{\int_\tau^0 \lambda_1(\sigma)\,d\sigma\right\} & 0 & \cdots & 0 \\ \omega_{21}(\tau) & \exp\left\{\int_\tau^0 \lambda_2(\sigma)\,d\sigma\right\} & \cdots & 0 \\ \vdots & \vdots & \ddots & \vdots \\ \omega_{\check{r}1}(\tau) & \omega_{\check{r}2}(\tau) & \cdots & \exp\left\{\int_\tau^0 \lambda_{\check{r}}(\sigma)\,d\sigma\right\} \end{pmatrix},$$

$$\Omega_0^\tau(\Phi) = \begin{pmatrix} \exp\left\{\int_0^\tau \mu_1(\sigma)\,d\sigma\right\} & 0 & \cdots & 0 \\ \bar{\omega}_{21}(\tau) & \exp\left\{\int_0^\tau \mu_2(\sigma)\,d\sigma\right\} & \cdots & 0 \\ \vdots & \vdots & \ddots & \vdots \\ \bar{\omega}_{\hat{r}1}(\tau) & \bar{\omega}_{\hat{r}2}(\tau) & \cdots & \exp\left\{\int_0^\tau \mu_{\hat{r}}(\sigma)\,d\sigma\right\} \end{pmatrix}.$$

In addition, taking the matrix $B_{33}(\tau)$ in the form $[I_{\check{r}}, 0]$, if $\check{r} < \hat{r}$, and $B_{33}(\tau) \equiv I_{\hat{r}}$ if $\check{r} = \hat{r}$, we get

$$\Omega_\tau^0(\check{Y}) B_{33}(\tau) \Omega_0^\tau(\hat{\Phi})$$

$$= \begin{pmatrix} e^{\int_0^\tau [\mu_1(\sigma)-\lambda_1(\sigma)]\,d\sigma} & 0 & \cdots & 0 \\ W_{21}(\tau) & e^{\int_0^\tau [\mu_2(\sigma)-\lambda_2(\sigma)]\,d\sigma} & \cdots & 0 \\ \vdots & \vdots & \ddots & \vdots \\ W_{\hat{r}1}(\tau) & W_{\hat{r}2}(\tau) & \cdots & e^{\int_0^\tau [\mu_{\hat{r}}(\sigma)-\lambda_{\hat{r}}(\sigma)]\,d\sigma} \end{pmatrix}.$$

Since the diagonal functions

$$\exp\left\{\int_0^\tau [\mu_i(\sigma) - \lambda_i(\sigma)]\, d\sigma\right\} > 0,$$

then, evidently, condition (1.6.14) is satisfied. This completes the proof of Theorem 1.6.1.

Remark 1.6.1 Though it is easy enough to find the quadratic form $V(t,x,y) = \langle S(t)y, y\rangle + 2\|S\|_0 \langle x, y\rangle$ possessing definite sign derivative along its solutions for system (1.5.1), it may prove to be difficult to find out such a form for system (1.6.1). This is shown by a simple example of the system

$$\frac{dx}{dt} = -(\tanh t)x, \quad \frac{dy}{dt} = bx + (\arctan t)y,$$

which is exponentially dichotomous on R for every $b = \text{const} \neq 0$.

It is easy to notice that, if all non-trivial solutions for the scalar equation

$$\frac{dx}{dt} = \hat{a}(t)x, \quad \hat{a}(t) \in C^0(R),$$

are exponentially damping on both sides, i.e.

$$\exp\left\{\int_\tau^t \hat{a}(\sigma)\, d\sigma\right\} \leq K_0 \left\{\begin{array}{ll} \exp\{-\gamma_0(t-\tau)\}, & 0 \leq \tau \leq t, \\ \exp\{\gamma_0(t-\tau)\}, & t \leq \tau \leq 0, \end{array}\right.$$

and all non-trivial solutions of the other scalar equation

$$\frac{dy}{dt} = \check{b}(t)y$$

increase exponentially on both sides

$$\exp\left\{\int_\tau^t \check{b}(\sigma)\, d\sigma\right\} \leq K_1 \left\{\begin{array}{ll} \exp\{-\gamma_1(t-\tau)\}, & 0 \leq t \leq \tau, \\ \exp\{\gamma_1(t-\tau)\}, & \tau \leq t \leq 0, \end{array}\right.$$

then the triangular system of the equations

$$\frac{dx}{dt} = \hat{a}(t)x, \quad \frac{dy}{dt} = px + \check{b}(t)y, \tag{1.6.18}$$

is exponentially dichotomous on R for every value of the parameter $p \neq 0$. Actually, having considered the corresponding inhomogeneous system of equations

$$\begin{aligned}\frac{dx}{dt} &= \hat{a}(t)x + f_1(t), \\ \frac{dy}{dx} &= px + \check{b}(t)y + f_2(t), \quad f_i(t) \in C^0(R),\end{aligned} \tag{1.6.19}$$

we verify that it has a unique solution bounded on R. All solutions to the first equation

$$x(t) = \exp\left\{\int_0^t \hat{a}(\sigma)\,d\sigma\right\}\left[x(0) + \int_0^t f_1(\tau)\exp\left\{-\int_0^\tau \hat{a}(\sigma)\,d\sigma\right\}d\tau\right]$$

are bounded on R. Substituting the solutions into the second equation, we get

$$\frac{dy}{dt} = \check{b}(t)y + px(0)\exp\left\{\int_0^t \hat{a}(\sigma)\,d\sigma\right\} + f_3(t).$$

For the solution to this equation which is bounded on R to exist, it is necessary that condition (1.3.18) be satisfied

$$\int_{-\infty}^{\infty}\exp\left\{\int_0^\tau [\hat{a}(\sigma) - \check{b}(\sigma)]\,d\sigma\right\}d\tau\, px(0)$$
$$+ \int_{-\infty}^{\infty} f_3(\tau)\exp\left\{\int_\tau^0 \check{b}(\sigma)\,d\sigma\right\}d\tau = 0. \tag{1.6.20}$$

Since

$$p\int_{-\infty}^{\infty}\exp\left\{\int_0^\tau [\hat{a}(\sigma) - \check{b}(\sigma)]\,d\sigma\right\}d\tau \neq 0,$$

a unique value of $x(0)$ is always found such that (1.6.20) is satisfied. Therefore, the system of equations (1.6.19) possesses a unique solution bounded on R for $p \neq 0$, and hence, system (1.6.18) is exponentially dichotomous on R. In this regard one might think that the system of equations

$$\frac{dx}{dt} = \hat{A}(t)x, \qquad \frac{dy}{dt} = B(t)x + \check{D}(t)y, \tag{1.6.21}$$

where $x \in R^n$ and $y \in R^n$ is exponentially dichotomous on R for every symmetrical definite sign matrix $B(t) \in C^0(R)$. In addition, all solutions to the n-dimensional system

$$\frac{dx}{dt} = \hat{A}(t)x,$$

are assumed to be uniformly exponentially damping over $\pm\infty$, and all non-trivial solutions of it are exponentially increasing over $\pm\infty$. We treat an example which shows this to be false.

Example 1.6.1 Let

$$\frac{dx_1}{dt} = -(\tanh t)x_1, \quad \frac{dx_2}{dt} = -3(\tanh t)x_2,$$

$$\frac{dy_1}{dt} = x_1 + \sqrt{\tfrac{5}{6}}\, x_2 + 3(\tanh t)y_1, \quad (1.6.22)$$

$$\frac{dy_2}{dt} = \sqrt{\tfrac{5}{6}}\, x_1 + x_2 + (\tanh t)y_2.$$

Here

$$B = \begin{pmatrix} 1 & \sqrt{\tfrac{5}{6}} \\ \sqrt{\tfrac{5}{6}} & 1 \end{pmatrix}$$

is a positive definite matrix,

$$\hat{A}(t) = \operatorname{diag}\{-\tanh t, -3\tanh t\},$$

$$\check{D}(t) = \operatorname{diag}\{3\tanh t, \tanh t\}.$$

Substituting the general solutions of the first two equations $x_1 = x_1(0)\cosh^{-1} t$ and $x_2 = x_2(0)\cosh^{-3} t$ into the last two equations we get

$$\frac{dy_1}{dt} = 3(\tanh t)y_1 + x_1(0)\cosh^{-1} t + \sqrt{\tfrac{5}{6}}\, x_2(0)\cosh^{-3} t,$$

$$\frac{dy_2}{dt} = (\tanh t)y_2 + \sqrt{\tfrac{5}{6}}\, x_1(0)\cosh^{-1} t + x_2(0)\cosh^{-3} t.$$

In this case the condition (1.3.18) is of the form

$$\int_{-\infty}^{\infty} \cosh^{-3} t \left[x_1(0)\cosh^{-1} t + \sqrt{\tfrac{5}{6}}\, x_2(0)\cosh^{-3} t \right] dt = 0,$$

$$\int_{-\infty}^{\infty} \cosh^{-1} t \left[\sqrt{\tfrac{5}{6}}\, x_1(0)\cosh^{-1} t + x_2(0)\cosh^{-3} t \right] dt = 0.$$

We find, computing the integrals,

$$\tfrac{4}{3} x_1(0) + \tfrac{16}{15}\sqrt{\tfrac{5}{6}}\, x_2(0) = 0,$$
$$2\sqrt{\tfrac{5}{6}}\, x_1(0) + \tfrac{4}{3} x_2(0) = 0. \quad (1.6.23)$$

The system of linear algebraic equations obtained relative to $x_1(0)$ and $x_2(0)$ has non-trivial solutions

$$x_1(0) = -\tfrac{4}{5}\sqrt{\tfrac{5}{6}}\, x_2(0).$$

Therefore, the system of the equations (1.6.22) has non-trivial solutions bounded on R, and hence, is not exponentially dichotomous on the whole R axis. It should be recalled that, if we set $\check{D}(t) = -[\hat{A}(t)]^*$ in (1.6.21), then this system is exponentially dichotomous on the R axis for every matrix $B(t)$ of definite sign.

1.7 Perturbation of the Block-Triangular Form Linear Systems which are Regular and Weakly Regular on the Whole R Axis

Let us clarify what perturbations of the system (1.6.1) preserve its regularity (or weak regularity) on the whole R axis. First, we consider a perturbed system of the form

$$\frac{dx}{dt} = [A(t) + p_1(t)I_n]x,$$
$$\frac{dy}{dt} = x - [A^*(t) + p_2(t)I_n]y, \tag{1.7.1}$$

where $p_i(t)$ are some scalar functions from the space $C^0(R)$. We assume condition (1.2.29) to be satisfied. Now, what functions $p_i(t) \in C^0(R)$ is a fixed value of $\lambda > 0$ found for, so that the derivative of the quadratic form (1.5.4) due to (1.7.1) is of definite sign?

We find the derivative of (1.5.4) due to (1.7.1)

$$\dot{V}(t,x,y) = -\lambda\langle(A + p_1 I_n)x, y\rangle - \lambda\langle x, x - (A^* + p_2 I_n)y\rangle$$
$$+ \langle \dot{S}y, y\rangle + \langle S[x - (A^* + p_2 I_n)y], y\rangle + \langle Sy, [x - (A^* + p_2 I_n)y]\rangle$$
$$= -\lambda\|x\|^2 + \lambda(p_2 - p_1)\langle x, y\rangle + 2\langle Sx, y\rangle - 2p_2\langle Sy, y\rangle$$
$$+ \langle[\dot{S} - SA^* - AS]y, y\rangle.$$

Hence

$$\dot{V}(t,x,y) = -\lambda\|x\|^2 + (\lambda|p_2 - p_1|_0 + 2\|S\|_0)\|x\|\,\|y\|$$
$$- (\beta - 2|p_2|_0\|S\|_0)\|y\|^2.$$

It is obvious that for the derivative $\dot{V}(t,x,y)$ to be negative definite, it is sufficient that the inequalities

$$\beta - 2|p_2|_0\|S\|_0 > 0,$$

$$\begin{vmatrix} -\lambda & \frac{1}{2}(\lambda|p_2 - p_1|_0 + 2\|S\|_0) \\ \frac{1}{2}(\lambda|p_2 - p_1|_0 + 2\|S\|_0) & -(\beta - 2|p_2|_0\|S\|_0) \end{vmatrix} > 0 \tag{1.7.2}$$

be satisfied. Analysing (1.7.2) we get the condition

$$|p_2|_0 + |p_2 - p_1|_0 < \beta(2\|S\|_0)^{-1} \qquad (1.7.3)$$

which ensures the existence of a constant value $\lambda > 0$ such that the derivative of the non-degenerate quadratic form (1.5.4) due to (1.7.1) is negative definite.

Under condition (1.2.29) we consider the system of equations

$$\begin{aligned}\frac{dx}{dt} &= [A(t) + A_1(t)]x, \\ \frac{dy}{dt} &= B(t)x - [A^*(t) + A_2(t)]y,\end{aligned} \qquad (1.7.4)$$

where $A_i(t), B(t) \in C^0(R)$, and the matrix $B(t)$, which is not necessarily symmetric, satisfies the condition

$$\langle B(t)x, x \rangle \geq \beta_0 \|x\|^2, \quad \beta_0 = \text{const} > 0. \qquad (1.7.5)$$

Note that the perturbed system (1.7.4) is of a more general form than (1.7.1). If we set $B(t) \equiv I_n$ and $A_i(t) \equiv p_i(t)I_n$ in (1.7.4) then we arrive at the system (1.7.1).

We designate by α_2 a constant, not necessarily a positive one, determined by

$$-2\langle S(t)A_2(t)y, y \rangle \leq \alpha_2 \|y\|^2, \qquad (1.7.6)$$

and prove the following assertion.

Theorem 1.7.1 *Let there exist a symmetric matrix $S(t) \in C^1(R)$ satisfying the condition (1.2.29) ($\det S(t)$ can take zero values for some $t \in R$). Then, if the inequalities*

$$\|A_2^*(t) - A_1(t)\|_0 < \frac{\beta_0(\beta - \alpha_2)}{2\|S(t)B(t)\|_0}, \quad \alpha_2 < \beta, \qquad (1.7.7)$$

are satisfied, the system of equations (1.7.4) is exponentially dichotomous on R.

Proof Computing the derivative of the quadratic form (1.5.4) due to (1.7.4), we find

$$\begin{aligned}\dot{V}(t, x, y) \leq &-\lambda \langle Bx, x \rangle + \lambda \langle (A_2^* - A_1)x, y \rangle - \beta \|y\|^2 \\ &+ 2\langle SBx, y \rangle - 2\langle SA_2 y_1, y \rangle.\end{aligned}$$

In view of the inequalities (1.7.5) and (1.7.6) we get

$$\dot{V}(t, x, y) \leq -\lambda \beta_0 \|x\|^2 + \lambda \|A_2^* - A_1\|_0 \|x\| \|y\| + 2\|SB\|_0 \|x\| \|y\| - (\beta - \alpha_2)\|y\|^2.$$

For the derivative $\dot{V}(t,x,y)$ to be negative definite it is sufficient that the inequality $\beta - \alpha_2 > 0$ is satisfied and the second order determinant is positive

$$\begin{vmatrix} -\lambda\beta_0 & \frac{1}{2}(\lambda|A_2^* - A_1|_0 + 2\|SB\|_0) \\ \frac{1}{2}(\lambda|A_2^* - A_1|_0 + 2\|SB\|_0) & -(\beta - \alpha_2) \end{vmatrix} > 0. \quad (1.7.8)$$

Uncovering the determinant in (1.7.8) and performing simple transformations we arrive at the inequality

$$\lambda\|A_2^* - A_1\|_0 - 2\sqrt{\lambda\beta_0(\beta - \alpha_2)} + 2\|SB\|_0 < 0,$$

which has positive solutions λ only in the case when the condition (1.7.7) is satisfied.

This complete the proof of Theorem 1.7.1.

Corollary 1.7.1 *Under the conditions (1.2.29) and (1.7.5) the system of equations*

$$\frac{dx}{dt} = [A(t) + S(t)W^*(t)]x,$$

$$\frac{dy}{dt} = B(t)x - [A^*(t) + W(t)S(t)]y$$

is exponentially dichotomous on R for every matrix function $W(t) \in C_0(R)$ satisfying the condition $\langle W(t)x, x \rangle \geq 0$, $\forall t \in R$, $x \in R^n$.

Remark 1.7.1 In the case of the system (1.7.1) $A_i(t) = p_i(t)I_n$ discussed above one can get the inequality (1.7.3) from the inequality (1.7.7).

Further we show that for the system (1.7.1) to be exponentially dichotomous on R the following inequality

$$|p_i(t)|_0 < \beta(2\|S\|_0)^{-1}, \quad i = 1, 2 \quad (1.7.9)$$

has to be satisfied instead of (1.7.3).

Theorem 1.7.2 *Let there exist a symmetric matrix $S(t) \in C^1(R)$ satisfying the condition (1.2.29). Then the system of equations*

$$\frac{dx}{dt} = [A(t) + p_1(t)I_n]x,$$
$$\frac{dy}{dt} = B(t)x - [A^*(t) + p_2(t)I_n]y, \quad (1.7.10)$$

is exponentially dichotomous on the whole R axis for every matrix $B(t) \in C^0(R)$ with the property

$$\langle B(t)x, x \rangle \geq \beta(t)\|x\|^2, \quad \beta(t) \geq 0, \quad \beta(t) \not\equiv 0 \quad (1.7.11)$$

and any scalar functions $p_i(t) \in C^0(R)$ satisfying (1.7.9).

Proof Note that the establishment of the quadratic form having a definite sign derivative along the solutions of system (1.7.10) is complicated by the weak assumptions (1.7.11) with regard to the matrix $B(t)$. Therefore we apply the decomposition proposed in Section 1.2. By means of the change of Lyapunov variables $x = L(t)z$ and $y = [L^{-1}(t)]^*v$ the system (1.7.10) is reduced to the form

$$\frac{dz_1}{dt} = [\Phi^+(t) + p_1(t)I_{r+}]z_1,$$

$$\frac{dz_2}{dt} = [\Phi^-(t) + p_2(t)I_{r-}]z_2,$$

$$\frac{dz_3}{dt} = \Phi_1(t)z_1 + \Phi_2(t)z_2 + [\hat{\Phi}(t) + p_1(t)I_{\hat{r}}]z_3,$$

$$\frac{dv_1}{dt} = \sum_{i=1}^{3} D_{1i}(t)z_i - [\Phi^+(t) + p_2(t)I_{r+}]^*v_1 - \Phi_1^*(t)v_3, \qquad (1.7.10a)$$

$$\frac{dv_2}{dt} = \sum_{i=1}^{3} D_{2i}(t)z_i - [\Phi^-(t) + p_2(t)I_{r-}]^*v_2 - \Phi_2^*(t)v_3,$$

$$\frac{dv_3}{dt} = \sum_{i=1}^{3} D_{3i}(t)z_i - [\hat{\Phi}(t) + p_2(t)I_{\hat{r}}]^*v_3.$$

Since $D(t) = L^*(t)B(t)L(t)$, then the matrix $D(t)$ has the property (1.7.11) too and, hence, its diagonal block $D_{33}(t)$ has the property

$$\langle D_{33}(t)z_3, z_3 \rangle \geq d(t)\|z_3\|^2, \quad d(t) \geq 0, \quad d \not\equiv 0. \qquad (1.7.11a)$$

Condition (1.2.29) and inequality (1.7.9) ensure the property of having a fixed sign of the derivative of the quadratic form $V(t,y) = \langle S(t)y, y \rangle$ due to the perturbed system

$$\frac{dy}{dt} = -[A^*(t) \pm p_i(t)I_n]y, \quad i = 1, 2.$$

Therefore, the size of the perturbation $p_i(t)$ of each of the diagonal subsystems of the system (1.7.10a) does not essentially influence the behaviour of its non-trivial solutions over infinity. If, for example, all non-trivial solutions of the subsystem

$$\frac{dv_3}{dt} = -\hat{\Phi}^*(t)v_3$$

are exponentially increasing on $\pm\infty$, then all non-trivial solutions of the perturbed system

$$\frac{dv_3}{dt} = -[\hat{\Phi}^*(t) + p_2(t)I_n]v_3$$

are also exponentially (with another exponent) increasing on $\pm\infty$.

Clearly, for the system (1.7.10) to be exponentially dichotomous on R, it is necessary and sufficient that the inhomogeneous system

$$\frac{dz_3}{dt} = [\hat{\Phi}(t) + p_1(t)I_{\hat{r}}]z_3 + \varphi_1(t),$$
$$\frac{dv_3}{dt} = D_{33}(t)z_3 - [\hat{\Phi}^*(t) + p_2(t)I_{\hat{r}}]v_3 + \varphi_2(t),$$
(1.7.12)

has a unique solution bounded on R for every vector function $(\varphi_1(t), \varphi_2(t)) \in C^0(R)$. The general solution of the first subsystem

$$z_3(t) = \Omega_0^t(\hat{\Phi}) \exp\left\{\int_0^t p_1(\sigma)\,d\sigma\right\}\left[z_3(0) + \int_0^t \Omega_\tau^0(\hat{\Phi}) \exp\left\{\int_\tau^0 p_1(\sigma)\,d\sigma\right\} \varphi_1(\tau)\,d\tau\right]$$

is substituted into the second subsystem (1.7.12), then

$$\frac{dv_3}{dt} = -[\hat{\Phi}^*(t) + p_2(t)I_{\hat{r}}]v_3$$
$$+ D_{33}(t)\Omega_0^t(\hat{\Phi}) \exp\left\{\int_0^t p_1(\sigma)\,d\sigma\right\} z_3(0) + \varphi(t),$$
(1.7.13)

where

$$\varphi(t) = D_{33}(t)\int_0^t \Omega_\tau^t(\hat{\Phi}) \exp\left\{\int_\tau^t p_1(\sigma)\,d\sigma\right\}\varphi_1(\tau)\,d\tau + \varphi_2(t).$$

For the solution to the system (1.7.13) which is bounded on R to exist, it is necessary and sufficient that

$$\int_{-\infty}^{\infty} [\Omega_\tau^t(\hat{\Phi})]^* \exp\left\{\int_0^t p_2(\sigma)\,d\sigma\right\}$$
$$\times \left[D_{33}(t)\Omega_0^t(\hat{\Phi}) \exp\left\{\int_0^t p_1(\sigma)\,d\sigma\right\} z_3(0) + \varphi(t)\right] dt = 0.$$
(1.7.14)

Let us show that the determinant of the matrix

$$M = \int_{-\infty}^{\infty} [\Omega_0^t(\hat{\Phi})]^* D_{33}(t)\Omega_0^t(\hat{\Phi}) \exp\left\{\int_0^t [p_1(\sigma) + p_2(\sigma)]\,d\sigma\right\} dt \qquad (1.7.15)$$

differs from zero. For this it is enough to verify that the quadratic form $\langle M\eta, \eta\rangle$ is positive definite. In view of (1.7.11a) we get

$$\langle M\eta,\eta\rangle = \int_{-\infty}^{\infty} \langle D_{33}(t)\Omega_0^t(\hat{\Phi})\eta, \Omega_0^t(\hat{\Phi})\eta\rangle \exp\left\{\int_0^t [p_1(\sigma)+p_2(\sigma)]\,d\sigma\right\} dt$$

$$\geq \int_{-\infty}^{\infty} d(t)\|\Omega_0^t(\hat{\Phi})\eta\|^2 \exp\left\{\int_0^t [p_1(\sigma)+p_2(\sigma)]\,d\sigma\right\} dt \geq \varepsilon_0 \|\eta\|^2,$$

$$\varepsilon_0 = \text{const} > 0.$$

Thus, the matrix (1.7.15) being non-degenerate ensures the unique solvability with respect to $z_3(0)$ of the system of algebraic equations (1.7.14). Therefore, the system (1.7.10a) together with (1.7.10) is exponentially dichotomous on the whole R axis.

Remark 1.7.2 If in the perturbed system (1.7.10) other matrices are substituted instead of the unique matrices I_n, the norm of which equals one, then for inequalities (1.7.5) and (1.7.9) the system (1.7.10) can not be exponentially dichotomous on R.

Let us show this by the example of (1.6.22). Comparing system (1.6.22) with system (1.7.10) we obtain

$$A(t) = -2(\tanh t)I_2, \quad B(t) = \begin{pmatrix} 1 & \sqrt{\frac{5}{6}} \\ \sqrt{\frac{5}{6}} & 1 \end{pmatrix},$$

$$p_i(t) = (-1)^{i+1} \tanh t, \quad i = 1, 2.$$

Here the matrix diag $\{1, -1\}$ is used instead of the identity matrix I_2. We take the matrix function $S(t) \in C^1(R)$ (with the smallest possible norm) satisfying the condition (1.2.29). To this end, we present the general solution to the equation

$$\frac{ds}{dt} = -4(\tanh t)s + 1:$$

$$s(t) = \frac{16 s(0)}{(e^t + e^{-t})^4} + \int_0^t \left(\frac{e^\tau + e^{-\tau}}{e^t + e^{-t}}\right)^4 d\tau$$

$$= \frac{16 s(0) + \frac{1}{4}(e^{4t} - e^{-4t}) + 2(e^{2t} - e^{-2t}) + 6t}{(e^t + e^{-t})^4},$$

where $s(0)$ is an arbitrary initial value. If we assume that $s(0) \geq 0$, then evidently, $\sup_{t\in R} |s(t)| = \sup_{t\geq 0} |s(t)|$. Immediately computing the derivative $ds(t)/dt$ we obtain

$$\frac{ds(t)}{dt} = \frac{-2(e^{3t} + e^{-3t}) + 18(e^t + e^{-t}) - 24(e^t - e^{-t}) - 64 s(0)(e^t - e^{-t})}{(e^t + e^{-t})^5}.$$

Some difficulties arise when solutions to the equation

$$\frac{ds(t)}{dt} = 0$$

are found. Therefore, we presume in this equation that it has the solution $t = \ln 2$ and find a concrete value $s(0) = s^*(0)$:

$$16s^*(0) = \frac{115}{24} - 6\ln 2 \approx 0.63 < 0.7.$$

Now by determining a value of the concrete function $s^*(t)$ at point $t = \ln 2$ we get that $\sup_{t \in R} |s(t)| = s^*(\ln 2) < 17/36$. Thus, if we take $-I_2 s^*(t)$ for the matrix $S(t)$, then the condition (1.2.29) is satisfied ($\beta = 1$), and, moreover $\|S\|_0 < 17/36$. Obviously, $|p_i|_0 = \sup_{t \in R} |\tanh t| = 1$. Therefore, inequality (1.7.9) is satisfied. Nevertheless, the system (1.6.22) was shown not to be exponentially dichotomous on the whole R axis.

Let us return to the system of equations (1.6.21) for different dimensions of x and y, $x \in R^{\hat{r}}$, $y \in R^{\check{r}}$, and under the assumption that it is weakly regular on the whole R axis. We recall that all solutions of the system

$$\frac{dx}{dt} = \hat{A}(t)x$$

are exponentially damping on $\pm\infty$, and all non-trivial solutions of the system

$$\frac{dx}{dt} = \check{D}(t)x$$

increase on both sides. We represent this in the form of the corresponding estimates for the matriciants

$$\|\Omega_\tau^t(\hat{A})\| \leq K_1 \begin{cases} \exp\{-\gamma_1(t-\tau)\}, & 0 \leq \tau \leq t, \\ \exp\{\gamma_1(t-\tau)\}, & t \leq \tau \leq 0, \end{cases}$$

$$\|\Omega_\tau^t(\check{D})\| \leq K_2 \begin{cases} \exp\{\gamma_2(t-\tau)\}, & 0 \leq t \leq \tau, \\ \exp\{-\gamma_2(t-\tau)\}, & \tau \leq t \leq 0, \end{cases} \quad (1.7.16)$$

$$K_i, \gamma_i = \text{const} > 0.$$

Obviously, for the system (1.6.21) to be weakly regular on R it is necessary and sufficient that the two conditions be satisfied (see (1.6.14))

$$\check{r} \leq \hat{r}, \quad \text{rank } M_0 = \check{r}, \quad (1.7.17)$$

where

$$M_0 = \int_{-\infty}^{\infty} \Omega_\tau^0(\check{D}) B(\tau) \Omega_0^\tau(\hat{A}) \, d\tau. \quad (1.7.17a)$$

Let us consider the perturbed system of equations

$$\begin{aligned}\frac{dx}{dt} &= [\hat{A}(t) + p_1 I_{\hat{r}}]x, \\ \frac{dy}{dt} &= B(t)x + [\check{D}(t) + p_2 I_{\check{r}}]y,\end{aligned} \quad (1.7.18)$$

where the parameters p_1 and p_2 do not depend on t and satisfy the inequalities

$$|p_i| < \gamma_i, \quad i = 1, 2. \quad (1.7.18a)$$

Evidently, the violation of one of the inequalities (1.7.18a) results in an essential violation of one of the estimates (1.7.16), and so, system (1.7.18) can not be weakly regular on R. This is easily seen by a simple example of two equations

$$\begin{aligned}\frac{dx}{dt} &= (-\tanh t + p_1)x, \\ \frac{dy}{dt} &= b(t)x + (\tanh t + p_2)y,\end{aligned} \quad (1.7.19)$$

where $b(t)$ is a scalar function from $C^0(R)$, $x \in R^1$, $y \in R^1$. Moreover, $\gamma_1 = \gamma_2 = 1$. If, for example, $p_i = 1$, then the system (1.7.19) is not regular on R any more. Note that even if the inequality $|p_i| < 1$ is satisfied, this does not ensure the regularity on R of the system (1.7.19). Thus, substituting $b(t) = \exp\{-\varepsilon t\}(\sinh t / \cosh 2t)$ in (1.7.19) we come to the conclusion that the system (1.7.19) is not regular on R for $p_1 = \frac{1}{2} + \varepsilon$, $p_2 = \frac{1}{2}$.

Theorem 1.7.3 *Let the estimates (1.7.16) and (1.7.17) be satisfied for the matriciants $\Omega_\tau^t(\hat{A})$ and $\Omega_\tau^t(\check{D})$. Then the perturbed system (1.7.18) is weakly regular on R for the inequalities (1.7.18a) and*

$$|p_1 - p_2| \leq \frac{2\gamma^2}{\lambda + \sqrt{\lambda^2 + 4\gamma^2} + \sqrt{2\lambda^2 + 2\lambda\sqrt{\lambda^2 + 4\gamma^2}}}, \quad (1.7.20)$$

where

$$\gamma = \gamma_1 + \gamma_2, \quad \lambda = \frac{K_1 K_2 \|B\|_0}{m_0}, \\ m_0 = \min_{\|y\|=1,\, y \in R^{\check{r}}} \|M_0^* y\|. \quad (1.7.21)$$

Proof When inequalities (1.7.18a) are satisfied, the system (1.7.18) is only weakly regular on R if
$$\text{rank } M_p = \check{r}, \qquad (1.7.22)$$
where
$$p = p_1 - p_2, \quad M_p = \int_{-\infty}^{\infty} \Omega_\tau^0(\check{D}) B(\tau) \Omega_0^\tau(\hat{A}) \exp\{p\tau\} d\tau.$$

Thus, for $p = 0$ the rank of the matrix M_0 is equal to $\check{r}$ and the small changes of the parameter p for which the rank of the matrix M_p is preserved remain to be determined. First, we assume $p > 0$ and consider the difference

$$M_p - M_0 = \int_{-\infty}^{\infty} \Omega_\tau^0(\check{D}) B(\tau) \Omega_0^\tau(\hat{A}) \left[\exp\{p\tau\} - 1\right] d\tau$$
$$= \int_{-\infty}^{\infty} \Omega_\tau^0(\check{D}) B(\tau) \Omega_0^\tau(\hat{A}) \exp\{\eta_p(\tau)\tau\} p\tau \, d\tau, \qquad (1.7.23)$$

where
$$\eta_p(\tau) = \begin{cases} \frac{1}{\tau} \ln \frac{\exp\{p\tau\}-1}{p\tau}, & \tau \neq 0, \\ \frac{1}{2} p, & \tau = 0. \end{cases} \qquad (1.7.24)$$

Note that the rectangular matrix M_p consists of $\check{r}$ lines and $\hat{r}$ columns ($\check{r} \leq \hat{r}$) and moreover, condition (1.7.22) is evidently equivalent to the following
$$\|M_p^* y\| > 0, \quad \forall y \in R^{\check{r}}, \quad y \neq 0. \qquad (1.7.25)$$

Setting $\|y\| = 1$, $y \in R^{\check{r}}$, and taking into account the estimates (1.7.16) and designations (1.7.21) we get from (1.7.23)
$$\|M_p^* y\| \geq m_0 - \lambda m_0 p \int_{-\infty}^{\infty} |\tau| \exp\{-\gamma|\tau| + \eta_p(\tau)\tau\} d\tau. \qquad (1.7.26)$$

Consider the integral
$$J = \int_{-\infty}^{\infty} |\tau| \exp\{-\gamma|\tau| + \eta_p(\tau)\tau\} d\tau$$
$$= \int_{-\infty}^{0} (-\tau) \exp\{\gamma\tau + \eta_p(\tau)\tau\} d\tau + \int_{0}^{\infty} \tau \exp\{-\gamma\tau + \eta_p(\tau)\tau\} d\tau \qquad (1.7.27)$$
$$= \int_{0}^{\infty} \tau \exp\{-\gamma\tau\} \left[\exp\{\eta_p(\tau)\tau\} + \exp\{-\eta_p(-\tau)\tau\}\right] d\tau.$$

Let us show that for all $\tau > 0$ the inequality

$$\exp\{\eta_p(\tau)\tau\} + \exp\{-\eta_p(-\tau)\tau\} < \exp\{p\tau\} + \exp\{-p\tau\} \tag{1.7.28}$$

is satisfied. In view of the structure of function (1.7.24) and designating $x = p\tau$, we reduce inequality (1.7.28) to the form

$$\frac{\exp x - 1}{x} - \frac{\exp\{-x\} - 1}{x} < \exp x + \exp\{-x\}. \tag{1.7.28a}$$

Hence for $x > 0$ we get

$$\varphi(x) = (x-1)\exp x + (x+1)\exp\{-x\} > 0. \tag{1.7.28b}$$

Since $\varphi(0) = 0$ and $\varphi'(x) = 2x \sinh x$ for $x > 0$, the inequality (1.7.28) becomes evident.

We estimate integral (1.7.27) using (1.7.28)

$$J < \int_0^\infty \tau[\exp\{-(\gamma-p)\tau\} + \exp\{-(\gamma+p)\tau\}]\,d\tau$$

$$= \frac{1}{(\gamma-p)^2} + \frac{1}{(\gamma+p)^2}. \tag{1.7.29}$$

Thus, in view of (1.7.29), the inequality (1.7.26) leads to the estimate

$$\|M_p^* y\| \geq m_0[1 - \lambda p((\gamma-p)^{-2} + (\gamma+p)^{-2})]. \tag{1.7.30}$$

Now the right-hand side of inequality (1.7.30) is required to be non-negative, $p(\gamma-p)^{-2} + p(\gamma+p)^{-2} \leq \lambda^{-1}$. Then we get $(p+\gamma^2/p-2\gamma)^{-1} + (p+\gamma^2/p+2\gamma)^{-1} \leq \lambda^{-1}$. Introducing the auxiliary designation $x = p + \gamma^2/p$ and analysing the inequality obtained, we conclude that for all $p \in [0, p_0]$, where

$$p_0 = \frac{\lambda + \sqrt{\lambda^2 + 4\gamma^2} - \sqrt{2\lambda^2 + 2\lambda\sqrt{\lambda^2 + 4\gamma^2}}}{2}$$

$$= \frac{2\gamma^2}{\lambda + \sqrt{\lambda^2 + 4\gamma^2} + \sqrt{2\lambda^2 + 2\lambda\sqrt{\lambda^2 + 4\gamma^2}}}, \tag{1.7.31}$$

the right-hand side of (1.7.30) is negative. Consequently, condition (1.7.25) is satisfied, which is equivalent to (1.7.22). For $p < 0$ we make the change of variables $p \to -p$ and $\tau \to -\tau$ in (1.7.23) and arrive at the same value of p_0 as in (1.7.31). Thus the variation of the parameter p within the limits from $-p_0$ to p_0, where p_0 is determined by (1.7.31) does not violate condition (1.7.22).

This completes the proof of Theorem 1.7.3.

1.8 Exponentially Dichotomous Linear Systems with Parameters

It is well known that if the right-hand side $f(x, t; p)$ of the system of ordinary differential equations

$$\frac{dx}{dt} = f(x, t; p), \quad x \in R^n,$$

has a certain smoothness with respect to the variables x and t (for example, satisfies Lipschitz conditions in x, is continuous in t) and is continuous with respect to the parameter p, the solution of the Cauchy problem

$$\frac{dx}{dt} = f(x, t; p), \quad x\big|_{t=0} = x_0,$$

depends continuously on the parameter p. Things are different with regard to solutions bounded on the whole R axis. Consider the example of the scalar equation

$$\frac{dx}{dt} = \frac{1 - p^2 t^2}{1 + p^2 t^2} x + 1, \quad p \in [-1, 1]. \tag{1.8.1}$$

Its bounded on R solution is

$$x(t; p) = \begin{cases} -1, & p = 0, \\ \int\limits_{-\infty}^{t} \exp\left\{\tau - t + \frac{2}{p}(\arctan pt - \arctan p\tau)\right\} d\tau, & 0 < |p| < 1. \end{cases} \tag{1.8.2}$$

Clearly, equation (1.8.1) has no other solutions bounded on R, but (1.8.2). It is evident from (1.8.2) that the solution $x(t; p)$ has a discontinuity with respect to the parameter p for $p = 0$.

Let us consider a system of linear differential equations with the vector of parameters $p = (p_1, \ldots, p_m)$

$$\frac{dx}{dt} = A(t; p)x, \tag{1.8.3}$$

where $x \in R^n$, $A(t; p)$ is an $n \times n$-dimensional matrix function continuous with respect to the totality of variables t and p, $t \in R$, $p = (p_1, \ldots, p_m) \in M$ (M is a simply connected compact set from R^m) and uniformly bounded, $\|A(t; p)\| \leq K_0 = \text{const} < \infty$. Let the system of the equations (1.8.3) be exponentially dichotomous on the whole R axis for every fixed value of $p \in M$, i.e. R^n can be represented in the form of a direct sum of two subspaces $E^+(p)$ and $E^-(p)$ depending on the vector of parameters p such that

$$\|\Omega_0^t(p; A)x^+\| \leq K(p)\|\Omega_0^\tau(p; A)x^+\| \exp\{-\gamma(p)(t - \tau)\},$$
$$\tau \leq t, \quad x^+ \in E^+(p),$$
$$\|\Omega_0^t(p; A)x^-\| \leq K(p)\|\Omega_0^\tau(p; A)x^-\| \exp\{\gamma(p)(t - \tau)\},$$
$$t \leq \tau, \quad x^- \in E^-(p), \tag{1.8.4}$$

where $\Omega_0^t(p; A)$ is a matriciant of the system (1.8.3). The question arises: where do the subspaces $E^+(p)$ and $E^-(p)$ depend continuously on the parameters p if the positive constants $K(p)$ and $\gamma(p)$ in (1.8.4) can be taken independently of p? The examples show this not to be so in the general case.

Example 1.8.1 Let

$$\frac{dx_i}{dt} = (-1)^i(1-p^2t^2)(1+p^2t^2)^{-1}x_i, \quad i=1,2, \quad p \in [0,1]. \tag{1.8.5}$$

Obviously, the coefficients of the present system depend continuously on the parameter p, though the subspaces $E^+(p)$ and $E^-(p)$ are not continuously dependent on p. For all $p \in (0,1]$ the subspace $E^+(p)$ consists of the points $\{(0, x_2)\}$, and for $p = 0$ $E^+(0) = \{(x_1, 0)\}$. We show that the constant $K(p)$ found in (1.8.4) cannot be taken in (1.8.5) to be one and the same for all $p \in [0,1]$. In fact, let a constant K_0 exist such that for all $p \in [0,1]$ the estimate

$$|x_{2t}| \leq K_0 |x_{2\tau}| \exp\{-\gamma(t-\tau)\}, \quad \tau \leq t, \quad \gamma > 0$$

is valid, where x_{2t} is a solution of the second equation (1.8.5)

$$x_{2t} = x_{2\tau} \exp\left\{-(t-\tau) + \frac{2}{p}(\arctan pt - \arctan p\tau)\right\}.$$

Then for all $p \in (0,1]$ and $\tau \leq t$ we get

$$\exp\left\{\frac{2}{p}(\arctan pt - \arctan p\tau)\right\} \leq K_0 \exp\{(1-\gamma)(t-\tau)\}.$$

Further we take and fix the values τ^*, t^* $(\tau^* < t^*)$ such that

$$\exp\{(1-\gamma)(t^* - \tau^*)\} > K_0.$$

On the one hand the assumed estimate

$$\exp\left\{\frac{2}{p}(\arctan pt^* - \arctan p\tau^*)\right\} \leq K_0 \exp\{(1-\gamma)(t^* - \tau^*)\}$$

is valid; on the other hand

$$\lim_{p \to 0} \exp\left\{(\gamma-1)(t^* - \tau^*) + \frac{2}{p}(\arctan pt^* - \arctan p\tau^*)\right\} > K_0.$$

The contradiction obtained proves the impossibility of choosing the constant $K(p)$ to be independent of the parameter $p \in [0,1]$. In this case it may be of the form

$$K(p) = \begin{cases} \exp\left\{\frac{2\pi}{p}\right\}, & p \in (0,1], \\ 1, & p = 0. \end{cases}$$

Example 1.8.2 Similarly, we state for the system

$$\frac{dx_i}{dt} = (-1)^i(1-p^3t^2)(1+p^2t^2)^{-1}x_i, \quad i=1,2, \quad p \in [0,1]$$

that not only $K(p) \to \infty$ but $\gamma(p) \to 0$ as $p \to +0$.

Theorem 1.8.1 *Let the system of the equations (1.8.3) be exponentially dichotomous on the whole R axis for every fixed vector of the parameters $p = (p_1, \ldots, p_m) \in M$, and for the continuous matrix function $A(t,p)$ the condition*

$$\|A(t,p) - A(t,\bar{p})\| \le \mu(\|p-\bar{p}\|) \tag{1.8.6}$$

is satisfied for all $p, \bar{p} \in M$, $\|p-\bar{p}\| \le d$ (d is a fixed constant, $\mu(u)$ is a continuous function defined on the segment $[0,d]$, $\mu(0) = 0$). Then the subspaces $E^+(p)$ and $E^-(p)$ depend continuously on the parameters $p \in M$ and the positive constants $K(p)$ and $\gamma(p)$ in the estimates (1.8.4) can be taken to be independent of the parameters p.

Proof Let us show that for every fixed value of the parameter $p_0 \in M$ there exists a positive number $d(p_0)$ such that for all $p \in M \cap \{p \mid \|p-\bar{p}\| \le d(p_0)\}$ the subspaces $E^+(p)$ and $E^-(p)$ do not change their dimensions and the constants $K(p)$ and $\gamma(p)$ in the estimates (1.8.4) can not be taken to be dependent on p, but only on p_0. To this end we consider the quadratic form

$$V(t,p_0,x) = \langle S(t,p_0)x, x \rangle, \tag{1.8.7}$$

where

$$\begin{aligned} S(t,p_0) = &\int_t^\infty [\Omega_0^\tau(p_0;A)C(p_0)\Omega_t^0(p_0;A)]^*[\Omega_0^\tau(p_0;A)C(p_0)\Omega_t^0(p_0;A)]\,d\tau \\ &-\int_{-\infty}^t \{\Omega_0^\tau(p_0;A)[I_n - C(p_0)]\Omega_t^0(p_0;A)\}^*\{\Omega_0^\tau(p_0;A)[I_n - C(p_0)] \\ &\qquad \times \Omega_t^0(p_0;A)\}\,d\tau, \end{aligned} \tag{1.8.7a}$$

and $C(p_0)$ is the projection matrix on the subspace $E^+(p_0)$ along $E^-(p_0)$. It is known that the condition of exponential dichotomy on the whole R axis of the system (1.8.3) is equivalent to the estimates

$$\begin{aligned} \|\Omega_0^t(p;A)C(p)\Omega_\tau^0(p;A)\| &\le K_2'(p)\exp\{-\gamma(p)(t-\tau)\}, \quad \tau \le t, \\ \|\Omega_0^t(p;A)[I_n - C(p)]\Omega_\tau^0(p;A)\| &\le K_2''(p)\exp\{\gamma(p)(t-\tau)\}, \quad t \le \tau. \end{aligned} \tag{1.8.8}$$

In addition it is easy to express the constants $K_2'(p)$ and $K_2''(p)$ via $K(p)$, $\gamma(p)$ and K_0. For example, letting $K_0 > \gamma$, it is possible to take $K_2'(p)$ as

$$K_2'(p) = K(p) \frac{\sqrt{K_0^2 - \gamma(p)}}{\gamma(p)} \left(\frac{K^2(p)(K_0 + \gamma(p))}{K_0 - \gamma(p)} \right)^{K_0/2\gamma(p)} \quad (1.8.9)$$

Immediately estimating the integrand expression in (1.8.7a) we verify the boundedness of the matrix function $S(t, p_0)$ for $t \in R$

$$\|S(t, p_0)\| \leq (2\gamma(p_0))^{-1}[(K_2'(p_0)) + (K_2''(p_0))^2] = K_3(p_0).$$

The derivative of the quadratic form (1.8.7) along solutions of the system (1.8.3) is negative definite for $p = p_0$

$$\langle [\dot{S}(t, p_0) + S(t, p_0)A(t, p_0) + A^*(t, p_0)S(t, p_0)]x, x \rangle \leq -\frac{1}{2}\|x\|^2.$$

In Section 1.1 it was proved that the number of positive eigenvalues of the matrix $S(t, p_0)$ determines the dimension of the subspace $E^+(p_0)$, and that of the negative eigenvalues specifies the dimension of $E^-(p_0)$. Let us compute the derivative of the quadratic form (1.8.7) along solutions to the system (1.8.3) where the values of the parameters $p \in M$ are close enough to p_0

$$\dot{V}(t, p_0, x) = \langle (\dot{S}(t, p_0) + S(t, p_0)A(t, p) + A^*(t, p)S(t, p_0))x, x \rangle$$
$$= \langle (\dot{S}(t, p_0) + S(t, p_0)A(t, p_0) + A^*(t, p_0)S(t, p_0))x, x \rangle$$
$$+ \langle [S(t, p_0)(A(t, p) - A(t, p_0)) + (A^*(t, p) - A^*(t, p_0))S(t, p_0)]x, x \rangle \quad (1.8.10)$$
$$\leq -\left(\frac{1}{2} - 2K_3(p_0)\mu(\|p - p_0\|)\right)\|x\|^2.$$

Now we take $d(p_0)$ from $\mu(d) \leq (8K_3(p_0))^{-1}$. We then find that the quadratic form (1.8.7) derivative along solutions of the system (1.8.3) is negative definite for all p close to p_0, and hence, the subspaces $E^+(p)$ and $E^-(p)$ do not change their dimensions when the parameters p are changed in the neighbourhood of p_0. Since M is a connected compact set, the subspaces $E^+(p)$ and $E^-(p)$ do not also change their dimensions for all $p \in M$. Let us suppose on the contrary, i.e. there exist two points $\tilde{p}, \tilde{\tilde{p}} \in M$ for which $\dim E^+(\tilde{p}) \neq \dim E^+(\tilde{\tilde{p}})$. We connect these points with a continuous curve belonging to M. There exist two sequences of points $p^{(n)}$ and $\bar{p}^{(n)}$ on this curve which have a common limit

$$p_0 = \lim_{n \to \infty} p^{(n)} = \lim_{n \to \infty} \bar{p}^{(n)} \quad \text{and} \quad \dim E^+(p^{(n)}) \neq \dim E^+(\bar{p}^{(n)}).$$

We find that the subspace $E^+(p)$ changes its dimension in the neighbourhood of the point p_0, and this contradicts the result proved above. Thus, the subspaces $E^+(p)$ and $E^-(p)$ do not change their dimensions for all $p \in M$.

Let us now show that the constants $K(p)$ and $\gamma(p)$ in (1.8.4) can be one and the same for all p from the neighbourhood of the point p_0: $\|p - p_0\| \leq d(p_0)$. To this end we estimate (1.8.10) after the choice of $d(p_0)$

$$\frac{d}{dt} V(t, p_0, x) \leq -\frac{1}{4} \|x\|^2. \tag{1.8.11}$$

The derivative is taken along solutions of the system (1.8.3) for all $p \in M$ which are close to p_0: $\|p - p_0\| \leq d(p_0)$. For all points $x \in E^+(p)$, $\|x\| \neq 0$ when $t \in R$ the estimate

$$\langle S(t, p_0)\Omega_0^t(p; A)x, \Omega_0^t(p; A)x \rangle > 0 \tag{1.8.12}$$

is valid. We designate

$$V_\varepsilon(t, p, p_0, x) = \langle S(t, p_0)\Omega_0^t(p; A)x, \Omega_0^t(p; A)x \rangle + \varepsilon \|\Omega_0^t(p; A)x\|^2.$$

In view of estimate (1.8.11) we have

$$\dot{V}_\varepsilon(t, p, p_0, x) \leq -\left(\frac{1}{4} - 2K_0\varepsilon\right)\|\Omega_0^t(p; A)x\|^2$$
$$\leq -\frac{1 - 8K_0\varepsilon}{4(K_3(p_0) + \varepsilon)} V_\varepsilon(t, p, p_0, x) = -\frac{1}{8K(p_0)} V_\varepsilon(t, p, p_0, x), \tag{1.8.13}$$

where $\varepsilon > 0$ is taken to be fixed, $\varepsilon = K_2(p_0)[16K_3(p_0)K_0 + 1]^{-1}$. From estimate (1.8.13) we obtain

$$V_\varepsilon(t, p, p_0, x) \leq V_\varepsilon(\tau, p, p_0, x) \exp\left\{-\frac{1}{8K_2(p_0)}(t - \tau)\right\}, \quad \tau \leq t.$$

Hence

$$\|\Omega_0^t(p; A)x\| \leq \left(\frac{K_2(p_0) + \varepsilon}{\varepsilon}\right)^{1/2} \|\Omega_0^\tau(p; A)\| \exp\left\{-\frac{1}{16K_2(p_0)}(t - \tau)\right\}, \quad \tau \leq t.$$

We find a similar estimate for solutions to the system (1.8.3) with the initial points from the subspace $E^-(p)$.

Thus, for every point $p_0 \in M$ there exists a neighbourhood $U_{p_0} \subset M$ such that for all $p \in U_{p_0}$ the positive constants K and γ in (1.8.4) can not be taken to be dependent on p, but only on p_0. Now, in view of the known Heine-Borel lemma we have that the constants K and γ in (1.8.4) can be taken independently of the parameters $p \in M$.

Now let us prove the continuous dependence of the subspaces $E^+(p)$ and $E^-(p)$ on the parameters $p \in M$. To this end we represent the Green function of the problem of bounded solutions for the system of the equations (1.8.3) for every fixed $p \in M$ in the form

$$G(t,\tau;p) = \begin{cases} \Omega_0^t(p;A)C(p)\Omega_\tau^0(p;A), & \tau \leq t, \\ \Omega_0^t(p;A)[C(p) - I_n]\Omega_\tau^0(p;A), & \tau > t. \end{cases} \quad (1.8.14)$$

This function satisfies the estimate

$$\|G(t,\tau;p)\| \leq K \exp\{-\gamma|t-\tau|\}, \quad (1.8.15)$$

with the constants K and γ being independent of the parameters $p \in M$. Due to the Green function (1.8.14) being unique the integral representation

$$G(t,\tau;p) - G(t,\tau;\bar{p})$$
$$= \int_{-\infty}^{\infty} G(t,t_1;p)[A(t_1;p) - A(t_1;\bar{p})]G(t_1,\tau;\bar{p})\,dt_1 \quad (1.8.16)$$

takes place for the difference $G(t,\tau;p) - G(t,\tau;\bar{p})$. Hence

$$\|G(t,\tau;p) - G(t,\tau;\bar{p})\|$$
$$\leq K^2 \int_{-\infty}^{\infty} \|A(t_1;p) - A(t_1;\bar{p})\| \exp\{-\gamma(|t-t_1| + |\tau - t_1|)\}\,dt_1.$$

In view of condition (1.8.6) we get the inequality

$$\|G(t,\tau;p) - G(t,\tau;\bar{p})\|$$
$$\leq K^2 \mu(\|p-\bar{p}\|) \int_{-\infty}^{\infty} \exp\{-\gamma(|t-t_1| + |\tau - t_1|)\}\,dt_1$$
$$= K^2 \mu(\|p-\bar{p}\|)(-\gamma^{-1} + |t-\tau|) \exp\{-\gamma|t-\tau|\},$$

which indicates the continuous dependence of the Green function $G(t,\tau,p)$ and by the same token of the projection matrix

$$C(p) \equiv \Omega_t^0(p,A)G(t,0;p), \quad t \geq 0$$

on the parameters $p \in M$.

1.9 Comments and References

Section 1.0 The properties of exponential dichotomy of linear systems of differential equations set out in the first chapter of the book have been thoroughly analysed and applied by Anosov [1], Baris and Fodchuk [1], Bronstein and Cherniy [1], Cheban [1], Cheresiz [1], Daletsky and Krein [1], Demidovich [1], Feshchenko, Shkil' and Nikolenko [1], Golets and Kulik [1], Grujić, Martynyuk and Ribbens-Pavella [1], Hartman [1], Izobov [1, 2], Krasnosel'sky, Burd and Kolesov [1], Krein [1], Majzel [1], Malkin [1], Massera and Schaeffer [1], Millionshchikov [1], Mitropolsky and Lykova [1], Molchanov [1], Nemytsky and Stepanov [1], Perov [1], Perov and Trubnikov [1], K.Persidsky [1], S.Persidsky [1], Pliss [1], Rejzin' [1], Rozenvasser [1], Shkil' [1], Starzhynsky and Yakubovich [1], Stepanov [1].

Section 1.1 The basic idea of the proof of Theorem 1.1.1 is due to Samoilenko and Kulik [3], who have discussed the linear extension of dynamical systems on a torus. For linear systems of differential equations this theorem and the converse Theorem 1.1.2 were proved by Mitropolsky, Samoilenko and Kulik [1].

Sections 1.2 – 1.5 The proof of Theorems 1.2.1 is set out by A.Kulik and V.Kulik [1]. A number of interesting results have been obtained in terms of Theorem 1.2.1. In particular Theorems 1.2.2, 1.2.3, 1.3.4, 1.3.6, 1.3.7, 1.4.1 – 1.4.3, 1.5.1, 1.5.5 that were cited by Kulik [7, 9, 12], Mitropolsky, Samoilenko and Kulik [2], Mitropolsky and Kulik [3] are proved.

Section 1.3 Theorems 1.3.1, 1.3.2 were first proved by Kulik [4], and then generalized by Coppel [2].

Section 1.6 The regularity of linear systems of block-triangular form was treated by Kulik [10] as was Theorem 1.6.1.

Section 1.7 The basic idea of the proof of Theorems 1.7.1 – 1.7.3 is due to Mitropolsky, Samoilenko and Kulik [2] and Mitropolsky and Kulik [2].

Section 1.8 The dependence of exponentially-dichotomous systems on parameters was discussed by Ordynskaya and Kulik [1] and in Theorem 1.8.1.

2 Linear Extension of Dynamical Systems on a Torus

2.0 Introduction

The investigations made in Chapter 1 are closely connected with the study of the theory of preserving invariant tori of dynamical systems under perturbations. When the closure of a set of values in the phase space R^k of the quasiperiodic solution $y = y(t)$, $t \in R$, $y \in R^k$, of the autonomous system

$$\frac{dy}{dt} = Y(y)$$

is considered, in many cases we get a surface homeomorphic to the m-dimensional torus $\mathcal{T}_m$. The investigation of changes of the toroidal surface $\mathcal{T}_m$ for small changes of the right-hand of the autonomous system is quite difficult. However it is an important problem in the theory of multifrequence oscillations. The introduction of cyclic coordinates $\varphi = (\varphi_1, \ldots, \varphi_m)$ and the normal ones $x \in R^n$ on the torus $\mathcal{T}_m$, $m + n = k$, allows the perturbed autonomous system to be written as

$$\frac{d\varphi}{dt} = a(\varphi, x); \quad \frac{dx}{dt} = A(\varphi, x)x + f(\varphi),$$

the linearization of which in the torus neighbourhood yields

$$\frac{d\varphi}{dt} = a(\varphi), \quad \frac{dx}{dt} = A(\varphi)x + f(\varphi). \qquad (2.0.1)$$

Chapter 2 deals with the investigation of the system (2.0.1). Here the existence conditions for both the unique and non-unique Green function $G_0(\tau, \varphi)$ of the problem on invariant tori are established by means of Lyapunov functions of variable sign. The problems of the smoothness of the Green function with respect to phase variables φ are also treated as is its dependence on the parameters. In the author's opinion, it is a result of interest that if the system

$$\frac{d\varphi}{dt} = a(\varphi), \quad \frac{dx}{dt} = A(\varphi)x$$

has many Green functions, then the extended system

$$\frac{d\varphi}{dt} = a(\varphi), \quad \frac{dx}{dt} = A(\varphi)x, \quad \frac{dy}{dt} = x - A^*(\varphi)y$$

has a unique $2n \times 2n$-dimensional Green function. This makes it possible to investigate the smoothness and dependence of the family of invariant tori of the system (2.0.1) on parameters. It is noted that similar results can also be obtained when the right-hand side of the system (2.0.1) is not periodic in φ. Essential differences are noticed in this case in the problems of smoothness of bounded invariant manifolds.

2.1 Necessary Existence Conditions for Invariant Tori

We designate the space of the functions $F(\varphi)$, continuous with respect to the totality of variables $\varphi_1, \ldots, \varphi_m$ and 2π-periodic in each variable φ_j, $j = \overline{1, m}$, i.e. being specified on the m-dimensional torus $\mathcal{T}_m$, by $C^0(\mathcal{T}_m)$. $C^r(\mathcal{T}_m)$ is a subspace of $C^0(\mathcal{T}_m)$ which consists of functions $F(\varphi)$ possessing continuous derivatives up to the order of r included.

Let $a(\varphi) = (a_1(\varphi), \ldots, a_m(\varphi)) \in C^1(\mathcal{T}_m)$. Then the system of equations

$$\frac{d\varphi}{dt} = a(\varphi) \tag{2.1.1}$$

specifies the dynamical system $\varphi \to \varphi_t(\varphi)$ on the torus $\mathcal{T}_m$. Moreover, $\varphi_t(\varphi)$ is the solution of the system (2.1.1) obeying the initial condition $\varphi_t(\varphi)|_{t=0} = \varphi$, the group property $\varphi_t(\varphi_\tau(\varphi)) \equiv \varphi_{t+\tau}(\varphi)$ and a periodicity condition

$$\varphi_t(\varphi + 2k\pi) = 2k\pi + \varphi_t(\varphi), \tag{2.1.2}$$

which is valid for arbitrary t, $\tau \in R = (-\infty, \infty)$, $\varphi \in \mathcal{T}_m$ and integer vectors $k = (k_1, \ldots, k_m)$. Also, $f = (f_1, \ldots, f_n) \in C^0(\mathcal{T}_m)$, $A = \{A_{ij}(\varphi)\}_{i,\,j=\overline{1,n}}$ is an n-dimensional square matrix, the elements $A_i = (A_{i1}, \ldots, A_{in})$, $i = \overline{1, n}$ of which belong to the space $C^0(\mathcal{T}_m)$.

The system of differential equations

$$\frac{d\varphi}{dt} = a(\varphi), \quad \frac{dx}{dt} = A(\varphi)x + f(\varphi) \tag{2.1.3}$$

is called a *linear extension of the dynamical system* (2.1.1) *on the torus* $\mathcal{T}_m$.

We designate the subspace $C^0(\mathcal{T}_m)$ formed by the functions $F(\varphi)$ for which the function $F(\varphi_t(\varphi))$ is continuously differentiable in t for all $t \in R$, $\varphi \in \mathcal{T}_m$ by $C'(\mathcal{T}_m; a)$. If $F(\varphi) \in C'(\mathcal{T}_m; a)$, then we designate by $\dot{F}(\varphi)$ the function determined by

$$\dot{F}(\varphi) = \left.\frac{dF(\varphi_t(\varphi))}{dt}\right|_{t=0}, \quad \varphi \in \mathcal{T}_m. \tag{2.1.4}$$

This definition and property (2.1.2) imply the inclusion $\dot{F}(\varphi) \in C^0(\mathcal{T}_m)$.

We call the surface

$$x = u(\varphi), \quad \varphi \in \mathcal{T}_m, \tag{2.1.5}$$

an *invariant torus of the system* (2.1.3), if $u(\varphi) \in C'(\mathcal{T}_m; a)$ and

$$\frac{du(\varphi_t(\varphi))}{dt} = A(\varphi_t(\varphi))u(\varphi_t(\varphi)) + f(\varphi_t(\varphi)), \tag{2.1.6}$$

for all $t \in R$, $\varphi \in \mathcal{T}_m$.

Then we define the linear operator $L \colon C'(\mathcal{T}_m; a) \to C^0(\mathcal{T}_m)$ on $C'(\mathcal{T}_m; a)$ setting

$$Lu(\varphi) = \dot{u}(\varphi) - A(\varphi)u(\varphi), \quad \varphi \in \mathcal{T}_m. \tag{2.1.7}$$

If the relation (2.1.5) prescribes the invariant torus of the dynamical system (2.1.3), then the function $u(\varphi)$ satisfies the obvious identity

$$Lu(\varphi) \equiv f(\varphi), \quad \varphi \in \mathcal{T}_m. \tag{2.1.8}$$

and vice versa. If some function $u(\varphi) \in C'(\mathcal{T}_m; a)$ satisfies the identity (2.1.8), then

$$\dot{u}(\varphi_\tau(\varphi)) = \left.\frac{du(\varphi_t(\varphi_\tau(\varphi)))}{dt}\right|_{t=0} = \left.\frac{du(\varphi_{t+\tau}(\varphi))}{dt}\right|_{t=0} = \frac{du(\varphi_\tau(\varphi))}{d\tau}, \tag{2.1.9}$$

for any $\tau \in R$, $\varphi \in \mathcal{T}_m$ which leads to (2.1.6) in view of (2.1.8), i.e. shows that the formula (2.1.5) thereby specifies the invariant torus of the system (2.1.3). This one-to-one correspondence of the invariant tori of the system (2.1.3) to the solutions of the equation

$$Lu = f, \tag{2.1.10}$$

renders the study of the solvability of the equation (2.1.10) urgent. Now let us establish necessary conditions for such solvability. We claim that the kernel of operator L is trivial and designate $\ker L = \{0 \mid |t| \to \infty\}$ whenever the correlation

$$\lim_{|t| \to \infty} u(\varphi_t(\varphi)) = 0 \quad \forall \varphi \in \mathcal{T}_m \tag{2.1.11}$$

holds for every function from the kernel $u(\varphi) \in \ker L$. We define the linear operator $L^* \colon C'(\mathcal{T}_m; a) \to C^0(\mathcal{T}_m)$ on $C'(\mathcal{T}_m; a)$ setting

$$L^* u(\varphi) = -\dot{u}(\varphi) - [A^*(\varphi) + \mu(\varphi) I_n] u(\varphi), \quad \varphi \in \mathcal{T}_m, \tag{2.1.12}$$

where A^* is a matrix conjugated with A, μ is a function from $C^0(\mathcal{T}_m)$ specified by

$$\mu(\varphi) = \sum_{i=1}^{m} \frac{\partial a_i(\varphi)}{\partial \varphi_i}. \tag{2.1.13}$$

The scalar product of functions $f(\varphi)$ and $g(\varphi)$ from $C^0(\mathcal{T}_m)$ is understood in terms of the product of the space $L_2(\mathcal{T}_m)$, i.e. we let

$$(f, g)_0 = \frac{1}{(2\pi)^m} \int_0^{2\pi} \cdots \int_0^{2\pi} \langle f(\varphi), g(\varphi) \rangle \, d\varphi_1 \ldots d\varphi_m, \tag{2.1.14}$$

where $\langle f, g \rangle = \sum_i f_i g_i$ is a scalar product in R^n. As usual, the function $f \in C^0(\mathcal{T}_m)$ is considered to be orthogonal to the set of functions M from $C^0(\mathcal{T}_m)$: $f \perp M$, if $(f, g)_0 = 0$, $\forall g \in M$. Then the necessary conditions of the solvability of equation (2.1.10) are expressed as

$$f \perp \ker L^*, \tag{2.1.15}$$

and the necessary conditions for the solvability of this equation for an arbitrary function f from $C^0(\mathcal{T}_m)$ are as follows

$$\ker L^* = \{0\}, \quad \ker L = \{0 \mid |t| \to \infty\}. \tag{2.1.16}$$

It should be noted that the latter of the correlations has not been known even for the contraction $L \colon C^r(\mathcal{T}_m) \to C^{r-1}(\mathcal{T}_m)$ $(r \geq 1)$ of the operator (2.1.7), when L is determined by the differential expression

$$L = \sum_{i=1}^{m} a_i(\varphi) \frac{\partial}{\partial \varphi_i} - A(\varphi), \tag{2.1.17}$$

and is a well-studied operator.

Now let us prove the above assertions. To this end we first establish an important correlation valid for the functions of the class $C'(\mathcal{T}_m; a)$.

Lemma 2.1.1 Let $u(\varphi) \in C'(\mathcal{T}_m; a)$. Then

$$\int_0^{2\pi} \cdots \int_0^{2\pi} [\dot{u}(\varphi) + \mu(\varphi)u(\varphi)] \, d\varphi_1 \ldots d\varphi_m = 0. \tag{2.1.18}$$

Proof The correlation (2.1.18) for function $u(\varphi) \in C'(\mathcal{T}_m)$ is obvious, since for such a function

$$\dot{u}(\varphi) + \mu(\varphi)u(\varphi) = \sum_{i=1}^m a_i(\varphi) \frac{\partial u(\varphi)}{\partial \varphi_i} + \mu(\varphi)u(\varphi)$$

$$= \sum_{i=1}^m \left[a_i(\varphi) \frac{\partial u(\varphi)}{\partial \varphi_i} + \frac{\partial a_i(\varphi)}{\partial \varphi_i} u(\varphi) \right]$$

$$= \sum_{i=1}^m \frac{\partial [a_i(\varphi)u(\varphi)]}{\partial \varphi_i}.$$

Consequently,

$$\int_0^{2\pi} \cdots \int_0^{2\pi} [\dot{u}(\varphi) + \mu(\varphi)u(\varphi)] \, d\varphi_1 \ldots d\varphi_m = \sum_{i=1}^m \int_0^{2\pi} \cdots \int_0^{2\pi} \{ [a_i(\varphi)u(\varphi)]|_{\varphi_i=2\pi}$$
$$- [a_i(\varphi)u(\varphi)]|_{\varphi_i=0} \} \, d\varphi_1 \ldots d\varphi_{i-1} d\varphi_{i+1} \ldots d\varphi_m = 0.$$

Now let $u(\varphi) \in C'(\mathcal{T}_m; a) \setminus C'(\mathcal{T}_m)$. We set

$$I(t) = \int_0^{2\pi} \cdots \int_0^{2\pi} u(\varphi_t(\varphi)) \, d\varphi_1 \ldots d\varphi_m, \quad t \in R. \tag{2.1.19}$$

As this function $u(\varphi_t(\varphi))$ is differentiable in t, then

$$\left. \frac{dI(t)}{dt} \right|_{t=0} = \int_0^{2\pi} \cdots \int_0^{2\pi} \left. \frac{du(\varphi_t(\varphi))}{dt} \right|_{t=0} d\varphi_1 \ldots d\varphi_m$$
$$= \int_0^{2\pi} \cdots \int_0^{2\pi} \dot{u}(\varphi) \, d\varphi_1 \ldots d\varphi_m. \tag{2.1.20}$$

In integral (2.1.19) we make the change of variables

$$\varphi = \varphi_{-t}(\psi), \tag{2.1.21}$$

viewing t as a parameter. Then

$$I(t) = \int_{K_t} u(\psi)|J(t,\psi)|\,d\psi, \qquad (2.1.22)$$

where J is a Jacobian of the transformation (2.1.21), K_t is a preimage of the cube $K_0 = \{0 \leq \varphi_i \leq 2\pi,\ i = \overline{1,m}\}$ of this transformation.

We find the Jacobian of the transformation (2.1.21)

$$J(t,\psi) = \det \frac{\partial \varphi_{-t}(\psi)}{\partial \psi}. \qquad (2.1.23)$$

Since $\frac{\partial \varphi_{-t}(\psi)}{\partial \psi}$ is a matriciant of the system of equations

$$\frac{dz}{dt} = -\frac{\partial a(\varphi_{-t}(\psi))}{\partial \varphi} z,$$

then by the *Ostrogradsky – Liouville relation*

$$\det \frac{\partial \varphi_{-t}(\psi)}{\partial \psi} = \exp\left\{-\int_0^t \sum_{i=1}^m \frac{\partial a_i(\varphi_{-t}(\psi))}{\partial \varphi_i}\,dt\right\},$$

which in view of (2.1.13) yields the equality

$$J(t,\psi) = \exp\left\{-\int_0^t \mu(\varphi_{-t}(\psi))\,dt\right\} = \exp\left\{\int_0^{-t} \mu(\varphi_\tau(\psi))\,dt\right\}.$$

Thus

$$I(t) = \int_{K_t} u(\psi) \exp\left\{\int_0^{-t} \mu(\varphi_\tau(\psi))\,d\tau\,d\psi\right\}. \qquad (2.1.24)$$

The change (2.1.21) implies $\psi = \varphi_t(\varphi)$, and so, when φ runs through the torus $\mathcal{T}_m$, ψ also runs through this torus for any fixed $t \in R$. However, this means that

$$K_t = K_0 \bmod 2\pi. \qquad (2.1.25)$$

Since for the function $f \in C^0(\mathcal{T}_m)$

$$f(\psi) = f(\psi \bmod 2\pi), \qquad (2.1.26)$$

the equalities (2.1.25) and (2.1.26) yield the identity

$$I(t) = \int_{K_t} u(\psi) \exp\left\{\int_0^{-t} \mu(\varphi_\tau(\psi))\, d\tau\, d\psi\right\}$$

$$= \int_{K_0} u(\psi) \exp\left\{\int_0^{-t} \mu(\varphi_\tau(\psi))\, d\tau\, d\psi\right\}$$

$$= \int_0^{2\pi}\cdots\int_0^{2\pi} u(\psi) \exp\left\{\int_0^{-t} \mu(\varphi_\tau(\psi))\, d\tau\, d\psi_1 \ldots d\psi_m\right\}.$$

Differentiating the latter in t we get the expression

$$\left.\frac{dI(t)}{dt}\right|_{t=0} = -\int_0^{2\pi}\cdots\int_0^{2\pi} u(\psi)\left|\mu(\varphi_{-t}(\psi))\exp\left\{\int_0^t \mu(\varphi_\tau(\psi))\, d\tau\right\}\right|_{t=0} d\psi_1 \ldots d\psi_m$$

$$= -\int_0^{2\pi}\cdots\int_0^{2\pi} \mu(\psi)u(\psi)\, d\psi_1 \ldots d\psi_m. \tag{2.1.27}$$

Substituting (2.1.27) into (2.1.20) yields (2.1.18).

Lemma 2.1.1 provides the proof of the conjugation of operators L and L^* relative to the product (2.1.14). Hence, its validity.

Lemma 2.1.2 *For any functions $u(\varphi)$, $v(\varphi) \in C'(\mathcal{T}_m; a)$ the equality*

$$(Lu, v)_0 = (u, L^*v)_0, \tag{2.1.28}$$

is valid.

Proof Actually, since

$$\left\langle \frac{du(\varphi_t(\varphi))}{dt} - A(\varphi_t(\varphi))u(\varphi_t(\varphi)), v(\varphi_t(\varphi))\right\rangle$$

$$= \frac{d}{dt}\langle u(\varphi_t(\varphi)), v(\varphi_t(\varphi))\rangle - \left\langle u(\varphi_t(\varphi)), \frac{dv(\varphi_t(\varphi))}{dt}\right\rangle$$

$$- \langle u(\varphi_t(\varphi)), A^*(\varphi_t(\varphi))v(\varphi_t(\varphi))\rangle,$$

then

$$(Lu, v)_0 = \frac{1}{(2\pi)^m}\int_0^{2\pi}\cdots\int_0^{2\pi} \dot{W}(\varphi)\, d\varphi_1 \ldots d\varphi_m - (u, \dot{v})_0 - (u, A^*v)_0,$$

where $W(\varphi)$ denotes $W(\varphi) = \langle u(\varphi), v(\varphi)\rangle$. As the functions $u(\varphi)$ and $v(\varphi)$ belong to the space $C'(\mathcal{T}_m; a)$, the function $W(\varphi)$ belongs to this space as well. Then, by Lemma 2.1.1,

$$\int_0^{2\pi}\cdots\int_0^{2\pi} \dot{W}(\varphi)\, d\varphi_1 \ldots d\varphi_m = -\int_0^{2\pi}\cdots\int_0^{2\pi} \mu(\varphi)W(\varphi)\, d\varphi_1 \ldots d\varphi_m,$$

from which equality (2.1.28) follows

$$(Lu, v)_0 = -(u, \dot{v} + A^*v + \mu v)_0.$$

We now consider equation (2.1.10). Let the function $u_0(\varphi)$ be its solution, so that $Lu_0(\varphi) \equiv f(\varphi)\ \forall \varphi \in \mathcal{T}_m$. Then by Lemma 2.1.2 for any function $v(\varphi)$ from $C'(\mathcal{T}_m; a)$ the relation

$$(f, v)_0 = (Lu_0, v)_0 = (u_0, L^*v)_0,$$

is satisfied, which implies that for any functions $v \in \ker L^*$

$$(f, v)_0 = 0.$$

Thus, we arrive at the following conclusion.

Theorem 2.1.1 *For the system of equations (2.1.10) to have a solution, it is necessary that*

$$(f, v)_0 = 0, \quad \forall v \in \ker L^*. \tag{2.1.29}$$

Equation (2.1.10), where $A = 0$, $a = \omega$ and the numbers $(\omega_1, \ldots, \omega_m) = \omega$ form a basis, and therefore $\langle k, \omega\rangle = 0$ for any non-zero integer vector $k = (k_1, \ldots, k_m)$, shows that the necessary conditions for the solvability of equation (2.1.10) are not sufficient in the general case. In fact, here $L^*u = -\dot{u}$, $\varphi_t(\varphi) = \omega t + \varphi$ and $\ker L^*$ consists of functions independent of φ: $\ker L^* = \{c\}$, c is an arbitrary constant. The orthogonality condition (2.1.29) requires the right-hand side of (2.1.10) to have zero average value

$$\int_0^{2\pi}\cdots\int_0^{2\pi} f(\varphi)\, d\varphi_1 \ldots d\varphi_m = 0. \tag{2.1.30}$$

The equation (2.1.10) is solvable, provided that the primitive of the function $f(\omega t)$ is almost periodic. For this it is known by Bohl's theorem, that it is not sufficient when only the equality (2.1.30) is satisfied.

We now clarify the conditions of the solvability of equation (2.1.10) for an arbitrary function $f(\varphi) \in C^0(\mathcal{T}_m)$. Designate the matricant of the linear system of equations

$$\frac{dx}{dt} = A(\varphi_t(\varphi))x, \tag{2.1.31}$$

by Ω_0^t, i.e. the fundamental matrix of solutions, where $\Omega_0^t|_{t=0} = I_n$, where I_n is an n-dimensional identity matrix.

Lemma 2.1.3 *Let $u_0(\varphi) \in \ker L$, $u_j(\varphi)$ is a solution of equation*

$$Lu = u_{j-1}, \quad j = 1, 2, \ldots. \qquad (2.1.32)$$

Then

$$u_j(\varphi_t(\varphi)) = \Omega_0^t(\varphi; A)\left[u_j(\varphi) + \frac{t}{1!}u_{j-1}(\varphi) + \cdots + \frac{t^j}{j!}u_0(\varphi)\right], \qquad (2.1.33)$$

$$\Omega_0^t(\varphi; A)u_j(\varphi) = u_j(\varphi_t(\varphi)) - \frac{t}{1!}u_{j-1}(\varphi_t(\varphi)) + \ldots (-1)^j \frac{t^j}{j!}u_0(\varphi_t(\varphi)).$$

Proof If $u_0(\varphi) \in \ker L$, then $x = u_0(\varphi_t(\varphi))$ is a solution of the system of equations (2.1.31) so that

$$u_0(\varphi_t(\varphi)) = \Omega_0^t(\varphi; A)u_0(\varphi). \qquad (2.1.34)$$

The function $x = u_1(\varphi_t(\varphi))$ then specifies the solution of the system

$$\frac{dx}{dt} = A(\varphi_t(\varphi))x + f(\varphi_t(\varphi)), \qquad (2.1.35)$$

for $f(\varphi) = u_0(\varphi)$. Therefore

$$\begin{aligned}
u_1(\varphi_t(\varphi)) &= \Omega_0^t(\varphi; A)u_1(\varphi) + \int_0^t \Omega_\tau^t(\varphi; A)u_0(\varphi_\tau(\varphi))\, d\tau \\
&= \Omega_0^t(\varphi; A)u_1(\varphi) + \int_0^t \Omega_\tau^t(\varphi; A)\Omega_0^\tau(\varphi; A)u_0(\varphi)\, d\tau \\
&= \Omega_0^t(\varphi; A)\left[u_1(\varphi) + tu_0(\varphi)\right],
\end{aligned} \qquad (2.1.36)$$

which proves the first identity (2.1.33) for $j = 1$. The function $x = u_2(\varphi_t(\varphi))$ prescribes the solution of the system (2.1.35) for $f = u_1$. Thus

$$\begin{aligned}
u_2(\varphi_t(\varphi)) &= \Omega_0^t(\varphi; A)u_2(\varphi) + \int_0^t \Omega_\tau^t(\varphi; A)u_1(\varphi_\tau(\varphi))\, d\tau \\
&= \Omega_0^t(\varphi; A)\left[u_2(\varphi) + \int_0^t (u_1(\varphi) + \tau u_0(\varphi))\, d\tau\right] \\
&= \Omega_0^t(\varphi; A)\left[u_2(\varphi) + \frac{t}{1!}u_1(\varphi) + \frac{t^2}{2!}u_0(\varphi)\right],
\end{aligned} \qquad (2.1.37)$$

which proves the first identity (2.1.33) for $j = 2$.

Arguments by induction prove this identity for any integer $j \geq 1$.

Let us prove the second identity (2.1.33). Relations (2.1.36) and (2.1.34) imply

$$u_1(\varphi_t(\varphi)) = \Omega_0^t(\varphi; A)u_1(\varphi) + tu_0(\varphi_t(\varphi)),$$

i.e. the second identity (2.1.33) is valid for $j = 1$. In order to prove this identity for any integer $j > 1$ note that the vector

$$\bar{u} = \operatorname{col}\{\Omega_0^t(\varphi; A)u_0(\varphi), \ldots, \Omega_0^t(\varphi; A)u_N(\varphi)\}$$

is related to the vector

$$u = \operatorname{col}\{u_0(\varphi_t(\varphi)), \ldots, u_N(\varphi_t(\varphi))\}$$

by the vector matrix identity

$$u = T\bar{u}, \tag{2.1.38}$$

the matrix T of which reads according to the first identity (2.1.33)

$$T = e^{zt} = \sum_{k=0}^{N} \frac{z^k t^k}{k!}, \quad z = \begin{pmatrix} 0 & 0 & \cdots & 0 & 0 \\ I_n & 0 & \cdots & 0 & 0 \\ 0 & I_n & \cdots & 0 & 0 \\ \vdots & \vdots & \ddots & \vdots & \vdots \\ 0 & 0 & \cdots & I_n & 0 \end{pmatrix}; \tag{2.1.39}$$

0 is a zero n-dimensional matrix. Solving identity (2.1.38) with respect to $\bar{u}$ we get

$$\bar{u} = T^{-1}u = e^{-zt}, \quad \bar{u} = \sum_{k=0}^{N} \frac{(-1)^k z^k t^k}{k!} u. \tag{2.1.40}$$

So, $\Omega_0^t(\varphi; A)u_j(\varphi)$ is expressed via $u_k(\varphi_t(\varphi))$, $k = 0, 1, \ldots, j$, according to the formula derived from the first identity (2.1.33) by the substitution by $\Omega_0^t(\varphi; A)u_k(\varphi)$ for $u_k(\varphi_k(\varphi))$, $-t$ for t and $u_j(\varphi_t(\varphi))$ for $\Omega_0^t(\varphi; A)u_j(\varphi)$. Then

$$\Omega_0^t(\varphi; A)u_j(\varphi) = u_j(\varphi_t(\varphi)) - \frac{t}{1!}u_{j-1}(\varphi_t(\varphi)) + \ldots (-1)^j \frac{t^j}{j!} u_0(\varphi_t(\varphi)).$$

We make use of Lemma 2.1.3 to prove the basic result of this section.

Theorem 2.1.2 *For the system (2.1.10) to have a solution for an arbitrary function $f(\varphi) \in C^0(\mathcal{T}_m)$, it is necessary that*

$$\ker L^* = \{0\}, \quad \ker L = \{0 \mid |t| \to \infty\}. \tag{2.1.41}$$

Proof For an arbitrary function $f(\varphi) \in C^0(\mathcal{T}_m)$ let the system (2.1.10) possess a solution. However, we assume either $\ker L^* \neq \{0\}$ or $\ker L \neq \{0 \mid |t| \to \infty\}$.

If $\ker L^*$ contains a function $v \neq 0$, then by Theorem 2.1.1 we have for $f = v$

$$0 = (f, v)_0 = (v, v)_0 = \|v\|_0^2 \neq 0.$$

The contradiction proves the first relation (2.1.41).

Now let $\ker L \neq \{0 \mid |t| \to \infty\}$; therefore the function $u_0(\varphi) \in \ker L$ exists for which the condition

$$\lim_{|t| \to \infty} u_0(\varphi_t(\varphi)) = 0 \tag{2.1.42}$$

does not hold for any $\varphi \in \mathcal{T}_m$. Consider the system of functions

$$u_0(\varphi), \ u_1(\varphi), \ \ldots, \ u_n(\varphi), \tag{2.1.43}$$

the function u_j of which is a solution of equation (2.1.32). Since the system of vectors (2.1.43) is linearly dependent for every fixed $\varphi \in \mathcal{T}_m$, we divide the set of points $\mathcal{T}_m$ into two subsets N_0 and $N_1 = \mathcal{T}_m \setminus N_0$, associating N_0 with all the points of $\mathcal{T}_m$ for which the vectors

$$u_0(\varphi), \ u_1(\varphi), \ \ldots, \ u_{n-1}(\varphi) \tag{2.1.44}$$

are linearly independent. Then, for a fixed $\varphi_0 \in N_0$, the constants $\alpha_0, \alpha_1, \alpha_2, \ldots, \alpha_{n-1}$ are found, by means of which the vector $u_n(\varphi_0)$ is decomposed by a system of vectors (2.1.44)

$$u_n(\varphi_0) = \sum_{i=0}^{n-1} \alpha_i u_i(\varphi_0). \tag{2.1.45}$$

Hence in view of (2.1.33) we get

$$\Omega_0^t(\varphi; A) u_n(\varphi_0) = \alpha_0 u_0(\varphi_t(\varphi_0)) + \alpha_1 \left[u_1(\varphi_t(\varphi_0)) - \frac{t}{1!} u_0(\varphi_t(\varphi_0)) \right]$$
$$+ \alpha_{n-1} \left[u_{n-1}(\varphi_t(\varphi_0)) - \frac{t}{1!} u_{n-2}(\varphi_t(\varphi_0)) + \cdots + (-1)^{n-1} \frac{t^{n-1}}{(n-1)!} u_0(\varphi_t(\varphi_0)) \right],$$

from which, again in view of (2.1.33), we find

$$(-1)^n \frac{t^n}{n!} u_0(\varphi_t(\varphi_0)) + (-1)^{n-1} \frac{t^{n-1}}{(n-1)!} u_1(\varphi_t(\varphi_0)) + \ldots$$

$$- \frac{t}{1!} u_{n-1}(\varphi_t(\varphi_0)) + u_n(\varphi_t(\varphi_0))$$

$$= \left[\alpha_0 - \frac{t}{1!} \alpha_1 + \cdots + (-1)^{n-1} \frac{t^{n-1}}{(n-1)!} \alpha_{n-1} \right] u_0(\varphi_t(\varphi_0)) \quad (2.1.46)$$

$$+ \left[\alpha_1 - \frac{t}{1!} \alpha_2 + \cdots + (-1)^{n-2} \frac{t^{n-2}}{(n-2)!} \alpha_{n-1} \right] u_1(\varphi_t(\varphi_0)) + \ldots$$

$$+ \alpha_{n-1} u_{n-1}(\varphi_t(\varphi_0)).$$

With regard to the boundedness of the functions $u_j(\varphi_t(\varphi_0))$ for all $t \in R$ this identity implies

$$\lim_{|t| \to \infty} \|u_0(\varphi_t(\varphi_0))\| = 0. \quad (2.1.47)$$

So, if $\varphi_0 \in N$, then the limiting correlation (2.1.47) holds true.

Let $\varphi_0 \in N$. Then the functions (2.1.44) are linearly dependent. We distinguish the points φ from N_1 for which the system of vectors

$$u_0(\varphi), \ u_1(\varphi), \ \ldots, \ u_{n-2}(\varphi) \quad (2.1.48)$$

is linearly dependent. The vector $u_{n-1}(\varphi_0)$ is decomposed into vectors (2.1.48) for every such point $\varphi = \varphi_0$ in the same way as the vector $u_n(\varphi_0)$ for $\varphi \in N_0$ into vectors (2.1.44). This yields an identity similar to (2.1.46) and proves the limiting correlation (2.1.47) for the points N_1, for which the vectors (2.1.48) are linearly independent.

Proceeding with the arguments, we conclude on the validity of (2.1.47) for all the points $\varphi \in \mathcal{T}_m$ for which at least one of the systems of vectors

$$u_0(\varphi), \ u_1(\varphi), \ \ldots, \ u_j(\varphi), \quad j = 1, 2, \ldots, n-1, \quad (2.1.49)$$

is linearly independent.

Let φ_0 be the point for which all systems of vectors (2.1.49) are linearly dependent. Then constants α_0 and α_1 are found such that $\alpha_0^2 + \alpha_1^2 \neq 0$ and $\alpha_0 u_0(\varphi_0) + \alpha_1 u_1(\varphi_0) = 0$.

If $\alpha_1 \neq 0$, then $u_1(\varphi_0) = -\frac{\alpha_0}{\alpha_1} u_0(\varphi_0)$ and the identity (2.1.33) yields for $j = 1$

$$-\frac{\alpha_0}{\alpha_1} u_0(\varphi_t(\varphi_0)) = u_1(\varphi_t(\varphi_0)) - t u_0(\varphi_t(\varphi_0)). \quad (2.1.50)$$

The last identity implies the relation (2.1.47). By this the relation is proved for all points $\varphi_0 \in \mathcal{T}_m$, for which

$$u_0(\varphi_0) = 0. \quad (2.1.51)$$

Let N be a set of such points, and so, consist only of those points of $\mathcal{T}_m$ for which the relation (2.1.51) is valid. The existence of a solution and uniqueness theorem for the Cauchy problem for the system of the equations (2.1.31) implies that the solution $x(t)$ of this system satisfies the identity $x(t) \equiv 0$ for $t \in R$ whenever $x(0) = 0$. Therefore, if $\varphi_0 \in N$, then $x = u(\varphi_t(\varphi_0)) = 0$ for $t \in R$. This also proves the limiting relation (2.1.47) for the points $\varphi_0 \in N$.

Thus, if in the conditions of Theorem 2.1.2 $u_0 \in \ker L$, then relation (2.1.47) holds true for any $\varphi_0 \in \mathcal{T}_m$. The latter contradicts the assumption of non-degeneracy of the operator L kernel. This contradiction completes the proof of the Theorem 2.1.2.

The example of the system of equations

$$\frac{d\varphi}{dt} = 0, \quad \frac{dx}{dt} = A(\varphi)x + f(\varphi), \tag{2.1.52}$$

for which the necessary conditions of the equation (2.1.10) solvability are reduced to the requirement

$$\det A(\varphi) \neq 0, \quad \forall \varphi \in \mathcal{T}_m, \tag{2.1.53}$$

demonstrates the accessibility of the latter, since conditions (2.1.53) are also sufficient for the solvability of the equation (2.1.10).

The example of system of the equations

$$\frac{d\varphi}{dt} = \sin \varphi, \quad \frac{dx}{dt} = (\cos \varphi)x + f(\varphi) \tag{2.1.54}$$

proves the independence of conditions of Theorem 2.1.2 and shows the degree of "unimprovability" revealing, in particular, the impossibility to replace these conditions by

$$\ker L^* = \ker L = \{0\}. \tag{2.1.55}$$

In fact, it is easy to verify that for the system in question

$$\ker L^* = \{0\}, \quad \ker L = \{c \sin \varphi\},$$

where c is an arbitrary constant. Since

$$\sin \varphi_t(\varphi) = \frac{2c_0}{1 + c_0^2 e^{2t}},$$

where

$$c_0 = \begin{cases} 0, & \varphi = 0, \pi, \\ \tan \frac{\varphi}{2}, & 0 < \varphi < \pi, \ \pi < \varphi < 2\pi, \end{cases}$$

then $\lim\limits_{|t| \to \infty} \sin \varphi_t(\varphi_0) = 0$, $\forall \varphi \in \mathcal{T}_1$, and consequently $\ker L = \{0 \mid |t| \to \infty\}$.

So, the triviality of the kernel of the operator L^* does not imply the triviality of the kernel of the operator L, allowing the latter to be degenerate. This means the independence of conditions of Theorem 2.1.2.

Let us now demonstrate the "unimprovability" of the conditions. To this end, we show the system (2.1.54) to be such that the equation (2.1.10) corresponding to it has a solution for any function $f \in C^0(\mathcal{T}_1)$. The function is specified by the relations

$$u_0(\varphi) = \begin{cases} -f(0), & \varphi = 0, \\ \sin\varphi \int_{\frac{\pi}{2}}^{\varphi} \frac{f(\psi)}{\sin^2\psi}\, d\psi, & 0 < \varphi < \pi, \\ f(\pi), & \varphi = \pi, \\ \sin\varphi \int_{\frac{3}{2}\pi}^{\varphi} \frac{f(\psi)}{\sin^2\psi}\, d\psi, & \pi < \varphi < 2\pi, \end{cases} \quad (2.1.56)$$

and extended in the rest of φ by periodicity. As

$$\lim_{\varphi \to +0} \sin\varphi \int_{\frac{\pi}{2}}^{\varphi} \frac{f(\psi)}{\sin^2\psi}\, d\psi = \lim_{\varphi \to +0} \frac{f(\varphi)}{-\cos\varphi} = -f(0),$$

$$\lim_{\varphi \to \pi \pm 0} \sin\varphi \int_{\frac{3}{2}\pi}^{\varphi} \frac{f(\psi)}{\sin^2\psi}\, d\psi = \lim_{\varphi \to \pi \pm 0} \frac{f(\varphi)}{-\cos\varphi} = f(\pi),$$

$$\lim_{\varphi \to 2\pi - 0} \sin\varphi \int_{\frac{3}{2}\pi}^{\varphi} \frac{f(\psi)}{\sin^2\psi}\, d\psi = \lim_{\varphi \to 2\pi - 0} \frac{f(\varphi)}{-\cos\varphi} = -f(2\pi) = -f(0),$$

then $u_0(\varphi) \in C^0(\mathcal{T}_1)$. Further

$$u_0(\varphi_t(0)) \equiv u_0(\varphi), \quad u_0(\varphi_t(\pi)) \equiv u_0(\pi)$$

and so the identity

$$\frac{du_0(\varphi_t(\varphi_0))}{dt} \equiv [\cos\varphi_t(\varphi_0)] u_0(\varphi_t(\varphi_0)) + f(\varphi_t(\varphi_0))$$

is satisfied for $\varphi_0 = 0$ and $\varphi = \pi$. Moreover by (2.1.56), for $\varphi_0 \in (0, \pi)$ or $\varphi_0 \in (\pi, 2\pi)$

$$\frac{du_0(\varphi_t(\varphi_0))}{dt} = \frac{\partial u_0(\varphi_t(\varphi_0))}{d\varphi} \frac{d(\varphi_t(\varphi_0))}{dt} \equiv [\cos\varphi_t(\varphi_0)] u_0(\varphi_t(\varphi_0)) + f(\varphi_t(\varphi_0)).$$

This means that $u_0 \in C'(\mathcal{T}_1; \sin\varphi)$ and $Lu_0 = f$. Thus, the equation (2.1.10) corresponding to the system (2.1.54) has a solution for any $f \in C^0(\mathcal{T}_1)$.

2.2 The Green Function. Sufficient Existence Conditions for an Invariant Torus

The fact that the functions $a(\varphi)$ and $A(\varphi)$ belong to the space $C^r(\mathcal{T}_m)$ implies immediately that $\varphi_t(\varphi) - \varphi$ and the matricant $\Omega_\tau^t(\varphi; A)$ also belong to the space $C^r(\mathcal{T}_m)$ for all $t, \tau \in R$ and for any $r \geq 1$. We now show that

$$\Omega_\tau^t(\varphi_\theta(\varphi); A) \equiv \Omega_{\tau+\theta}^{t+\theta}(\varphi; A), \qquad (2.2.1)$$

for any $t, \tau, \theta \in R$. To this end, in the identity specifying $\Omega_\tau^t(\varphi; A)$

$$\frac{d}{dt}\Omega_\tau^t(\varphi; A) = A(\varphi_t(\varphi))\Omega_\tau^t(\varphi; A),$$

we replace φ by $\varphi_\theta(\varphi)$ and, in view of the group property of function $\varphi_t(\varphi)$ expressed by $\varphi_t(\varphi_\theta(\varphi)) \equiv \varphi_{t+\theta}(\varphi)$, get

$$\frac{d}{dt}\Omega_\tau^t(\varphi_\theta(\varphi); A) = A(\varphi_{t+\theta}(\varphi))\Omega_\theta^t(\varphi_\theta(\varphi); A).$$

The latter identity is valid for arbitrary $t, \tau, \theta \in R$ and means that $\Omega_\tau^t(\varphi_\theta(\varphi); A)$ is a fundamental matrix of the solutions of the system

$$\frac{dx}{dt} = A(\varphi_{t+\theta}(\varphi))x, \qquad (2.2.2)$$

taking the value I_n for $t = \tau$. The matrix $\Omega_{\tau+\theta}^{t+\theta}(\varphi; A)$ has the same property. This is only possible when the matrices $\Omega_\tau^t(\varphi_\theta(\varphi); A)$ and $\Omega_{\tau+\theta}^{t+\theta}(\varphi; A)$ coincide. This proves the desired identity (2.2.1).

Definition 2.2.1 Let $C(\varphi)$ be the matrix belonging to the space $C^0(\mathcal{T}_m)$. We set

$$G_0(\tau, \varphi) = \begin{cases} \Omega_\tau^0(\varphi; A)C(\varphi_\tau(\varphi)), & \text{for } \tau \leq 0, \\ \Omega_\tau^0(\varphi; A)[C(\varphi_\tau(\varphi)) - I_n], & \text{for } \tau > 0 \end{cases} \qquad (2.2.3)$$

and refer to $G_0(\tau, \varphi)$ as *Green function* of the system of equations

$$\frac{d\varphi}{dt} = a(\varphi), \quad \frac{dx}{dt} = A(\varphi)x \qquad (2.2.4)$$

whenever the integral $\int_{-\infty}^{\infty} \|G_0(\tau, \varphi)\| d\tau$ converges uniformly in $\varphi \in \mathcal{T}_m$. Moreover

$$\int_{-\infty}^{\infty} \|G_0(\tau, \varphi)\| d\tau \leq K < \infty. \qquad (2.2.5)$$

Let us indicate the simplest properties of the Green function (2.2.3). By its definition it follows that $G_0(\tau,\varphi) \in C^0(\mathcal{T}_m)$ for every $\tau \neq 0$, and

$$G_0(-0,\varphi) - G_0(+0,\varphi) = I_n.$$

Let $G_t(\tau,\varphi)$ be a matrix defined by

$$G_t(\tau,\varphi) = \begin{cases} \Omega_\tau^t(\varphi;A)C(\varphi_\tau(\varphi)), & \text{for } t \geq \tau, \\ \Omega_\tau^t(\varphi;A)[C(\varphi_\tau(\varphi)) - I_n], & \text{for } t < \tau. \end{cases} \quad (2.2.6)$$

Then

$$G_0(\tau,\varphi_t(\varphi)) \equiv G_t(\tau+t,\varphi). \quad (2.2.7)$$

The last relation follows immediately from the matrix $\Omega_\tau^t(\varphi;A)$ property expressed by (2.2.1)

$$G_0(\tau,\varphi_t(\varphi)) = \begin{cases} \Omega_\tau^0(\varphi_t(\varphi);A)C(\varphi_\tau(\varphi_t(\varphi))), & \text{for } \tau \leq 0, \\ \Omega_\tau^0(\varphi_t(\varphi);A)[C(\varphi_\tau(\varphi_t(\varphi))) - I_n], & \text{for } \tau > 0, \end{cases}$$

$$= \begin{cases} \Omega_{\tau+t}^t(\varphi;A)C(\varphi_{\tau+t}(\varphi)), & \text{for } \tau+t \leq t, \\ \Omega_{\tau+t}^t(\varphi;A)[C(\varphi_{\tau+t}(\varphi)) - I_n], & \text{for } \tau+t > t, \end{cases}$$

$$= G_t(\tau+t,\varphi).$$

Equation (2.2.7) shows that the matrix

$$G_0(-t,\varphi_t(\varphi)) = G_t(0,\varphi) = \begin{cases} \Omega_0^t(\varphi;A)C(\varphi), & \text{for } t \geq 0, \\ \Omega_0^t(\varphi;A)[C(\varphi) - I_n], & \text{for } t < 0, \end{cases} \quad (2.2.3a)$$

consists of solutions of the homogeneous system of the equations (2.1.31) treated for $t \geq 0$ and $t < 0$ respectively.

Let $f \in C^0(\mathcal{T}_m)$. Consider the integral

$$\int_{-\infty}^{\infty} G_0(\tau,\varphi)f(\varphi_\tau(\varphi))\,d\tau, \quad (2.2.8)$$

which is majorized by the convergent integral according to the estimate

$$\left\| \int_{-\infty}^{\infty} G_0(\tau,\varphi)f(\varphi_\tau(\varphi))\,d\tau \right\| \leq \int_{-\infty}^{\infty} \|G_0(\tau,\varphi)\|\,d\tau\,\|f\|_0.$$

We set

$$u(\varphi) = \int_{-\infty}^{\infty} G_0(\tau,\varphi)f(\varphi_\tau(\varphi)\,d\tau$$

and show that $u(\varphi) \in C^0(\mathcal{T}_m)$. Actually, let

$$u_i(\varphi) = \int_{-i}^{i} G_0(\tau, \varphi) f(\varphi_\tau(\varphi)) \, d\tau, \quad i = 1, 2, \ldots.$$

The sequence of functions $u_i(\varphi)$ evidently belongs to the space $C^0(\mathcal{T}_m)$ and satisfies the estimate

$$\|u(\varphi) - u_t(\varphi)\| \leq \left[\int_{-\infty}^{-i} \|G_0(\tau, \varphi)\, d\tau\| + \int_{i}^{\infty} \|G_0(\tau, \varphi)\| \, d\tau\right] \|f\|_0.$$

The uniform in φ convergence of the integral (2.2.5) ensures the limiting relation $\lim_{i \to \infty} u_i(\varphi) = u(\varphi)$ which is uniform in $\varphi \in \mathcal{T}_m$. The latter denotes the convergence of the sequence $u_i(\varphi)$, $i = 1, 2, \ldots$ with respect to the norm of the space $C^0(\mathcal{T}_m)$. Since $C^0(\mathcal{T}_m)$ is a complete space, then $u(\varphi) \in C^0(\mathcal{T}_m)$. The integral (2.2.8) prescribes some operator T in the space of functions $C^0(\mathcal{T}_m)$:

$$Tf(\varphi) = \int_{-\infty}^{\infty} G_0(\tau, \varphi) f(\varphi_\tau(\varphi)) \, d\tau. \tag{2.2.9}$$

Let us show that the set

$$x = u(\varphi) = Tf(\varphi), \quad \varphi \in \mathcal{T}_m, \tag{2.2.10}$$

determines the invariant torus of the inhomogeneous system (2.1.3). To this end the function $x(t, \varphi) = u(\varphi_t(\varphi))$ is considered. By the inequalities (2.2.7) and (2.2.9) we have the representation for $u(\varphi_t(\varphi))$:

$$u(\varphi_t(\varphi)) = \int_{-\infty}^{\infty} G_0(\tau, \varphi_t(\varphi)) f(\varphi_\tau(\varphi_t(\varphi))) \, d\tau$$

$$= \int_{-\infty}^{\infty} G_t(\tau + t, \varphi) f(\varphi_{\tau+t}(\varphi)) \, d\tau = \int_{-\infty}^{\infty} G_t(\tau, \varphi) f(\varphi_\tau(\varphi)) \, d\tau,$$

the differentiation of which in t yields

$$\frac{du(\varphi_t(\varphi))}{dt} = C(\varphi_t(\varphi)) f(\varphi_t(\varphi)) + [I_n - C(\varphi_t(\varphi))] f(\varphi_t(\varphi))$$

$$+ A(\varphi_t(\varphi)) \int_{-\infty}^{\infty} G_t(\tau, \varphi) f(\varphi_\tau(\varphi)) \, d\tau = A(\varphi_t(\varphi)) u(\varphi_t(\varphi)) + f(\varphi_t(\varphi)),$$

for any $t \in R$. The latter proves that $u(\varphi) \in C'(\mathcal{T}_m; a)$ and the set (2.2.10) is an invariant torus of the system (2.1.3).

Thus, the existence of the Green function of the system (2.2.4) ensures the existence of the invariant torus of the system (2.1.3) for an arbitrary function $f \in C^0(\mathcal{T}_m)$.

Note that in the above arguments the fact that the functions $a(\varphi)$ and $A(\varphi)$ belong to the space $C^r(\mathcal{T}_m; a)$ was only used to determine the continuity of the matrix $\Omega_\tau^t(\varphi; A)$ in φ. Therefore all the above arguments remain valid for $a(\varphi) \in C_{\text{Lip}}(\mathcal{T}_m)$, $A(\varphi) \in C^0(\mathcal{T}_m)$. Also it follows here from the assertion determining sufficient existence conditions for the invariant torus of the system (2.1.3).

Theorem 2.2.1 *Let the right-hand side of the system of equations (2.2.4) satisfy the conditions $a(\varphi) \in C_{\text{Lip}}^r(\mathcal{T}_m)$, $A(\varphi) \in C^r(\mathcal{T}_m)$, $r \geq 0$ and the system itself have the Green function $G_0(\tau, \varphi)$.*

Then for any $f \in C^r(\mathcal{T}_m)$ the system of equations (2.1.3) has an invariant torus defined by relation (2.2.10) and satisfying the estimate

$$\|u\|_0 = \|Tf\|_0 \leq K\|f\|_0, \qquad K = \max_{\varphi \in \mathcal{T}_m} \int_{-\infty}^{\infty} \|G_0(\tau, \varphi)\| \, d\tau.$$

We show by example that the invariant torus discussed in the cited theorem cannot be other that continuous even when the functions a, A and f are analytical in φ.

We consider the system of equations

$$\frac{d\varphi}{dt} = -\sin\varphi, \qquad \frac{dx}{dt} = -x + \sin\varphi.$$

By definition (2.2.3) the Green function for this system with the matrix $C(\varphi) \equiv 1$ has the form

$$G_0(\tau, \varphi) = \begin{cases} e^\tau, & \text{for } \tau \leq 0, \\ 0, & \text{for } \tau > 0. \end{cases}$$

The invariant torus (2.2.10) is determined by

$$x = u(\varphi) = \int_{-\infty}^{0} e^\tau \sin\varphi_\tau(\varphi) \, d\tau, \qquad \varphi \in \mathcal{T}_1,$$

where

$$\sin\varphi_\tau(\varphi) = \begin{cases} 0, & \text{for } \varphi = k\pi, \quad k = 0, \pm 1, \ldots, \\ \frac{2e^\tau \tan\frac{\varphi}{2}}{e^{2\tau} + \tan^2\frac{\varphi}{2}}, & \text{for } \varphi \neq k\pi. \end{cases}$$

Computing the function u yields

$$u(\varphi) = \begin{cases} 0, & \text{for } \varphi = k\pi, \\ \tan\frac{\varphi}{2} \ln \sin^2\frac{\varphi}{2}, & \text{for } \varphi \neq k\pi, \end{cases}$$

which shows that for $\varphi = 0$ the functions $u(\varphi)$ have no finite derivative

$$\lim_{\varphi \to 0} \frac{du(\varphi)}{d\varphi} = \lim_{\varphi \to 0} \left(\frac{\ln \sin^2 \frac{\varphi}{2}}{2 \cos^2 \frac{\varphi}{2}} + 1 \right) = -\infty.$$

2.3 Existence Conditions for an Exponentially Stable Invariant Torus

Assume that the matriciant $\Omega_0^t(\varphi; A)$ of the system (2.1.31) satisfies the inequality

$$\|\Omega_0^t(\varphi; A)\| \leq K e^{-\gamma \tau}, \quad \text{for } t \geq 0 \tag{2.3.1}$$

for all $\varphi \in \mathcal{T}_m$ and some positive constants K and γ independent of φ. Let us show that in this case the function

$$G_0(\tau, \varphi) = \begin{cases} \Omega_\tau^0(\varphi; A), & \text{for } \tau \leq 0, \\ 0, & \text{for } \tau > 0, \end{cases} \tag{2.3.2}$$

is the Green function of the system (2.2.4).

Actually, inequality (2.3.1) implies

$$\|\Omega_\tau^0(\varphi; A)\| = \|\Omega_0^{-\tau}(\varphi_\tau(\varphi); A)\| \leq K e^{\gamma t}, \quad \text{for } \tau \leq 0.$$

Therefore, for function (2.3.2) the inequality

$$\|G_0(\tau, \varphi)\| \leq K e^{-\gamma |\tau|}, \quad \tau \in R,$$

is satisfied, ensuring the estimate (2.2.5) for the integral of $\|G_0(\tau, \varphi)\|$, with this function being the Green function of the system (2.2.4).

The set

$$x = u(\varphi) = Tf(\varphi), \quad \varphi \in \mathcal{T}_m, \tag{2.3.3}$$

where T is the operator (2.2.9) is the invariant torus of the system (2.1.3). We show that this torus is exponentially stable.

Designate the general solution of the system (2.1.31) by $x = x(t, \varphi, x_0) = \Omega_0^t(\varphi; A) x_0$. Incorporating properties of the matrix $\Omega_0^t(\varphi; A)$ for this solution yields the estimate

$$\begin{aligned} \|x(t, \varphi, x_0)\| &= \|\Omega_0^t(\varphi; A) x_0\| = \|\Omega_\tau^t(\varphi; A) \Omega_0^\tau(\varphi; A) x_0\| \\ &\leq \|\Omega_0^{t-\tau}(\varphi_\tau(\varphi); A)\| \, \|x(\tau, \varphi, x_0)\| \leq K e^{-\gamma(t-\tau)} \|x(\tau, \varphi, x_0)\|, \end{aligned} \tag{2.3.4}$$

valid for all $t \geq \tau$ and arbitrary $\varphi \in \mathcal{T}_m$. This estimate is sufficient for the invariant torus (2.3.3) to be exponentially stable. Consequently, if the matrix $\Omega_0^t(\varphi; A)$ satisfies the inequality (2.3.1), the system (2.1.3) has an exponentially stable invariant torus which is prescribed by

$$x = Tf(\varphi) = \int_{-\infty}^{0} \Omega_\tau^0(\varphi; A) f(\varphi_\tau(\varphi)) \, d\tau, \quad \varphi \in \mathcal{T}_m, \quad (2.3.5)$$

and satisfies the estimate

$$\|Tf\|_0 \leq K\gamma^{-1}\|f\|_0. \quad (2.3.6)$$

Introduce an index allowing us to verify if the inequality (2.3.1) is satisfied for the matriciant $\Omega_0^t(\varphi; A)$ of the system (2.1.31). We designate the set of $n \times n$-dimensional positive definite symmetric matrices $S(\varphi)$ that belong to the space $C^1(\mathcal{T}_m; a)$ by $\mathcal{N}_0$ and as above $\dot{S}(\varphi)$ denotes the matrix

$$\lim_{t \to 0} \frac{d}{dt} S(\varphi_t(\varphi)) = \dot{S}(\varphi).$$

We set

$$\inf_{S \in \mathcal{N}_0} \max_{\|x\|=1} \frac{\langle [S(\varphi)A(\varphi) + (\frac{1}{2}\dot{S}(\varphi)]x, x\rangle}{\langle S(\varphi)x, x\rangle} \leq -\beta(\varphi),$$

and determine β_0 by the relation

$$\beta_0 = \inf_{\varphi \in \mathcal{T}_m} \beta(\varphi). \quad (2.3.7)$$

Lemma 2.3.1 *For any $\mu > 0$ a number $K = K(\mu) > 0$ can be found such that the solution $x(t, \varphi, x_0) = \Omega_0^t(\varphi; A)x_0$ of the system (2.1.31) satisfies the inequality*

$$\|x(t, \varphi, x_0)\| \leq Ke^{-(\beta_0 - \mu)(t-\tau)}\|x(\tau, \varphi, x_0)\|,$$

for all $t \geq \tau$ and arbitrary $\varphi \in \mathcal{T}_m$.

This lemma implies that inequality (2.3.1) is valid whenever

$$\beta_0 > 0. \quad (2.3.8)$$

Proof If an infimum is reached on the set $\mathcal{N}_0$, then a matrix $S(\varphi) \in \mathcal{N}_0$ can be found such that

$$\max_{\|x\|=1} \frac{\langle [S(\varphi)A(\varphi) + (1/2)\dot{S}(\varphi)]x, x\rangle}{\langle S(\varphi)x, x\rangle} \leq -\beta(\varphi)$$

Then, however,

$$\left\langle S(\varphi_t(\varphi)) \frac{dx(t,\varphi,x_0)}{dt}, x(t,\varphi,x_0) \right\rangle = \langle S(\varphi_t(\varphi))A(\varphi_t(\varphi))x(t,\varphi,x_0), x(t,\varphi,x_0) \rangle,$$

and hence

$$\frac{d}{dt}\langle S(\varphi_t(\varphi))x(t,\varphi,x_0), x(t,\varphi,x_0) \rangle \leq -2\beta(\varphi_t(\varphi))\langle S(\varphi_t(\varphi))x(t,\varphi,x_0), x(t,\varphi,x_0) \rangle.$$

Integrating the last inequality yields, for all $t \geq \tau$,

$$\langle S(\varphi_t(\varphi))x(t,\varphi,x_0), x(t,\varphi,x_0) \rangle$$
$$\leq \exp\left\{-2\int_\tau^t \beta(\varphi_s(\varphi))\,ds\right\} \langle S(\varphi_\tau(\varphi))x(t,\varphi,x_0), x(t,\varphi,x_0) \rangle. \quad (2.3.9)$$

The periodicity and positive definiteness of the matrix $S(\varphi)$ provide the estimate

$$K_0 \langle x, x \rangle \leq \langle S(\varphi)x, x \rangle \leq K^0 \langle x, x \rangle$$

with some positive constants K_0 and K^0 independent of $\varphi \in \mathcal{T}_m$ and $x \in R^n$. Therefore, the inequality (2.3.9) leads to the estimate

$$K_0 \|x(t,\varphi,x_0)\|^2 \leq \exp\left\{-2\int_\tau^t \beta(\varphi_s(\varphi))\,ds\right\} K^0 \|x(\tau,\varphi,x_0)\|^2, \quad (2.3.10)$$

for $t \geq \tau$, which implies the desired estimate

$$\|x(t,\varphi,x_0)\| \leq K e^{-\beta_0(t-\tau)} \|x(\tau,\varphi,x_0)\|, \quad t \geq \tau,$$

where $K^2 = K^0/K_0$ does not depend on φ, x_0 and τ.

Provided that the infimum is not reached on the set $\mathcal{N}_0$, a sequence of the matrices $S_\nu(\varphi) \in \mathcal{N}_0$, $\nu = 1, 2, \ldots$, is found such that

$$\lim_{\nu \to \infty} \max_{\|x\|=1} \frac{\langle [S_\nu(\varphi)A(\varphi) + (1/2)\dot{S}_\nu(\varphi)]x, x \rangle}{\langle S_\nu(\varphi)x, x \rangle} \leq -\beta(\varphi).$$

For arbitrary μ a matrix $S_\mu(\varphi) \in N_0$ exists such that

$$\max_{\|x\|=1} \frac{\langle [S_\mu(\varphi)A(\varphi) + (1/2)\dot{S}_\mu(\varphi)]x, x \rangle}{\langle S_\mu(\varphi)x, x \rangle} \leq -\beta(\varphi) + \mu.$$

Replacing $S(\varphi)$ by $S_\mu(\varphi)$ and $\beta(\varphi)$ by $\beta(\varphi) - \mu$ in the above arguments we come to the required estimate for $x(t,\varphi,x_0)$.

This completes the proof of Lemma 2.3.1.

Corollary 2.3.1 *Inequality (2.3.8) is sufficient for the existence of an exponentially stable invariant torus of system (2.1.3) for an arbitrary function $f(\varphi) \in C'(\mathcal{T}_m)$.*

Let us prove the contrary. Let the invariant torus $x = 0$, $\varphi \in \mathcal{T}_m$ of the system of equations (2.2.4) be exponentially stable. Then

$$\|x(t,\varphi,x_0)\| \le K e^{-\gamma(t-\tau)} \|x(\tau,\varphi,x_0)\|, \quad t \ge \tau, \tag{2.3.11}$$

for arbitrary $\varphi \in \mathcal{T}_m$, $x_0 \in R^n$, $\tau \in R$ and some positive K and γ independent of φ, x_0 and τ.

Since $x(t,\varphi,x_0) = \Omega_0^t(\varphi;A)x_0$, then (2.3.11) implies the estimate (2.3.1) for the matrix $\Omega_0^t(\varphi;A)$. We apply this estimate to prove the inequality (2.3.8). Designate by $S(\varphi)$ the matrix determined by

$$S(\varphi) = \int_0^\infty (\Omega_0^\tau(\varphi;A))^* \Omega_0^\tau(\varphi;A)\, d\tau. \tag{2.3.12}$$

We show that $S \in \mathcal{N}_0$ and

$$\max_{\|x\|=1} \frac{\langle [S(\varphi)A(\varphi) + (1/2)\dot S(\varphi)]x, x\rangle}{\langle S(\varphi)x, x\rangle} \le -\beta,$$

where β is a positive constant. As $\|(\Omega_0^\tau(\varphi;A))^*\| = \|\Omega_0^\tau(\varphi;A)\|$, the inequality (2.3.1) ensures uniform convergence of the integral (2.3.12) and the matrix $S(\varphi)$ belonging to the space $C^0(\mathcal{T}_m)$. The matrix under the integral sign being symmetric in (2.3.12) implies that the matrix $S(\varphi)$ is symmetric. Further we have

$$\begin{aligned}
S(\varphi_t(\varphi)) &= \int_0^\infty (\Omega_t^{t+\tau}(\varphi;A))^* \Omega_t^{t+\tau}(\varphi;A)\, d\tau \\
&= (\Omega_t^0(\varphi;A))^* \int_t^\infty (\Omega_0^\tau(\varphi;A))^* \Omega_0^\tau(\varphi;A)\, d\tau\, \Omega_t^0(\varphi;A),
\end{aligned} \tag{2.3.13}$$

that proves differentiability in t for $t \in R$ of the function $S(\varphi_t(\varphi))$. This means that $S(\varphi) \in C'(\mathcal{T}_m; a)$. We show that the matrix $S(\varphi)$ is positive definite. In view of the inequality

$$\langle S(\varphi)x, x\rangle = \frac{\langle S(\varphi)x, x\rangle}{\langle x, x\rangle} \|x\|^2 \ge \min_{\varphi \in \mathcal{T}_m, \|\xi\|=1} \langle S(\varphi)x, x\rangle \|x\|^2 = \langle S(\varphi_0)\xi_0, \xi_0\rangle \|x\|^2,$$

where φ_0 and ξ_0 are fixed values from the set $\mathcal{T}_m$, $\|\xi\| = 1$, we set out the estimate

$$\langle S(\varphi)x, x \rangle \geq \langle S(\varphi_0)\xi_0, \xi_0 \rangle \|x\|^2$$

$$= \left\langle \int_0^\infty (\Omega_0^\tau(\varphi_0; A))^* \Omega_0^\tau(\varphi_0; A) \, d\tau \, \xi_0, \xi_0 \right\rangle \|x\|^2$$

$$= \int_0^\infty \langle \Omega_0^\tau(\varphi_0; A)\xi_0, \Omega_0^\tau(\varphi_0; A)\xi_0 \rangle \, d\tau \, \|x\|^2 \quad (2.3.14)$$

$$= \int_0^\infty \|\Omega_0^\tau(\varphi_0; A)\xi_0\|^2 \, d\tau \, \|x\|^2 = \lambda_1 \|x\|^2,$$

where $\lambda_1 = \int_0^\infty \|\Omega_0^\tau(\varphi_0; A)\xi_0\|^2 \, d\tau$.

Since $\Omega_0^t(\varphi_0; A)\xi_0$ is a solution of the homogeneous system of equations (2.1.31) taking the value $\xi_0 \neq 0$ for $t = 0$, then $\|\Omega_0^\tau(\varphi_0; A)\xi_0\|^2 > 0$ for all $\tau \in R$. Therefore $\lambda_1 > 0$, which that proves the positive definiteness of the matrix $S(\varphi)$. This also proves that $S(\varphi) \in \mathcal{N}_0$.

We establish $\dot{S}(\varphi)$. Differentiating (2.3.13) in t we obtain

$$\frac{dS(\varphi_t(\varphi))}{dt} = -A^*(\varphi_t(\varphi))S(\varphi_t(\varphi)) - I_n - S(\varphi_t(\varphi))A(\varphi_t(\varphi)).$$

Now it follows from

$$\left\langle \left[S(\varphi)A(\varphi) + \frac{1}{2}\dot{S}(\varphi)\right]x, x \right\rangle = \frac{1}{2} \langle [S(\varphi)A(\varphi) + A^*(\varphi)S(\varphi)$$

$$- A^*(\varphi)S(\varphi) - S(\varphi)A(\varphi) - I_n]x, x \rangle = -\frac{1}{2}\|x\|^2,$$

in view of (2.3.14), that

$$\max_{\|x\|=1} \frac{\langle [S(\varphi)A(\varphi) + (1/2)\dot{S}(\varphi)]x, x \rangle}{\langle S(\varphi)x, x \rangle} = \max_{\|x\|=1} \frac{-\|x\|^2}{2\langle S(\varphi)x, x \rangle}$$

$$= -\frac{1}{2} \frac{1}{\max_{\|x\|=1} \langle S(\varphi)x, x \rangle} = -\frac{1}{2\lambda_2},$$

where $\lambda_2 \geq \lambda_1 > 0$. Consequently,

$$\max_{\|x\|=1} \frac{\langle [S(\varphi)A(\varphi) + (1/2)\dot{S}(\varphi)]x, x \rangle}{\langle S(\varphi)x, x \rangle} \leq -\beta, \quad (2.3.15)$$

where $\beta = \frac{1}{2\lambda_2}$ is a positive constant. Inequality (2.3.15) leads to the estimate $\beta_0 \geq \beta > 0$ for (2.3.7).

The above discussion is summarized as follows.

Theorem 2.3.1 *Let the right-hand side of the system of equations (2.2.4) satisfy the conditions $a(\varphi) \in C^r_{\text{Lip}}(\mathcal{T}_m)$, $A(\varphi) \in C^r(\mathcal{T}_m)$, $r \geq 0$. The system of equations (2.1.3) has an exponentially invariant torus for any $f \in C^r(\mathcal{T}_m)$ if and only if (2.3.7) is positive.*

When the positiveness of β_0 is verified, S is chosen from $\mathcal{N}_0$ so that the derivative of $V = \langle S(\varphi)x, x \rangle$ along the solutions of the system (2.2.4) is negative definite. Since the mentioned derivative is

$$\dot{V} = 2\langle [S(\varphi)A(\varphi) + (1/2)\dot{S}(\varphi)]x, x \rangle = \langle \hat{S}(\varphi)x, x \rangle,$$

where $\hat{S}(\varphi)$ is a symmetric matrix

$$\hat{S}(\varphi) = S(\varphi)A(\varphi) + A^*(\varphi)S(\varphi) + \dot{S}(\varphi),$$

then $\dot{V}$ is negative definite, while $-\hat{S}(\varphi)$ is a positive definite matrix. Applying the well-known Sylvester criterion to verify the definiteness of the symmetric matrix positive we deduce on the positiveness of β_0 whenever a symmetric matrix $S(\varphi) \in C'(\mathcal{T}_m; a)$ is found, such that its main minors and the main minors of the matrix $\hat{S}(\varphi)$ corresponding to it are positive for all $\varphi \in \mathcal{T}_m$. These conditions are satisfied, particularly, when the main minors of the constant symmetric matrix S and the matrix $\hat{S}(\varphi) = -[SA(\varphi) + A^*(\varphi)S]$ are positive for all $\varphi \in \mathcal{T}_m$.

2.4 Uniqueness Conditions for the Green Function and its Properties

The necessary existence conditions for the invariant torus of the system (2.3.1) for an arbitrary function $f \in C^n(\mathcal{T}_m)$ imply by Theorem 2.2.1 that the absence of a non-degenerate invariant torus of the homogeneous system of equations (2.2.4) is a necessary condition for the existence of the Green function of this system. We show that the absence of a degenerate invariant torus of the system (2.2.4) ensures the uniqueness of the Green function.

Theorem 2.4.1 *Let $a(\varphi) \in C_{\text{Lip}}(\mathcal{T}_m)$, $A(\varphi) \in C^0(\mathcal{T}_m)$ and the system (2.2.4) have the Green function $G_0(\tau, \varphi)$. The torus $x = 0$, $\varphi \in \mathcal{T}_m$, is a unique invariant torus of the system (2.2.4), if and only if this system has no other Green function, but $G_0(\tau, \varphi)$.*

Proof We assume that the system (2.2.4) has a Green function $G_1(\tau, \varphi)$ other than $G_0(\tau, \varphi)$. We set

$$u_0(\varphi) = \int_{-\infty}^{\infty} G_1(\tau, \varphi) f(\varphi_\tau(\varphi)) \, d\tau - \int_{-\infty}^{\infty} G_0(\tau, \varphi) f(\varphi_\tau(\varphi)) \, d\tau, \qquad (2.4.1)$$

where $f(\varphi)$ is an arbitrary function from $C^0(\mathcal{T}_m)$. Evidently the set $x = u_0(\varphi)$, $\varphi \in \mathcal{T}_m$ determines the invariant torus of system the (2.2.4). Let us prove that

$$u_0(\varphi) \not\equiv 0, \qquad (2.4.2)$$

for some function $f(\varphi) \in C^0(\mathcal{T}_m)$.

By definition of the Green function and formula (2.4.1)

$$u_0(\varphi) = \int_{-\infty}^{\infty} \Omega_\tau^0(\varphi; A) R(\varphi_\tau(\varphi)) f(\varphi_\tau(\varphi)) \, d\tau,$$

where $R(\varphi)$ is a non-zero matrix from $C^0(\mathcal{T}_m)$. In order to prove inequality (2.4.2) it is sufficient to show that the operator T_o determined by

$$\begin{aligned}T_0 f(\varphi) &= \int_{-\infty}^{\infty} G_1(\tau, \varphi) f(\varphi_\tau(\varphi)) \, d\tau - \int_{-\infty}^{\infty} G_0(\tau, \varphi) f(\varphi_\tau(\varphi)) \, d\tau \\ &= \int_{-\infty}^{\infty} \Omega_\tau^0(\varphi; A) R(\varphi_\tau(\varphi)) f(\varphi_\tau(\varphi)) \, d\tau,\end{aligned} \qquad (2.4.3)$$

on the space of functions $C^0(\mathcal{T}_m)$ is not a zero-operator of this space. Suppose on the contrary, i.e.

$$T_0 f \equiv 0, \quad \forall f(\varphi) \in C^0(\mathcal{T}_m). \qquad (2.4.4)$$

We show that (2.4.4) should imply

$$R(\varphi) \equiv 0, \quad \forall \varphi \in \mathcal{T}_m, \qquad (2.4.5)$$

which contradicts the assumption that $G_0(\tau, \varphi) \not\equiv G_1(\tau, \varphi)$, implying the inequality $R(\varphi) \neq 0$, $\forall \varphi \in \mathcal{T}_m$.

Passing to the proof of equality (2.4.5) we first consider the value of the matrix $R(\varphi)$ at the points belonging to the trajectories of the system (2.2.4) which are periodic on $\mathcal{T}_m$. Let φ_0 be such a point, $\varphi = \varphi_{t+T}(\varphi_0) = \varphi_t(\varphi_0) \bmod 2\pi$, $t \in R$, is a trajectory corresponding to it, T is its period. By the Floquet–Lyapunov theory

$$\Omega_0^t(\varphi_0; A) = \Phi(t) e^{At},$$

where $\Phi(t)$ is a periodic matrix with the period T, A is a constant matrix. Then, however

$$\Omega_\tau^0(\varphi_0; A) = (\Omega_0^\tau(\varphi_0; A))^{-1} = e^{-A\tau} \Phi^{-1}(\tau),$$

so that

$$T_0 f(\varphi_0) = \int_{-\infty}^{\infty} e^{-A\tau} \Phi^{-1}(\tau) R(\varphi_\tau(\varphi_0)) f(\varphi_\tau(\varphi_0))\, d\tau$$

$$= \sum_{p=-\infty}^{\infty} \int_{pT}^{(p+1)T} e^{-A\tau} \Phi^{-1}(\tau) R(\varphi_\tau(\varphi_0)) f(\varphi_\tau(\varphi_0))\, d\tau$$

$$= \sum_{p=-\infty}^{\infty} \int_0^T e^{-ApT} e^{-A\tau} \Phi^{-1}(\tau) R(\varphi_\tau(\varphi_0)) f(\varphi_\tau(\varphi_0))\, d\tau = \sum_{p=-\infty}^{\infty} \left(e^{-AT}\right)^p B_f,$$

where

$$B_f = \int_0^T e^{-A\tau} \Phi^{-1}(\tau) R(\varphi_\tau(\varphi_0)) f(\varphi_\tau(\varphi_0))\, d\tau.$$

Since the series $\sum_{p=-\infty}^{\infty} \left(e^{-AT}\right)^p$ diverges for any matrix A, then $T_0 f(\varphi_0) = 0$ only if

$$B_f = 0. \tag{2.4.6}$$

We consider the equality (2.4.6). The case when φ_0 is a rest point of the dynamical system on $\mathcal{T}_m$ and the case when φ_0 is not such a point are distinguished. In the first case, equality (2.4.6) takes place for any number $T > 0$. The latter is possible only when

$$e^{-A\tau} \Phi^{-1}(\tau) R(\varphi_\tau(\varphi_0)) f(\varphi_\tau(\varphi_0))\, d\tau \equiv 0.$$

Then, however $R(\varphi_\tau(\varphi_0)) f(\varphi_\tau(\varphi_0)) \equiv 0$, which is possible in view of the identity $\varphi_\tau(\varphi_0) \equiv \varphi_0$ and the arbitrariness of the function f only provided that

$$R(\varphi_0) = 0. \tag{2.4.7}$$

In the second case the number $T > 0$ is found, which is a real period of the trajectory periodic on the torus, such that

$$\varphi_{t+T}(\varphi_0) = \varphi_t(\varphi_0) \bmod 2\pi, \quad \varphi_{t+\tau}(\varphi) \neq \varphi_t(\varphi_0) \bmod 2\pi,$$

for any $0 < \tau < T$. The point $\psi_t = \nu t$, $\nu = 2\pi/T$, of the circle $\mathcal{T}_1$ is set to correspond one-to-one to the point $\varphi_t(\varphi_0)$ of the torus $\mathcal{T}_m$. This correspondence enables the function $F \in C^0(\mathcal{T}_1)$ related to F by $f(\varphi_t(\varphi_0)) = F(\nu t)$, $t \in R$ to be put in a correspondence with the arbitrary function $f \in C^0(\mathcal{T}_m)$. In this case equality (2.4.6) becomes

$$\int_0^T e^{-A\tau}\,\Phi^{-1}(\tau)R(\varphi_\tau(\varphi_0))F(\nu\tau)\,d\tau = 0. \tag{2.4.8}$$

Since the equality (2.4.8) is satisfied for an arbitrary function F from $C^0(\mathcal{T}_1)$, it is only possible when

$$e^{-A\tau}\,\Phi^{-1}(\tau)R(\varphi_\tau(\varphi_0)) \equiv 0.$$

This results, with regard to the non-degeneracy of the matrices $e^{-A\tau}$ and $\Phi^{-1}(\tau)$, in the identity

$$R(\varphi_t(\varphi_0)) \equiv 0, \quad \tau \in [0,T]. \tag{2.4.9}$$

In fact, we set $F(\nu t) = \{0, \ldots, F_j(\nu t), \ldots, 0\}$, $r_{ij}(t) = \{e^{-A\tau}\,\Phi^{-1}(\tau)R(\varphi_\tau(\varphi_0))\}_{ij}$, $i, j = 1, \ldots, n$, and rewrite (2.4.8) for the chosen function $F(\nu t)$:

$$\int_0^T r_{ij}(\tau) F_j(\nu\tau)\,d\tau = 0. \tag{2.4.10}$$

Replacing F_j in (2.4.10) by the function from $C^0(\mathcal{T}_1)$ determined for $\psi \in [0, 2\pi]$ by $F_j(\psi) = \bar{r}_{ij}(\psi/\nu)f_0(\psi)$, where $\bar{r}_{ij}$ is a complex function conjugated with r_{ij}, f_0 is a function from $C^0(\mathcal{T}_1)$ equal to zero for $\psi = 0$ and larger than zero for $\psi \in (0, 2\pi)$, yields

$$\int_0^T |r_{ij}(\tau)|^2 f_0(\nu\tau)\,d\tau = 0.$$

This equality is possible provided that $|r_{ij}(\tau)|^2 f_0(\nu\tau) \equiv 0$, $\tau \in (0,T)$, which is equivalent to

$$r_{ij}(\tau) \equiv 0, \quad \tau \in [0,T]. \tag{2.4.11}$$

Since r_{ij} is an arbitrary element of the matrix $e^{-A\tau}\,\Phi^{-1}(\tau)R(\varphi_\tau(\varphi_0))$, then (2.4.11) is equivalent to (2.4.9). This proves the equality (2.4.7) for any point φ_0 belonging to the trajectory of the system (2.2.4) periodic on $\mathcal{T}_m$.

We consider the values of the matrix $R(\varphi)$ at the other points of the torus $\mathcal{T}_m$. Let φ_0 be such a point, $\varphi = \varphi_t(\varphi_0)$, $t \in R$ is a trajectory corresponding to it. The uniform boundedness of the integrals

$$\int_{-\infty}^{\infty} \|G_i(\tau, \varphi)\|\,d\tau, \quad i = 1, 0,$$

implies the uniform boundedness of the integral

$$\int_{-\infty}^{\infty} \|\Omega_\tau^0(\varphi; A) R(\varphi_t(\varphi_0))\| \, d\tau.$$

Then a monotone vanishing sequence of positive numbers ε_n, $\varepsilon_{n+1} < \varepsilon_n$, $\lim\limits_{n\to\infty} \varepsilon_n = 0$, and a monotone sequence of positive numbers increasing to $+\infty$ t_n, $t_{n+1} > t_n$, $\lim\limits_{n\to\infty} t_n = +\infty$ can be found such that

$$\int_{-\infty}^{-t_n} \|\Omega_\tau^0(\varphi_0; A) R(\varphi_\tau(\varphi_0))\| \, d\tau + \int_{t_n}^{\infty} \|\Omega_\tau^0(\varphi_0; A) R(\varphi_\tau(\varphi_0))\| \, d\tau < \varepsilon_n.$$

As $T_0 f(\varphi_0) = 0$, then

$$\begin{aligned}
&\left\| \int_{-t_n}^{t_n} \Omega_\tau^0(\varphi_0; A) R(\varphi_\tau(\varphi_0)) f(\varphi_\tau(\varphi_0)) \, d\tau \right\| \\
&\leq \left\| \int_{-\infty}^{\infty} \Omega_\tau^0(\varphi_0; A) R(\varphi_\tau(\varphi_0)) f(\varphi_\tau(\varphi_0)) \, d\tau \right\| \\
&\quad + \int_{-\infty}^{-t_n} \|\Omega_\tau^0(\varphi_0; A) R(\varphi_\tau(\varphi_0))\| \, d\tau \, \|f\|_0 \\
&\quad + \int_{t_n}^{\infty} \|\Omega_\tau^0(\varphi_0; A) R(\varphi_\tau(\varphi_0))\| \, d\tau \, \|f\|_0 \leq \varepsilon_n \|f\|_0.
\end{aligned} \quad (2.4.12)$$

The arch $\varphi = \varphi_t(\varphi_0)$, $-t_n \leq t \leq t_n$ of the trajectory on $\mathcal{T}_m$ has no self-intersections, and so, for any continuous function $F(t)$, determined for $t \in R$, a function $f_n \in C^0(\mathcal{T}_m)$ is found which is related to F by $f_n(\varphi_t(\varphi_0)) = F(t)$ for $-t_n \leq t \leq t_n$.

Assume that $R(\varphi_t(\varphi_0)) \not\equiv 0$ for $t \in R$. Then $\Omega_0^t(\varphi_0; A) R(\varphi_t(\varphi_0)) \not\equiv 0$ for $t \in R$. Let $r_{ij} = \{\Omega_t^0(\varphi_0; A) R(\varphi_t(\varphi_0))\}_{ij}$ be the element of the matrix $\Omega_t^0(\varphi_0; A) R(\varphi_t(\varphi_0))$ for which $r_{ij} \not\equiv 0$, $t \in R$. We set

$$\hat{r}_{ij}(t) = \begin{cases} \bar{r}_{ij}(t), & \text{for } |r_{ij}(t)| \leq 1, \\ \operatorname{sign} \bar{r}_{ij}(t), & \text{for } |r_{ij}(t)| > 1. \end{cases}$$

Evidently, the function $\hat{r}_{ij}(t)$ is continuous for $t \in R$ and $\max\limits_{t \in R} |\hat{r}_{ij}(t)| \leq 1$.

We take the scalar functions $f_n \in C^0(\mathcal{T}_m)$ so that

$$f_n(\varphi_t(\varphi_0)) = \hat{r}_{ij}(t) \quad \text{for} \quad -t_n \leq t \leq t_n, \quad \|f_n\|_0 \leq 1,$$

and consider the inequality (2.4.12) for the functions $f_n(\varphi)e_j$, where $e_j = \{0, \ldots, 1, \ldots, 0\}$ is the jth unique basis vector. Then we have

$$\left\| \int_{-t_n}^{t_n} \Omega_\tau^0(\varphi_0; A) R(\varphi_\tau(\varphi_0)) f(\varphi_\tau(\varphi_0)) e_j \, d\tau \right\| = \left| \int_{-t_n}^{t_n} r_{ij}(\tau) \hat{r}_{ij}(\tau) \, d\tau \right|$$

$$= \int_{-t_n}^{t_n} r_{ij}(\tau) \hat{r}_{ij}(\tau) \, d\tau \leq \varepsilon_n. \tag{2.4.13}$$

Since $r_{ij}(\tau)\hat{r}_{ij}(\tau) \neq 0$ for $t \in R$, then

$$\int_{-t_N}^{t_N} r_{ij}(\tau)\hat{r}_{ij}(\tau) \, d\tau = \delta_1 > 0$$

for some N sufficiently large. Then, however, non-positiveness of the function $r_{ij}(t)\hat{r}_{ij}(t)$ yields

$$\int_{-t_n}^{t_n} r_{ij}(\tau)\hat{r}_{ij}(\tau) \, d\tau \geq \delta_1 \quad \forall n \geq N,$$

which contradicts the inequality (2.4.13) and the condition $\lim_{n \to \infty} \varepsilon_n = 0$. This contradiction caused by the assumption that $r_{ij} \neq 0$ for $t \in R$ proves that

$$\Omega_t^0(\varphi_0; A) R(\varphi_t(\varphi_0)) \equiv 0, \quad \tau \in R. \tag{2.4.14}$$

This equality proves relation (2.4.7) for any point $\varphi_0 \in \mathcal{T}_m$ which does not belong to the trajectory of system (2.2.4) periodic on $\mathcal{T}_m$.

Summing up the above arguments we make sure that equality (2.4.5) is proved. As has been noted, the latter contradicts the assumption on the Green function being non unique. Thus, the non-uniqueness of the Green function necessarily implies the existence of the invariant torus of the system (2.2.4) other than the trivial $x = 0$, $\varphi \in \mathcal{T}_m$.

To complete the proof of Theorem 2.4.1 the uniqueness of the Green function is assumed not to ensure the uniqueness of the invariant torus of the system (2.2.4). We first establish two important equalities resulting from the assumption on the uniqueness of the Green function.

Lemma 2.4.1 *Let the function $G_0(\tau,\varphi)$ be the sole Green function of the system (2.2.4). Then the matrix $C(\varphi) = G_0(-0,\varphi)$ satisfies the equalities*

$$C(\varphi_t(\varphi)) \equiv \Omega_0^t(\varphi;A)C(\varphi)\Omega_t^0(\varphi;A), \quad \forall\, t \in R, \quad \varphi \in \mathcal{T}_m; \qquad (2.4.15)$$

$$C^2(\varphi) \equiv C(\varphi), \quad \forall\, \varphi \in \mathcal{T}_m. \qquad (2.4.16)$$

Proof Actually, let the first one of the relations not be valid, i.e. for some $t_1 \neq 0$

$$C(\varphi_{t_1}(\varphi)) \neq \Omega_0^{t_1}(\varphi;A)C(\varphi)\Omega_{t_1}^0(\varphi;A), \quad \varphi \in \mathcal{T}_m.$$

We set

$$R(\varphi) = \Omega_{t_1}^0(\varphi;A)C(\varphi_{t_1}(\varphi)) - C(\varphi)\Omega_{t_1}^0(\varphi;A)$$

and consider the function

$$G_1(\tau,\varphi) = G_0(\tau,\varphi) + \Omega_\tau^0(\varphi;A)R(\varphi_\tau(\varphi)).$$

For this function

$$\int_{-\infty}^{\infty} \|G_1(\tau,\varphi)\|\, d\tau \leq \int_{-\infty}^{\infty} \|G_0(\tau,\varphi)\|\, d\tau + \int_{-\infty}^{\infty} \|\Omega_\tau^0(\varphi;A)R(\varphi_\tau(\varphi))\|\, d\tau$$

$$\leq K + \int_{-\infty}^{\infty} \|\Omega_{\tau+t_1}^0(\varphi;A)C(\varphi_{\tau+t_1}(\varphi)) - \Omega_\tau^0(\varphi;A)C(\varphi_\tau(\varphi))\Omega_{t_1}^0(\varphi_\tau(\varphi);A)\|\, d\tau$$

$$\leq K + \int_{-\infty}^{-t_1} \|\Omega_{\tau+t_1}^0(\varphi;A)C(\varphi_{\tau+t_1}(\varphi))\|\, d\tau$$

$$+ \int_{-t_1}^{\infty} \|\Omega_{\tau+t_1}^0(\varphi;A)[I_n - C(\varphi_{\tau+t_1}(\varphi))]\|\, d\tau + \int_{-\infty}^{-t_1} \Omega_\tau^0(\varphi;A)C(\varphi_\tau(\varphi))\|\, d\tau\, \|\Omega_{t_1}^0\|_0$$

$$+ \int_{-t_1}^{\infty} \|\Omega_{\tau+t_1}^0(\varphi;A) - \Omega_\tau^0(\varphi;A)C(\varphi_\tau(\varphi))\Omega_{t_1}^0(\varphi_\tau(\varphi);A)\|\, d\tau$$

$$\leq 2K + \Bigg\{ \int_{-\infty}^{-t_1} \|\Omega_\tau^0(\varphi;A)C(\varphi_\tau(\varphi))\|\, d\tau$$

$$+ \int_{-t_1}^{\infty} \|\Omega_\tau^0(\varphi;A)[I_n - C(\varphi_\tau(\varphi))]\|\, d\tau \Bigg\} \|\Omega_{t_1}^0\|_0$$

$$\leq 2K + K\|\Omega^0_{t_1}(\varphi; A)\|_0$$

$$+ \left\{ \left| \int_{-t_1}^{0} \|\Omega^0_\tau(\varphi; A)C(\varphi_\tau(\varphi))\|\, d\tau \right| + \left| \int_{0}^{-t_1} \|\Omega^0_\tau(\varphi; A)[I_n - C(\varphi_\tau(\varphi))]\|\, d\tau \right| \right\} \|\Omega^0_{t_1}\|_0$$

$$\leq 2K + (K + \bar{K})\|\Omega^0_{t_1}\|_0 = K_1.$$

In terms of these relations the inequality

$$\int_{-\infty}^{\infty} \|G_1(\tau, \varphi)\|\, d\tau \leq K_1,$$

is satisfied uniformly in $\varphi \in \mathcal{T}_m$. Then the function $G_1(\tau, \varphi)$ satisfies all the conditions which the Green function of system (2.2.4) must satisfy. This contradicts the assumption on the uniqueness of the Green function. The contradiction proves the equality (2.4.15).

Now let $C^2(\varphi) \neq C(\varphi)$, $\varphi \in \mathcal{T}_m$. The function

$$G_1(\tau, \varphi) = G_0(\tau, \varphi) + \Omega^0_\tau(\varphi; A)[C^2(\varphi_\tau(\varphi)) - C(\varphi_\tau(\varphi))]$$

then satisfies the chain of inequalities

$$\int_{-\infty}^{\infty} \|G_1(\tau, \varphi)\|\, d\tau \leq K + \int_{-\infty}^{0} \|\Omega^0_\tau(\varphi; A)C(\varphi_\tau(\varphi))\|\, d\tau\, \|C - I_n\|_0$$

$$+ \int_{0}^{\infty} \|\Omega^0_\tau(\varphi; A)[I_n - C(\varphi_\tau(\varphi))]\|\, d\tau\, \|C\|_0 \leq \tilde{K}_1,$$

due to which $G_1(\tau, \varphi)$ is the Green function of the system (2.2.4) as well. The latter contradicts the assumption on the uniqueness of the Green function and proves the equality (2.4.16).

We employ (2.4.15) to complete the proof of Theorem 2.4.1. Let, in spite of the uniqueness of the Green function, the torus

$$x = u_0(\varphi), \quad \varphi \in \mathcal{T}_m, \tag{2.4.17}$$

be an invariant torus of the system (2.2.4) other than the trivial one. Then

$$u_0(\varphi) \not\equiv 0, \quad \varphi \in \mathcal{T}_m. \tag{2.4.18}$$

Consider the function $u_1(\varphi) \in C'(\mathcal{T}_m; a)$ which determines the invariant torus $x = u_1(\varphi)$, $\varphi \in \mathcal{T}_m$ of the system

$$\frac{d\varphi}{dt} = a(\varphi), \quad \frac{dx}{dt} = A(\varphi)x + u_0(\varphi). \tag{2.4.19}$$

The invariance of the torus (2.4.17) of the homogeneous system of equations corresponding to (2.4.19) implies $u_0(\varphi_t(\varphi)) = \Omega_0^t(\varphi; A)u_0(\varphi)$. Therefore

$$u_1(\varphi) = \int_{-\infty}^{\infty} G_0(\tau, \varphi) u_0(\varphi_\tau(\varphi)) \, d\tau = \int_{-\infty}^{\infty} G_0(\tau, \varphi) \Omega_0^\tau(\varphi; A) u_0(\varphi) \, d\tau$$

$$= \int_{-\infty}^{0} \Omega_\tau^0(\varphi; A) C(\varphi_\tau(\varphi)) \Omega_0^\tau(\varphi; A) u_0(\varphi) \, d\tau \qquad (2.4.20)$$

$$- \int_{0}^{\infty} \Omega_\tau^0(\varphi; A) [I_n - C(\varphi_\tau(\varphi))] \Omega_0^\tau(\varphi; A) u_0(\varphi) \, d\tau,$$

which, leads, in view of (2.4.15), to

$$u_1(\varphi) = \int_{-\infty}^{0} C(\varphi) u_0(\varphi) \, d\tau - \int_{0}^{\infty} [I_n - C(\varphi)] u_0(\varphi) \, d\tau. \qquad (2.4.21)$$

The latter is possible provided that

$$C(\varphi) u_0(\varphi) \equiv 0, \quad C(\varphi) u_0(\varphi) \equiv u_0(\varphi)$$

simultaneously. Then $u_0(\varphi) \equiv 0$, $\forall \varphi \in \mathcal{T}_m$, which contradicts the inequality (2.4.18).

The main properties of the Green function following from the assumption on its uniqueness are specified by (2.4.15) and (2.4.16). The former means that the matrix $C(\varphi)$ belongs to the space $C'(\mathcal{T}_m; a)$ and is a solution to the matrix equation

$$\frac{d}{dt} C = A(\varphi) C - CA;$$

the latter shows the matrix $C(\varphi)$ to be a projector.

We demonstrate that the equality (2.4.15) is a characteristic property of the uniqueness of the Green function. Namely, if the system (2.2.4) has more than one Green function, then neither of them satisfies the equality (2.4.15). Let us formulate the above in the form of a separate assertion.

Lemma 2.4.2 *Let $a(\varphi) \in C_{\text{Lip}}(\mathcal{T}_m)$, $A(\varphi) \in C^0(\mathcal{T}_m)$ and the system (2.2.4) have an invariant torus other than the trivial $x = 0$, $\varphi \in \mathcal{T}_m$. Then neither of the Green functions of system (2.2.4) satisfies the relation (2.4.15).*

Proof We consider the torus (2.4.17) to be an invariant torus of the system (2.2.4) other than the trivial one. Let $G_0(\tau, \varphi)$ be the Green function of the system (2.2.4) satisfying (2.4.15). Then the system (2.4.19) has an invariant torus determined by (2.4.20) which by (2.4.21) can only be zero if $u_1(\varphi) \equiv 0$, $\varphi \in \mathcal{T}_m$. Moreover, $\dot{u}_1(\varphi) \equiv 0$, $\varphi \in \mathcal{T}_m$ and the second equation (2.4.19) implies $u_0(\varphi) \equiv 0$, $\varphi \in \mathcal{T}_m$. The latter contradicts the non-triviality of the torus (2.4.17). The contradiction proves that the function $G_0(\tau, \varphi)$ cannot satisfy the relation (2.4.15).

We note, finally, that the absence of degenerate invariant tori of the system (2.2.4) is a sufficient condition of the uniqueness of the Green function of this system. Therefore, the quasiperiodic system, i.e. system (2.2.4) can only have one Green function for $a(\varphi) = \omega = \text{const}$, such that the relations (2.4.15) and (2.4.16) are satisfied.

2.5 Sufficient Conditions for Exponential Dichotomy of the Invariant Torus

Definition 2.5.1 The invariant torus $x = 0$ of the system (2.2.4) is called *exponential dichotomous* as well as the system (2.2.4) whenever the Euclidean space R^n can be represented as a direct sum of two subspaces E^+ and E^- of mutually complementary dimensions r and $n - r$ for any $\varphi_0 \in \mathcal{T}_m$, so that any solution $\varphi_t(\varphi_0)$, $x_t(\varphi_0, x_0)$ of the system (2.2.4), where $x_0 \in E^+$ satisfies the inequality

$$\|x_t(\varphi_0, x_0)\| \leq K e^{-\gamma(t-\tau)} \|x_\tau(\varphi_0, x_0)\|, \quad t \geq \tau,$$

and any solution $\varphi_t(\varphi_0)$, $x_t(\varphi_0, x_0)$ of the system (2.2.4), where $x_0 \in E^-$, satisfies the inequality

$$\|x_t(\varphi_0, x_0)\| \leq K_1 e^{\gamma_1(t-\tau)} \|x_\tau(\varphi_0, x_0)\|, \quad t \leq \tau,$$

for arbitrary $\tau \in R$, $\varphi_0 \in \mathcal{T}_m$ and some positive constants K, γ, K_1 and γ_1, independent of φ_0, x_0 and τ.

In order to determine the conditions of exponential dichotomy of the trivial torus $x = 0$, $\varphi \in \mathcal{T}_m$ of the system (2.2.4) it is reasonable to apply the method of sign-variable Lyapunov functions which are quadratic forms with respect to variables x. This leads to a result giving sufficient conditions for the exponential dichotomy of the invariant torus.

Theorem 2.5.1 *Let $a(\varphi) \in C_{\text{Lip}}(\mathcal{T}_m)$, $A(\varphi) \in C^0(\mathcal{T}_m)$ and there exist non-degenerate symmetric matrix $S(\varphi) \in C'(\mathcal{T}_m; a)$ such that the matrix*

$$\hat{S}(\varphi) = \dot{S}(\varphi) + S(\varphi)A(\varphi) + A^*(\varphi)S(\varphi)$$

is negative definite for all $\varphi \in \mathcal{T}_m$. Then the trivial torus $x = 0$, $\varphi \in \mathcal{T}_m$ of the system (2.2.4) is exponentially dichotomous.

Proof We consider that the matrix $S(\varphi)$ has r positive and $n - r$ negative eigenvalues. Let

$$S_t = (\Omega_0^t(\varphi; A))^* S(\varphi_t(\varphi)) \Omega_0^t(\varphi; A),$$

and K_t be a set of points x, the space R^n satisfying the inequality

$$\langle S_t x, x \rangle \geq 0, \tag{2.5.1}$$

where φ and t are fixed values $\varphi \in \mathcal{T}_m$, $t \geq 0$. We set $K^+ = \bigcap\limits_{t \geq 0} K_t$ and prove that K^+ is an r-dimensional linear subspace R^n standing for the space E^+ in the definition of the dichotomy for a trivial torus of the system (2.2.4). In view of the expressions of the matrices S_t and $\hat{S}(\varphi)$ we find

$$\frac{d}{dt} S_t = (\Omega_0^t(\varphi; A))^* \hat{S}(\varphi_t(\varphi; A)) \Omega_0^t(\varphi; A).$$

Therefore, for every $x \in R^n$, $x \neq 0$,

$$\frac{d}{dt} \langle S_t x, x \rangle = \langle \hat{S}(\varphi_t(\varphi)) \Omega_0^t(\varphi; A) x, \Omega_0^t(\varphi; A) x \rangle = \langle \hat{S}(\varphi_t(\varphi)) x_t(\varphi, x), x_t(\varphi, x) \rangle,$$

which results, because of the negative definiteness of the matrix $\hat{S}(\varphi)$ in the inequality

$$\langle S_t x, x \rangle < \langle S_\tau x, x \rangle \quad \forall t > \tau, \quad \text{and} \quad \forall \tau \in R. \tag{2.5.2}$$

This inequality implies the inclusion $K_t \subset K_\tau$ for $t > \tau$ and the fact that the intersection of the boundary ∂K_t of the set K_t with the boundary ∂K_τ of the set K_τ contains a zero point only means that $\partial K_t \cap \partial K_\tau = \{0\}$.

Thus, K^+ is the intersection of embedded sets K_t. We show that K^+ contains r linearly independent vectors. Since the matrix S_t is continuous in t and non-degenerate, it has the same number of positive eigenvalues as the matrix $S_0 = S(\varphi)$. Being symmetric, the matrix S_t is orthogonal similar to its *Jordan form* J_t which can be represented as

$$J_t = \text{diag}\{D_1(t), -D_2(t)\},$$

where $D_1(t) = \text{diag}\{\lambda_1(t), \ldots, \lambda_r(t)\}$, $D_2(t) = \text{diag}\{\lambda_{r+1}(t), \ldots, \lambda_n(t)\}$, $\lambda_j(t)$ are positive values. Then

$$S_t = O^*(t) J_t O(t) = O_1^*(t) \, \text{diag}\{I_r, -I_{n-r}\} O_1(t), \tag{2.5.3}$$

where $O(t)$ is an orthogonal matrix, $O_1(t) = O(t) \, \text{diag}\{D_1^{1/2}(t), D_2^{1/2}(t)\}$.

For a linear transformation of the coordinates by the formula

$$x = By, \quad \det B \neq 0,$$

the quadratic form $\langle S_t x, x \rangle$ is evidently transformed into the form $\langle \tilde{S}_t y, y \rangle$ with the matrix $\tilde{S}_t = B^* S_t B$. So, an appropriate choice of the system of coordinates in R^n can reduce the quadratic form with the matrix $S_0 = S(\varphi)$ to the form with the matrix $\tilde{S}_0 = \text{diag}\{I_r, -I_{n-r}\}$. The system of coordinates x can be considered to be already taken so that the quadratic form $\langle S(\varphi)x, x\rangle$ in it is of the form $x_1^2 + \cdots + x_r^2 - x_{r+1}^2 - \cdots - x_n^2$. In other words, without loss of generality, one can consider that $S(\varphi) = \text{diag}\{I_r, -I_{n-r}\}$ (see Kulik [7] and Letov [1]).

Consider now the cross-section of the set K_t by the plane P_0 specified as

$$x_1 = 0, \quad \ldots, \quad x_j = 1, \quad \ldots, \quad x_r = 0. \tag{2.5.4}$$

The set $K_0 \cap P_0$ consists, evidently, of the points satisfying the inequality $x_{r+1}^2 + \cdots + x_n^2 \leq 1$ and is a closed ball $\mathcal{M}_0$ in the $n-r$-dimensional subspace R^n as the plane P_0 is.

The set $K_t \cap P_0$ consists of the points satisfying the inequality (2.5.1) and (2.5.4) simultaneously. We show that the set $K_t \cap P_0$ is non-empty. For this we represent inequality (2.5.1), specifying K_t in the system of coordinates $y = O(t)x$, where $O(t)$ is the matrix from the representation of S_t, in the form of (2.5.3)

$$\langle S_t x, x\rangle = \langle J_t O(t)x, O(t)x\rangle = \langle J_t y, y\rangle$$
$$= \lambda_1(t)y_1^2 + \cdots + \lambda_r(t)y_r^2 - \lambda_{r+1}(t)y_{r+1}^2 - \cdots - \lambda_n(t)y_n^2 \geq 0,$$

or

$$\lambda_1(t)y_1^2 + \cdots + \lambda_r(t)y_r^2 \geq \lambda_{r+1}(t)y_{r+1}^2 + \cdots + \lambda_n(t)y_n^2. \tag{2.5.5}$$

Since the functions $\lambda_j(t)$ are positive, then the values $y_1 = c_1, \ldots, y_r = c_r, y_{r+1} = \cdots = y_n = 0$, satisfy the inequality (2.5.5), where $c_1, \ldots, c_r$ are arbitrary constants. Therefore the set K_t contains the subspace R_r determined by

$$x = O(t)\begin{bmatrix} C \\ 0 \end{bmatrix}, \quad C = \text{col}\{c_1, \ldots, c_r\},$$

or, what is the same, by

$$x = \begin{bmatrix} O_1(t) \\ O_{21}(t) \end{bmatrix} C,$$

where O_1 is an $r \times r$-dimensional matrix and O_{21} is an $(n-r) \times r$-dimensional matrix formed by the first r columns of the matrix $O(t)$. With regard to the inclusion $K_t \subset K_0$ we get that the linear subspace R_r belongs to the set

$$x_1^2 + \cdots + x_r^2 \geq x_{r+1}^2 + \cdots + x_n^2.$$

Then the matrix $O_1(t)$ is non-degenerate: let $\det O_1(t) \neq 0$, since otherwise the point R^r determined by the zero value of C taken from $O_1 C = 0$, which in view of $O(t)\begin{bmatrix} C \\ 0 \end{bmatrix} \neq 0$ yields the inequality $O_{21}(t)C \neq 0$, does not belong to the set

K_0: $0 \geq \sum_{j=1}^{n-r}(O_{21}(t)C)_j^2$, which is impossible. The non-degeneracy of the matrix $O_1(t)$ ensures the existence of the vector $O_1^{-1}(t)e_j$, where $e_j = (0,\ldots,1,\ldots,0)$ is an r-dimensional jth basis vector and by the same token the existence of the point

$$x = \begin{bmatrix} O_1(t) \\ O_{21}(t) \end{bmatrix} O_1^{-1}(t)e_j = \begin{bmatrix} e_j \\ O_{21}(t)O_1^{-1}(t)e_j \end{bmatrix},$$

belonging to the intersection of planes P_0 and R^r and therefore also to the set $K_t \cap P_0$ as well.

So, the set $K_t \cap P_0$ is non-empty. We show that $K_t \cap P_0$ either consists of one point, or is a homeomorphic closed ball of an $n-r$-dimensional subspace R^n, as the plane P_0 is. To this end we represent the inequality (2.5.2) for $\tau = 0$ as

$$x_1^2 + \cdots + x_r^2 - x_{r+1}^2 - \cdots - x_n^2 \geq \langle S_t x, x \rangle, \quad t > 0. \tag{2.5.6}$$

If S_t is applied in the block form

$$S_t = \begin{bmatrix} S_1 & S_{12} \\ S_{21} & S_2 \end{bmatrix},$$

when S_1 is referred to as a $r \times r$-dimensional matrix, then in view of (2.5.6) we get the negative definiteness of the matrix $\frac{1}{2}(S_2 + S_2^*)$:

$$\langle S_2 \bar{x}_2, \bar{x}_2 \rangle \leq -\|\bar{x}_2\|^2,$$

for every vector $\bar{x}_2 = (x_{r+1},\ldots,x_n) \neq 0$. Then the matrix $\frac{1}{2}(S_2 + S_2^*)$ can be represented in the form

$$\frac{1}{2}(S_2 + S_2^*) = -O_2^* O_2,$$

where O_2 is a non-degenerate matrix. The change of variables

$$y_2 = O_2 \bar{x}_2, \quad y_2 = (y_{r+1},\ldots,y_n), \tag{2.5.7}$$

reduces the quadratic form $\langle S_2 \bar{x}_2, \bar{x}_2 \rangle$ to the form

$$\langle S_2 \bar{x}_2, \bar{x}_2 \rangle = -\langle O_2 \bar{x}_2, O_2 \bar{x}_2 \rangle = -\|y_2\|^2.$$

Setting $x = (\bar{x}_1, \bar{x}_2)$, $\bar{x}_1 = (x_1, \ldots, x_r)$, we represent the inequality determining K_t in the form

$$\langle S_1 \bar{x}_1, \bar{x}_1 \rangle + \langle (S_{12}^* + S_{21})\bar{x}_1, \bar{x}_2 \rangle + \langle S_2 \bar{x}_2, \bar{x}_2 \rangle \geq 0.$$

Therefore, the set $K_t \cap P_0$ consists of points satisfying the inequality

$$\langle S_1 e_j, e_j \rangle + \langle (S_{12}^* + S_{21})e_j, \bar{x}_2 \rangle + \langle S_2 \bar{x}_2, \bar{x}_2 \rangle \geq 0. \tag{2.5.8}$$

In the system of coordinates with P_0 chosen by (2.5.7), the inequality (2.5.8) becomes
$$\|y_2\|^2 - \langle (S_{12}^* + S_{21})e_j, O_2^{-1}y_2 \rangle - \langle S_1 e_j, e_j \rangle \leq 0,$$
or
$$(y_{r+1} - \beta_1)^2 + \cdots + (y_n - \beta_{n-r})^2 \leq f, \qquad (2.5.9)$$
where $\beta_1, \ldots, \beta_{n-r}$, f are the functions t which are dependent on e_j. Since the set $K_t \cap P_0$ is non-empty, the inequality (2.5.9) is satisfied by at least one value of y_2, which is possible only if $f \geq 0$.

Let $f = 0$ for some $t = t_0$. The set $K_{t_0} \cap P_0$ then consists of one point
$$\bar{x}_1 = e_j, \quad y_{r+\nu} = \beta_\nu, \quad \beta_\nu = \beta_\nu(t_0), \quad \nu = 1, \ldots, n - r. \qquad (2.5.10)$$

Since $K_t \subset K_{t_0}$ for $t > t_0$, then $(K_t \cap P_0) \subset (K_{t_0} \cap P_0)$ for $t > t_0$, and consequently for $t > t_0$ the set $K_t \cap P_0$ contains the point (2.5.10) as well, and moreover, only this point.

For $f > 0$ the set (2.5.9) is a closed ball in the space P_0, and $K_t \cap P_0$ is a homeomorphic closed ball $\mathcal{M}_t$ of the $n - r$-dimensional subspace R^n, as the plane P_0 is.

Since $\mathcal{M}_t \subset \mathcal{M}_\tau$ for $t > \tau$, then the set $\mathcal{M}_t$ for $t > 0$ either consists of one point (2.5.10) or is an intersection of the closed embedded non-empty "balls" and consequently contains at least one point $\bar{x}_1 = e_j$, $\bar{x}_2 = x_2^0$. In both cases the set $K^+ \cap P_0$ contains at least one point. Changing the basis vector e_j and subsequently setting $j = 1, 2, \ldots, r$ we get r points belonging to the set K^+.

Let
$$x_1, \ldots, x_r \qquad (2.5.11)$$
be the indicated points. As $x_j = (e_j, x_{2j}^0)$, $j = 1, \ldots, r$, the vectors directed from the origin to the points (2.5.11) are linearly independent. Moreover, since the segments connecting the origin with the points (2.5.11) are formed by the points of the form μx_j, $0 \leq \mu \leq 1$, then they are located in the set K^+.

Let x be the point of K^+. We prove that the function $x_t = \Omega_0^t(\varphi; A)$ satisfies the inequality
$$\|x_t\| \leq K e^{-\gamma(t-\tau)} \|x_\tau\|, \quad \forall t > \tau, \quad \text{and} \quad \forall \tau \in R, \qquad (2.5.12)$$
where K and γ are positive constants independent of the choice of point φ. The inclusion $x \in K^+$ implies $\langle S_t x, x \rangle \geq 0$, $\forall t > 0$. The latter inequality together with (2.5.2) means that $\langle S_t x, x \rangle \geq 0$, $\forall t \in R$.

To obtain estimate (2.5.12) we set
$$V_\varepsilon(t) = \langle S_t x, x \rangle + \varepsilon \langle x_t, x_t \rangle,$$

where ε is considered to be a small positive constant. Then for the derivative of function $V_\varepsilon(t)$ we have

$$\frac{dV_\varepsilon(t)}{dt} = \langle \hat{S}(\varphi_t(\varphi))x_t, x_t \rangle + \varepsilon \langle [A(\varphi_t(\varphi)) + A^*(\varphi_t(\varphi))]x_t, x_t \rangle$$
$$\leq -\gamma_1 \|x_t\|^2 + \varepsilon M \|x_t\|^2 = -(\gamma_1 - \varepsilon M)\|x_t\|^2 = -\gamma_2 \|x_t\|^2,$$

where

$$M = \max_{\varphi \in \mathcal{T}_m, \|x\|=1} |\langle [A(\varphi) + A^*(\varphi)]x, x \rangle|, \quad \gamma_2 = \gamma_1 - \varepsilon M > 0.$$

Since moreover

$$V_\varepsilon(t) \leq \langle S_t x, x \rangle + \varepsilon \|x_t\|^2 = \langle S(\varphi_t(\varphi))x_t, x_t \rangle + \varepsilon \|x_t\|^2 \leq (M_1 + \varepsilon)\|x_t\|^2,$$
$$\|x_t\|^2 \leq \frac{\langle S_t x, x \rangle + \varepsilon \|x_t\|^2}{\varepsilon} = \frac{V_\varepsilon(t)}{\varepsilon},$$
(2.5.13)

where $M_1 = \max\limits_{\varphi \in \mathcal{T}_m, \|x\|=1} |\langle S(\varphi)x, x \rangle|$, then

$$\frac{d}{dt} V_\varepsilon(t) \leq -\gamma_2 \|x_t\|^2 \leq -\frac{\gamma_2}{M_1 + \varepsilon} V_\varepsilon(t) = -2\gamma V_\varepsilon(t).$$

Integrating this inequality yields

$$V_\varepsilon(t) \leq V_\varepsilon(\tau) e^{-2\gamma(t-\tau)}, \quad \forall t > \tau,$$

or, in view of (2.5.13)

$$\|x_t\|^2 \leq \frac{V_\varepsilon(t)}{\varepsilon} \leq \left(\frac{V_\varepsilon(t)}{\varepsilon}\right) e^{-2\gamma(t-\tau)}$$
$$\leq \left(1 + \frac{M_1}{\varepsilon}\right) e^{-2\gamma(t-\tau)} \|x_\tau\|^2 \leq K^2 e^{-2\gamma(t-\tau)} \|x_\tau\|^2,$$
$$\forall t > \tau, \quad \text{and} \quad \forall \tau \in R.$$

As M_1 and γ are positive constants independent of the choice of φ, the latter of the inequalities is the desired one (2.5.12).

We designate a linear subspace of R^n stretched on the vectors (2.5.11) by R^r. Let us show that $R^r \subset K^+$. Suppose this to be false and that $y = \sum\limits_{j=1}^{r} \mu_j x_j$ is a linear combination of vectors (2.5.11) which does not belong to the set K^+. The fact that $y \notin K^+$ implies the existence of $t_0 > 0$ such that $\langle S_t y, y \rangle < 0$, $\forall t \geq t_0$. Then in view of inequality (2.5.2)

$$\langle -S_t y, y \rangle > \langle -S_{t_0} y, y \rangle = \text{const} > 0, \quad \forall t > t_0. \qquad (2.5.14)$$

On the other hand

$$\langle S_t y, y \rangle = \sum_{j=1}^{r} \mu_j \langle S_t x_j, x_j \rangle = \sum_{j=1}^{r} \mu_j \langle S(\varphi_t(\varphi))x_t(x_j), x_t(x_j) \rangle,$$

where $x_t(x_j) = \Omega_0^t(\varphi; A)x_j$ due to $x_j \in K^+$ satisfying the estimate (2.5.12) based on which

$$\lim_{t \to \infty} x_t(x_j) = 0. \tag{2.5.15}$$

For $t > 0$ the boundedness of the matrix $S(\varphi_t(\varphi))$ and relation (2.5.13) imply $\lim_{t \to +\infty} \langle -S_t y, y \rangle = 0$. The latter contradicts the inequality (2.5.14). The contradiction proves that $R^r \subset K^+$.

Since $R^r \subset K^+$, then the solution $\varphi_t(\varphi)$, $x_t(\varphi, x) = \Omega_0^t(\varphi; A)x$ of the system (2.2.4) with $x \in R^r$ satisfies the inequality (2.5.12). Owing to this E^+ relative to the solutions of (2.2.4) stands for the space E^+ mentioned in the definition of exponential dichotomy of the trivial torus of the system (2.2.4).

Proceeding with arguments similar to the above we establish the existence of an n-dimensional space R^{n-r} in the intersection K^- of the sets $\tilde{K}_t$ formed by the points $x \in R^n$ for which $\langle S_t x, x \rangle \leq 0$, $t \leq 0$ for the function $-\langle S_t x, x \rangle$.

For every point $x \in K^-$ the function $x_t = \Omega_0^t(\varphi; A)$ satisfies the inequality

$$\|x_t\| \leq K e^{\gamma(t-\tau)} \|x_\tau\|, \quad \forall t \leq \tau, \quad \text{and} \quad \forall \tau \in R. \tag{2.5.16}$$

The latter is implied by the fact that the derivative of the function

$$V_\varepsilon^1(t) = -\langle S_t x, x \rangle + \varepsilon \|x_t\|^2$$

satisfies, for sufficiently small positive ε the inequality

$$\frac{dV_\varepsilon^1(t)}{dt} = -\langle \hat{S}(\varphi_t(\varphi))x_t, x_t \rangle + \varepsilon \langle [A(\varphi_t(\varphi)) + A^*(\varphi_t(\varphi))]x_t, x_t \rangle$$

$$\geq (\gamma_1 - \varepsilon M)\|x_t\|^2 = \gamma_2 \|x_t\|^2 \geq \frac{\gamma_2}{M_1 + \varepsilon} V_\varepsilon^1(t) = 2\gamma V_\varepsilon^1(t), \quad \forall t \in R.$$

Integrating this inequality yields the estimate

$$V_\varepsilon^1(t) \geq e^{2\gamma(\tau-t)} V_\varepsilon^1(t), \quad \forall t \leq \tau, \quad \text{and} \quad \forall \tau \in R,$$

which in its turn implies

$$\|x_t\|^2 \leq \frac{V_\varepsilon^1(t)}{\varepsilon} \leq \left(\frac{V_\varepsilon^1(t)}{\varepsilon}\right) e^{2\gamma(t-\tau)}$$

$$\leq \left(1 + \frac{M_1}{\varepsilon}\right) e^{2\gamma(t-\tau)} \|x_\tau\|^2 \leq K^2 e^{2\gamma(t-\tau)} \|x_\tau\|^2,$$

$$\forall t \leq \tau, \quad \text{and} \quad \forall \tau \in R.$$

Due to the inequality (2.5.16) the space R^{n-r} relative to the solutions of the system (2.2.4) stands for the space E^- mentioned in the definition of the exponential dichotomy of the trivial torus of the system (2.2.4).

Since the inequality $\langle S(\varphi)x, x\rangle > 0$ is valid for the points $x \in R^r$, $x \neq 0$, and the inequality $\langle S(\varphi)x, x\rangle < 0$ is satisfied for the points $x \in R^{n-r}$, $x \neq 0$, the sets R^r and R^{n-r} intersect at one point only $R^r \cap R^{n-r} = \{0\}$. Then the expressions $K^+ = R^r$, $K^- = R^{n-r}$, $R^n = K^+ \oplus K^-$ follow from the equality $R^n = R^r \oplus R^{n-r}$, inequalities (2.5.12), (2.5.16) and the correlation $\lim\limits_{|t|\to\infty} \|\Omega_0^t(\varphi; A)x\| = \infty$, $x \notin R^r \cup R^{n-r}$ which is deduced by the same arguments as in Section 2.4

$$\lim_{|t|\to\infty} \|x_t(\varphi_0, x_0)\| = \infty, \quad \text{for} \quad (\varphi_0, x_0) \notin M^+ \bigcup M^-.$$

The arbitrariness of the choice of $\varphi \in \mathcal{T}_m$ and the independence of the constants K and γ of $\varphi \in \mathcal{T}_m$ prove exponential dichotomy of the invariant torus $x = 0$, $\varphi \in \mathcal{T}_m$ of the system (2.2.4).

This completes the proof of the Theorem 2.5.1.

Remark 2.5.1 The proof of the theorem implies that the manifold $M^+\big|_{\varphi=\text{const}}$ belongs to the "cone" of the space R^n, determined by $\langle S(\varphi)x, x\rangle > 0$ for each $\varphi \in \mathcal{T}_m$, and the manifold $M^-\big|_{\varphi=\text{const}}$ belongs to the "cone" R^n determined by $\langle S(\varphi)x, x\rangle < 0$.

2.6 Necessary Conditions for Exponential Dichotomy of the Invariant Torus

Let us show that the sufficient conditions cited in the above section are also necessary for the exponential dichotomy of the trivial torus $x = 0$, $\tau \in \mathcal{T}_m$ of the system (2.2.4). In addition, it is clear that the behaviour of the solutions of (2.2.4) characterizes the linear system

$$\frac{dx}{dt} = A(\varphi_t(\varphi))x \tag{2.6.1}$$

as exponentially dichotomous on the whole axis $R = (-\infty, \infty)$.

We designate the projectors corresponding to the expansion of R^n into direct sum of the subspaces $R^r(\varphi)$ and $R^{n-r}(\varphi)$ by $C(\varphi)$ and $C_1(\varphi) = I_n - C(\varphi)$. The function

$$G_t(\tau, \varphi) = \begin{cases} \Omega_0^t(\varphi; A)C(\varphi)\Omega_\tau^0(\varphi; A), & t \geq \tau, \\ -\Omega_0^t(\varphi; A)C_1(\varphi)\Omega_\tau^0(\varphi; A), & t < \tau, \end{cases} \tag{2.6.2}$$

then satisfies the estimate

$$\|G_t(\tau, \varphi)\| \leq K_1 e^{-\gamma|t-\tau|}, \quad t, \tau \in R. \tag{2.6.3}$$

We apply this function to establish the properties of the matrix $C(\varphi)$.

Lemma 2.6.1 *Let $a(\varphi) \in C_{\text{Lip}}(\mathcal{T}_m)$, $A(\varphi) \in C_{\text{Lip}}(\mathcal{T}_m)$ and the function (2.6.2) satisfies the inequality (2.6.3). Then the matrix $C(\varphi)$ belongs to the space $C^0(\mathcal{T}_m)$.*

Proof Consider the matrix

$$Z_t(\varphi, \bar\varphi) = \begin{cases} \Omega_0^t(\varphi; A)C(\varphi) - \Omega_0^t(\bar\varphi; A)C(\bar\varphi), & t \geq 0, \\ -\Omega_0^t(\varphi; A)C_1(\varphi) - \Omega_0^t(\bar\varphi; A)C_1(\bar\varphi), & t < 0, \end{cases} \quad (2.6.4)$$

equal to the matrix $C(\varphi) - C(\bar\varphi)$ for $t = 0$. It is continuous in t, satisfies the inequality

$$\|Z_t(\varphi, \bar\varphi)\| \leq 2K_1 e^{-\gamma|t|}, \quad t \in R, \quad (2.6.5)$$

and is a solution of the differential equation

$$\frac{dz}{dt} = A(\varphi_t(\bar\varphi))Z + [A(\varphi_t(\varphi)) - A(\varphi_t(\bar\varphi))]G_t(0, \varphi). \quad (2.6.6)$$

The exponential dichotomy of the system (2.6.1) implies the uniqueness of the solution of the system (2.6.6) which is bounded on R and its representation in the form of the integral

$$Z_t(\varphi, \bar\varphi) = \int_{-\infty}^{\infty} G_t(\tau, \bar\varphi)[A(\varphi_\tau(\varphi)) - A(\varphi_\tau(\bar\varphi))]G_\tau(0, \varphi)\,d\tau. \quad (2.6.7)$$

Hence

$$C(\varphi) - C(\bar\varphi) = \int_{-\infty}^{\infty} G_t(\tau, \bar\varphi)[A(\varphi_\tau(\varphi)) - A(\varphi_\tau(\bar\varphi))]G_\tau(0, \varphi)\,d\tau.$$

We find estimate of the function under integral sign of (2.6.7). Let the estimates

$$\|a(\varphi) - a(\bar\varphi)\| \leq \alpha\|\varphi - \bar\varphi\|, \quad \|A(\varphi) - A(\bar\varphi)\| \leq \beta\|\varphi - \bar\varphi\|$$

be satisfied for any $\varphi, \bar\varphi \in \mathcal{T}_m$ and some positive α and β. From system (2.2.4) it is easy to derive the estimates

$$\|\varphi_t(\varphi) - \varphi_t(\bar\varphi)\| \leq e^{\alpha|t|}\|\varphi - \bar\varphi\|, \quad \|\varphi_t(\varphi) - \varphi_t(\bar\varphi)\| \leq \|\varphi - \bar\varphi\| + 2\|a\|_0|t|,$$

due to which

$$\|\varphi_t(\varphi) - \varphi_t(\bar\varphi)\| \leq e^{\alpha|t|/(\nu+1)}(\|\varphi - \bar\varphi\| + 2\|a\|_0|t|)^{\nu/(\nu+1)}\|\varphi - \bar\varphi\|^{1/(\nu+1)},$$

for all $t \in R$ and arbitrary $\nu \geq 0$. This together with the inequality (2.6.3) for the function $G_t(\tau, \varphi)$ yields

$$\|Z_t(\varphi, \bar\varphi)\| \leq \beta K_1^2 \|\varphi - \bar\varphi\|^{1/(\nu+1)} \int_{-\infty}^{\infty} \exp\left\{-\gamma(|t-\tau| + |\tau|) + \frac{\alpha}{\nu+1}|\tau|\right\} \qquad (2.6.8)$$
$$\times (2\|a\|_0|\tau| + \|\varphi - \bar\varphi\|)^{\nu/(\nu+1)}\, d\tau,$$

which implies the convergence of the integral (2.6.7) for sufficiently large ν: $\nu+1 > \alpha/\gamma$ and the satisfaction of the inequality for $Z_0(\varphi, \bar\varphi) = C(\varphi) - C(\bar\varphi)$ of the form

$$\|C(\varphi) - C(\bar\varphi)\| \leq \beta K_1^2 M \|\varphi - \bar\varphi\|^{1/(\nu+1)}. \qquad (2.6.9)$$

Here M is a positive constant equal to the value of the integral in inequality (2.6.8) for $t = 0$ and $\|\varphi - \bar\varphi\| = 2\pi\sqrt{m}$. The inequality (2.6.9) proves the continuity of the matrix $C(\varphi)$ in φ for $\varphi \in \mathcal{T}_m$. The periodicity of the matrix $C(\varphi)$ in φ follows from the definition of this matrix. Thus $C(\varphi) \in C^0(\mathcal{T}_m)$.

Consider now the function

$$G_0(\tau, \varphi) = \begin{cases} \Omega_\tau^0(\varphi; A) C(\varphi_\tau(\varphi)), & \tau \leq 0, \\ -\Omega_\tau^0(\varphi; A) C_1(\varphi_\tau(\varphi)), & \tau > 0, \end{cases} \qquad (2.6.10)$$

viewing $C(\varphi)$ as the matrix found in (2.6.2). The formulas (2.6.2) and (2.6.10) show that $G_0(\tau, \varphi) = G_t(0, \varphi_\tau(\varphi))\big|_{t=-\tau}$, and because to this the inequality

$$\|G_0(\tau, \varphi)\| \leq K_1 e^{-\gamma|\tau|}, \quad \tau \in R. \qquad (2.6.11)$$

is valid. Since $C(\varphi) \in C^0(\mathcal{T}_m)$, the function (2.6.10) is the Green function of the system (2.2.4). The absence of the solutions to the system (2.6.1) which are bounded on R other than the trivial one ensures the uniqueness of this function and the validity of the inequality

$$G_0(\tau, \varphi) = \begin{cases} C(\varphi)\Omega_\tau^0(\varphi; A), & \tau \leq 0 \\ -C_1(\varphi)\Omega_\tau^0(\varphi; A), & \tau > 0 \end{cases} = G_t(\tau, \varphi)\big|_{t=0},$$

proving that $G_0(\tau, \varphi)$ is derived from the function (2.6.2) for $t = 0$. This gives the coinciding designations of (2.6.2) and (2.6.10). Thus, the exponential dichotomy of the trivial torus $x = 0$, $\varphi \in \mathcal{T}_m$ of the system (2.2.4) ensures the existence and uniqueness of the Green function of this system, which is exponentially damping as $|\tau| \to \infty$ according to (2.6.11).

We incorporate the properties of the functions $G_t(\tau, \varphi)$ and $G_0(\tau, \varphi)$ to prove the main result of this section.

Theorem 2.6.1 Let $a(\varphi) \in C_{\text{Lip}}(\mathcal{T}_m)$, $A(\varphi) \in C_{\text{Lip}}(\mathcal{T}_m)$ and the trivial torus of the system (2.2.4) be exponentially dichotomous. Then there exists a non-degenerate symmetric matrix $S(\varphi) \in C'(\mathcal{T}_m; a)$ such that the matrix $\hat{S}(\varphi) = \dot{S}(\varphi) + S(\varphi)A(\varphi) + A^*(\varphi)S(\varphi)$ is negative definite for all $\varphi \in \mathcal{T}_m$.

Proof We define the matrix

$$S(\varphi) = S_1(\varphi) - S_2(\varphi), \qquad (2.6.12)$$

setting

$$S_1(\varphi) = \int_0^\infty C^*(\varphi)(\Omega_0^\tau(\varphi; A))^* \Omega_0^\tau(\varphi; A) C(\varphi) \, d\tau,$$

$$S_2(\varphi) = \int_{-\infty}^0 C_1^*(\varphi)(\Omega_0^\tau(\varphi; A))^* \Omega_0^\tau(\varphi; A) C_1(\varphi) \, d\tau. \qquad (2.6.13)$$

Inequality (2.6.3) ensures uniform convergence of the integrals (2.6.4). Taking the matrices $C(\varphi)$ and $\Omega_0^\tau(\varphi; A)$ belonging to the space $C^0(\mathcal{T}_m)$ into account, we find that for every fixed $\tau \in R$, $S_i(\varphi) \in C^0(\mathcal{T}_m)$, $i = 1, 2$.

Let us show that $S_i \in C'(\mathcal{T}_m; a)$. To this end we use the fact that the matrix $C(\varphi)$ satisfies the equality (2.4.15) (the latter is based on the uniqueness of the Green function $G_0(\tau, \varphi)$) and transform the expression for $S_1(\varphi_t(\varphi))$ as

$$S_1(\varphi_t(\varphi)) = \int_0^\infty C^*(\varphi_t(\varphi))(\Omega_t^{t+\tau}(\varphi; A))^* \Omega_t^{t+\tau}(\varphi; A) C(\varphi_t(\varphi)) \, d\tau$$

$$= \int_0^\infty [\Omega_0^{t+\tau}(\varphi; A)) C(\varphi) \Omega_t^0(\varphi; A)]^* \Omega_0^{t+\tau}(\varphi; A) C(\varphi) \Omega_t^0(\varphi; A) \, d\tau \quad (2.6.14)$$

$$= \int_t^\infty (\Omega_t^0(\varphi; A))^* C^*(\varphi)(\Omega_0^\tau(\varphi; A))^* \Omega_0^\tau(\varphi; A) C(\varphi) \Omega_t^0(\varphi; A) \, d\tau.$$

This formula implies the continuous differentiability of the function $S_1(\varphi_t(\varphi))$ in t and therefore the function $S_1(\varphi)$ belonging to the space $C'(\mathcal{T}_m; a)$. For the derivative of this function we get the expression

$$\frac{d}{dt} S_1(\varphi_t(\varphi)) = -C^*(\varphi_t(\varphi)) C(\varphi_t(\varphi)) - A^*(\varphi_t(\varphi)) S_1(\varphi_t(\varphi)) - S_1(\varphi_t(\varphi)) A(\varphi_t(\varphi)),$$

from which it follows that

$$\dot{S}_1(\varphi) = -C^*(\varphi) C(\varphi) - A^*(\varphi) S_1(\varphi) - S_1(\varphi) A(\varphi). \qquad (2.6.15)$$

Similarly, $S_2(\varphi_t(\varphi))$ is transformed to an expression which implies

$$S_2(\varphi_t(\varphi)) = \int_{-\infty}^{t} (\Omega_t^0(\varphi;A))^* C_1^*(\varphi)(\Omega_0^\tau(\varphi;A))^* \Omega_0^\tau(\varphi;A) C_1(\varphi) \Omega_t^0(\varphi;A) \, d\tau,$$

that $S_2(\varphi) \in C'(\mathcal{T}_m;a)$ and

$$\dot{S}_2(\varphi) = -C_1^*(\varphi)C_1(\varphi) - A^*(\varphi)S_2(\varphi) - S_2(\varphi)A(\varphi). \quad (2.6.16)$$

The equalities (2.6.15) and (2.6.16) imply negative definiteness of the matrix $\hat{S}(\varphi)$. Actually, as

$$\hat{S} = S + SA + A^*S = \dot{S}_1 - \dot{S}_2 + SA + A^*S$$
$$= -C^*C - A^*S_1 - S_1A - C_1^*C_1 + A^*S_2 + S_2A + SA + A^*S$$
$$= -C^*C - C_1^*C_1 - A^*(S_1 - S_2) - (S_1 - S_2)A + A^*S + SA = -C^*C - C_1^*C_1,$$

then

$$\langle \hat{S}x, x \rangle = -\langle Cx, Cx \rangle - \langle C_1 x, C_1 x \rangle \leq -(\|Cx\|^2 + \|C_1 x\|^2) \leq -\frac{1}{2}\|x\|^2,$$

which is equivalent to the negative definiteness of the matrix $\hat{S}(\varphi)$ for all $\varphi \in \mathcal{T}_m$.

To complete the proof of the theorem we demonstrate the non-degeneracy of the matrix $S(\varphi)$ for $\varphi \in \mathcal{T}_m$. In view that $C^2(\varphi) = C(\varphi)$ and

$$\|C(\varphi)\| = \|\Omega_\tau^0(\varphi;A)\Omega_0^\tau(\varphi;A)C(\varphi)\|$$
$$\leq \|\Omega_\tau^0(\varphi;A)\| \|\Omega_0^\tau(\varphi;A)C(\varphi)\| \leq \|\Omega_\tau^0(\varphi;A)\| K e^{-\gamma\tau},$$

for $\tau \geq 0$, we get for $\langle S_1(\varphi)x, x \rangle$

$$\langle S_1(\varphi)x, x \rangle = \int_0^\infty \|\Omega_0^\tau(\varphi;A)C(\varphi)x\|^2 \, d\tau$$

$$\leq \int_0^\infty \|\Omega_0^\tau(\varphi;A)C(\varphi)\|^2 \, d\tau \, \|C(\varphi)x\|^2 \leq \frac{K^2}{2\gamma}\|C(\varphi)x\|^2 \leq \lambda_2 \|C(\varphi)x\|^2,$$

$$\langle S_1(\varphi)x, x \rangle \geq \int_0^\infty \frac{d\tau}{\|\Omega_\tau^0(\varphi;A)\|^2} \|C(\varphi)x\|^2$$

$$\geq \int_0^1 \frac{d\tau}{\|\Omega_\tau^0(\varphi;A)\|^2} \|C(\varphi)x\|^2 \geq \lambda_1 \|C(\varphi)x\|^2.$$

Putting together these estimates one obtains the inequality

$$\lambda_1 \|C(\varphi)x\|^2 \leq \langle S_1(\varphi)x, x\rangle \leq \lambda_2 \|C(\varphi)x\|^2, \quad (2.6.17)$$

valid for all $x \in R^n$, $\varphi \in \mathcal{T}_m$ and some positive λ_1 and λ_2.

The inequalities for $\langle S_2(\varphi)x, x\rangle$ are obtained in the same way in the form

$$\lambda_3 \|C_1(\varphi)x\|^2 \leq \langle S_2(\varphi)x, x\rangle \leq \lambda_4 \|C_1(\varphi)x\|^2,$$

valid for all $x \in R^n$, $\varphi \in \mathcal{T}_m$ and some positive λ_3 and λ_4.

Let $T(\varphi)$ be the matrix which reduces $C(\varphi)$ to the diagonal form $C(\varphi) = T(\varphi)JT^{-1}(\varphi)$, $J = \text{diag}\{i_r, 0\}$. In (2.6.17) we make the change of variables $x = T(\varphi)y$. In view of

$$\|Jy\| = \|T^{-1}CTy\| \leq \|T^{-1}\|\,\|CTy\| \leq \|T^{-1}\|\,\|T\|\,\|Jy\|,$$

we get new inequalities from (2.6.17):

$$\lambda_1 \|T^{-1}\|^{-2}\|Jy\|^2 \leq \lambda_1 \|CTy\|^2 \leq \langle T^*S_1Ty, y\rangle \leq \lambda_2\|CTy\|^2 \leq \lambda_2\|T\|^2\|Jy\|^2,$$

which indicates that the matrix T^*S_1T has a block-diagonal form $T^*S_1T = \text{diag}\{D_1, 0\}$, where D_1 is an $r \times r$-dimensional positive definite matrix for any variable.

Similar arguments for the matrix $S_2(\varphi)$ result in the inequalities

$$\lambda_3\|T^{-1}\|^{-2}\|(I_n - J)y\|^2 \leq \langle T^*S_2Ty, y\rangle \leq \lambda_4\|T\|^2\|(I_n - J)y\|^2,$$

which show that the matrix T^*S_2T also has a block-diagonal form $T^*S_2T = \text{diag}\{0, D_2\}$, where D_2 is an $(n-r) \times (n-r)$-dimensional positive definite matrix for any $\varphi \in \mathcal{T}_m$. Then the matrix $S(\varphi) = S_1(\varphi) - S_2(\varphi)$ can be represented as

$$S(\varphi) = (T^{-1}(\varphi))^* \,\text{diag}\{D_1(\varphi), -D_2(\varphi)\}T^{-1}(\varphi), \quad (2.6.18)$$

and for every $\varphi \in \mathcal{T}_m$ has precisely r positive and $n - r$ negative eigenvalues. Hence, the non-degeneracy of the matrix $S(\varphi)$ is proved.

Remark 2.6.1 The above presentation shows that when the trivial torus of the system (2.2.4) is exponentially dichotomous, the matrix $C(\varphi)$ determining the Green function $G_0(\tau, \varphi)$ and the matrix $S(\varphi)$ determined by (2.6.12) and (2.6.13) are related by the fact that any matrix $T(\varphi)$ reducing $C(\varphi)$ to the Jordan form ensures the representation of $S(\varphi)$ in the form of (2.6.18) with positive definite blocks $D_1(\varphi)$ and $D_2(\varphi)$ of the dimensions $r \times r$ and $(n-r) \times (n-r)$ respectively.

Now we combine the results of this section and the previous one in one assertion on exponential dichotomy of the trivial torus of system (2.2.4).

Theorem 2.6.2 *Let $a(\varphi) \in C_{\text{Lip}}(\mathcal{T}_m)$ and $A(\varphi) \in C_{\text{Lip}}(\mathcal{T}_m)$. The trivial torus of the system (2.2.4) is exponentially dichotomous, if and only if there exists a non-degenerate symmetric matrix $S(\varphi) \in C'(\mathcal{T}_m; a)$, for which the matrix $\hat{S}(\varphi) = \dot{S}(\varphi) + S(\varphi)A(\varphi) + A^*(\varphi)S(\varphi)$ is negative definite for all $\varphi \in \mathcal{T}_m$, or if there exists a unique the Green function (2.6.10), (2.6.11) of system (2.2.4).*

2.7 Existence Criterion for the Green Function

In addition, we assume that the vector-function $a(\varphi)$ satisfies a less strict condition than $a(\varphi) \in C_{\text{Lip}}(\mathcal{T}_m)$, namely, the vector function $a(\varphi)$ is such that the Cauchy problem

$$\frac{d\varphi}{dt} = a(\varphi), \quad \varphi|_{t=0} = \varphi_0$$

has a unique solution $\varphi_t(\varphi_0)$ depending continuously on φ_0 for every fixed $\varphi_0 \in R^n$. For this it is sufficient (the well-known Osgood theorem) that the condition $\|a(\varphi) - a(\bar{\varphi})\| \leq \omega(\|\varphi - \bar{\varphi}\|)$, $\varphi, \bar{\varphi} \in R^m$, where $\omega(\sigma)$ is a continuous monotone increasing scalar function such that $\omega(0) = 0$ and $\int_{+0} \frac{d\sigma}{\omega(\sigma)} = \infty$. In the matrix function $A(\varphi)$ we assume that $A(\varphi) \in C^0(\mathcal{T}_m)$.

Now we prove necessary and sufficient existence conditions for the Green functions (2.6.10) and (2.6.11). These conditions are formulated in terms of the sign variable quadratic forms of Lyapunov functions and allow one to establish the existence both of the unique Green function and the non-unique one.

Theorem 2.7.1 *Let the system (2.2.4) have the Green function of the problem on invariant tori (2.6.10) with the estimate (2.6.11). Then there exists an n-dimensional symmetric matrix $S(\varphi) \in C'(\mathcal{T}_m; a)$ satisfying the condition*

$$\langle [\dot{S}(\varphi) - S(\varphi)A^*(\varphi) - A(\varphi)S(\varphi)]x, x \rangle \leq -\|x\|^2. \tag{2.7.1}$$

In addition, if the Green function is unique, then

$$\det S(\varphi) \neq 0, \quad \forall \varphi \in R^m. \tag{2.7.2}$$

Proof Let us show that for such a matrix $S(\varphi)$ the following matrix

$$S(\varphi) = 2[S_1(\varphi) - S_2(\varphi)], \tag{2.7.3}$$

can be taken, where

$$S_1(\varphi) = \int_0^\infty \Omega_\tau^0(\varphi; A)[C(\varphi_\tau(\varphi)) - I_n]\{\Omega_\tau^0(\varphi; A)[C(\varphi_\tau(\varphi)) - I_n]\}^* d\tau; \tag{2.7.4}$$

$$S_2(\varphi) = \int_{-\infty}^{0} \Omega_\tau^0(\varphi; A) C(\varphi_\tau(\varphi)) [\Omega_\tau^0(\varphi; A) C(\varphi_\tau(\varphi))]^* \, d\tau. \tag{2.7.4a}$$

Obviously, due to the uniform convergence of the integrals (2.7.4), (2.7.4a) in $\varphi \in R^m$, $S_i(\varphi) \in C^0(\mathcal{T}_m)$. In view of (2.2.1), we find

$$S_1(\varphi_t(\varphi)) = \int_{t}^{\infty} \Omega_\sigma^t(\varphi; A)[C(\varphi_\sigma(\varphi)) - I_n]\{\Omega_\sigma^t(\varphi; A)[C(\varphi_\sigma(\varphi)) - I_n]\}^* \, d\sigma; \tag{2.7.5}$$

$$S_2(\varphi_t(\varphi)) = \int_{-\infty}^{t} \Omega_\sigma^t(\varphi; A) C(\varphi_\sigma(\varphi)) [\Omega_\sigma^t(\varphi; A) C(\varphi_\sigma(\varphi))]^* \, d\sigma. \tag{2.7.5a}$$

From here it is clear that the functions $S_i(\varphi_t(\varphi))$, $i = 1, 2$ are continuously differentiable in t, i.e. $S_i(\varphi) \in C'(\mathcal{T}_m; a)$. In addition, we get from (2.7.5) and (2.7.5a)

$$\dot{S}_1(\varphi) = -[C(\varphi) - I_n][C(\varphi) - I_n]^* + A(\varphi) S_1(\varphi) + S_1(\varphi) A^*(\varphi),$$
$$\dot{S}_2(\varphi) = C(\varphi) C^*(\varphi) + A(\varphi) S_2(\varphi) + S_2(\varphi) A^*(\varphi).$$

Thus, we make sure that the matrix function (2.7.3) satisfies the condition (2.7.1)

$$\langle [\dot{S}(\varphi) - S(\varphi) A^*(\varphi) - A(\varphi) S(\varphi)] x, x \rangle$$
$$= -2\{\|[C(\varphi) - I_n]^* x\|^2 + \|C^*(\varphi) x\|^2\} \leq -\|x\|^2.$$

Now let us show that in the case of the unique Green function (2.6.10) the eigenvalues of the matrix (2.7.3) differ from zero for all $\varphi \in R^n$. We prove the inequalities

$$\begin{aligned}\lambda_1 \|[C^*(\varphi) - I_n] x\|^2 &\leq \langle S_1(\varphi) x, x \rangle \leq \lambda_2 \|[C^*(\varphi) - I_n] x\|^2, \\ \lambda_3 \|C^*(\varphi) x\|^2 &\leq \langle S_2(\varphi) x, x \rangle \leq \lambda_4 \|C^*(\varphi) x\|^2,\end{aligned} \tag{2.7.6}$$

for all $x \in R^n$, $\varphi \in R^m$; $i = \overline{1,4}$ are positive constants. Since the matrix $C(\varphi)$ satisfies condition (2.4.15), (2.4.16) we get

$$\langle S_1(\varphi) x, x \rangle = \int_{0}^{\infty} \|\{\Omega_\tau^0(\varphi; A)[C(\varphi_\tau(\varphi)) - I_n]\}^* x\|^2 \, d\tau$$

$$= \int_{0}^{\infty} \|[\Omega_\tau^0(\varphi; A)]^* [C^*(\varphi) - I_n]^2 x\|^2 \, d\tau$$

$$\leq \int_0^\infty \|[\Omega_\tau^0(\varphi;A)]^*[C^*(\varphi) - I_n]\|^2 \, d\tau \, \|[C^*(\varphi) - I_n]x\|^2,$$

$$\langle S_1(\varphi)x, x \rangle \geq \int_0^\infty \frac{d\tau}{\|\Omega_0^\tau(\varphi;A)\|^2} \|[C^*(\varphi) - I_n]x\|^2.$$

The second of the estimates (2.7.6) is proved similarly. Let $L(\varphi_0)$ be a non-degenerate matrix reducing the matrix $C^*(\varphi_0)$ to the diagonal form:

$$L^{-1}(\varphi_0)C^*(\varphi_0)L(\varphi_0) = J_r = \mathrm{diag}\,\{I_r, 0\}.$$

We make the change of variables $x = L(\varphi_0)y$ for $\varphi = \varphi_0$ in (2.7.6):

$$\langle L^*(\varphi_0)S_1(\varphi)L(\varphi_0)y, y\rangle \leq \lambda_2 \|[C^*(\varphi_0) - I_n]L(\varphi_0)y\|^2$$

$$\leq \lambda_2 \|L(\varphi_0)\|^2 \|J_{n-r}\,y\|^2,$$

$$\langle L^*(\varphi_0)S_1(\varphi)L(\varphi_0)y, y\rangle \geq \lambda_1 \|[C^*(\varphi_0) - I_n]L(\varphi_0)y\|^2$$

$$\geq \lambda_1 \|L^{-1}(\varphi_0)\|^{-2} \|J_{n-r}\,y\|^2,$$

$$J_{n-r} = \mathrm{diag}\,\{0, J_{n-r}\},$$

$$\langle L^*(\varphi_0)S_2(\varphi_0)L(\varphi_0)y, y\rangle \leq \lambda_4 \|L(\varphi_0)\|^2 \|J_r\,y\|^2,$$

$$\langle L^*(\varphi_0)S_2(\varphi_0)L(\varphi_0)y, y\rangle \geq \lambda_3 \|L^{-1}(\varphi_0)\|^{-2} \|J_r\,y\|^2.$$

The inequalities obtained imply that the matrices $\bar{S}_i(\varphi_0) = L^*(\varphi_0)S_i(\varphi)L(\varphi_0)$, $i = 1, 2$, have the block diagonal form $\bar{S}_1(\varphi_0) = \mathrm{diag}\,\{0, \tilde{S}_1(\varphi_0)\}$, $\bar{S}_2(\varphi_0) = \mathrm{diag}\,\{\tilde{S}_2(\varphi_0), 0\}$ and moreover $\det \tilde{S}_i(\varphi_0) \neq 0$, $i = 1, 2$. Thus,

$$\det S(\varphi_0) = \det\left[(L^*(\varphi_0))^{-1} \mathrm{diag}\,\{-\tilde{S}_2(\varphi_0), \tilde{S}_1(\varphi_0)\}L^{-1}(\varphi_0)\right] \neq 0.$$

Since $\varphi = \varphi_0$ was taken arbitrarily from R^m, the inequality (2.7.2) is obtained.

Theorem 2.7.2 *Let there exist an n-dimensional symmetric matrix function $S(\varphi) \in C'(\mathcal{T}_m; a)$ satisfying condition (2.7.1). Then there exists the Green function (2.6.10) which satisfies estimate (2.6.11). If we require additionally from $S(\varphi)$ that the condition (2.7.2) is satisfied, then the Green function is unique, and the identities (2.4.15) and (2.4.16) are satisfied for it. In the case of degeneracy of the matrix $S(\varphi)$ for some $\varphi = \varphi_0 \in R^m$ there exist an infinite number of the Green functions (2.6.10) with estimate (2.6.11).*

Proof First consider, the case when the determinant of the matrix $S(\varphi)$ differs from zero for all $\varphi \in R^m$. Condition (2.7.1) being satisfied means that the derivative of the quadratic form $\langle S(\varphi_t(\varphi_0)x, x\rangle$ by virtue of the system conjugated with (2.6.1):

$$\frac{dx}{dt} = -A^*(\varphi_t(\varphi_0))x \qquad (2.7.7)$$

is negative definite for every fixed $\varphi_0 \in R^m$. Consequently we conclude from the conditions of Theorems 1.1.1–1.1.3, that the systems (2.6.1) and (2.7.7) are exponentially dichotomous on the whole R axis for every fixed $\varphi_0 \in R^m$. In addition, K, γ, K_0 found in estimates (1.1.2), (1.1.3) and (1.1.57) of Chapter 1 can be taken to be independent of $\varphi_0 \in R^m$. We demonstrate that this is sufficient for the continuity of the Green function $G_t(\tau, \varphi_0)$ of the problem for solutions to the system (2.6.1) bounded on R in parameters φ_0.

Consider the difference $G_t(\tau, \varphi_0) - G_t(\tau, \bar{\varphi}_0)$ which can be represented due to the uniqueness of the Green function $G_t(\tau, \varphi_0)$ as

$$G_t(\tau, \varphi_0) - G_t(\tau, \bar{\varphi}_0) = \int_{-\infty}^{\infty} G_t(\sigma, \varphi_0)[A(\varphi_\sigma(\varphi_0)) - A(\varphi_\sigma(\bar{\varphi}_0))] G_\sigma(\tau, \bar{\varphi}_0) \, d\sigma. \quad (2.7.8)$$

Actually, assuming that the difference $G_t(\tau, \varphi_0) - G_t(\tau, \bar{\varphi}_0) = R_t$ is a function of the variable $t \in R$ for fixed values $\tau \in R$, $\varphi_0 \in R^m$ we note that the inequality

$$\frac{d}{dt} R_t = A(\varphi_t(\varphi_0)) G_t(\tau, \varphi_0) - A(\varphi_t(\bar{\varphi}_0)) G_t(\tau, \bar{\varphi}_0)$$
$$= A(\varphi_t(\varphi_0)) R_t + [A(\varphi_t(\varphi_0)) - A(\varphi_t(\bar{\varphi}_0))] G_t(\tau, \bar{\varphi}_0)$$

holds for $t \ne \tau$. Hence, the value R_t, $t \ne \tau$ is a solution which is bounded on the whole R axis of the linear inhomogeneous system of equations

$$\frac{dR}{dt} = A(\varphi_t(\varphi_0)) R + F_t,$$

where F_t is a matrix function which is bounded on R which has a unique discontinuity of the first type for $t = \tau$. Therefore due to the uniqueness of the Green function $G_t(\tau, \varphi)$ we have the equality

$$R_t = \int_{-\infty}^{\infty} G_t(\tau, \varphi_0) F_\tau \, d\tau,$$

which leads to (2.7.8).

From (2.7.8) we get the inequality

$$\|G_t(\tau, \varphi_0) - G_t(\tau, \bar{\varphi}_0)\| \le K_0^2 \int_{-\infty}^{\infty} \|A(\varphi_\sigma(\varphi_0)) - A(\varphi_\sigma(\bar{\varphi}_0))\| \quad (2.7.9)$$
$$\times \exp\{-\gamma_0(|\sigma - t| + |\sigma - \tau|)\} \, d\sigma.$$

In view of the estimate $|t - \sigma| + |\tau - \sigma| \geq |\sigma| - |t|$ we proceed with the inequality (2.7.9) in the form

$$\|G_t(\tau, \varphi_0) - G_t(\tau, \bar{\varphi}_0)\| \leq K_0^2 \int_{-\infty}^{\infty} \|A(\varphi_\sigma(\varphi_0)) - A(\varphi_\sigma(\bar{\varphi}_0))\| \qquad (2.7.10)$$

$$\times \exp\{-\gamma_0 |\sigma|\} d\sigma \cdot \exp\{\gamma_0 |t|\}.$$

Note that the function

$$L_{\gamma_0}(\varphi_0, \bar{\varphi}_0) = \int_{-\infty}^{\infty} \|A(\varphi_\sigma(\varphi_0)) - A(\varphi_\sigma(\bar{\varphi}_0))\| \exp\{-\gamma_0 |\sigma|\} d\sigma,$$

which depends on $2m$ variables $\varphi_{10}, \ldots, \varphi_{m0}, \bar{\varphi}_{10}, \ldots, \bar{\varphi}_{m0}$, is continuous with respect to the whole totality of these variables (by virtue of the uniform convergence of the integral) and $L\gamma_0(\varphi_0, \varphi_0) \equiv 0$. Therefore due to the estimate (2.7.10) the Green function $G_t(\tau, \gamma_0)$, $t \neq \tau$, is continuous with respect to the whole totality of the parameters $\varphi_{10}, \ldots, \varphi_{m0}$, and hence, the unique Green function $G_0(\tau, \varphi)$ of the problem on invariant tori for the system (2.2.4) is also continuous in φ.

Now consider the case when the determinant of the matrix $S(\varphi)$ vanishes for some value $\varphi = \varphi_0 \in R^m$. We show that the extended system

$$\frac{d\varphi}{dt} = a(\varphi), \quad \frac{dx}{dt} = A(\varphi)x, \quad \frac{dy}{dt} = -x - A^*(\varphi)y, \quad y \in R^n, \qquad (2.7.11)$$

has a unique the Green function $\bar{G}_0(\tau, \varphi)$ of the problem on invariant tori. For this it is sufficient to verify that the derivative of the non-degenerate quadratic form $V(\varphi, x, y) = \lambda \langle x, y \rangle + \langle S(\varphi)y, y \rangle$, computed along the solution of system (2.7.11) is negative definite for a sufficiently large value of the parameter $\lambda > 0$. In fact, when computing this derivative we get

$$\dot{V}(\varphi, x, y) = \lambda \langle Ax, y \rangle + \lambda \langle x, -x - A^*y \rangle + \langle \dot{S}y, y \rangle$$
$$+ \langle S(-x - A^*y), y \rangle \langle Sy, -x - A^*y \rangle \leq -\lambda \|x\|^2 - 2\langle Sx, y \rangle - \|y\|^2.$$

It is obvious that the derivative $\dot{V}(\varphi, x, y)$ is negative definite for a sufficiently large value of λ: $\lambda > \max_{\varphi \in R^m} \|S(\varphi)\|^2 = \|S_0\|^2$. Thus, the system (2.7.11) has the unique

Green function

$$\bar{G}_0(\tau,\varphi) = \begin{cases} \begin{bmatrix} \Omega_\tau^0(\varphi;A) & 0 \\ -\int_\tau^0 [\Omega_0^\sigma(\varphi;A)]^*\Omega_\tau^\sigma(\varphi;A)\,d\sigma & [\Omega_0^\tau(\varphi;A)]^* \end{bmatrix} \\ \qquad \times \begin{bmatrix} C_{11}(\varphi_\tau(\varphi)) & C_{12}(\varphi_\tau(\varphi)) \\ C_{21}(\varphi_\tau(\varphi)) & C_{22}(\varphi_\tau(\varphi)) \end{bmatrix}, \qquad \tau < 0, \\ \begin{bmatrix} \Omega_\tau^0(\varphi;A) & 0 \\ -\int_\tau^0 [\Omega_0^\sigma(\varphi;A)]^*\Omega_\tau^\sigma(\varphi;A)\,d\sigma & [\Omega_0^\tau(\varphi;A)]^* \end{bmatrix} \\ \qquad \times \begin{bmatrix} C_{11}(\varphi_\tau(\varphi)) - I_n & C_{12}(\varphi_\tau(\varphi)) \\ C_{21}(\varphi_\tau(\varphi)) & C_{22}(\varphi_\tau(\varphi)) - I_n \end{bmatrix}, \qquad \tau > 0, \end{cases} \quad (2.7.12)$$

which satisfies the estimate

$$\|\bar{G}_0(\tau,\varphi)\| \leq \bar{K}_0 \exp\{-\bar{\gamma}|\tau|\}, \quad \bar{K}_0, \bar{\gamma}-\text{const} > 0. \tag{2.7.13}$$

From here it follows that the block of the matrix function (2.7.12) located at the left-side upper corner

$$G_0(\tau,\varphi) = \begin{cases} \Omega_\tau^0(\varphi;A)C_{11}(\varphi_\tau(\varphi)), & \tau \leq 0, \\ \Omega_\tau^0(\varphi;A)[C_{11}(\varphi_\tau(\varphi)) - I_n], & \tau > 0, \end{cases} \tag{2.7.14}$$

satisfies the similar estimate $\|\bar{G}_0(\tau,\varphi)\| \leq \bar{K}_0 \exp\{-\bar{\gamma}|\tau|\}$.

We show that Green functions exist other than (2.7.14). The identify (2.4.15) is represented for the projection matrix $C(\varphi) = \{C_{ij}(\varphi)\}_{i,j=1,2}$ as

$$\begin{bmatrix} C_{11}(\varphi_\tau(\varphi)) & C_{12}(\varphi_\tau(\varphi)) \\ C_{21}(\varphi_\tau(\varphi)) & C_{22}(\varphi_\tau(\varphi)) \end{bmatrix}$$

$$\equiv \begin{bmatrix} \Omega_\tau^0(\varphi;A) & 0 \\ -\int_0^\tau [\Omega_\tau^\sigma(\varphi;A)]^*\Omega_0^\sigma(\varphi;A)\,d\sigma & [\Omega_\tau^0(\varphi;A)]^* \end{bmatrix} \begin{bmatrix} C_{11}(\varphi) & C_{12}(\varphi) \\ C_{21}(\varphi) & C_{22}(\varphi) \end{bmatrix} \quad (2.7.15)$$

$$\times \begin{bmatrix} \Omega_\tau^0(\varphi;A) & 0 \\ -\int_\tau^0 [\Omega_0^\sigma(\varphi;A)]^*\Omega_\tau^\sigma(\varphi;A)\,d\sigma & [\Omega_0^\tau(\varphi;A)]^* \end{bmatrix}.$$

From here we get, in particular,

$$C_{11}(\varphi_\tau(\varphi)) \equiv \Omega_0^\tau(\varphi;A)C_{11}(\varphi)\Omega_\tau^0(\varphi;A)$$

$$- \Omega_0^\tau(\varphi;A)C_{12}(\varphi)\int_\tau^0 [\Omega_0^\sigma(\varphi;A)]^*\Omega_\tau^\sigma(\varphi;A)\,d\sigma, \tag{2.7.16}$$

$$C_{12}(\varphi_\tau(\varphi)) \equiv \Omega_0^\tau(\varphi;A)C_{12}(\varphi)[\Omega_0^\tau(\varphi;A)]^*.$$

Moreover, the estimate

$$\|\Omega_0^t(\varphi; A)C_{12}(\varphi)[\Omega_\tau^0(\varphi; A)]^*\| \leq \bar{K}_0 \exp\{-\bar{\gamma}|t-\tau|\}, \qquad (2.7.17)$$

is valid, which is implied by (2.7.13) and the second relation of (2.7.16). Clearly, $C_{12} \neq 0$. Otherwise the first of the identities (2.7.16) transforms into (2.4.15) which is only possible in the case of the non-degeneracy of the matrix $S(\varphi)$ (see Theorem 1.4.2).

Thus, by virtue of (2.7.17) the Green functions other than (2.7.14) can be represented as

$$G_0(\tau, \varphi) = \begin{cases} \Omega_0^\tau(\varphi; A)[C_{11}(\varphi_\tau(\varphi)) + pC_{12}(\varphi_\tau(\varphi))], & \tau \geq 0, \\ \Omega_0^\tau(\varphi; A)[C_{11}(\varphi_\tau(\varphi)) + pC_{12}(\varphi_\tau(\varphi)) - I_n], & \tau > 0, \end{cases} \qquad (2.7.18)$$

where p is an arbitrarily fixed parameter from R, $p \neq 0$.

This completes the proof of Theorem 2.7.2.

We establish further the symmetric matrices $C_{12}(\varphi)$, $C_{21}(\varphi)$, which are found in the structure of the Green function (2.7.12). First, consider an auxiliary assertion.

Lemma 2.7.1 *Let two systems of equations*

$$\begin{cases} \dfrac{d\varphi}{dt} = a(\varphi), \\ \dfrac{dz}{dt} = \mathcal{P}_1(\varphi)z, \end{cases} \qquad \begin{cases} \dfrac{d\varphi}{dt} = a(\varphi), \\ \dfrac{dz}{dt} = \mathcal{P}_2(\varphi)z, \end{cases} \quad z \in R^{2n}, \qquad (2.7.19)$$

have the Green functions of the problem on invariant tori

$$G_0(\tau, \varphi; \mathcal{P}_i) = \begin{cases} \Omega_\tau^0(\varphi; \mathcal{P}_i)C(\varphi_\tau(\varphi); \mathcal{P}_i), & \tau \leq 0, \\ \Omega_\tau^0(\varphi; \mathcal{P}_i)[C(\varphi_\tau(\varphi); \mathcal{P}_i) - I_{2n}], & \tau > 0, \quad i = 1, 2, \end{cases} \qquad (2.7.20)$$

which satisfy the estimates $\|G_0(\tau, \varphi; \mathcal{P}_i)\| \leq K \exp\{-\gamma|\tau|\}$, K, $\gamma - \text{const} > 0$, *and the $2n \times 2n$-dimensional matrices* $\mathcal{P}_i(\varphi) \in C^0(\mathcal{T}_m)$, $i = 1, 2$, *are related by*

$$\mathcal{P}_2(\varphi) \equiv J\mathcal{P}_1^*(\varphi)J, \quad J = \begin{pmatrix} 0 & I_n \\ -I_n & 0 \end{pmatrix}. \qquad (2.7.21)$$

Then the Green functions (2.7.20) are unique and related by

$$G_0(\tau, \varphi; \mathcal{P}_2) \equiv JG_\tau^*(0, \varphi; \mathcal{P}_1)J. \qquad (2.7.22)$$

Also, for the projection matrices $C(\varphi; \mathcal{P}_i)$, $i = 1, 2$ *the identities*

$$C(\varphi; \mathcal{P}_2) \equiv JC^*(\varphi; \mathcal{P}_1)J + I_{2n}. \qquad (2.7.23)$$

are valid.

Proof The existence of the two Green functions (2.7.20) yields the existence of two $2n$-dimensional symmetric matrices $S_i(\varphi) \in C'(\mathcal{T}_m; a)$, $i = 1, 2$ which satisfy the condition

$$\langle [\dot{S}_i(\varphi) - S_i(\varphi)\mathcal{P}_i^*(\varphi) - \mathcal{P}_i(\varphi)S_i(\varphi)]z, z \rangle \leq -\|z\|^2. \qquad (2.7.24)$$

Making the change $z \to Jz$ and in view of $J^* = -J$ we get from (2.7.24)

$$\langle J[\dot{S}_i(\varphi) - S_i(\varphi)\mathcal{P}_i^*(\varphi) - \mathcal{P}_i(\varphi)S_i(\varphi)]Jz, z \rangle \geq \|z\|^2,$$

or

$$\langle [J\dot{S}_i(\varphi)J + JS_i(\varphi)JJ\mathcal{P}_i^*(\varphi)J + J\mathcal{P}_i(\varphi)JJS_i(\varphi)J]z, z \rangle \geq \|z\|^2.$$

With regard to (2.7.21) which implies $\mathcal{P}_1 \equiv J\mathcal{P}_2^*J$, we have

$$\langle \dot{\bar{S}}_i(\varphi) + \bar{S}_i(\varphi)\mathcal{P}_j(\varphi) + \mathcal{P}_j^*(\varphi)\bar{S}_i(\varphi)]z, z \rangle \geq \|z\|^2, \qquad (2.7.25)$$
$$i \neq j, \quad i, j = 1, 2,$$

where $\bar{S}_i(\varphi) = JS_i(\varphi)J$. The simultaneous validity of the inequalities (2.7.24), (2.7.25) implies the exponential dichotomy on the whole R axis of the two linear systems

$$\frac{dz}{dt} = \mathcal{P}_i(\varphi_t(\varphi))z, \quad i = 1, 2. \qquad (2.7.26)$$

Consider now the Green function $G_t(\tau, \varphi; \mathcal{P}_1)$ of the problem on solutions which are bounded on R for system (2.7.26) for $i = 1$ and cite the following expression

$$JG_\tau^*(t, \varphi; \mathcal{P}_1)J = \begin{cases} J[\Omega_0^\tau(\varphi; \mathcal{P}_1)C(\varphi; \mathcal{P}_1)\Omega_t^0(\varphi; \mathcal{P}_1)]^*J, & t \leq \tau, \\ J[\Omega_0^\tau(\varphi; \mathcal{P}_1)(C(\varphi; \mathcal{P}_1) - I_{2n})\Omega_t^0(\varphi; \mathcal{P}_1)]^*J, & \tau < t. \end{cases} \qquad (2.7.27)$$

In view of the equalities

$$J^2 = -I_{2n}, \quad \Omega_0^t(\varphi; \mathcal{P}_2) \equiv -J[\Omega_t^0(\varphi; \mathcal{P}_1)]^*J, \qquad (2.7.28)$$

we transform the right-side part of (2.7.27) as

$$JG_\tau^*(t, \varphi; \mathcal{P}_1)J$$
$$= \begin{cases} J[\Omega_t^0(\varphi; \mathcal{P}_1)]^*JJC^*(\varphi; \mathcal{P}_1)JJ[\Omega_0^\tau(\varphi; \mathcal{P}_1)]^*J, & t \leq \tau, \\ J[\Omega_t^0(\varphi; \mathcal{P}_1)]^*JJ[C^*(\varphi; \mathcal{P}_1) - I_{2n}]JJ[\Omega_0^\tau(\varphi; \mathcal{P}_1)]^*J, & \tau < t, \end{cases} \qquad (2.7.29)$$
$$= \begin{cases} \Omega_0^t(\varphi; \mathcal{P}_2)[JC^*(\varphi; \mathcal{P}_1)J]\Omega_\tau^0(\varphi; \mathcal{P}_2), & t \leq \tau, \\ \Omega_0^t(\varphi; \mathcal{P}_2)[JC^*(\varphi; \mathcal{P}_1)J + I_{2n}]\Omega_\tau^0(\varphi; \mathcal{P}_2), & \tau < t. \end{cases}$$

Note that the matrix $C = JC^*(\varphi; \mathcal{P}_1)J + I_{2n}$ is projecting, $C^2 = [JC^*(\varphi; \mathcal{P}_1)J + I_{2n}]^2 = -JC^*(\varphi; \mathcal{P}_1)J + JC^*(\varphi; \mathcal{P}_1)J + JC^*(\varphi; \mathcal{P}_1)J + I_{2n} = C$. The equality (2.7.29) is represented as

$$JG_\tau^*(t, \varphi; \mathcal{P}_1)J = \begin{cases} \Omega_0^t(\varphi; \mathcal{P}_2)C\Omega_\tau^0(\varphi; \mathcal{P}_2), & \tau \le t, \\ \Omega_0^t(\varphi; \mathcal{P}_2)[C - I_n]\Omega_\tau^0(\varphi; \mathcal{P}_2), & \tau > t. \end{cases} \quad (2.7.30)$$

As the function (2.7.30) has the same structure as the Green function $G_t(\tau, \varphi; \mathcal{P}_2)$ and the Green function is unique (due to the exponential dichotomy on R of the system (2.7.26), then

$$G_t(\tau, \varphi; \mathcal{P}_2) = JG_\tau^*(t, \varphi; \mathcal{P}_1)J. \quad (2.7.31)$$

Setting $t = 0$ in (2.7.31) we come to (2.7.22). The comparison of the corresponding projection matrices yields (2.7.23).

Applying the lemma it is easy to establish the following results.

Theorem 2.7.3 *Let the system of equations*

$$\frac{d\varphi}{dt} = a(\varphi),$$

$$\frac{dx}{dt} = A(\varphi)x, \quad (2.7.32)$$

$$\frac{dy}{dt} = B(\varphi)x - A^*(\varphi)y, \quad y \in R^n,$$

have the Green function on invariant tori for some symmetric matrix $B(\varphi) \in C^0(\mathcal{T}_m)$:

$$\bar{G}_0(\tau, \varphi) = \begin{cases} \begin{bmatrix} \Omega_\tau^0(\varphi; A) & 0 \\ \int_\tau^0 [\Omega_0^\sigma(\varphi; A)]^* B(\varphi_\sigma(\varphi))\Omega_\tau^\sigma(\varphi; A)\, d\sigma & [\Omega_0^\tau(\varphi; A)]^* \end{bmatrix} \\ \times \begin{bmatrix} C_{11}(\varphi_\tau(\varphi)) & C_{12}(\varphi_\tau(\varphi)) \\ C_{21}(\varphi_\tau(\varphi)) & C_{22}(\varphi_\tau(\varphi)) \end{bmatrix}, \quad \tau \le 0, \\[2ex] \begin{bmatrix} \Omega_\tau^0(\varphi; A) & 0 \\ \int_\tau^0 [\Omega_0^\sigma(\varphi; A)]^* B(\varphi_\sigma(\varphi))\Omega_\tau^\sigma(\varphi; A)\, d\sigma & [\Omega_0^\tau(\varphi; A)]^* \end{bmatrix} \\ \times \begin{bmatrix} C_{11}(\varphi_\tau(\varphi)) - I_n & C_{12}(\varphi_\tau(\varphi)) \\ C_{21}(\varphi_\tau(\varphi)) & C_{22}(\varphi_\tau(\varphi)) - I_n \end{bmatrix}, \quad \tau > 0. \end{cases}$$

$$(2.7.33)$$

Then this function is unique and the $n \times n$-dimensional blocks $C_{ij}(\varphi)$, $i, j = 1, 2$ of the projection matrix

$$C(\varphi) = \begin{bmatrix} C_{11}(\varphi) & C_{12}(\varphi) \\ C_{21}(\varphi) & C_{22}(\varphi) \end{bmatrix}, \quad C^2(\varphi) \equiv C(\varphi), \qquad (2.7.34)$$

have the following properties

$$C_{12}(\varphi) \equiv C_{12}^*(\varphi), \quad C_{21}(\varphi) \equiv C_{21}^*(\varphi), \quad C_{11}(\varphi) + C_{22}^*(\varphi) \equiv I_n. \qquad (2.7.35)$$

Proof We designate by $\mathcal{P}(\varphi)$ the matrix

$$\mathcal{P}(\varphi) = \begin{bmatrix} A(\varphi) & 0 \\ B(\varphi) & -A^*(\varphi) \end{bmatrix} \qquad (2.7.36)$$

and note that an identify similar to (2.7.21) holds for it

$$\mathcal{P}(\varphi) \equiv J\mathcal{P}^*(\varphi)J. \qquad (2.7.37)$$

Therefore, by the above lemma the Green function (2.7.33) is unique. Moreover, (2.7.23) for the projection matrix (2.7.34) is

$$\begin{bmatrix} C_{11}(\varphi) & C_{12}(\varphi) \\ C_{21}(\varphi) & C_{22}(\varphi) \end{bmatrix} \equiv \begin{bmatrix} 0 & I_n \\ -I_n & 0 \end{bmatrix} \begin{bmatrix} C_{11}^*(\varphi) & C_{12}^*(\varphi) \\ C_{21}^*(\varphi) & C_{22}^*(\varphi) \end{bmatrix}$$
$$\times \begin{bmatrix} 0 & I_n \\ -I_n & 0 \end{bmatrix} + \begin{bmatrix} I_n & 0 \\ 0 & I_n \end{bmatrix}. \qquad (2.7.38)$$

Proceeding with the expressions obtained we arrive at (2.7.35).

Note that (2.7.37) can be satisfied not only for the matrix (2.7.36) but for the matrix of a more general form

$$\mathcal{P}(\varphi) = \begin{bmatrix} A(\varphi) & D(\varphi) \\ B(\varphi) & -A^*(\varphi) \end{bmatrix}, \qquad (2.7.39)$$

where $B(\varphi), D(\varphi) \in C^0(\mathcal{T}_m)$ are n-dimensional symmetric matrices. The identity (2.7.37) indicates the possibility to pass from the system

$$\frac{d\varphi}{dt} = a(\varphi), \quad \frac{dz}{dt} = \mathcal{P}(\varphi)z, \quad z = \begin{pmatrix} x \\ y \end{pmatrix} \in R^{2n}, \qquad (2.7.40)$$

to the conjugated system

$$\frac{d\varphi}{dt} = a(\varphi), \quad \frac{d\bar{z}}{dt} = -\mathcal{P}^*(\varphi)\bar{z}, \qquad (2.7.41)$$

by means of a non-degenerate change of variables $\bar{z} = Jz$. Thus, the existence of the quadratic form $V(\varphi, z)$ which has a sign-definite derivative along the solution of the system (2.7.40) induces the existence of the same form for the conjugated system (2.7.41) $\bar{V}(\varphi, \bar{z}) = V(\varphi, J^{-1}\bar{z})$ and the presence of the two forms excludes its degeneracy. Now we assume, in spite of (2.7.37), that the system (2.7.40) is transformed by means of the non-degenerate change of variables $\bar{z} = X(\varphi)z$, $\det X(\varphi) \neq 0$, $X(\varphi) \in C'(\mathcal{T}_m; a)$ to the conjugated system (2.7.41). Then, evidently, the matrix function $X(\varphi)$ satisfies the equation $\dot{X} + X\mathcal{P}(\varphi) + \mathcal{P}^*(\varphi)X = 0$. Therefore the solution of the matrix equation

$$X + XA(\varphi) + A^*(\varphi)X = 0 \qquad (2.7.42)$$

is of a special interest.

Theorem 2.7.4 *Let the matrix equation (2.7.42) have the non-degenerate solution $X = L(\varphi) \in C'(\mathcal{T}_m; a)$, $\det L(\varphi) \neq 0$, $\forall \varphi \in R^m$, where the $n \times n$-dimensional matrix $L(\varphi)$ is not necessarily symmetric and there exist an n-dimensional symmetric matrix $S(\varphi) \in C'(\mathcal{T}_m; a)$ satisfying the inequality*

$$\langle [\dot{S}(\varphi) + S(\varphi)A(\varphi) + A^*(\varphi)S(\varphi)]x, x \rangle \leq -\|x\|^2 \qquad (2.7.43)$$

for all $x \in R^n$, $\varphi \in R^m$. Then $\det S(\varphi) \neq 0$, $\forall \varphi \in R^m$ and the system of equations (2.2.4) has a unique Green function (2.6.10) with the estimate (2.6.11). In addition n is even and the identify

$$G_0(\tau, \varphi) \equiv -L^{-1}(\varphi)G^*_\tau(0, \varphi)L(\varphi_\tau(\varphi)) \qquad (2.7.44)$$

holds for the Green function $G_0(\tau, \varphi)$.

Proof Let us verify that the derivative of the quadratic form $\langle S(\varphi_t(\varphi_0))L^{-1}(\varphi_t(\varphi_0))x, L^{-1}(\varphi_t(\varphi_0))x \rangle$ computed along the solution of the system (2.7.7) is negative definite. Taking into account (2.7.43), we have

$$\langle (L^{-1})^*(\dot{S} + SA^* + AS)L^{-1}x, x \rangle \leq -\|L\|^{-2}\|x\|^2. \qquad (2.7.45)$$

Designating $\bar{S} = (L^{-1})^*SL^{-1}$ and in view of the identify $\dot{L}^{-1}(\varphi) \equiv -L^{-1}(\varphi)\dot{L}(\varphi)L^{-1}(\varphi)$, we transform the left-hand side of (2.7.45) to the form

$$\langle [(L^{-1})^*\dot{S}L^{-1} + (L^{-1})^*SL^{-1}LAL^{-1} + (L^{-1})^*A^*L^*(L^{-1})^*SL^{-1}]x, x \rangle$$
$$= \langle [\dot{\bar{S}} - (L^{-1})^*\dot{L}^*\bar{S} - \bar{S}\dot{L}L^{-1} + \bar{S}LAL^{-1} + (LAL^{-1})^*\bar{S}]x, x \rangle$$
$$= \langle [\dot{\bar{S}} + \bar{S}(LAL^{-1} + \dot{L}L^{-1}) + (\dot{L}L^{-1} + LAL^{-1})^*\bar{S}]x, x \rangle$$
$$= \langle (\dot{\bar{S}} - \bar{S}A^* - A\bar{S})x, x \rangle.$$

Thus, the simultaneous satisfaction of the inequalities (2.7.45) and (2.7.43) is only possible if the matrix $S(\varphi)$ is non-degenerate for all $\varphi \in R^m$. Hence system (2.2.4) has a unique Green function (2.6.10) satisfying the estimate (2.6.11)

The solutions $x = x(t)$, $y = y(t)$ of the two mutually conjugated linear systems

$$\frac{dx}{dt} = A(\varphi_t(\varphi_0))x, \quad \frac{dy}{dt} = -A^*(\varphi_t(\varphi_0))y$$

are related by $x(t) = L^{-1}(\varphi_t(\varphi_0))y(t)$, i.e. if $y = y(t)$ is a solution of the second system, then multiplying $y(t)$ by the matrix function $L^{-1}(\varphi_t(\varphi_0))$ yields the solution of the first system. From here it follows that if a solution of the second system damps to $+\infty$, then the corresponding solution of the first system also damps to $+\infty$. However, on the other hand, for all $t \in R$, for any solutions of the first system $\langle L(\varphi_t(\varphi_0))x(t), x(t)\rangle \equiv \text{const}$. Thus, the first system has as many independent solutions damping on $+\infty$ as the increasing ones. Consequently, n is an even valued.

We prove identify (2.7.44). Consider the Green function $G_t(\tau, \varphi_0)$ of the problem on solutions to system (2.6.1) bounded on R and the expression

$$-L^{-1}(\varphi_t(\varphi_0))G_\tau^*(t, \varphi_0)L(\varphi_\tau(\varphi_0)) = G.$$

We have

$$G(\tau, \varphi) = \begin{cases} -L^{-1}(\varphi_t(\varphi_0))[\Omega_t^0(\varphi_0; A)]^* L(\varphi_0) L^{-1}(\varphi_0)[C^*(\varphi_0) - I_n] \\ \times L(\varphi_0)L^{-1}(\varphi_0)[\Omega_0^\tau(\varphi_0; A)]^* L(\varphi_\tau(\varphi_0)), & \tau \leq t; \\ -L^{-1}(\varphi_t(\varphi_0))[\Omega_t^0(\varphi_0; A)]^* L(\varphi_0) L^{-1}(\varphi_0) C^*(\varphi_0) \\ \times L(\varphi_0)L^{-1}(\varphi_0)[\Omega_0^\tau(\varphi_0; A)]^* L(\varphi_\tau(\varphi_0)), & \tau > t. \end{cases} \quad (2.7.46)$$

In view of the interconnection between the matriciants

$$L^{-1}(\varphi_t(\varphi_0))[\Omega_t^\tau(\varphi_0; A)]^* L(\varphi_t(\varphi_0)) = \Omega_\tau^t(\varphi_0; A)$$

we represent (2.7.46) as

$$G(\tau, \varphi) = \begin{cases} \Omega_0^t(\varphi_0; A)P(\varphi_0)\Omega_\tau^0(\varphi_0; A), & \tau \leq t, \\ \Omega_0^t(\varphi_0; A)[P(\varphi_0) - I_n]\Omega_\tau^0(\varphi_0; A), & \tau > t, \end{cases} \quad (2.7.47)$$

where $P(\varphi_0) = -[L^{-1}(\varphi_0)C^*(\varphi_0)L(\varphi_0) - I_n]$. Due to the uniqueness of the Green function equality (2.7.47) determines the Green function $G_t(\tau, \varphi_0)$. Therefore $G_t(\tau, \varphi_0) \equiv -L^{-1}(\varphi_t(\varphi_0))G_\tau^*(t, \varphi_0)L(\varphi_\tau(\varphi_0))$ and the projection matrix $C(\varphi)$ satisfies the identify

$$L(\varphi)C(\varphi) + C^*(\varphi)L(\varphi) \equiv L(\varphi), \quad (2.7.47a)$$

This completes the proof of Theorem 2.7.4.

Corollary 2.7.1 *If there exists an n-dimensional symmetric matrix function $S(\varphi) \in C'(\mathcal{T}_m; a)$ satisfying the condition (2.7.1) and for some $\varphi = \varphi_0 \in R^m$ det $S(\varphi_0) = 0$, then the matrix equation (2.7.42) has no non-degenerate solutions, but possesses non-trivial degenerating solutions $X = C_{12}(\varphi)$, $C_{12}(\varphi) \neq 0$, det $C_{12}(\varphi^0) = 0$ for some $\varphi^0 \in R^m$.*

Remark 2.7.1 Letting $L(\varphi) = J$ in (2.7.42a), we arrive at (2.7.23).

As system (2.7.11) has the unique Green function (2.7.12), a question arises as to what properties the matrices $B(\varphi)$, $D(\varphi) \in C^0(\mathcal{T}_m)$, which are not necessarily symmetric ones, should possess for the system

$$\frac{d\varphi}{dt} = a(\varphi),$$
$$\frac{dx}{dt} = A(\varphi)x + B(\varphi)y, \qquad (2.7.48)$$
$$\frac{dy}{dt} = D(\varphi)x - A^*(\varphi)y, \quad y \in R^n,$$

to have a unique Green function.

Theorem 2.7.5 *Let there exist an n-dimensional symmetric matrix $S(\varphi) \in C'(\mathcal{T}_m; a)$ satisfying condition (2.2.39) and the matrix $B(\varphi) \in C^0(\mathcal{T}_m)$ satisfying the inequality*

$$\langle B(\varphi)y, y \rangle \geq \beta_0 \|y\|^2, \quad \beta_0 = \text{const} > 0, \qquad (2.7.49)$$

for all $y \in R^n$. Then, for every matrix $D(\varphi) \in C^0(\mathcal{T}_m)$ satisfying the inequality

$$\langle D(\varphi)x, x \rangle \geq 0, \quad \forall x \in R^n, \quad \varphi \in R^m, \qquad (2.7.50)$$

the system (2.7.48) has the unique Green function of the problem on invariant tori.

Proof We take the quadratic form as

$$V(\varphi, x, y) = \langle S(\varphi)x, x \rangle - \lambda \langle x, y \rangle, \quad \lambda > 0, \qquad (2.7.50a)$$

and compute the derivative along solutions of (2.7.48)

$$\dot{V}(\varphi, x, y) = \langle \dot{S}(\varphi)x, x \rangle + \langle S(Ax + By), x \rangle + \langle Sx, Ax + By \rangle$$
$$- \lambda \langle Ax + By, y \rangle - \lambda \langle x, Dx - A^*y \rangle$$
$$\leq -\|x\|^2 + 2\langle SBy, x \rangle - \lambda \langle By, y \rangle - \lambda \langle Dx, x \rangle$$
$$\leq -\|x\|^2 + \beta_0 \lambda \|y\|^2 + 2\langle SBy, x \rangle.$$

We see that the derivative $\dot{V}(\varphi, x, y)$ of the non-degenerate quadratic form (2.7.50a) computed along solutions of the system (2.7.48) is negative definite for sufficiently large values of λ. Thus, the system (2.7.48) has the unique Green function of the problem on invariant tori.

Remark 2.7.2 In the cited theorem the matrices $B(\varphi)$ and $D(\varphi)$ are not necessarily symmetric. It is sufficient to assume that all eigenvalues of the matrix $\hat{B}(\varphi) = B(\varphi) + B^*(\varphi)$ are positive (negative), and eigenvalues of the matrix $\hat{D}(\varphi) = D(\varphi) + D^*(\varphi)$ are non-negative (non-positive).

Remark 2.7.3 The condition (2.7.50) is not necessary. It can be replaced by a weaker requirement $\lambda \langle D(\varphi)y, y \rangle \geq -\varepsilon_0 \|y\|^2$, where $\varepsilon_0 > 0$ is a sufficiently small fixed number. Actually, with regard to the above estimate for $\dot{V}$ we have

$$\dot{V}(\varphi, x, y) \leq -(1 - \lambda \varepsilon_0)\|x\|^2 + 2\|S\|_0 \|B\|_0 \|x\| \|y\| - \beta_0 \lambda \|y\|^2. \tag{2.7.51}$$

It is clear then, that for the negative definiteness of the derivative $\dot{V}(\varphi, x, y)$ it is necessary that a positive value λ exists satisfying the inequality

$$\begin{vmatrix} 1 - \lambda \varepsilon_0 & -\|S\|_0 \|B\|_0 \\ -\|S\|_0 \|B\|_0 & \beta_0 \lambda \end{vmatrix} > 0. \tag{2.7.52}$$

Expending the determinant and multiplying all its terms by $-\beta_0^{-1}$ one obtains

$$\varepsilon_0 \lambda^2 - \lambda + \frac{\|S\|_0^2 \|B\|_0^2}{\beta} < 0. \tag{2.7.53}$$

It is necessary and sufficient for the existence of the positive solution to this inequality that $\varepsilon_0 < \beta_0 \left(4\|S\|_0^2 \|B\|_0^2\right)^{-1}$.

Returning to the system (2.7.48), we note that if both matrices $B(\varphi)$ and $D(\varphi)$ are positive definite, the system (2.7.48) has a unique the Green function for any matrix $A(\varphi) \in C^0(\mathcal{T}_m)$. We formulate this as follows.

Theorem 2.7.6 *Let $n \times n$-dimensional matrices $B(\varphi)$, $D(\varphi) \in C^0(\mathcal{T}_m)$ (not necessarily symmetric) satisfy the condition*

$$\langle B(\varphi)y, y \rangle \langle D(\varphi)x, x \rangle \geq \alpha_0 \|y\|^2 \|x\|^2, \quad \alpha_0 = \text{const} > 0, \tag{2.7.54}$$

for all independent $x, y \in R^n$. Then for every $n \times n$-dimensional matrix $A(\varphi) \in C^0(\mathcal{T}_m)$ system (2.7.48) has the unique Green function of the problem on invariant tori.

Proof Obviously, condition (2.7.54) denotes either negative definiteness of both the matrices

$$\langle B(\varphi)y, y \rangle \leq -\alpha_0^{(1)} \|y\|^2, \quad \langle D(\varphi)x, x \rangle \leq -\alpha_0^{(2)} \|x\|^2, \tag{2.7.55}$$

or its positive definiteness

$$\langle B(\varphi)y, y\rangle \geq \alpha_0^{(1)}\|y\|^2, \quad \langle D(\varphi)x, x\rangle \geq \alpha_0^{(2)}\|x\|^2. \tag{2.7.56}$$

Let the second case take place. Then we make sure that the derivative of the non-degenerate quadratic form

$$V(x,y) = -\langle x, y\rangle, \tag{2.7.57}$$

computed along solutions to system (2.7.48) is negative definite

$$\dot{V}(x,y) = -[\langle A(\varphi)x + B(\varphi)y, y\rangle + \langle x, D(\varphi)x - A^*(\varphi)y\rangle]$$
$$= -[\langle B(\varphi)y, y\rangle + \langle D(\varphi)x, x\rangle] \leq -\alpha_0^{(1)}\|y\|^2 - \alpha_0^{(2)}\|x\|^2.$$

In the case when inequalities (2.7.55) are satisfied the quadratic form is taken as $V(x,y) = \langle x, y\rangle$.

This completes the proof of Theorem 2.7.6.

We make the change of variables $x = u - v$, $y = u + v$ in the conditions of Theorem 2.7.6 in system (2.7.48) and in the quadratic form (2.7.57). The system (2.7.48) becomes

$$\frac{d\varphi}{dt} = a(\varphi),$$

$$\frac{du}{dt} = \frac{1}{2}[A(\varphi) + B(\varphi) + C(\varphi) - A^*(\varphi)]u + \frac{1}{2}[B(\varphi) - A(\varphi) - C(\varphi) - A^*(\varphi)]v,$$

$$\frac{dv}{dt} = \frac{1}{2}[C(\varphi) - A(\varphi) - A^*(\varphi) - B(\varphi)]u + \frac{1}{2}[A(\varphi) - C(\varphi) - A^*(\varphi) - B(\varphi)]v,$$

and the quadratic form (2.7.57) is $\|u\|^2 - \|v\|^2$. We designate

$$L(\varphi) = \frac{1}{2}[B(\varphi) + C(\varphi)], \quad M(\varphi) = \frac{1}{2}[A(\varphi) - A^*(\varphi)],$$

$$N(\varphi) = \frac{1}{2}[A(\varphi) + A^*(\varphi)], \quad R(\varphi) = \frac{1}{2}[C(\varphi) - B(\varphi)],$$

and note that the matrices L, M, N, R have the following properties

$$M(\varphi) + M^*(\varphi) \equiv 0; \tag{2.7.58}$$

$$\langle [L(\varphi) \pm R(\varphi)]x, x\rangle \geq \gamma\|x\|^2, \tag{2.7.59}$$

or

$$\langle [L(\varphi) \pm R(\varphi)]x, x\rangle \leq -\gamma\|x\|^2, \quad \gamma = \text{const} > 0. \tag{2.7.60}$$

Theorem 2.7.7 *In the system of equations*

$$\frac{d\varphi}{dt} = a(\varphi),$$
$$\frac{dx}{dt} = [L(\varphi) + M(\varphi)]x + [N(\varphi) - R(\varphi)]y, \qquad (2.7.61)$$
$$\frac{dy}{dt} = [L(\varphi) + R(\varphi)]x + [M(\varphi) - L(\varphi)]y$$

let the matrix functions $M(\varphi)$, $L(\varphi)$, $R(\varphi) \in C^0(\mathcal{T}_m)$ satisfy the conditions (2.7.58), (2.7.59) or (2.7.60). Then for every n-dimensional symmetric matrix $N(\varphi) \in C^0(\mathcal{T}_m)$ the system (2.7.61) has a unique Green function $\bar{G}_0(\tau, \varphi)$ of the problem on invariant tori.

Proof Computing the derivative of the quadratic form $V(x,y) = \|x\|^2 - \|y\|^2$ along solutions of (2.7.61), one obtains

$$\begin{aligned}\dot{V}(x,y) &= 2\langle[(L+M)x + (N-R)y], x\rangle - 2\langle y, [(N+R)x + (M-L)y]\rangle \\ &= 2\langle(L+M)x, x\rangle + 2\langle(N-R)y, x\rangle - 2\langle(N+R)x, y\rangle - 2\langle y, (M-L)y\rangle.\end{aligned} \qquad (2.7.62)$$

In view of the equivalence of (2.7.58) to $\langle M(\varphi)x, xt\rangle \equiv 0$, we proceed with (2.7.62) as

$$\begin{aligned}\dot{V}(x,y) &= 2\langle Lx, x\rangle + 2\langle Ly, y\rangle - 2\langle Ry, x\rangle - 2\langle Rx, y\rangle \\ &= \langle(L+R)(x+y), (x+y)\rangle + \langle(L-R)(x-y), (x-y)\rangle \\ &\geq \gamma(\|x+y\|^2 + \|x-y\|^2) = 2\gamma(\|x\|^2 + \|y\|^2).\end{aligned}$$

The estimate obtained completes the proof of Theorem 2.7.7.

2.8 The Non-unique Green Function and the Properties of the System Implied by its Existence

We note that for the non-unique Green function (2.6.10) the identities (2.4.15) and (2.4.16) are not satisfied. The first of the identities is replaced by (2.7.16). Moreover, the following result is valid.

Theorem 2.8.1 *Let there exist two $n \times n$-dimensional matrix functions $C(\varphi)$, $H(\varphi) \in C^0(\mathcal{T}_m)$ satisfying the identity*

$$\Omega_t^0(\varphi; A)C(\varphi_t(\varphi))\Omega_0^t(\varphi; A) \equiv C(\varphi) - \int_t^0 H^*(\varphi)[\Omega_0^\sigma(\varphi; A)]^*\Omega_0^\sigma(\varphi; A)\,ds \qquad (2.8.1)$$

and the estimates

$$\|\Omega_0^t(\varphi; A)C(\varphi)\| \leq K \exp\{-\gamma t\}, \quad t \geq 0,$$
$$\|\Omega_0^t(\varphi; A)[C(\varphi) - I_n]\| \leq K \exp\{\gamma t\}, \quad t \leq 0, \quad (2.8.2)$$
$$\|\Omega_0^t(\varphi; A)H(\varphi)\| \leq K \exp\{-\gamma |t|\}, \quad t \in R, \quad K, \gamma - \text{const} > 0.$$

These matrix functions $C(\varphi)$, $H(\varphi)$ are then unique.

Proof Let there exist matrices $\bar{C}(\varphi)$, $\bar{\bar{C}}(\varphi)$ and $\bar{H}(\varphi)$, $\bar{\bar{H}}(\varphi)$ satisfying (2.8.1) and (2.8.2). Then for $\tilde{C}(\varphi) = \bar{C}(\varphi) - \bar{\bar{C}}(\varphi)$ and $\tilde{H}(\varphi) = \bar{H}(\varphi) - \bar{\bar{H}}(\varphi)$ we get an identity similar to (2.8.1)

$$\Omega_t^0(\varphi; A)\tilde{C}(\varphi_t(\varphi))\Omega_0^t(\varphi; A) \equiv \tilde{C}(\varphi) - \int_t^0 \tilde{H}^*(\varphi)[\Omega_0^\sigma(\varphi; A)]^*\Omega_0^\sigma(\varphi; A)\,d\sigma \quad (2.8.3)$$

By virtue of (2.8.2)

$$\|\Omega_0^t(\varphi; A)\tilde{C}(\varphi)\| \leq 2K \exp\{-\gamma |t|\},$$
$$\|\Omega_0^t(\varphi; A)\tilde{H}(\varphi)\| \leq 2K \exp\{-\gamma |t|\}, \quad t \in R. \quad (2.8.4)$$

Since the constants K, γ do not depend on $\varphi \in R^m$, then the first inequality (2.8.4) yields

$$\|\Omega_\tau^t(\varphi; A)\tilde{C}(\varphi_\tau(\varphi))\| \leq 2K \exp\{-\gamma |t - \tau|\}, \quad (2.8.5)$$

for all $t, \tau \in R$. Multiplying (2.8.3) from the right by the matrix $H(\varphi)$

$$\Omega_t^0(\varphi; A)\tilde{C}(\varphi_t(\varphi))\Omega_0^t(\varphi; A) \equiv \tilde{C}(\varphi)\tilde{H}(\varphi)$$
$$- \int_t^0 [\Omega_0^\sigma(\varphi; A)\tilde{H}(\varphi)]^*\Omega_0^\sigma(\varphi; A)\tilde{H}(\varphi)\,d\sigma, \quad (2.8.6)$$

and taking into account the estimates (2.8.4), (2.8.5) we get, passing to the limit as $t \to -\infty$

$$0 \equiv \tilde{C}(\varphi)\tilde{H}(\varphi) - \int_{-\infty}^0 [\Omega_0^\sigma(\varphi; A)\tilde{H}(\varphi)]^*\Omega_0^\sigma(\varphi; A)\tilde{H}(\varphi)\,d\sigma. \quad (2.8.7)$$

Now we pass to the limit as $t \to +\infty$ in (2.8.6)

$$0 \equiv \tilde{C}(\varphi)\tilde{H}(\varphi) + \int_0^\infty [\Omega_0^\sigma(\varphi; A)\tilde{H}(\varphi)]^*\Omega_0^\sigma(\varphi; A)\tilde{H}(\varphi)\,d\sigma. \quad (2.8.8)$$

Identities (2.8.7) and (2.8.8) imply

$$\int_{-\infty}^{\infty} [\Omega_0^\sigma(\varphi; A)\tilde{H}(\varphi)]^* \Omega_0^\sigma(\varphi; A)\tilde{H}(\varphi)\, d\sigma \equiv 0. \tag{2.8.9}$$

Hence for an arbitrary numerical vector $\eta \in R^n$ we get

$$\int_{-\infty}^{\infty} \|\Omega_0^\sigma(\varphi; A)\tilde{H}(\varphi)\eta\|^2\, d\sigma \equiv 0. \tag{2.8.10}$$

This is only possible when $\tilde{H}(\varphi) \equiv 0$. Thus, (2.8.3) becomes

$$\Omega_t^0(\varphi; A)\tilde{C}(\varphi_t(\varphi))\Omega_0^t(\varphi; A) \equiv \tilde{C}(\varphi).$$

From this we find for every fixed $\varphi = \varphi_0 \in R^m$

$$[\Omega_0^t(\varphi_0; A)\tilde{C}(\varphi_t(\varphi_0))]^* \equiv [\Omega_t^0(\varphi_0; A)]^* \tilde{C}^*(\varphi_0). \tag{2.8.11}$$

Note that the left-hand side of (2.8.9) is exponentially damping as $t \to \pm\infty$ due to the estimate (2.8.5) and the right-hand side of it is a solution of the conjugated system (2.7.7) which evidently has no non-trivial solutions bounded on R. Thus, there is one possibility left, i.e. that the identity $\tilde{C}(\varphi) \equiv 0$ is satisfied.

This completes the proof of Theorem 2.8.1.

In the assumption on satisfying the condition of Theorem 2.7.2 we consider the inhomogeneous system of equations of the type

$$\begin{aligned}\frac{d\varphi}{dt} &= a(\varphi), \\ \frac{dx}{dt} &= A(\varphi)x + f(\varphi), \\ \frac{dy}{dt} &= -x - A^*(\varphi)y + g(\varphi),\end{aligned} \tag{2.8.12}$$

where $f(\varphi), g(\varphi) \in C^0(\mathcal{T}_m)$. Obviously, this system has a unique invariant torus for every fixed vector function $f(\varphi), g(\varphi) \in C^0(\mathcal{T}_m)$ and this torus can be represented via the Green function (2.7.12) as

$$\begin{pmatrix} x \\ y \end{pmatrix} = \begin{pmatrix} u(\varphi) \\ v(\varphi) \end{pmatrix} = \int_{-\infty}^{\infty} \bar{G}_0(\tau, \varphi) \begin{pmatrix} f(\varphi_\tau(\varphi)) \\ g(\varphi_\tau(\varphi)) \end{pmatrix} d\tau. \tag{2.8.13}$$

We set $f(\varphi) \equiv 0$ in this inequality and get for the first n coordinates

$$x = u(\varphi) = \int_{-\infty}^{\infty} \Omega_\tau^0(\varphi; A) C_{12}(\varphi_\tau(\varphi)) g(\varphi_\tau(\varphi)) \, d\tau$$

$$= \int_{-\infty}^{\infty} C_{12}(\varphi) [\Omega_0^\tau(\varphi; A)]^* g(\varphi_\tau(\varphi)) \, d\tau.$$
(2.8.14)

Clearly, equality (2.8.14) determines the invariant torus of system (2.2.4) no matter what the vector function $g(\varphi) \in C^0(\mathcal{T}_m)$ is.

Proposition 2.8.1 *Equality (2.8.14) determines all invariant tori of the system (2.2.4).*

Actually, let the equality $x = u_0(\varphi)$ define some non-trivial invariant torus of the system (2.2.4). Then the system of equations

$$\frac{d\varphi}{dt} = a(\varphi),$$

$$\frac{dx}{dt} = A(\varphi)x, \qquad (2.8.15)$$

$$\frac{dy}{dt} = -x - A^*(\varphi)y + u_0(\varphi),$$

has a unique torus $x = u_0(\varphi)$, $y = 0$ which is represented by (2.8.13), where $f(\varphi) \equiv 0$, $g(\varphi) = u_0(\varphi)$. Therefore

$$u_0(\varphi) \equiv \int_{-\infty}^{\infty} C_{12}(\varphi) [\Omega_0^\tau(\varphi; A)]^* u_0(\varphi_\tau(\varphi)) \, d\tau. \qquad (2.8.16)$$

The representation of non-trivial invariant tori of system (2.2.4) in the form of (2.8.14) and the identity (2.8.16) yield the following assertion.

Proposition 2.8.2 *The operator $\mathfrak{M}$:*

$$\mathfrak{M}g = \int_{-\infty}^{\infty} C_{12}(\varphi) [\Omega_0^\tau(\varphi; A)]^* g(\varphi_\tau(\varphi)) \, d\tau \qquad (2.8.17)$$

acting from the space $C^0(\mathcal{T}_m)$ into the space $C'(\mathcal{T}_m; a)$ is projecting, i.e. $\mathfrak{M}^2 = \mathfrak{M}$.

Note that Theorem 1.5.3 yields the following.

Theorem 2.8.2 *Let there exist an n-dimensional symmetric matrix $S(\varphi) \in C'(\mathcal{T}_m; a)$ which satisfies the condition (2.7.1) and the determinant of the matrix $S(\varphi_0)$ equal to zero at some point $\varphi = \varphi_0 \in R^m$. Then there exists a unique n-dimensional symmetric matrix $H(\varphi) \in C^0(\mathcal{T}_m^-)$ satisfying the two conditions:*

(1) *the equality $x = \Omega_0^t(\varphi_0; A) H(\varphi_0) \eta$ determines all bounded on R solutions of system (2.6.1) for the changing constant vector $\eta \in R^n$ and every fixed $\varphi_0 \in R^m$;*
(2) *the identity*

$$H(\varphi) = \int_{-\infty}^{\infty} H(\varphi) [\Omega_0^\tau(\varphi; A)]^* \Omega_0^\tau(\varphi; A) H(\varphi) \, d\tau \qquad (2.8.18)$$

holds for all $\varphi \in R^m$. In addition the matrix $H(\varphi)$ coincides with the matrix $C_{12}(\varphi)$ found in (2.8.14) – (2.8.17).

Remark 2.8.1 Identity (2.8.18) allows us to determine the matrix function $H(\varphi)$ unambiguously in some cases.

Let us verify this by example (2.1.54). We recall that

$$\Omega_0^t(\varphi; A) = \left[\cos^2 \frac{\varphi}{2} \exp -t + \sin^2 \frac{\varphi}{2} \exp t \right]^{-1}.$$

Then (2.8.18) becomes

$$H(\varphi) \equiv \int_{-\infty}^{\infty} H^2(\varphi) \left[\cos^2 \frac{\varphi}{2} \exp -t + \sin^2 \frac{\varphi}{2} \exp t \right]^{-2} dt.$$

From here we get, letting $\varphi \neq \pi k, \ k \in Z$

$$H(\varphi) = \left\{ \int_{-\infty}^{\infty} \left[\cos^2 \frac{\varphi}{2} \exp -t + \sin^2 \frac{\varphi}{2} \exp t \right]^{-2} dt \right\}^{-1} = \frac{1}{2} \sin^2 \varphi. \qquad (2.8.19)$$

Evidently, it is necessary to let $H(\pi k) = 0$ at the points $\varphi = \pi k$. Thus, in view of the above representation of one of the tori of system (2.1.54) and function (2.8.19) all invariant tori of system (2.1.54) are represented

$$x = \sin^2 \frac{\varphi}{2} \int_{-\infty}^{0} \exp\{\tau\} f(\varphi_\tau(\varphi)) \, d\tau - \cos^2 \frac{\varphi}{2} \int_{0}^{\infty} \exp\{-\tau\} f(\varphi_\tau(\varphi)) \, d\tau$$

$$+ \frac{1}{2}\int_{-\infty}^{\infty} \frac{g(\varphi_\tau(\varphi)) \sin^2 \varphi}{\cos^2 \frac{\varphi}{2} \exp\{-\tau\} + \sin^2 \frac{\varphi}{2} \exp\{\tau\}} d\tau,$$

where $g(\varphi)$ is an arbitrary scalar function from the space $C^0(\mathcal{T}_m)$.

Applying Theorem 1.3.4 we conclude as follows.

Theorem 2.8.3 *Let there exist two symmetric matrix functions $S_1(\varphi), S_2(\varphi) \in C'(\mathcal{T}_m; a)$ satisfying the conditions*

$$\langle [\dot{S}_1(\varphi) + S_1(\varphi)A(\varphi) + A^*(\varphi)S_1(\varphi)]x, x\rangle \leq -\beta_1\|x\|^2; \qquad (2.8.20)$$

$$\langle [\dot{S}_2(\varphi) - S_2(\varphi)A^*(\varphi) - A(\varphi)S_2(\varphi)]x, x\rangle \leq -\beta_2\|x\|^2, \qquad (2.8.21)$$

where $\beta_i = \mathrm{const} > 0$. Then $\det S_i(\varphi) \neq 0$, $i = 1, 2$, for all $\varphi \in R^m$ and the system (2.2.4) and the system

$$\frac{d\varphi}{dt} = a(\varphi), \quad \frac{dx}{dt} = -A^*(\varphi)x, \qquad (2.8.22)$$

conjugated with (2.2.4) have unique Green functions of the problem on invariant tori.

Note that if the condition (2.8.20) is satisfied and $\det S_1(\varphi) \neq 0$, $\forall \varphi \in R^m$, then the matrix $-S_1^{-1}(\varphi) = S_2(\varphi)$ can be taken for $S_2(\varphi)$. If the matrix function $S_1(\varphi)$ degenerates for some values $\varphi = \varphi_0 = R^m$ under condition (2.8.20), then the matrix function $S_2(\varphi) \in C'(\mathcal{T}_m; a)$ satisfying (2.8.21) no longer exists. Also, system (2.8.22) has a set of the Green functions, and system (2.2.4) does not have the Green function. There arises a question: what additional properties should the vector function $f(\varphi)$ possess so that the system (2.1.3) has an invariant torus.

Theorem 2.8.4 *Let there exist an n-dimensional symmetric matrix $S_1(\varphi) \in C'(\mathcal{T}_m; a)$ satisfying the condition (2.8.20) and for some $\varphi = \varphi_0 \in R^m$ the equality $\det S_1(\varphi_0) = 0$ holds true. Then the necessary and sufficient condition for the existence of the invariant torus of the system (2.1.3) is the identity*

$$\int_{-\infty}^{\infty} \langle f(\varphi_\sigma(\varphi)), v(\varphi_\sigma(\varphi))\rangle \, d\sigma \equiv 0, \quad \forall \varphi \in R^m, \qquad (2.8.23)$$

being satisfied for every function $x = v(\varphi)$ which determines the non-trivial torus of the system (2.8.22).

Proof We first note that for every non-trivial invariant torus $x = v(\varphi)$ of the system (2.8.22) the value of $v(\varphi_t(\varphi_0))$, which is the function of t, vanishes exponentially as $t \to \pm\infty$. Therefore the integral (2.8.23) converges for any vector function $f(\varphi) \in C^0(\mathcal{T}_m)$. Evidently the extended system

$$\frac{d\varphi}{dt} = a(\varphi),$$
$$\frac{dx}{dt} = A(\varphi)x - y, \qquad (2.8.24)$$
$$\frac{dy}{dt} = -A^*(\varphi)y,$$

has a unique Green function

$$\bar{G}_0(\tau, \varphi) = \begin{cases} \begin{bmatrix} \Omega_\tau^0(\varphi; A) & \omega(\tau, \varphi) \\ 0 & [\Omega_0^\tau(\varphi; A)]^* \end{bmatrix} \\ \times \begin{bmatrix} C_{11}(\varphi_\tau(\varphi)) & C_{12}(\varphi_\tau(\varphi)) \\ C_{21}(\varphi_\tau(\varphi)) & C_{22}(\varphi_\tau(\varphi)) \end{bmatrix}, & \tau \leq 0, \\ \begin{bmatrix} \Omega_\tau^0(\varphi; A) & \omega(\tau, \varphi) \\ 0 & [\Omega_0^\tau(\varphi; A)]^* \end{bmatrix} \\ \times \begin{bmatrix} C_{11}(\varphi_\tau(\varphi)) - I_n & C_{12}(\varphi_\tau(\varphi)) \\ C_{21}(\varphi_\tau(\varphi)) & C_{22}(\varphi_\tau(\varphi)) - I_n \end{bmatrix}, & \tau > 0, \end{cases} \qquad (2.8.25)$$

where

$$\omega(\tau, \varphi) = -\int_\tau^0 \Omega_\sigma^0(\varphi; A)[\Omega_\sigma^0(\varphi; A)]^* \, d\sigma; \qquad (2.8.26)$$

$C(\varphi) = \{C_{ij}(\varphi)\}$, $i, j = 1, 2$, is the projection matrix. For the Green function (2.8.25) the estimate (2.7.13) is also valid.

For any fixed vector function $f(\varphi) \in C^0(\mathcal{T}_m)$ the system

$$\frac{d\varphi}{dt} = a(\varphi),$$
$$\frac{dx}{dt} = A(\varphi)x - y + f(\varphi), \qquad (2.8.27)$$
$$\frac{dy}{dt} = -A^*(\varphi)y,$$

has a unique invariant torus

$$\begin{pmatrix} x \\ y \end{pmatrix} = \begin{pmatrix} u(\varphi) \\ v(\varphi) \end{pmatrix} = \int_{-\infty}^{\infty} \bar{G}_0(\tau, \varphi) \begin{pmatrix} f(\varphi_\tau(\varphi)) \\ 0 \end{pmatrix} d\tau.$$

Moreover

$$u(\varphi) = \int_{-\infty}^{0} [\Omega_\tau^0(\varphi; A) C_{11}(\varphi_\tau(\varphi)) + \omega(\tau, \varphi) C_{21}(\varphi_\tau(\varphi))] f(\varphi_\tau(\varphi))\, d\tau$$

$$+ \int_{0}^{\infty} \Omega_\tau^0(\varphi; A) [C_{11}(\varphi_\tau(\varphi)) - I_n + \omega(\tau, \varphi) C_{21}(\varphi_\tau(\varphi))] f(\varphi_\tau(\varphi))\, d\tau, \qquad (2.8.28)$$

$$v(\varphi) = \int_{-\infty}^{\infty} [C_{21}^*(\varphi_\tau(\varphi)) \Omega_0^\tau(\varphi; A)]^* f(\varphi_\tau(\varphi))\, d\tau.$$

Due to the uniqueness of the Green function (2.8.25) the identity

$$\begin{bmatrix} \Omega_\tau^0(\varphi; A) & \omega(\tau, \varphi) \\ 0 & [\Omega_0^\tau(\varphi; A)]^* \end{bmatrix} \begin{bmatrix} C_{11}(\varphi_\tau(\varphi)) & C_{12}(\varphi_\tau(\varphi)) \\ C_{21}(\varphi_\tau(\varphi)) & C_{22}(\varphi_\tau(\varphi)) \end{bmatrix}$$

$$\equiv \begin{bmatrix} C_{11}(\varphi) & C_{12}(\varphi) \\ C_{21}(\varphi) & C_{22}(\varphi) \end{bmatrix} \begin{bmatrix} \Omega_\tau^0(\varphi; A) & \omega(\tau, \varphi) \\ 0 & [\Omega_0^\tau(\varphi; A)]^* \end{bmatrix}$$

is valid for all $\varphi \in R^m$, $\tau \in R$. Therefore

$$C_{21}(\varphi_\tau(\varphi)) \equiv [\Omega_\tau^0(\varphi; A)]^* C_{21}(\varphi_\tau(\varphi)) \Omega_\tau^0(\varphi; A), \qquad (2.8.29)$$

and the second of the equalities (2.8.28) can be represented as

$$v(\varphi) = \int_{-\infty}^{\infty} C_{21}(\varphi) \Omega_\tau^0(\varphi; A) f(\varphi_\tau(\varphi))\, d\tau. \qquad (2.8.30)$$

Obviously, if the system (2.1.3) has the invariant torus $x = u_0(\varphi)$, then the system (2.8.27) also has the invariant torus $x = u_0(\varphi)$, $y = 0$. Consequently

$$\int_{-\infty}^{\infty} C_{21}(\varphi) \Omega_\tau^0(\varphi; A) f(\varphi_\tau(\varphi))\, d\tau \equiv 0. \qquad (2.8.31)$$

On the other hand, if for some vector-function $f(\varphi) \in C^0(\mathcal{T}_m)$ the identity (2.8.31) is valid, then due to (2.8.28) and (2.8.30) the system (2.8.27) has the invariant torus $x = u(\varphi)$, $y = 0$ and so, the system (2.1.3) has the torus $x = u(\varphi)$. Thus, condition (2.8.31) is necessary and sufficient for the existence of the invariant torus of the system (2.1.3).

LINEAR EXTENSION OF DYNAMICAL SYSTEMS ON A TORUS 161

Now we show that (2.8.31) is equivalent to (2.8.23). For some vector-function $f(\varphi) \in C^0(\mathcal{T}_m)$ let the identity (2.8.23) be satisfied. Every non-trivial torus of the system (2.8.22) can be represented as

$$y = v(\varphi) = \int_{-\infty}^{\infty} C_{21}(\varphi)\Omega_\tau^0(\varphi; A)g(\varphi_\tau(\varphi))\, d\tau,$$

where $g(\varphi)$ is an arbitrary vector-function from the space $C^0(\mathcal{T}_m)$. In view of (2.8.29) we get

$$
\begin{aligned}
v(\varphi_\sigma(\varphi)) &= \int_{-\infty}^{\infty} [\Omega_\sigma^0(\varphi; A)]^* C_{21}(\varphi)\Omega_\sigma^0(\varphi; A)\Omega_{\tau+\sigma}^\sigma(\varphi; A) g(\varphi_{\tau+\sigma}(\varphi))\, d\tau \\
&= \int_{-\infty}^{\infty} [\Omega_\sigma^0(\varphi; A)]^* C_{21}(\varphi)\Omega_{t_1}^0(\varphi; A) g(\varphi_{t_1}(\varphi))\, dt_1.
\end{aligned}
\qquad (2.8.32)
$$

Substituting the representation of $v(\varphi_\sigma(\varphi))$ obtained into the left-hand side of (2.8.23) yields

$$
\begin{aligned}
&\int_{-\infty}^{\infty} f^*(\varphi_\sigma(\varphi)) \int_{-\infty}^{\infty} [\Omega_\sigma^0(\varphi; A)]^* C_{21}(\varphi)\Omega_{t_1}^0(\varphi; A) g(\varphi_{t_1}(\varphi))\, dt_1\, d\sigma \\
&= \int_{-\infty}^{\infty} \left\{ \int_{-\infty}^{\infty} C_{21}(\varphi)\Omega_\sigma^0(\varphi; A) f(\varphi_\sigma(\varphi))\, d\sigma \right\}^* \Omega_{t_1}^0(\varphi; A) g(\varphi_{t_1}(\varphi))\, dt_1.
\end{aligned}
\qquad (2.8.33)
$$

To justify the replacement of the integrals in (2.8.33) we show that for every fixed $\varphi = \varphi_0 \in R^m$ estimate

$$\|[\Omega_\sigma^0(\varphi_0; A)]^* C_{21}(\varphi_0)\Omega_t^0(\varphi_0; A)\| \le K(\varphi_0)\exp\{-\gamma|\sigma|-\gamma|t|\}, \qquad (2.8.34)$$

is satisfied, where the positive values $K(\varphi_0)$, γ do not depend on t, $\sigma \in R$, but only on $\varphi_0 \in R^m$. We fix a value $\varphi = \varphi_0 \in R^m$ and reduce the symmetric matrix $C_{21}(\varphi_0)$ to the Jordan form $QC_{21}(\varphi_0)Q^{-1} = \text{diag}\{\Lambda, 0\}$. The satisfying of estimate (2.7.13) by the Green function (2.8.25), with the constants $\bar{K}_0$, $\bar{\gamma}$ being independent of φ means that the inequality

$$\|[\Omega_\sigma^0(\varphi; A)]^* C_{21}(\varphi)\Omega_t^0(\varphi; A)\| \le \bar{K}_0 \exp\{-\bar{\gamma}|\sigma - t|\} \qquad (2.8.35)$$

is satisfied for all $\sigma, t \in R$, $\varphi \in R^m$. Thus, for the fixed $\varphi = \varphi_0$

$$
\begin{aligned}
\|[\Omega_\sigma^0(\varphi_0; A)]^* C_{21}(\varphi_0)\Omega_t^0(\varphi_0; A)\| &= \|[\Omega_\sigma^0(\varphi_0; A)]^* Q^{-1}\,\text{diag}\{\Lambda, 0\} Q\Omega_t^0(\varphi_0; A)\| \\
&= \|[\Omega_\sigma^0(\varphi_0; A)]^* Q^{-1}\,\text{diag}\{I_r, 0\} Q Q^{-1}\,\text{diag}\{\Lambda, 0\} Q Q^{-1}\,\text{diag}\{I_r, 0\} Q\Omega_t^0(\varphi_0; A)\| \\
&\le \|[\Omega_\sigma^0(\varphi_0; A)]^* Q^{-1}\,\text{diag}\{I_r, 0\}Q\|\, \|C_{21}(\varphi_0)\|\, \|Q^{-1}\,\text{diag}\{I_r, 0\} Q\Omega_t^0(\varphi_0; A)\| \\
&\le K_1(\varphi_0)\exp\{-\gamma|\sigma|\}\, \|C_{21}(\varphi_0)\|\, K_2(\varphi_0)\exp\{-\gamma|t|\} \\
&= K(\varphi_0)\exp\{-\gamma|\sigma|-\gamma|t|\}.
\end{aligned}
$$

Since the matrix $C_{21}(\varphi)$ is symmetric and the identity (2.8.29) is satisfied for the vector-function $f(\varphi)$, then (2.8.32) implies (2.8.23).

Now assume that the identity (2.8.23) is satisfied for some vector-function $f(\varphi) \in C^0(\mathcal{T}_m)$. Then substituting

$$g(\varphi) = \int_{-\infty}^{\infty} C_{21}(\varphi)\Omega_\sigma^0(\varphi; A)f(\varphi_\sigma(\varphi))\, d\sigma,$$

in (2.8.32) gives

$$\int_{-\infty}^{\infty} \|g(\varphi_\sigma(\varphi))\|^2 \, d\sigma \equiv 0.$$

Hence it follows that $g(\varphi) \equiv 0$.

This completes the proof of Theorem 2.8.4.

Further we consider a system of differential equations of the type

$$\frac{d\varphi}{dt} = a(\varphi),$$
$$\frac{dx_1}{dt} = b_1(\varphi)x_1, \qquad (2.8.36)$$
$$\frac{dx_2}{dt} = b_{21}(\varphi)x_1 + b_2(\varphi)x_2,$$

where $\varphi \in R^m$, $x_1, x_2 \in R$, $b_1(\varphi)$, $b_{21}(\varphi)$, $b_2(\varphi)$ are scalar functions belonging to the space $C^1(\mathcal{T}_m)$, i.e. continuously differentiable and 2π-periodic in every variable φ_j, $j = 1, \ldots, m$. We assume that there exist scalar functions $s_1(\varphi)$, $s_2(\varphi) \in C^1(\mathcal{T}_m)$ satisfying the conditions

$$\sum_{i=1}^{m} \frac{\partial s_1(\varphi)}{\partial \varphi_i} a_i(\varphi) - 2b_1(\varphi)s_1(\varphi) \leq -1; \qquad (2.8.37)$$

$$\sum_{i=1}^{m} \frac{\partial s_2(\varphi)}{\partial \varphi_i} a_i(\varphi) + 2b_2(\varphi)s_2(\varphi) \leq -1. \qquad (2.8.38)$$

We wonder which properties of the functions $b_1(\varphi)$, $b_{21}(\varphi)$, $b_2(\varphi)$ ensure that the system (2.8.36) has the Green function of the problem on invariant tori. In spite of the fact that the inequalities (2.8.37) and (2.8.38) provides the existence of the Green function for each of the systems

$$\frac{d\varphi}{dt} = a(\varphi), \quad \frac{dx_1}{dt} = b_1(\varphi)x_1; \qquad (2.8.39)$$

$$\frac{d\varphi}{dt} = a(\varphi), \quad \frac{dx_2}{dt} = -b_2(\varphi)x_2, \qquad (2.8.40)$$

the Green function may not exist for the system (2.8.36). We can verify this by the example of the system

$$\frac{d\varphi_1}{dt} = \sin \varphi_1, \quad \frac{dx_1}{dt} = (\cos \varphi_1)x_1,$$
$$\frac{d\varphi_2}{dt} = \sin \varphi_2, \quad \frac{dx_2}{dt} = b_{21}(\varphi)x_1 - (\cos \varphi_2)x_2, \quad (2.8.41)$$

where $b_{21}(\varphi) \in C^1(\mathcal{T}_m)$.

We introduce the following designations

$$\hat{M}_1 = \left(\varphi_0 \in R^m \ \Big| \ \exp\left\{\int_0^t b_1(\varphi_\sigma(\varphi_0))\,d\sigma\right\} \xrightarrow[|t|\to\infty]{} 0\right),$$
$$\hat{M}_2 = \left(\varphi_0 \in R^m \ \Big| \ \exp\left\{-\int_0^t b_2(\varphi_\sigma(\varphi_0))\,d\sigma\right\} \xrightarrow[|t|\to\infty]{} 0\right). \quad (2.8.42)$$

Note that in the case when $s_2(\varphi) \neq 0$, $\forall \varphi \in R^m$ the system (2.8.36) has a Green function, since for the conjugated system

$$\frac{d\varphi}{dt} = a(\varphi),$$
$$\frac{dx_1}{dt} = -b_1(\varphi)x_1 - b_{21}(\varphi)x_2, \quad (2.8.43)$$
$$\frac{dx_2}{dt} = -b_2(\varphi)x_2$$

the quadratic form

$$V(\varphi, x_1, x_2) = s(\varphi)x_1^2 - \frac{k}{s_2(\varphi)} x_2^2, \quad (2.8.44)$$

exists, having a sign-definite derivative along the solutions of (2.8.43) for a sufficiently large value of $k > 0$. Besides, an additional assumption on the non-degeneracy of $s_1(\varphi)$ immediately implies the non-degeneracy of the form (2.8.44) and, consequently, the existence of the unique Green function for the system (2.8.36). Thus, it is of interest when both the functions $s_1(\varphi)$ and $s_2(\varphi)$ vanish for some, maybe various, values $\varphi \in R^m$.

Theorem 2.8.5 *Let there exist functions $s_i(\varphi) \in C^1(\mathcal{T}_m)$, $i = 1, 2$, which satisfy the conditions (2.8.37) and (2.8.38).*

Then for the Green function $G_0(\tau, \varphi)$ of the system (2.8.36) to exist, which satisfies estimate (2.6.11) it is necessary and sufficient that two conditions be satisfied

$$\hat{M}_2 \subseteq \hat{M}_1, \quad (2.8.45)$$

and
$$\int_{-\infty}^{\infty} b_{21}(\varphi_\tau(\varphi_0)) \exp\left\{\int_0^\tau [b_1(\varphi_\sigma(\varphi_0)) - b_2(\varphi_\sigma(\varphi_0))]\,d\sigma\right\} d\tau \neq 0, \quad (2.8.46)$$

for all $\varphi_0 \in \hat{M}_2$.

Proof We show that under the conditions (2.8.45) and (2.8.46) the system
$$\frac{dx_1}{dt} = -b_1(\varphi_t(\varphi_0))x_1 - b_{21}(\varphi_t(\varphi_0))x_2,$$
$$\frac{dx_2}{dt} = -b_2(\varphi_t(\varphi_0))x_2, \quad (2.8.47)$$

has no non-trivial solutions bounded on the whole R axis for any fixed value $\varphi_0 \in R^m$. This is a necessary and sufficient condition for the existence of the Green function $G_0(\tau, \varphi)$ for the system (2.8.36) (see Bronstein [1]). If $\varphi_0 \notin \hat{M}_2$, then all non-trivial solutions of the second equation of (2.8.47) are not bounded on R. Assume that $\varphi_0 \in \hat{M}_2$. Substituting the solution of the second equation of system (2.8.47)
$$x_2 = \eta_2 \exp\left\{-\int_0^t b_2(\varphi_\sigma(\varphi_0))\,d\sigma\right\},$$

into the first one yields
$$\frac{dx_1}{dt} = -b_1(\varphi_t(\varphi_0))x_1 - \eta_2 b_{21}(\varphi_t(\varphi_0)) \exp\left\{-\int_0^t b_2(\varphi_\sigma(\varphi_0))\,d\sigma\right\}, \quad (2.8.48)$$

where η_2 is an arbitrary constant. The general solution to (2.8.48) is
$$x_1(t;\varphi_0) = \exp\left\{-\int_0^t b_1(\varphi_\sigma(\varphi_0))\,d\sigma\right\}\left(\eta_1 - \eta_2 \int_0^t b_{21}(\varphi_\tau(\varphi_0))\right.$$
$$\left.\times \exp\left\{\int_0^\tau [b_1(\varphi_\sigma(\varphi_0)) - b_2(\varphi_\sigma(\varphi_0))]\,d\sigma\right\} dt\right), \quad (2.8.49)$$

where η_1 and η_2 are arbitrary constants. We see that if $\varphi_0 \notin \hat{M}_1$, then, for every fixed $\eta_2 \in R$, equation (2.8.48) has a unique solution bounded on R:

$$x_1(t;\varphi_0) = \eta_2 \begin{cases} \int_t^\infty b_{21}(\varphi_\tau(\varphi_0)) \exp\left\{\int_t^\tau [b_1(\varphi_\sigma(\varphi_0)) - b_2(\varphi_\sigma(\varphi_0))]\,d\sigma\right\} d\tau, \\ \qquad\qquad\qquad\qquad \text{if} \quad s_1(\varphi_t(\varphi_0)) < 0, \\ -\int_{-\infty}^t b_{21}(\varphi_\tau(\varphi_0)) \exp\left\{\int_t^\tau [b_1(\varphi_\sigma(\varphi_0)) - b_2(\varphi_\sigma(\varphi_0))]\,d\sigma\right\} d\tau, \\ \qquad\qquad\qquad\qquad \text{if} \quad s_1(\varphi_t(\varphi_0)) > 0. \end{cases}$$

As it is required that equation (2.8.48) has no non-trivial solutions bounded on R, it is necessary that $\varphi_0 \in \hat{M}_1$. Thus, if $\varphi_0 \in \hat{M}_2$, then $\varphi_0 \in \hat{M}_1$. Consequently, condition (2.8.45) is necessary and sufficient for the absence of non-trivial solutions bounded on R to the system (2.8.47).

We note further that in the case when $\varphi_0 \in \hat{M}_2$, the equation

$$\frac{dx_1}{dt} = -b_1(\varphi_t(\varphi_0))x_1 + f(t),$$

does not have a solution bounded on R for every function $f(t)$ continuous bounded on R. For such a solution to exist it is necessary and sufficient that

$$\int_{-\infty}^{\infty} f(\tau) \exp\left\{\int_0^{\tau} b_1(\varphi_\sigma(\varphi_0))\,d\sigma\right\} d\tau = 0. \qquad (2.8.50)$$

For

$$f(t) = -\eta_2 b_{21}(\varphi_t(\varphi_0)) \exp\left\{-\int_0^t b_2(\varphi_\sigma(\varphi_0))\,d\sigma\right\},$$

this equality becomes

$$-\eta_2 \int_{-\infty}^{\infty} b_{21}(\varphi_\tau(\varphi_0)) \exp\left\{\int_0^{\tau} [b_1(\varphi_\sigma(\varphi_0)) - b_2(\varphi_\sigma(\varphi_0))]\,d\sigma\right\} d\tau = 0.$$

Therefore, the condition (2.8.46) ensures the absence of non-trivial solutions to equation (2.8.48) bounded on R for every $\varphi = \varphi_0 \in R^m$ and hence, the absence of such solutions of the system (2.8.47) for every fixed $\varphi_0 \in R^m$.

This completes the proof of Theorem 2.8.5.

Remark 2.8.2 If the inclusion of sets (2.8.45) in the condition of Theorem 2.8.5 is replaced by

$$\hat{M}_2 = \hat{M}_1, \qquad (2.8.51)$$

then the condition of the uniqueness of the Green function $G_0(\tau, \varphi)$ is satisfied.

Actually, under conditions (2.8.51) and (2.8.46) not only the system (2.8.47) has no non-trivial solutions bounded on R, but neither does the system conjugated with it

$$\frac{dx_1}{dt} = b_1(\varphi_t(\varphi_0))x_1,$$

$$\frac{dx_2}{dt} = b_{21}(\varphi_t(\varphi_0))x_1 + b_2(\varphi_t(\varphi_0))x_2,$$

Consequently the Green functions exist for the system (2.8.36) and (2.8.43). Hence its uniqueness.

Remark 2.8.3 In the example of the system (2.8.41) the conditions (2.8.37), (2.8.38) are satisfied for $s_1(\varphi) = \cos\varphi_1$, $s_2(\varphi) = \cos\varphi_2$, and the condition (2.8.45) is not satisfied, since $\hat{M}_1 = \{(\varphi_1, \varphi_2) \mid \varphi_1 \neq \pi n, \varphi_2 \in R\}$, $\hat{M}_2 = \{(\varphi_1, \varphi_2) \mid \varphi_1 \in R, \varphi_2 \neq \pi n\}$. Clearly, a sufficient condition for the equality (2.8.51) to be satisfied is the validity of the identity $s_1(\varphi) \equiv s_2(\varphi)$ for all $\varphi \in R^m$.

Theorem 2.8.6 *Let there exist two scalar functions $s_i(\varphi) \in C^1(\mathcal{T}_m)$, $i = 1, 2$, satisfying the inequalities (2.8.37), (2.8.38) and the conditions (2.8.45), (2.8.46) are satisfied. Then*

$$2\alpha_0 < \min\left\{|s_1|_0^{-1}, |s_2|_0^{-1}\right\}, \tag{2.8.52}$$

where

$$\alpha_0 = \max_{\varphi \in R^m}\left(\max_{\|\eta\|=1}\left|\left\langle\frac{\partial a(\varphi)}{\partial\varphi}\eta, \eta\right\rangle\right|\right), \tag{2.8.53}$$

$$|s_i|_0 = \max_{\varphi \in R^m}|s_i(\varphi)|,$$

ensures the existence of a smooth invariant torus $x_i = u_i(\varphi)$, $i = 1, 2$, $u_i \in C^1(\mathcal{T}_m)$ of the perturbed system

$$\frac{d\varphi}{dt} = a(\varphi),$$

$$\frac{dx_1}{dt} = b_1(\varphi)x_1 + f_1(\varphi), \tag{2.8.54}$$

$$\frac{dx_2}{dt} = b_{21}(\varphi)x_1 + b_2(\varphi)x_2 + f_2(\varphi),$$

for any scalar functions $f_1(\varphi), f_2(\varphi) \in C^1(\mathcal{T}_m)$.

Proof We designate $\nu_0 = 2^{-1}\min\left\{|s_1|_0^{-1}, |s_2|_0^{-1}\right\}$ and show that for every fixed $\delta \in (-\nu_0, \nu_0)$ the system

$$\frac{d\varphi}{dt} = a(\varphi),$$

$$\frac{dx_1}{dt} = b_1(\varphi)x_1 + \delta x_1,$$

$$\frac{dx_2}{dt} = b_{21}(\varphi)x_1 + b_2(\varphi)x_2 + \delta x_2, \tag{2.8.55}$$

$$\frac{dy_1}{dt} = x_1 - b_1(\varphi)y_1 - b_{21}(\varphi)y_2 + \delta y_1,$$

$$\frac{dy_2}{dt} = x_2 - b_2(\varphi)y_2 + \delta y_2,$$

has the unique Green function of the problem on invariant tora

$$\bar{G}_0(\tau,\varphi;\delta) = \begin{cases} \bar{C}(\varphi)\bar{\Omega}_\tau^0(\varphi)\exp\{-\delta\tau\}, & \tau \leq 0, \\ [\bar{C}(\varphi) - I_4]\bar{\Omega}_\tau^0(\varphi)\exp\{-\delta\tau\}, & \tau > 0. \end{cases} \quad (2.8.56)$$

For this it is sufficient to show the absence of non-trivial solutions bounded on R of two mutually conjugated systems of differential equations

$$\begin{aligned} \frac{dx_1}{dt} &= b_1(\varphi_t(\varphi_0))x_1 + \delta x_1, \\ \frac{dx_2}{dt} &= b_{21}(\varphi_t(\varphi_0))x_1 + b_2(\varphi_t(\varphi_0))x_2 + \delta x_2, \\ \frac{dy_1}{dt} &= x_1 - b_1(\varphi_t(\varphi_0))y_1 - b_{21}(\varphi_t(\varphi_0))y_2 + \delta y_1, \\ \frac{dy_2}{dt} &= x_2 - b_2(\varphi_t(\varphi_0))y_2 + \delta y_2, \end{aligned} \quad (2.8.57)$$

and

$$\begin{aligned} \frac{d\bar{x}_1}{dt} &= -[b_1(\varphi_t(\varphi_0)) + \delta]\bar{x}_1 - b_{21}(\varphi_t(\varphi_0))\bar{x}_2 - \bar{y}_1, \\ \frac{d\bar{x}_2}{dt} &= -[b_2(\varphi_t(\varphi_0)) + \delta]\bar{x}_2 - \bar{y}_2, \\ \frac{d\bar{y}_1}{dt} &= [b_1(\varphi_t(\varphi_0)) - \delta]\bar{y}_1, \\ \frac{d\bar{y}_2}{dt} &= b_{21}(\varphi_t(\varphi_0))\bar{y}_1 + [b_2(\varphi_t(\varphi_0)) - \delta]\bar{y}_2. \end{aligned} \quad (2.8.58)$$

Note that the system (2.8.58) can be derived from (2.8.57) by the change of variables $x_1 = -\bar{y}_1$, $x_2 = -\bar{y}_2$, $y_1 = \bar{x}_1$, $y_2 = \bar{x}_2$ and the change of parameter $\delta \to -\delta$. Thus, it is sufficient to confirm the absence of non-trivial solutions bounded on R of the system (2.8.57) only. Representing all the solutions of (2.8.57) in explicit form and analysing them for $\varphi_0 \in \hat{M}_2$, $\varphi_0 \in \hat{M}_1 \setminus \hat{M}_2$, $\varphi_0 \notin \hat{M}_1$, we verify that system (2.8.57) has no non-trivial solutions bounded on R.

In terms of the structure of the Green function (2.3.49) we get the estimate

$$\|\bar{G}_0(\tau,\varphi;0)\| \leq K_1(\varepsilon)\exp\{-(\nu_0-\varepsilon)|\tau|\}, \quad (2.8.59)$$

for every $\varepsilon > 0$ and all $\tau \in R$, $\varphi \in R^m$; $K_1(\varepsilon)$ is a constant which depends on ε (the smaller $\varepsilon > 0$ is, the larger $K_1(\varepsilon)$ can be). If the estimate (2.8.59) is not satisfied for some $\varepsilon = \varepsilon_0 > 0$, then a sequence of values $\tau^{(n)}$, $\varphi^{(n)}$ exists such that

$$\exp\left\{(\nu_0 - \varepsilon_0)|\tau^{(n)}|\right\} \left\|\bar{G}_0(\tau^{(n)}, \varphi^{(n)}; 0)\right\| \xrightarrow[n\to\infty]{} \infty. \tag{2.8.60}$$

Clearly, one of the equalities $|\tau^{(n)}| = -\tau^{(n)}$, $|\tau^{(n)}| = \tau^{(n)}$ can be considered to be valid for all $n = 1, 2, \ldots$, since otherwise there exists a subsequence $\tau^{(n_k)}$ for which one of these equalities should hold. Setting $|\tau^{(n)}| = \tau^{(n)}$ in (2.8.60) we have

$$\exp\left\{-(\nu_0 - \varepsilon_0)\tau^{(n)}\right\} \left\|\bar{G}_0(\tau^{(n)}, \varphi^{(n)}; 0)\right\| = \left\|\bar{G}_0(\tau^{(n)}, \varphi^{(n)}; -(\nu_0 - \varepsilon_0))\right\| \xrightarrow[n\to\infty]{} \infty.$$

This contradicts the existence of the Green function (2.8.56) for $\delta = -(\nu_0 - \varepsilon_0)$. Thus, for every $\varepsilon > 0$ estimate (2.8.59) is valid. It is well known that the inequality $\nu_0 - \varepsilon_0 > \alpha_0$ ensures differentiability of the Green function (2.8.56) in parameters $\varphi_1, \ldots, \varphi_m$. As $\varepsilon > 0$ can be taken to be arbitrarily small, the inequality (2.8.52) is sufficient for the continuous differentiability of the Green function $\bar{G}_0(\tau, \varphi; 0)$ in φ. In addition, for the derivatives $\frac{\partial}{\partial \varphi_i} \bar{G}_0(\tau, \varphi; 0)$ the estimate

$$\left\|\frac{\partial}{\partial \varphi_i} \bar{G}_0(\tau, \varphi; 0)\right\| \leq K_2(\varepsilon) \exp\left\{-(\nu_0 - \alpha_0 - \varepsilon)|\tau|\right\}, \tag{2.8.61}$$

takes place, where α_0 is determined by (2.8.53), $K_2(\varepsilon) = \text{const} > 0$. In view of (2.8.59), (2.8.61) and the possibility of choosing $\varepsilon > 0$ to be arbitrarily small, we make sure that the extended system

$$\frac{d\varphi}{dt} = a(\varphi),$$

$$\frac{dx_1}{dt} = b_1(\varphi)x_1 + f_1(\varphi),$$

$$\frac{dx_2}{dt} = b_{21}(\varphi)x_1 + b_2(\varphi)x_2 + f_2(\varphi),$$

$$\frac{dy_1}{dt} = x_1 - b_1(\varphi)y_1 - b_{21}(\varphi)y_2 + f_3(\varphi),$$

$$\frac{dy_2}{dt} = x_2 - b_2(\varphi)y_2 + f_4(\varphi),$$

has a unique invariant torus $x_i = u_i(\varphi)$, $y_i = v_i(\varphi)$, $u_i(\varphi), v_i(\varphi) \in C^1(\mathcal{T}_m)$, $i = 1, 2$ for any functions $f_i \in C^1(\mathcal{T}_m)$.

Remark 2.8.4 The sets $\hat{M}_1$ and $\hat{M}_2$ determined by (2.8.42) can be immediately characterized via the functions $s_i(\varphi_t(\varphi_0))$ as follows. The set $\hat{M}_i$ consists of the values $\varphi_0 \in R^m$, for which the scalar function $s_i(\varphi_t(\varphi_0))$ vanishes for some $t = t_0 \in R$.

2.9 Invariant Tori of Linear Extensions with Slowly Changing Phase

Consider the system of differential equations with a small parameter

$$\frac{d\varphi}{dt} = \varepsilon a(\varphi),$$
$$\frac{dx}{dt} = A(\varphi)x + f(\varphi), \qquad (2.9.1)$$

where $\varphi \in R^m$, $x \in R^n$, $a(\varphi), A(\varphi), f(\varphi) \in C^1(\mathcal{T}_m)$. Note that for $\varepsilon = 0$ the system (2.9.1) has an invariant torus for every vector function $f(\varphi) \in C^1(\mathcal{T}_m)$ only if $\det A(\varphi) \neq 0$, $\forall \varphi \in R^m$. Clearly, this is not sufficient for the perturbed system (2.9.1) ($\varepsilon > 0$) to have an invariant torus for every vector-function $f(\varphi) \in C^1(\mathcal{T}_m)$. Therefore it was assumed (see Birkhoff [1]) that the real parts of all eigenvalues of the matrix $A(\varphi)$ differ from zero. Moreover, it is necessary that the matrix $A(\varphi)$ can be presented in the form

$$A(\varphi) = S(\varphi) \, \text{diag}\, \{D_1(\varphi), -D_2(\varphi)\} S^{-1}(\varphi), \qquad (2.9.2)$$

where $S(\varphi) \in C^1(\mathcal{T}_m)$. It is to be noted here, that not every matrix for which the real parts of its eigenvalues differ from zero, can be represented in the form of (2.9.2) (see Bylov *et al.* [1]). The application of sign-variable Lyapunov functions enables us to get rid of condition (2.9.2).

Theorem 2.9.1 *For every fixed $\varphi \in R^m$ let the real parts of all eigenvalues of the matrix $A(\varphi)$ differ from zero. Then there exists a number $\varepsilon > 0$ such that for all $\varepsilon \in [0, \varepsilon_0]$ the system (2.9.1) has an invariant torus $x = u_\varepsilon(\varphi) \in C^1(\mathcal{T}_m)$ for every vector-function $f(\varphi) \in C^1(\mathcal{T}_m)$.*

Proof We fix a value $\varphi = \varphi_0 \in R^m$ and consider the system of linear differential equations with constant coefficients

$$\frac{dx}{dt} = A(\varphi_0)x. \qquad (2.9.3)$$

Since the real parts of all the eigenvalues of the matrix $A(\varphi_0)$ differ from zero, then the system (2.9.3) is exponentially dichotomous on the whole R axis. Hence, (see Daletsky and Krein [1]) there exists a constant non-degenerate symmetric matrix S_{φ_0} satisfying the inequality

$$\langle [S_{\varphi_0} A(\varphi_0) + A^*(\varphi_0) S_{\varphi_0}] x, x \rangle \leq -\|x\|^2. \qquad (2.9.4)$$

Now we show that the derivative of the quadratic form

$$V_{\varphi_0}(x) = \langle S_{\varphi_0} x, x \rangle, \qquad (2.9.5)$$

computed along the solutions of the system

$$\frac{dx}{dt} = A(\varphi)x, \tag{2.9.6}$$

where the value φ is taken to be constant and close to φ_0, is also negative definite. Obviously, computing the derivative yields

$$\dot{V}_{\varphi_0}(x) = \langle [S_{\varphi_0}A(\varphi) + A^*(\varphi)S_{\varphi_0}]x, x \rangle$$
$$= \langle [S_{\varphi_0}A(\varphi_0) + A^*(\varphi_0)S_{\varphi_0}]x, x \rangle + 2\langle S_{\varphi_0}[A(\varphi) - A(\varphi_0)]x, x \rangle \tag{2.9.7}$$
$$\leq -(1 - 2\|S_{\varphi_0}\| \|A(\varphi) - A(\varphi_0)\|)\|x\|^2.$$

Since the matrix function $A(\varphi)$ is continuous with respect to the whole totality of variables $\varphi_1, \ldots, \varphi_m$, then there exists a number $\delta(\varphi_0) > 0$ such that for all $\varphi \in R^m$ satisfying the inequality

$$\|\varphi - \varphi_0\| \leq \delta(\varphi_0), \tag{2.9.8}$$

the condition

$$\|S_{\varphi_0}\| \|A(\varphi) - A(\varphi_0)\| \leq \frac{1}{4}, \tag{2.9.9}$$

holds true. Therefore, the derivative of the non-degenerate quadratic form (2.9.5) along the solutions of system (2.9.6) for all $\varphi \in R^m$ satisfying inequality (2.9.8), is negative definite $\dot{V}_{\varphi_0}(x) \leq -\frac{1}{2}\|x\|^2$. In view of the result of Theorem 1.1.3 one can claim that the Green function of the problem of solutions for the system (2.9.6) on bounded on R can be estimated for all $\varphi \in R^m$ satisfying (2.9.8) as follows

$$\|G(t, \tau, \varphi)\| \leq K(\varphi_0) \exp\{-\gamma(\varphi_0)|t - \tau|\}, \quad t, \tau \in R, \tag{2.9.10}$$

where $K(\varphi_0) = (2 + \sqrt{2})(2\|A\|_0 \|S_{\varphi_0}\|)^{3/2}$, $\gamma(\varphi_0) = (4\|S_{\varphi_0}\|)^{-1}$.

Thus, for every fixed value $\varphi \in R^m$ there exists a neighbourhood $U_{\varphi_0} \subset R^m$ such that for all $\varphi \in U_{\varphi_0}$ the constants $K(\varphi_0)$ and $\gamma(\varphi_0)$ in the estimate (2.9.10) are one and the same, i.e. they only depend on the point φ_0. Since the matrix function $A(\varphi)$ is 2π-periodic in φ_j, $j = 1, \ldots, m$, it is sufficient to consider it only in cube $T = \{\varphi \in R^m \mid |\varphi_0| \leq \pi, \ i = 1, \ldots, m\}$. A finite number of neighbourhoods can be taken from the infinite cover of the cube T by the neighbourhoods U_{φ_0} (the well-known Heine–Borel lemma). So, for the positive constants K and γ independent of $\varphi \in R^m$ the estimate

$$\|G(t, \tau, \varphi)\| \leq K \exp\{-\gamma|t - \tau|\}, \quad t, \tau \in R, \tag{2.9.11}$$

is to be satisfied. From this follows the existence of the continuous symmetric matrix function $S(\varphi)$ satisfying the inequality for all $\varphi \in R^m$ and $x \in R^n$:

$$\langle [S(\varphi)A(\varphi) + A^*(\varphi)S(\varphi)]x, x \rangle \leq -\|x\|^2.$$

Besides, $\det S(\varphi) \neq 0$, $\forall \varphi \in R^m$. For such a matrix $S(\varphi)$ the following one

$$S(\varphi) = 2 \int_0^\infty G^*(t,0,\varphi) G(t,0,\varphi)\, dt - 2 \int_{-\infty}^0 G^*(t,0,\varphi) G(t,0,\varphi)\, dt, \qquad (2.9.12)$$

can be taken.

Evidently, there exists a continuously differentiable matrix function $\bar{S}(\varphi) \equiv \bar{S}^*(\varphi)$ ($\det \bar{S}(\varphi) \neq 0$), satisfying the inequality

$$\langle [\bar{S}(\varphi) A(\varphi) + A^*(\varphi) \bar{S}(\varphi)] x, x \rangle \leq -\frac{1}{2} \|x\|^2.$$

Now taking the value $\bar{\varepsilon}_0 > 0$ from

$$\bar{\varepsilon}_0 \left\| \frac{\partial \bar{S}(\varphi)}{\partial \varphi} a(\varphi) \right\| \leq \frac{1}{4}, \qquad (2.9.13)$$

we get the following

$$\left\langle \left[\varepsilon \frac{\partial \bar{S}(\varphi)}{\partial \varphi} a(\varphi) + \bar{S}(\varphi) A(\varphi) + A^*(\varphi) \bar{S}(\varphi) \right] x, x \right\rangle \leq -\frac{1}{4} \|x\|^2, \qquad (2.9.14)$$

for all $\varepsilon \in [0, \bar{\varepsilon}_0]$, $x \in R^n$, $\varphi \in R^m$. Also, $\det \bar{S}(\varphi) \neq 0$, $\forall \varphi \in R^m$.

Thus, the perturbed system

$$\frac{d\varphi}{dt} = \varepsilon a(\varphi), \quad \frac{dx}{dt} = A(\varphi) x,$$

for every $\varepsilon \in [0, \bar{\varepsilon}_0]$ has a unique Green function of the problem on invariant tori satisfying the estimate

$$\|G_0(\tau, \varphi; \varepsilon)\| \leq K \exp\{-\gamma |\tau|\}, \qquad (2.9.15)$$

where $K = (2 + \sqrt{2}) (4\|\bar{S}\|_0 \|A\|_0)^{3/2}$, $\gamma = (\delta \|S\|)^{-1}$. Hence it follows that the system (2.9.1) has the invariant torus for every $\varepsilon \in [0, \bar{\varepsilon}_0]$, where $\bar{\varepsilon}_0$ is determined by (2.9.13):

$$x = u_\varepsilon(\varphi) = \int_{-\infty}^\infty G_0(\tau, \varphi; \varepsilon) f(\varphi_\tau^\varepsilon(\varphi))\, d\tau, \qquad (2.9.16)$$

where $\varphi_t^\varepsilon(\varphi)$ is a solution of the Cauchy problem

$$\frac{d\varphi}{dt} = \varepsilon a(\varphi), \quad \varphi_t^\varepsilon(\varphi)\big|_{t=0} = \varphi.$$

For the function (2.9.16) to be continuously differentiable in φ it is necessary, in addition to $a(\varphi)$, $A(\varphi)$, $f(\varphi) \in C^1(\mathcal{T}_m)$, that the inequality $\gamma > \varepsilon \|\partial a/\partial \varphi\|_0$ be satisfied. Therefore, taking $\varepsilon_0 = \min\{\bar{\varepsilon}_0, \bar{\bar{\varepsilon}}_0\}$, where $\bar{\bar{\varepsilon}}$ is determined by

$$8\bar{\bar{\varepsilon}}\|\bar{S}(\varphi)\|\left\|\frac{\partial a}{\partial \varphi}\right\|_0 < 1,$$

and the $\bar{\varepsilon}_0$ is specified by (2.9.13), we ensure the existence of the torus $x = u_\varepsilon(\varphi)$ of the system (2.9.1) for $\varepsilon \in [0, \bar{\varepsilon}_0]$ and the result that it belongs to the space of continuously differentiable functions $C^1(\mathcal{T}_m)$. This completes the proof of this theorem.

Theorem 2.9.1 leads easily to the following result.

Theorem 2.9.1 *Let the right-hand side of the nonlinear system of differential equations*

$$\frac{d\varphi}{dt} = \varepsilon a(\varphi, x, \varepsilon),$$

$$\frac{dx}{dt} = X(\varphi, x, \varepsilon),$$

satisfy the following conditions:

(1) *the functions $a(\varphi, x, \varepsilon)$, $X(\varphi, x, \varepsilon)$ are definite for all $\varphi \in R^m$, $x \in D$, $D = \{x \in R^n : \|x\| \leq d\}$ and $\varepsilon \in [0, \varepsilon_0]$, 2π-periodic in φ_j, $j = 1, \ldots, m$, and continuous with respect to the totality of variables φ, x, ε together with its partial derivatives in φ, x up to the order $l \geq 2$ included;*

(2) *the system of equations $X(\varphi, x, 0) = 0$ has an isolated solution $x = x_0(\varphi) \in C^1(\mathcal{T}_m)$ belonging to the domain D together with its ρ-neighbourhood;*

(3) *the real parts of all eigenvalues of the matrix $A(\varphi) = \partial X(\varphi, x_0(\varphi), 0)/\partial x$ differ from zero.*

Then a number $\varepsilon^0 > 0$ ($\varepsilon^0 \leq \varepsilon_0$) can be indicated such that for all $\varepsilon \in [0, \varepsilon^0]$ the initial nonlinear system of the equations has the invariant torus $x = u(\varphi, \varepsilon) \in C^{l-2}(\mathcal{T}_m)$ satisfying the condition $\lim_{\varepsilon \to 0} \|u(\varphi, \varepsilon) - x_0(\varphi)\|_{l-2} = 0$.

Now consider the system of differential equations of the type

$$\frac{d\varphi}{dt} = \omega + \varepsilon a(\varphi, x, \varepsilon), \quad \frac{dx}{dt} = [A(\varphi) + \varepsilon B(\varphi)]x + f(\varphi), \tag{2.9.17}$$

where $a(\varphi)$, $A(\varphi), B(\varphi), f(\varphi) \in C^l(\mathcal{T}_m)$, $l \geq 1$.

Theorem 2.9.2 *Let the linear system of differential equations*

$$\frac{dx}{dt} = A(\omega t + \varphi)x, \tag{2.9.18}$$

be exponentially dichotomous on the whole R axis for every fixed $\varphi \in R^m$. Then, there exists a $\varepsilon_0 > 0$ such that for all $\varepsilon \in [0, \varepsilon_0]$ the system (2.9.17) has the invariant torus $x = u_\varepsilon(\varphi)$, $u_\varepsilon(\varphi) \in C^l(\mathcal{T}_m)$ for every vector-function $f(\varphi) \in C^l(\mathcal{T}_m)$.

Remark 2.9.1 If ω is a vector with incommensurable coordinates, then in the cited theorem it is sufficient to require exponential dichotomy on the semiaxis of the system of equations

$$\frac{dx}{dt} = A(\omega t)x.$$

If $\omega = 0$, it is sufficient that the real parts of eigenvalues of the matrix $A(\varphi)$ differ from zero for all $\varphi \in R^m$.

Proof of the Theorem 2.9.2 Designating the modulus of continuity of the matrix function $A(\varphi)$ by $\mu(A; \sigma)$:

$$\sup_{\|\varphi - \bar\varphi\| \le \sigma} \|A(\varphi) - A(\bar\varphi)\| = \mu(A; \sigma),$$

we note that for any $\varphi, \bar\varphi \in R^m$, when all $t \in R$, the estimate

$$\|A(\omega t + \varphi) - A(\omega t + \bar\varphi)\| \le \mu(A; \|\varphi - \bar\varphi\|),$$

is valid. Therefore the system

$$\frac{d\varphi}{dt} = \omega, \quad \frac{dx}{dt} = A(\varphi)x,$$

has a unique Green function $G_0(\tau, \varphi)$ for which the estimate (2.6.11) is true with constants K, γ independent of $\varphi \in R^m$. Evidently the non-degenerate matrix function

$$S(\varphi) = \int_0^\infty [\Omega_0^\tau(\varphi; A)C(\varphi)]^* \Omega_0^\tau(\varphi; A)C(\varphi)\, d\tau$$

$$- \int_{-\infty}^0 \{\Omega_0^\tau(\varphi; A)[I_n - C(\varphi)]\}^* \Omega_0^\tau(\varphi; A)[I_n - C(\varphi)]\, d\tau, \tag{2.9.19}$$

belongs to the space $C^l(\mathcal{T}_m)$ and for it

$$\left\langle \left[\frac{\partial S(\varphi)}{\partial \varphi}\omega + S(\varphi)A(\varphi) + A^*(\varphi)S(\varphi)\right]x, x\right\rangle \le -\frac{1}{2}\|x\|^2.$$

Hence

$$\left\langle \left[\frac{\partial S(\varphi)}{\partial \varphi}[\omega + \varepsilon a(\varphi)] + S(\varphi)[A(\varphi) + \varepsilon B(\varphi)]\right.\right.$$
$$\left.\left. + [A^*(\varphi) + \varepsilon B^*(\varphi)]S(\varphi)\right]x, x\right\rangle \leq -\left(\frac{1}{2} - \varepsilon K\right)\|x\|^2.$$

Now letting $\varepsilon = \frac{1}{3} K^{-1}$, where

$$K = \max_{\varphi \in R^m} \left[2\|B(\varphi)\| \|S(\varphi)\| + \|a(\varphi)\| \left(\sum_{i=1}^{m} \left\|\frac{\partial S(\varphi)}{\partial \varphi_i}\right\|^2\right)^{1/2}\right],$$

we get that the function $S(\varphi)$ prescribed by (2.9.19) is also suitable for the perturbed system (2.9.17). This immediately implies the existence of the invariant torus $x = u_\varepsilon(\varphi)$ of the system (2.9.17) for every vector-function $f(\varphi) \in C^l(\mathcal{T}_m)$ and the fact that $\varepsilon > 0$ is sufficiently small allows us to claim that $u_\varepsilon(\varphi) \in C^l(\mathcal{T}_m)$.

Lemma 2.9.1 *Let the scalar function $u(\varphi)$ belongs to the space $C^l(\mathcal{T}_m; a)$ and, in addition $a(\varphi) \in C^1(\mathcal{T}_m)$. Then there exists a sequence of functions $u_n(\varphi) \in C^1(\mathcal{T}_m)$ such that*

$$|u_n(\varphi) - u(\varphi)| \xrightarrow[n\to\infty]{} 0, \quad \text{and} \quad |\dot{u}_n(\varphi) - \dot{u}(\varphi)| \xrightarrow[n\to\infty]{} 0.$$

Proof Consider an auxiliary system of equations

$$\frac{d\varphi}{dt} = a(\varphi), \quad \frac{dx}{dt} = 2\alpha_0 x + c(\varphi), \tag{2.9.20}$$

where $c(\varphi) = \dot{u}(\varphi) - 2\alpha_0 u(\varphi)$, $\alpha_0 = \text{const} > \left\|\frac{\partial a(\varphi)}{\partial \varphi}\right\|$. Clearly, the system (2.9.20) has the invariant torus $x = u(\varphi)$, and the function $u(\varphi)$ can be represented as

$$u(\varphi) = -\int_0^\infty e^{-2\alpha_0 \tau} c(\varphi_\tau(\varphi))\, d\tau.$$

Now we approximate the function $c(\varphi)$ by the sequence of functions $c_n(\varphi) \in C^1(\mathcal{T}_m)$, $|c_n(\varphi) - c(\varphi)| \xrightarrow[n\to\infty]{} 0$ and consider the systems of differential equations

$$\frac{d\varphi}{dt} = a(\varphi), \quad \frac{dx}{dt} = 2\alpha_0 x + c_n(\varphi), \quad n = 1, 2, \ldots. \tag{2.9.21}$$

For every $c_n(\varphi)$ the system (2.9.21) has the invariant torus described by

$$x = u_n(\varphi) = -\int_0^\infty e^{-2\alpha_0 \tau} c_n(\varphi_\tau(\varphi))\, d\tau.$$

The estimation of the difference $u_n(\varphi) - u(\varphi)$ is

$$|u_n(\varphi) - u(\varphi)| \leq \frac{1}{2\alpha_0} |c(\varphi) - c_n(\varphi)| \xrightarrow[n\to\infty]{} 0.$$

Since the functions $x = u(\varphi)$ and $x = u_n(\varphi)$ determine the invariant tori of systems (2.9.20) and (2.9.21) we have

$$\frac{du(\varphi_t(\varphi))}{dt} \equiv 2\alpha_0 u(\varphi_t(\varphi)) + c(\varphi_t(\varphi)),$$

$$\frac{du_n(\varphi_t(\varphi))}{dt} \equiv 2\alpha_0 u_n(\varphi_t(\varphi)) + c_n(\varphi_t(\varphi)).$$

Thus

$$\left|\frac{du(\varphi_t(\varphi))}{dt} - \frac{du_n(\varphi_t(\varphi))}{dt}\right| \leq 2|c(\varphi) - c_n(\varphi)|_0 \xrightarrow[n\to\infty]{} 0.$$

Theorem 2.9.3 *Let the vector-function $a(\varphi)$ and the matrix-function $A(\varphi)$ belong to the space $C^1(\mathcal{T}_m)$ and the trivial torus $x = 0$ of the system (2.2.4) be e-dichotomous. Then the torus $x = 0$ of the perturbed system*

$$\frac{d\varphi}{dt} = a(\varphi) + a_1(\varphi), \quad \frac{dx}{dt} = [A(\varphi) + A_1(\varphi)]x \qquad (2.9.22)$$

is also e-dichotomous for any functions $a_1(\varphi), A_1(\varphi) \in C^1(\mathcal{T}_m)$ which are sufficiently small with respect to the Euclidean norm.

Proof Consider the matrix function $S(\varphi)$ of the form of (2.9.19). In this case the function is, generally speaking, continuous in φ, although $S(\varphi_t(\varphi))$ is continuously differentiable in t. The fact that the function $S(\varphi)$ belongs to the space $C^1(\mathcal{T}_m)$ is ensured by the inequality $\gamma > \max_{\varphi \in R^m} \left\|\frac{\partial a(\varphi)}{\partial \varphi}\right\|$ which is not always satisfied. Immediate verification shows that

$$\dot{S}(\varphi) = \left.\frac{dS(\varphi_t(\varphi))}{dt}\right|_{t=0} \in C^0(\mathcal{T}_m),$$

and the estimate

$$\langle [\dot{S}(\varphi_t) + S(\varphi_t)A(\varphi_t) + A^*(\varphi_t)S(\varphi_t)]x, x\rangle \leq -\frac{1}{2}\|x\|^2, \qquad (2.9.23)$$

is satisfied. By virtue of the cited lemma the matrix function $S(\varphi)$ can be approximated by the continuously differentiable functions $\tilde{S}(\varphi)$ so that the derivatives $\dot{S}(\varphi_t)$ and $\dot{\tilde{S}}(\varphi_t)$ are close. Then the estimate (2.9.23) is valid and for the functions $\tilde{S}$:

$$\left\langle [\dot{\tilde{S}}(\varphi_t) + \tilde{S}(\varphi_t)A(\varphi_t) + A^*(\varphi_t)\tilde{S}(\varphi_t)]x, x \right\rangle \leq -\left(\frac{1}{2} - \varepsilon\right)\|x\|^2, \qquad (2.9.24)$$

where $\varepsilon > 0$ is a sufficiently small fixed value depending on the approximation of the functions $\tilde{S}(\varphi)$, $S(\varphi)$. The estimate (2.9.24) is replaced by

$$\left\langle \left[\frac{\partial \tilde{S}(\varphi)}{\partial \varphi} a(\varphi) + \tilde{S}(\varphi)A(\varphi) + A^*(\varphi)\tilde{S}(\varphi)\right]x, x \right\rangle \leq -\left(\frac{1}{2} - \varepsilon\right)\|x\|^2. \qquad (2.9.25)$$

Thus, the small changes of the functions $a(\varphi)$ and $A(\varphi)$ do not influence the satisfaction of (2.9.25).

This proves the Theorem 2.9.3.

The application of quadratic forms having a sign-definite derivative along the solutions of systems of linear differential equations results in the following assertion.

Theorem 2.9.4 *Let the following conditions hold for the system (2.2.4), where $a(\varphi)$, $A(\varphi) \in C^l(\mathcal{T}_m)$, $l \geq 1$,*

(1) $\left\|\frac{\partial}{\partial \varphi} A(\varphi_t(\varphi))\right\| \leq M = \text{const} < \infty;$

(2) *for every fixed $\varphi \in R^m$ there exists a non-degenerate symmetric matrix function $S_\varphi(t)$ which is continuously differentiable and bounded in t and satisfying the condition*

$$\left\langle [\dot{S}_\varphi(t) + S_\varphi(t)A(\varphi_t) + A^*(\varphi_t)S_\varphi(t)]x, x \right\rangle \leq -\gamma(\varphi)\|x\|^2, \quad \gamma(\varphi) > 0.$$

Then the trivial torus $x = 0$ of the system (2.2.4) is e-dichotomous.

Remark 2.9.2 In the case when $a(\varphi) \equiv \omega = \text{const}$ the first condition of Theorem 2.9.4 is obviously satisfied, since $\varphi_t(\varphi) = \omega t + \varphi$ and

$$\frac{\partial}{\partial \varphi_i} A(\varphi_t(\varphi)) = \sum_{j=1}^m \frac{\partial A(\varphi_t(\varphi))}{\partial \varphi_{tj}} \frac{\partial \varphi_{tj}}{\partial \varphi_i} = \frac{\partial}{\partial \varphi_i} A(\varphi)\bigg|_{\varphi = \omega t + \varphi}.$$

Remark 2.9.3 The condition of non-degeneracy of the symmetric matrix functions $S_\varphi(t)$ can be replaced by the equivalent existence condition for two families of symmetric matrix functions $\tilde{S}_\varphi(t)$ and $S_\varphi(t)$ which are continuously differentiable and bounded in t and satisfy the conditions

$$\left\langle [\dot{S}_\varphi(t) + S_\varphi(t)A(\varphi_t) + A^*(\varphi_t)S_\varphi(t)]x, x \right\rangle \leq -\gamma(\varphi)\|x\|^2,$$

$$\left\langle [\dot{\tilde{S}}_\varphi(t) - \tilde{S}_\varphi(t)A^*(\varphi_t) - A(\varphi_t)\tilde{S}_\varphi(t)]x, x \right\rangle \leq -\gamma(\varphi)\|x\|^2, \quad \gamma(\varphi) > 0.$$

2.10 Preserving the Green Function under Small Perturbations of Linear Expansions on a Torus

We assume that the system (2.2.4) has the Green function of the problem on the invariant tori (2.6.10) (unique or not unique) satisfying the estimate (2.6.11). There arises a question as to whether the perturbed system of equations

$$\frac{d\varphi}{dt} = a(\varphi) + a_1(\varphi), \quad \frac{dx}{dt} = [A(\varphi) + A_1(\varphi)]x, \qquad (2.10.1)$$

will have the Green function for sufficiently small norms of the vector function $a_1(\varphi) \in C^1(\mathcal{T}_m)$ and the matrix function $A_1(\varphi) \in C^0(\mathcal{T}_m)$. We discuss this question in the present section.

First consider the case when only the matrix function $A(\varphi)$ is perturbed, i.e. $a_1(\varphi) \equiv 0$. The presence of the Green function (2.6.10) induces the existence of a symmetric matrix (2.7.3) which satisfies the condition (2.7.1). We note that the small changes of the matrix function $A(\varphi)$ do not essentially affect (2.7.1) being satisfied. Actually, the immediate calculation yields

$$\langle \{\dot{S}(\varphi) - S(\varphi)[A^*(\varphi) + A_1^*(\varphi)] - [A(\varphi) + A_1(\varphi)]S(\varphi)\}x, x\rangle$$
$$= \langle [\dot{S}(\varphi) - S(\varphi)A^*(\varphi) - A(\varphi)S(\varphi)]x, x\rangle - \langle [S(\varphi)A_1^*(\varphi) + A_1(\varphi)S(\varphi)]x, x\rangle$$
$$\leq -\|x\|^2 + 2\|S(\varphi)\|\,\|A_1(\varphi)\|\,\|x\|^2 \leq -(1 - 2\|S\|_0\|A_1\|_0)\|x\|^2.$$

Hence, it is clear that if

$$\|A_1\|_0 < (2\|S\|_0)^{-1}, \qquad (2.10.2)$$

the inequality (2.2.1) is satisfied provided that $A(\varphi)$ is replaced by $A(\varphi) + A_1(\varphi)$ and $S(\varphi)$ by $-kS(\varphi)$, where $k = (1 - 2\|S\|_0\|A_1\|_0)^{-1}$. So, if the system (2.2.4) has the Green function (2.6.10), then the perturbed system (2.10.1), where $a_1(\varphi) \equiv 0$ and the matrix $A_1(\varphi)$ satisfies inequality (2.10.2) also has the Green function.

In (2.7.1) let the matrix function $S(\varphi)$ be continuously differentiable in every variable φ_j, $j = 1, \ldots, m$, i.e. $S(\varphi) \in C^1(\mathcal{T}_m)$. Then, the derivative is

$$\dot{S}(\varphi) = \sum_{j=1}^{m} \frac{\partial S(\varphi)}{\partial \varphi_j} a_j(\varphi) \qquad (2.10.3)$$

and the inequality (2.7.1) can be represented as

$$\left\langle \left[\sum_{j=1}^{m} \frac{\partial S(\varphi)}{\partial \varphi_j} a_j(\varphi) - S(\varphi)A^*(\varphi) - A(\varphi)S(\varphi)\right]x, x\right\rangle \leq -\|x\|^2. \qquad (2.10.4)$$

Hence, it is clear that small changes of the vector function $a(\varphi)$ do not affect essentially the estimate (2.10.4) being satisfied.

Note that the function $S(\varphi)$ cannot be continuously differentiable in φ. In the case of the Green function (2.6.10) the existence of one such matrix function $S(\varphi)$ can be represented in the form of (2.7.3). In this regard there appears to be a problem in the possibility of approximating the functions $S(\varphi) \in C^1(\mathcal{T}_m; a)$ by the differentiable functions $\tilde{S}(\varphi) \in C^1(\mathcal{T}_m; a)$ so as to keep the closeness of its derivatives $\dot{S}(\varphi)$ and $\dot{\tilde{S}}(\varphi)$ as well. Here the derivatives are computed along the solutions of the system

$$\frac{d\varphi}{dt} = a(\varphi).$$

Lemma 2.10.1 *Let the continuous vector function $a(\varphi)$ be such that the Cauchy problem*

$$\frac{d\varphi}{dt} = a(\varphi), \quad \varphi|_{t=0} = \varphi_0$$

has a unique solution $\varphi_t(\varphi)$ for every fixed $\varphi_0 \in R^m$ and let some scalar function $u(\varphi)$ belong to the space $C'(\mathcal{T}_m; a)$. Then the condition

$$\lim_{\sigma \to +0} \frac{\omega(\sigma; a)\omega(\sigma; u)}{\sigma} = 0, \quad (2.10.5)$$

where $\omega(\sigma; a)$, $\omega(\sigma; u)$ are the continuity moduli of functions $a(\varphi)$ and $u(\varphi)$, is sufficient for the sequence of the differentiable functions $u_n(\varphi) \in C'(\mathcal{T}_m)$ to exist, such that

$$\lim_{n \to \infty} (|u(\varphi) - u_n(\varphi)|_0 + |\dot{u}(\varphi) - \dot{u}_n(\varphi)|_0) = 0. \quad (2.10.6)$$

Proof We designate by $\mu_i(\sigma)$, $i = 1, 2, 3$, some (so far arbitrary) continuous scalar functions determined on $[0, 1]$, with $\mu_i(0) = 0$, $i = 1, 2, 3$. Evidently there exist continuously differentiable functions $a_\sigma(\varphi)$, $u_\sigma(\varphi)$, $g_\sigma(\varphi) \in C^1(\mathcal{T}_m)$ which satisfying the inequalities

$$\|a(\varphi) - a_\sigma(\varphi)\| \leq \mu_1(\sigma),$$
$$|u(\varphi) - u_\sigma(\varphi)| \leq \mu_2(\sigma), \quad (2.10.7)$$
$$|\dot{u}(\varphi) - g_\sigma(\varphi)| \leq \mu_3(\sigma).$$

We designate

$$K(\sigma) = \max\left\{1, \max_{\varphi \in R^m} \left\|\frac{\partial a_\sigma(\varphi)}{\partial \varphi}\right\|\right\} \quad (2.10.8)$$

and show that for the functions $u_n(\varphi)$ satisfying condition (2.10.6) functions of the type $u_n(\varphi) = u^{1/n}(\varphi)$ can stand, where

$$u^{(\sigma)}(\varphi) = \int_0^\infty \exp\{-2K(\sigma)\tau\} [2K(\sigma)u_\sigma(\varphi_\tau^\sigma(\varphi)) - g_\sigma(\varphi_\tau^\sigma(\varphi))] \, d\tau, \quad (2.10.9)$$

$\sigma = \frac{1}{n}$, $\varphi_t^\sigma(\varphi)$ is a solution of the system

$$\frac{d\varphi}{dt} = a_\sigma(\varphi),$$

under the initial condition $\varphi_t^\sigma(\varphi)|_{t=0} = \varphi$. We estimate the difference $\varphi_t(\varphi) - \varphi_t^\sigma(\varphi)$ for positive t. Based on the integral representation

$$\varphi_t(\varphi) - \varphi_t^\sigma(\varphi) = \int_0^t [a(\varphi_\tau(\varphi)) - a_\sigma(\varphi_\tau^\sigma(\varphi))] \, d\tau$$

$$= \int_0^t [a(\varphi_\tau(\varphi)) - a_\sigma(\varphi_\tau(\varphi))] \, d\tau + \int_0^t [a_\sigma(\varphi_\tau(\varphi)) - a_\sigma(\varphi_\tau^\sigma(\varphi))] \, d\tau,$$

we have

$$\|\varphi_t(\varphi) - \varphi_t^\sigma(\varphi)\| \leq \int_0^t \|a(\varphi_\tau(\varphi)) - a_\sigma(\varphi_\tau^\sigma(\varphi))\| \, d\tau$$

$$+ \int_0^t \|a_\sigma(\varphi_\tau(\varphi)) - a_\sigma(\varphi_\tau^\sigma(\varphi))\| \, d\tau \qquad (2.10.10)$$

$$\leq \mu_1(\sigma)t + K(\sigma) \int_0^t \|\varphi_\tau(\varphi) - \varphi_\tau^\sigma(\varphi)\| \, d\tau.$$

Designating the right-hand side of (2.10.10) by $\Delta(t)$ and differentiating it yields

$$\frac{d\Delta(t)}{dt} = \mu_1(\sigma) + K(\sigma)\|\varphi_t(\varphi) - \varphi_t^\sigma(\varphi)\| \leq \mu_1(\sigma)t + K(\sigma)\Delta(t).$$

Hence it follows that

$$\frac{d}{dt}[\Delta(t) \exp\{-K(\sigma)t\}] \leq \mu_1(\sigma) \exp\{-K(\sigma)t\}.$$

Integrating both parts of the inequality obtained within the limits from 0 to t we find

$$\Delta(t) \exp\{-K(\sigma)t\} - \Delta(0) \leq \frac{\mu_1(\sigma)}{K(\sigma)}[1 - \exp\{-K(\sigma)t\}].$$

Thus the estimate

$$\|\varphi_t(\varphi) - \varphi_t^\sigma(\varphi)\| \leq \frac{\mu_1(\sigma)}{K(\sigma)}[\exp\{K(\sigma)t\} - 1], \quad t \geq 0, \qquad (2.10.11)$$

is valid.

The immediate integration by parts demonstrates that the identity

$$u(\varphi) \equiv \int_0^\infty \exp\{-2K(\sigma)\tau\}\left[2K(\sigma)u(\varphi_\tau(\varphi)) - \dot{u}(\varphi_\tau(\varphi))\right]d\tau, \qquad (2.10.12)$$

holds for the function $u(\varphi)$. Therefore, applying the designation $\partial_a u(\varphi) = \dot{u}(\varphi)$ we get the following representation for the difference $u^{(\sigma)}(\varphi) - u(\varphi)$:

$$u^{(\sigma)}(\varphi) - u(\varphi) = \int_0^\infty \exp\{-2K(\sigma)\tau\}\left[2K(\sigma)(u_\sigma(\varphi_\tau^\sigma(\varphi)) - u(\varphi_\tau^\sigma(\varphi)))\right.$$

$$\left. + (\partial_a u(\varphi_\tau(\varphi)) - g_\sigma(\varphi_\tau^\sigma(\varphi)))\right] d\tau$$

$$= \int_0^\infty \exp\{-2K(\sigma)\tau\}\{2K(\sigma)[u_\sigma(\varphi_\tau^\sigma(\varphi)) - u(\varphi_\tau^\sigma(\varphi))]\}\, d\tau$$

$$+ \int_0^\infty \exp\{-2K(\sigma)\tau\}\{2K(\sigma)[u(\varphi_\tau^\sigma(\varphi)) - u(\varphi_\tau(\varphi))]\}\, d\tau \qquad (2.10.13)$$

$$+ \int_0^\infty \exp\{-2K(\sigma)\tau\}[\partial_a u(\varphi_\tau(\varphi)) - \partial_a u(\varphi_\tau^\sigma(\varphi))]\, d\tau$$

$$+ \int_0^\infty \exp\{-2K(\sigma)\tau\}[\partial_a u(\varphi_\tau^\sigma(\varphi)) - g_\sigma(\varphi_\tau^\sigma(\varphi))]\, d\tau$$

$$= J_1 + J_2 + J_3 + J_4.$$

We estimate each of the additives J_j, $j = 1, \ldots, 4$, by means of (2.10.7) and (2.10.11). For J_1 we have

$$|J_1| \leq \int \exp\{-2K(\sigma)\tau\}\, d\tau\, 2K(\sigma)\mu_2(\sigma) = \mu_2(\sigma). \qquad (2.10.14)$$

Incorporating the continuity modulus $\omega(\sigma; u)$ of the function $u(\varphi)$ we get the estimate for J_2:

$$|J_2| \leq \int_0^\infty \exp\{-2K(\sigma)\tau\}\, |u(\varphi_\tau^\sigma(\varphi)) - u(\varphi_\tau(\varphi))|\, d\tau\, 2K(\sigma)$$

$$\leq 2K(\sigma)\int_0^\infty \exp\{-2K(\sigma)\tau\}\, \omega(\|\varphi_\tau^\sigma(\varphi) - \varphi_\tau(\varphi)\|; u)\, d\tau. \qquad (2.10.15)$$

Since the continuity modulus $\omega(\sigma; u)$ is always a semiadditive function $\omega(\sigma_1 + \sigma_2; u) \leq \omega(\sigma_1; u) + \omega(\sigma_2; u)$, then $\omega(n\sigma; u) \leq n\omega(\sigma; u)$, $n \in N$. Consequently, in view of (2.10.11) we find

$$\omega(\|\varphi_\tau^\sigma(\varphi) - \varphi_\tau(\varphi)\|; u) \leq \omega\left(\frac{\mu_1(\sigma)}{K(\sigma)}[\exp\{K(\sigma)\tau\} - 1]; u\right)$$

$$\leq \omega\left(\frac{\mu_1(\sigma)}{K(\sigma)}[\exp\{K(\sigma)\tau\}]; u\right) \leq [\exp\{K(\sigma)\tau\}]\omega\left(\frac{\mu_1(\sigma)}{K(\sigma)}; u\right)$$

$$\leq \exp\{K(\sigma)\tau\}\omega\left(\frac{\mu_1(\sigma)}{K(\sigma)}; u\right).$$

Applying the inequality obtained to (2.10.15) results in

$$|J_2| \leq 2K(\sigma)\int_0^\infty \exp\{-K(\sigma)\tau\}\,d\tau\, \omega\left(\frac{\mu_1(\sigma)}{K(\sigma)}; u\right) = 2\omega\left(\frac{\mu_1(\sigma)}{K(\sigma)}; u\right). \quad (2.10.16)$$

We arrive at the following estimates in the same way

$$|J_3| \leq \int_0^\infty \exp\{-2K(\sigma)\tau\}\,|\partial_a u(\varphi_\tau(\varphi)) - \partial_a u(\varphi_\tau^\sigma(\varphi))|\,d\tau$$

$$\leq \int_0^\infty \exp\{-2K(\sigma)\tau\}\,\omega(\|\varphi_\tau(\varphi) - \varphi_\tau^\sigma(\varphi)\|; \partial_a u)\,d\tau \quad (2.10.17)$$

$$\leq \int_0^\infty \exp\{-2K(\sigma)\tau\}\,\omega\left(\frac{\mu_1(\sigma)}{K(\sigma)}[\exp\{K(\sigma)\tau\} - 1]; \partial_a u\right)d\tau$$

$$\leq \int_0^\infty \exp\{-K(\sigma)\tau\}\,\omega\left(\frac{\mu_1(\sigma)}{K(\sigma)}; \partial_a u\right)d\tau = \frac{1}{K(\sigma)}\omega\left(\frac{\mu_1(\sigma)}{K(\sigma)}; \partial_a u\right),$$

where $\omega(\sigma; \partial_a u)$ is a continuity modulus of the function $\partial_a u(\varphi) \stackrel{\text{def}}{=} \dot{u}(\varphi) \in C^0(\mathcal{T}_m)$;

$$|J_4| \leq \int_0^\infty \exp\{-2K(\sigma)\tau\}\,|\partial_a u(\varphi_\tau^\sigma(\varphi)) - g_\sigma(\varphi_\tau^\sigma(\varphi))|\,d\tau$$

$$\leq \int_0^\infty \exp\{-2K(\sigma)\tau\}\,\mu_3(\sigma)\,d\tau = \frac{\mu_3(\sigma)}{2K(\sigma)}. \quad (2.10.18)$$

We return to the representation (2.10.13) and take into account the estimate (2.10.14), (2.10.16)–(2.10.18)

$$|u^{(\sigma)}(\varphi) - u(\varphi)| \leq \mu_2(\sigma) + 2\omega\left(\frac{\mu_1(\sigma)}{K(\sigma)}; u\right)$$
$$+ \frac{1}{K(\sigma)}\omega\left(\frac{\mu_1(\sigma)}{K(\sigma)}; \partial_a u\right) + \frac{\mu_3(\sigma)}{2K(\sigma)}.$$
(2.10.19)

It is clear that the difference $u^{(\sigma)}(\varphi) - u(\varphi)$ uniformly vanishes as $\sigma \to 0$.

Undertaking an estimate of the difference of the derivatives $\partial_a u(\varphi) - \partial_a u^{(\sigma)}(\varphi)$ we note that

$$\partial_a u(\varphi) \equiv 2K(\sigma)u(\varphi) + (-2K(\sigma)u(\varphi) + \partial_a u(\varphi)),$$
$$\partial_{a_\sigma} u^{(\sigma)}(\varphi) \equiv 2K(\sigma)u^{(\sigma)}(\varphi) + (-2K(\sigma)u_\sigma(\varphi) + g_\sigma(\varphi)).$$

Therefore

$$\partial_a u(\varphi) - \partial_{a_\sigma} u^{(\sigma)}(\varphi) = 2K(\sigma)[u(\varphi) - u^{(\sigma)}(\varphi)] + 2K(\sigma)[u_\sigma(\varphi) - u(\varphi)]$$
$$+ \partial_a u(\varphi) - g_\sigma(\varphi).$$

Hence it follows that

$$|\partial_a u(\varphi) - \partial_{a_\sigma} u^{(\sigma)}(\varphi)| \leq 2K(\sigma)|u(\varphi) - u^{(\sigma)}(\varphi)|$$
$$+ 2K(\sigma)|u_\sigma(\varphi) - u(\varphi)| + |\partial_a u(\varphi) - g_\sigma(\varphi)|$$
$$\leq 2K(\sigma)\left[\mu_2(\sigma) + 2\omega\left(\frac{\mu_1(\sigma)}{K(\sigma)}; u\right) + \frac{1}{K(\sigma)}\omega\left(\frac{\mu_1(\sigma)}{K(\sigma)}; \partial_a u\right)\right]$$
$$+ 2K(\sigma)\mu_2(\sigma) + \mu_3(\sigma) = 4K(\sigma)\mu_2(\sigma) + 4K\omega\left(\frac{\mu_1(\sigma)}{K(\sigma)}; u\right)$$
$$+ 2\omega\left(\frac{\mu_1(\sigma)}{K(\sigma)}; \partial_a u\right) + \mu_3(\sigma).$$
(2.10.20)

We designate by $K_2(\sigma)$ the value which is dependent on $\sigma \in [0,1]$ and satisfies the inequality

$$\left\|\frac{\partial}{\partial \varphi} u^{(\sigma)}(\varphi)\right\| \leq K_2(\sigma),$$
(2.10.21)

where

$$\frac{\partial}{\partial \varphi} u^{(\sigma)}(\varphi) = \left(\frac{\partial}{\partial \varphi_1} u^{(\sigma)}(\varphi), \ldots, \frac{\partial}{\partial \varphi_m} u^{(\sigma)}(\varphi)\right).$$

Estimating the difference $\partial_{a_\sigma} u^{(\sigma)}(\varphi) - \partial_a u^{(\sigma)}(\varphi)$ we get

$$|\partial_{a_\sigma} u^{(\sigma)}(\varphi) - \partial_a u^{(\sigma)}(\varphi)| = \left|\sum_{j=1}^{m} \left(\frac{\partial}{\partial \varphi_j} u^{(\sigma)}(\varphi)\right)(a_{j\sigma}(\varphi) - a_j(\varphi))\right| \quad (2.10.22)$$

$$\leq \left\|\frac{\partial}{\partial \varphi} u^{(\sigma)}(\varphi)\right\| \|a_\sigma(\varphi) - a(\varphi)\| \leq K_2(\sigma)\mu_1(\sigma).$$

Applying the estimates (2.10.20), (2.10.22) we arrive at

$$|\partial_a u(\varphi) - \partial_a u^{(\sigma)}(\varphi)| = |\partial_a u(\varphi) - \partial_{a_\sigma} u^{(\sigma)}(\varphi)$$
$$+ \partial_{a_\sigma} u^{(\sigma)}(\varphi) - \partial_a u^{(\sigma)}(\varphi)|$$
$$\leq |\partial_a u(\varphi) - \partial_{a_\sigma} u^{(\sigma)}(\varphi)| + |\partial_{a_\sigma} u^{(\sigma)}(\varphi) - \partial_a u^{(\sigma)}(\varphi)|$$
$$\leq 4K(\sigma)\mu_2(\sigma) + 4K(\sigma)\omega\left(\frac{\mu_1(\sigma)}{K(\sigma)}; u\right) \quad (2.10.23)$$
$$+ 2\omega\left(\frac{\mu_1(\sigma)}{K(\sigma)}; \partial_a u\right) + \mu_3(\sigma) + K_2(\sigma)\mu_1(\sigma).$$

In the right-hand side of the obtained inequality the quantities $2\omega\left(\frac{\mu_1(\sigma)}{K(\sigma)}; \partial_a u\right)$, $\mu_3(\sigma)$ vanish as $\sigma \to +0$. Thus, for the whole of the right-hand side of (2.10.23) to vanish, it is necessary that

$$K(\sigma)\mu_2(\sigma) + K_2(\sigma)\mu_1(\sigma) \xrightarrow[\sigma \to +0]{} 0; \quad (2.10.24)$$

$$K(\sigma)\omega\left(\frac{\mu_1(\sigma)}{K(\sigma)}; u\right) \xrightarrow[\sigma \to +0]{} 0. \quad (2.10.25)$$

Now we take Steklov functions for the approximate functions $a_\sigma(\varphi), g_\sigma(\varphi), u_\sigma(\varphi)$ in the estimates (2.10.7). For example, taking

$$u_\sigma(\varphi) = (2\sigma)^{-m} \int_{\varphi_m - \sigma}^{\varphi_m + \sigma} \cdots \int_{\varphi_1 - \sigma}^{\varphi_1 + \sigma} u(\psi_1, \ldots, \psi_m)\, d\psi_1 \ldots d\psi_m,$$

we get for the difference $u_\sigma(\varphi) - u(\varphi)$

$$|u_\sigma(\varphi) - u(\varphi)| \leq (2\sigma)^{-m} \int_{\varphi_m - \sigma}^{\varphi_m + \sigma} \cdots \int_{\varphi_1 - \sigma}^{\varphi_1 + \sigma} |u(\psi_1, \ldots, \psi_m)$$
$$- u(\varphi_1, \ldots, \varphi_m)|\, d\psi_1 \ldots d\psi_m \quad (2.10.26)$$
$$\leq (2\sigma)^{-m} \int_{\varphi_m - \sigma}^{\varphi_m + \sigma} \cdots \int_{\varphi_1 - \sigma}^{\varphi_1 + \sigma} \omega(\|\psi - \varphi\|; u)\, d\psi_1 \ldots d\psi \leq m\omega(\sigma; u).$$

Estimating the derivative $\frac{\partial}{\partial \varphi_m} u_\sigma(\varphi)$ we find

$$\left|\frac{\partial}{\partial \varphi_m} u_\sigma(\varphi)\right| = (2\sigma)^{-m} \left| \int_{\varphi_m-\sigma}^{\varphi_m+\sigma} \cdots \int_{\varphi_1-\sigma}^{\varphi_1+\sigma} (u(\psi_1,\ldots,\psi_{m-1},\varphi_m+\sigma) \right.$$

$$\left. - u(\psi_1,\ldots,\psi_{m-1},\varphi_m-\sigma)) \, d\psi_1 \ldots d\psi_{m-1} \right| \leq \frac{\omega(\sigma;u)}{\sigma}. \qquad (2.10.27)$$

Thus in the estimates (2.10.7) one can set

$$\mu_1(\sigma) = m\omega(\sigma;a),$$
$$\mu_2(\sigma) = m\omega(\sigma;u), \qquad (2.10.28)$$
$$\mu_3(\sigma) = m\omega(\sigma;\partial_a u);$$

and, in addition

$$K(\sigma) = \sqrt{m}\,\frac{\omega(\sigma;u)}{\sigma}. \qquad (2.10.29)$$

We calculate the value of $K_2(\sigma)$. To this end the equality (2.10.9) is differentiated in φ_i

$$\frac{\partial u^{(\sigma)}(\varphi)}{\partial \varphi_i} = \int_0^\infty \exp\{-2K(\sigma)\tau\} \left[2K(\sigma) \sum_{j=1}^m \frac{\partial u_\sigma(\varphi_\tau^\sigma(\varphi))}{\partial \varphi_{\tau j}^\sigma} \frac{\partial \varphi_{\tau j}^\sigma(\varphi)}{\partial \varphi_i} \right.$$

$$\left. - \sum_{j=1}^m \frac{\partial g_\sigma(\varphi_\tau^\sigma(\varphi))}{\partial \varphi_{\tau j}^\sigma} \frac{\partial \varphi_{\tau j}^\sigma(\varphi)}{\partial \varphi_i} \right] d\tau. \qquad (2.10.30)$$

The following inequality can be established in the same way as the estimate (2.10.27)

$$\left\|\frac{\partial}{\partial \varphi_i} g_\sigma(\varphi)\right\| \leq \frac{\omega(\sigma;\partial_a u)}{\sigma}. \qquad (2.10.31)$$

Since the derivative $\frac{\partial \varphi_\tau^\sigma(\varphi)}{\partial \varphi_i}$ can be estimated as

$$\left\|\frac{\partial \varphi_\tau^\sigma(\varphi)}{\partial \varphi_i}\right\| \leq \exp\{K(\sigma)\tau\}, \quad t \geq 0,$$

we find from (2.10.30)

$$\left\|\frac{\partial u^{(\sigma)}(\varphi)}{\partial \varphi_i}\right\| \leq (c)\left[\frac{\omega(\sigma;u)}{\sigma} + \frac{\omega(\sigma;\partial_a u)}{K(\sigma)\sigma}\right] = (c)\left[\frac{\omega(\sigma;u)}{\sigma} + \frac{\omega(\sigma;\partial_a u)}{\sqrt{m}\,\omega(\sigma;a)}\right]. \qquad (2.10.32)$$

Thus, the value of $K_2(\sigma)$ is represented as

$$K_2(\sigma) = (c)\left[\frac{\omega(\sigma;u)}{\sigma} + \frac{\omega(\sigma;\partial_a u)}{\sqrt{m}\,\omega(\sigma;a)}\right], \qquad (2.10.33)$$

where (c) is a constant.

Calculating the left-hand side in (2.10.24) we get

$$K(\sigma)\mu_2(\sigma) + K_2(\sigma)\mu_1(\sigma)$$

$$= m^{3/2}\frac{\omega(\sigma;a)\omega(\sigma;u)}{\sigma} + (c)\left[\frac{\omega(\sigma;u)\omega(\sigma;a)}{\sigma} + \frac{\omega(\sigma;\partial_a u)}{\sqrt{m}}\right].$$

Hence it is clear that the condition (2.10.5) ensures the vanishing of the expression obtained for $\sigma \to +0$. We now have for the left-hand side of (2.10.25)

$$K(\sigma)\omega\left(\frac{\mu_1(\sigma)}{K(\sigma)}; u\right) = \sqrt{m}\,\frac{\omega(\sigma;u)}{\sigma}\,\omega(\sqrt{m}\,\sigma; u)$$

$$\leq \sqrt{m}\,([\sqrt{m}]+1)\frac{\omega(\sigma;a)\omega(\sigma;u)}{\sigma} \xrightarrow[\sigma\to+0]{} 0.$$

This completes the proof of the lemma.

Theorem 2.10.1 *Let the continuous vector-function $a(\varphi)$ be such that its continuity modulus $\omega(\sigma;a)$ satisfies the condition*

$$\lim_{\sigma\to+0}\frac{\omega(\sigma;a)}{\sigma}\int_\sigma^d \exp\left\{-\gamma_0\int_\sigma^z\frac{dt}{\omega(t;a)}\right\}dz = 0, \qquad (2.10.34)$$

and the matrix function $A(\varphi)$ belong to $C^1(\mathcal{T}_m)$ and the system (2.2.4) has the Green function of the problem on the invariant tori (2.6.10) (either unique or not), satisfying estimate (2.6.11). There then exists a number $,\varepsilon_0 > 0$ such that for every vector-function $a_1(\varphi) \in C^1(\mathcal{T}_m)$ satisfying the inequality

$$\|a_1(\varphi)\| \leq \varepsilon_0, \qquad (2.10.35)$$

the system of equations (2.10.1), where $A_1(\varphi) \equiv 0$, has the Green function $\tilde{G}_0(\tau,\varphi)$ which satisfies the estimate

$$\|\tilde{G}_0(\tau,\varphi)\| \leq K(\varepsilon_0)\exp\{-\gamma(\varepsilon_0)|\tau|\}, \qquad (2.10.36)$$

where the positive constants $K(\varepsilon_0)$, $\gamma(\varepsilon_0)$ only depend on ε_0.

Proof Consider the matrix function (2.7.3) and estimate its continuity modulus. To this end we represent the difference $S(\varphi) - S(\bar{\varphi})$ as

$$S(\varphi) - S(\bar{\varphi}) = 2 \int_0^\infty [G_0(\sigma, \varphi) G_0^*(\sigma, \varphi) - G_0(\sigma, \bar{\varphi}) G_0^*(\sigma, \bar{\varphi})] d\sigma$$

$$+ 2 \int_{-\infty}^0 [G_0(\sigma, \bar{\varphi}) G_0^*(\sigma, \bar{\varphi}) - G_0(\sigma, \varphi) G_0^*(\sigma, \varphi)] d\sigma$$

$$= 2 \int_0^\infty [G_0(\sigma, \varphi) - G_0(\sigma, \bar{\varphi})] G_0^*(\sigma, \varphi) d\sigma + 2 \int_0^\infty G_0(\sigma, \bar{\varphi}) [G_0^*(\sigma, \varphi) - G_0^*(\sigma, \bar{\varphi})] d\sigma$$

$$+ 2 \int_{-\infty}^0 [G_0(\sigma, \bar{\varphi}) - G_0(\sigma, \varphi)] G_0^*(\sigma, \bar{\varphi}) d\sigma + 2 \int_{-\infty}^0 G_0(\sigma, \varphi) [G_0^*(\sigma, \bar{\varphi}) - G_0^*(\sigma, \varphi)] d\sigma.$$

In view of (2.6.11) and the above representation we find that

$$\|S(\varphi) - S(\bar{\varphi})\| \leq 4K_0 \int_{-\infty}^\infty \|G_0(\sigma, \varphi) - G_0(\sigma, \bar{\varphi})\| \exp\{-\gamma_0 |\sigma|\} d\sigma. \quad (2.10.37)$$

Making the assumption that the Green function $G_0(\tau, \varphi)$ is unique, estimate (2.7.9) is valid for the difference $G_t(\sigma, \varphi) - G_t(\sigma, \bar{\varphi})$, which yields

$$\|G_0(\tau, \varphi) - G_0(\tau, \bar{\varphi})\| \leq K_0^2 \int_{-\infty}^\infty \omega(\|\varphi_\sigma(\varphi) - \varphi_\sigma(\bar{\varphi})\|; A) \exp\{-\gamma_0 |\sigma|\} d\sigma, \quad (2.10.38)$$

where $\omega(\sigma; A)$ is a continuity modulus of the matrix function $A(\varphi)$. Since $A(\varphi) \in C^1(\mathcal{T}_m)$, then evidently, the function $\omega(\sigma; A)$ can be estimated as

$$\omega(\sigma; A) \leq M \begin{cases} \sigma, & \sigma \in [0, d], \\ d, & \sigma \in [d, \infty), \end{cases} \quad (2.10.39)$$

where M and d are some positive constants.

We estimate the difference of solutions $\varphi_t(\varphi) - \varphi_t(\bar{\varphi})$. Setting $t \geq 0$ in the integral equation

$$\varphi_t(\varphi) - \varphi_t(\bar{\varphi}) = \varphi - \bar{\varphi} + \int_0^t [a(\varphi_\sigma(\varphi)) - a(\varphi_\sigma(\bar{\varphi}))] d\sigma,$$

one obtains

$$\|\varphi_t(\varphi) - \varphi_t(\bar{\varphi})\| \leq \|\varphi - \bar{\varphi}\| + \int_0^t \omega(\|\varphi_\sigma(\varphi) - \varphi_\sigma(\bar{\varphi})\|)\, d\sigma. \tag{2.10.40}$$

Applying the well-known Bihary lemma, we find from (2.10.40)

$$\|\varphi_t(\varphi) - \varphi_t(\bar{\varphi})\| \leq \Phi^{-1}(\Phi(\|\varphi - \bar{\varphi}\|) + t), \tag{2.10.41}$$

where

$$\Phi(u) = \int_d^u \frac{dt}{\omega(t; a)}, \tag{2.10.42}$$

$\Phi^{-1}(v)$ is an inverse function. Proceeding with the estimate (2.10.38), in view of (2.10.39), (2.10.41), we have

$$\|G_0(\tau, \varphi) - G_0(\tau, \bar{\varphi})\| \leq 2K_0^2 \int_0^\infty \omega(\Phi^{-1}(\Phi(\|\varphi - \bar{\varphi}\|) + \sigma); A) \exp\{-\gamma_0 \sigma\}\, d\sigma$$

$$\leq 2K_0^2 M \int_{\|\varphi-\bar{\varphi}\|}^d \frac{z}{\omega(z;a)} \exp\{-\gamma_0(\Phi(z) - \Phi(\|\varphi - \bar{\varphi}\|))\}\, dz$$

$$+ 2K_0^2 M d \int_d^\infty \frac{1}{\omega(z;a)} \exp\{-\gamma_0(\Phi(z) - \Phi(\|\varphi - \bar{\varphi}\|))\}\, dz.$$

We integrate the first integral by parts and compute the second. Then

$$\|G_0(\tau, \varphi) - G_0(\tau, \bar{\varphi})\|$$
$$\leq \frac{2K_0^2 M}{\gamma}\left(\|\varphi - \bar{\varphi}\| + \int_{\|\varphi-\bar{\varphi}\|}^d \exp\{-\gamma_0(\Phi(z) - \Phi(\|\varphi - \bar{\varphi}\|))\}\, dz\right). \tag{2.10.43}$$

The estimation of the difference of the Green functions (2.10.43) allows us to proceed with (2.10.37) as

$$\|S(\varphi) - S(\bar{\varphi})\|$$
$$\leq \frac{16K_0^2 M}{\gamma_0^2}\left(\|\varphi - \bar{\varphi}\| + \int_{\|\varphi-\bar{\varphi}\|}^d \exp\{-\gamma_0(\Phi(z) - \Phi(\|\varphi - \bar{\varphi}\|))\}\, dz\right).$$

Thus, the continuity modulus $\omega(\sigma; S)$ of the function $S(\varphi)$ can be estimated from the above

$$\omega(\sigma; S) \leq K_1\left(\sigma + \int_\sigma^d \exp\left\{-\gamma_0 \int_\sigma^z \frac{dt}{\omega(t;a)}\right\} dz\right). \quad (2.10.44)$$

We first verify that (2.10.5) is satisfied. We substitute $\omega(\sigma; S)$ for $\omega(\sigma; u)$ and estimate it

$$\frac{\omega(\sigma;a)\omega(\sigma;S)}{\sigma} \leq K_1\omega(\sigma;a) + K_1 \frac{\omega(\sigma;a)}{\sigma} \int_\sigma^d \exp\left\{-\gamma_0 \int_\sigma^z \frac{dt}{\omega(t;a)}\right\} dz. \quad (2.10.45)$$

In view of (2.10.34) we note that the right-hand side of (2.10.5) vanishes as $\delta \to +0$. So, in the conditions of the proved lemma one can claim that for every $\delta \in (0,1)$ there exists a symmetric matrix function $S_\delta(\varphi) \in C^1(\mathcal{T}_m)$ satisfying the inequality

$$\|\dot{S}_\delta(\varphi) - \dot{S}(\varphi)\| + 2\|A\|_0\|S_\delta(\varphi) - S(\varphi)\| \leq \delta. \quad (2.10.46)$$

Now, in the left-hand side of (2.7.1) we replace $S(\varphi) \in C'(\mathcal{T}_m;a)$ by $S_\delta(\varphi)$ which has continuous partial derivatives in φ_j, $j = 1, \ldots, m$

$$\left\langle \left[\sum_{i=1}^m \frac{\partial S_\delta(\varphi)}{\partial \varphi_i} a_i(\varphi) - S_\delta(\varphi)A^*(\varphi) - A(\varphi)S_\delta(\varphi)\right]x, x\right\rangle$$
$$= \left\langle [\dot{S}(\varphi) - S(\varphi)A^*(\varphi) - A(\varphi)S(\varphi)]x, x\right\rangle$$
$$+ \left\langle \{[\dot{S}_\delta(\varphi) - \dot{S}(\varphi)] + [S(\varphi) - S_\delta(\varphi)]A^*(\varphi)\right.$$
$$\left. + A(\varphi)[S(\varphi) - S_\delta(\varphi)]\}x, x\right\rangle \leq -\|x\|^2 + [\|\dot{S}_\delta(\varphi) - \dot{S}(\varphi)\|$$
$$+ 2\|A\|_0\|S_\delta(\varphi) - S(\varphi)\|]\|x\|^2 \leq -(1-\delta)\|x\|^2.$$

Adding the coordinates $a_i^{(1)}(\varphi)$ of $a_1(\varphi)$ to the coordinates a_i, $i = 1, \ldots, m$, of the vector functions $a(\varphi)$ we get

$$\left\langle \left[\sum_{i=1}^m \frac{\partial S_\delta(\varphi)}{\partial \varphi_i}(a_i(\varphi) + a_i^{(1)}(\varphi)) - S_\delta(\varphi)A^*(\varphi) - A(\varphi)S_\delta(\varphi)\right]x, x\right\rangle$$
$$\leq -(1-\delta)\|x\|^2 + \left\|\sum_{i=1}^m \frac{\partial S_\delta(\varphi)}{\partial \varphi_i} a_i^{(1)}(\varphi)\right\|\|x\|^2 \quad (2.10.47)$$
$$\leq -\left[1 - \delta - \|a_1(\varphi)\|\left(\sum_{i=1}^m \left\|\frac{\partial S_\delta(\varphi)}{\partial \varphi_i}\right\|^2\right)^{1/2}\right]\|x\|^2.$$

Hence, it is clear that the choice of $\varepsilon_0 > 0$ which satisfies the inequality
$$\bar{\beta} = 1 - \delta - \varepsilon_0 P_\delta > 0, \tag{2.10.48}$$
where
$$P_\delta = \left(\sum_{i=1}^{m} \left\| \frac{\partial S_\delta(\varphi)}{\partial \varphi_i} \right\|_0^2 \right)^{1/2}, \tag{2.10.49}$$
ensures negative definiteness of the derivative of the quadratic form
$$V_\delta(\varphi, x) = \langle S_\delta(\varphi)x, x \rangle, \tag{2.10.50}$$
computed along the solutions of the perturbed system
$$\frac{d\varphi}{dt} = a(\varphi) + a_1(\varphi), \quad \frac{dx}{dt} = -A^*(\varphi)x$$
under the condition (2.10.35).

Now the constants $K(\varepsilon_0)$, $\gamma(\varepsilon_0)$ found in (2.10.36) are immediately represented via the quadratic form (2.10.50) and the matrix function $A(\varphi)$. To this end we recall (see Theorem 1.1.3) that if there exists a non-degenerate matrix function $S_1(\varphi) = S_1^*(\varphi) \in C'(\mathcal{T}_m; a)$ which satisfies condition (2.7.43), then in the estimate of the Green function (2.6.11) the constants K_0, γ_0 can be represented as
$$K_0 = (2 + \sqrt{2})\left(\frac{\|A\|_0 \|S_1\|_0}{\beta_1} \right)^{3/2}, \quad \gamma_0 = \frac{\beta_1}{2\|S_1\|_0}. \tag{2.10.51}$$
However, in this case the quadratic form (2.10.50) can be degenerate at some points $\varphi \in R^m$. That is why we consider the extended system
$$\frac{d\varphi}{dt} = a(\varphi) + a_1(\varphi),$$
$$\frac{dx}{dt} = A(\varphi)x, \tag{2.10.52}$$
$$\frac{dy}{dt} = -x - A^*(\varphi)y,$$
and take the non-degenerate quadratic form having a sign-definite derivative along its solutions. This form is looked for in the form
$$V_{\delta,\lambda}(\varphi, x, y) = \lambda \langle x, y \rangle + \langle S_\delta(\varphi)y, y \rangle, \tag{2.10.53}$$
where λ is a so far unknown positive parameter. Incorporating (2.10.47) and (2.10.48) we estimate the derivative of the quadratic form (2.10.53) computed along solutions of the system (2.10.52)
$$\dot{V}_{\delta,\lambda}(\varphi, x, y) \leq -\lambda \|x\|^2 + 2\|S_\delta\|_0 \|x\| \|y\| - \bar{\beta}_1 \|y\|^2$$
$$\leq -\left(\lambda - \frac{2}{\bar{\beta}_1} \|S_\delta\|_0^2 \right) \|x\|^2 - \frac{\bar{\beta}_1}{2} \|y\|^2.$$

Letting
$$\lambda = \frac{\bar{\beta}_1}{2} + \frac{2}{\bar{\beta}_1}\|S_\delta\|_0^2, \qquad (2.10.54)$$
here we come to the inequality
$$\dot{V}_{\delta,\lambda}(\varphi,x,y) \le -\frac{\bar{\beta}_1}{2}(\|x\|^2 + \|y\|^2). \qquad (2.10.55)$$

Estimating the norms of the matrices
$$\bar{S}(\varphi) = \begin{pmatrix} 0 & \frac{\lambda}{2}I_n \\ \frac{\lambda}{2}I_n & S_\delta(\varphi) \end{pmatrix}, \quad \bar{A}(\varphi) = \begin{pmatrix} A(\varphi) & 0 \\ -I_n & -A^*(\varphi) \end{pmatrix},$$
we easily obtain
$$\|\bar{S}\|_0 \le \frac{\lambda}{2} + \|S_\delta\|_0, \quad \|\bar{A}\|_0 \le 1 + \sqrt{2}\,\|A\|_0.$$

We now use the representation of constants (2.10.51). Then
$$\gamma_0 = \frac{\bar{\beta}_1/2}{2\|\bar{S}\|_0} \ge \frac{\bar{\beta}_1}{2\lambda + 4\|S_\delta\|_0} = \frac{\bar{\beta}_1}{\bar{\beta}_1 + 4\bar{\beta}_1^{-1}\|S_\delta\|_0^2 + 4\|S_\delta\|_0}; \qquad (2.10.56)$$

$$K_0 = (2+\sqrt{2})\left(\frac{\|\bar{A}\|_0\|\bar{S}\|_0}{\bar{\beta}_1/2}\right)^{3/2}$$
$$\le (2+\sqrt{2})\left(\frac{(1+\sqrt{2}\,\|A\|_0)(\lambda + 2\|S_\delta\|_0)}{\bar{\beta}_1}\right)^{3/2} \qquad (2.10.57)$$
$$= (2+\sqrt{2})\left[(1+\sqrt{2}\,\|A\|_0)\left(\frac{1}{2} + \frac{2}{\bar{\beta}_1^2}\|S_\delta\|_0^2 + \frac{2}{\bar{\beta}_1}\|S_\delta\|_0\right)\right]^{3/2},$$

where $\bar{\beta}_1$ is determined by (2.10.48).

Thus, having fixed $\delta = 1/2$ we obtain the following representations of the constants $\gamma(\varepsilon_0)$ and $K(\varepsilon_0)$ from (2.10.56), (2.10.57)

$$\gamma(\varepsilon_0) = \left[1 + \frac{4\|S_{1/2}\|_0^2}{\left(\frac{1}{2} - \varepsilon_0 p_{1/2}\right)^2} + \frac{4\|S_{1/2}\|_0}{\frac{1}{2} - \varepsilon_0 p_{1/2}}\right]^{-1},$$

$$K(\varepsilon_0) = (2+\sqrt{2})(1+\sqrt{2}\,\|A\|_0)\left[\frac{1}{2} + \frac{2\|S_{1/2}\|_0^2}{\left(\frac{1}{2} - \varepsilon_0 p_{1/2}\right)^2} + \frac{2\|S_{1/2}\|_0}{\frac{1}{2} - \varepsilon_0 p_{1/2}}\right]^{3/2}.$$

This completes the proof of Theorem 2.10.1.

Remark 2.10.1 Condition (2.10.34) is satisfied, if the vector function $a(\varphi)$ satisfies the Lipschitz condition.

We verify the that of (2.10.34) is satisfied in the case when $\omega(\sigma; a) = \alpha\sigma|\ln \sigma|$, where α is a constant. We compute the corresponding integral

$$\int_\sigma^z \frac{dt}{\omega(t;a)} = -\int_\sigma^z \frac{dt}{\alpha t \ln t} = -\frac{1}{\alpha} \ln |\ln t| \Big|_\sigma^z = \frac{1}{\alpha} \ln \left|\frac{\ln \sigma}{\ln z}\right|,$$

$$\exp\left\{-\gamma_0 \int_\sigma^z \frac{dt}{\omega(t;a)}\right\} = \exp\left\{-\frac{\gamma_0}{\alpha} \ln \left|\frac{\ln \sigma}{\ln z}\right|\right\} = \left|\frac{\ln z}{\ln \sigma}\right|^{\gamma_0/\alpha}.$$

Thus, setting $d = e^{-1}$ we have

$$\frac{\omega(\sigma;a)}{\sigma} \int_\sigma^d \exp\left\{-\gamma_0 \int_\sigma^z \frac{dt}{\omega(t;a)}\right\} dz = \frac{\alpha|\ln \sigma|}{|\ln \sigma|^{\gamma_0/\alpha}} \int_\sigma^{\exp\{-1\}} |\ln \sigma|^{\gamma_0/\alpha} dz.$$

Since the integral $\int_\sigma^{\exp\{-1\}} |\ln \sigma|^{\gamma_0/\alpha} dz$ is convergent at zero, then $\alpha < \gamma_0$ is sufficient for (2.10.34) be satisfied.

2.11 On the Smoothness of an Exponentially Stable Invariant Torus

When the invariant torus

$$x = u(\varphi), \qquad (2.11.1)$$

of the system of equations

$$\frac{d\varphi}{dt} = a(\varphi),$$
$$\frac{dx}{dt} = A(\varphi)x + f(\varphi), \qquad (2.11.2)$$

is considered a question arises on the dependence of its smoothness on that of the right-hand side of (2.11.2). As noted above, this dependence is not obvious, and so is to be studied separately.

Theorem 2.11.1 *Let $a, A \in C^l(\mathcal{T}_m)$, $l \geq 1$. Then if the Green function of the system (2.11.1) for an arbitrary function $f \in C^l(\mathcal{T}_m)$ satisfies the inequality*

$$|G_0(\tau, \varphi)f(\varphi_\tau(\varphi))|_l \leq K e^{-\gamma|\tau|} |f|_l, \quad t \in R, \qquad (2.11.3)$$

where $\varphi_t(\varphi)$ is the solution of the first equation of (2.11.2) for arbitrary $\varphi \in \mathcal{T}_m$, K and γ are positive constants independent of φ, then the invariant torus (2.11.1) of the system (2.11.2) belongs to the space $C^l(\mathcal{T}_m)$ and satisfies the inequality

$$|u|_l \leq 2K\gamma^{-1}|f|_l. \tag{2.11.4}$$

The proof of the theorem follows obviously from estimate (2.11.3) and the integral representation of the function $u(\varphi)$ which determines the torus (2.11.1)

$$u(\varphi) = \int_{-\infty}^{\infty} G_0(\tau, \varphi) f(\varphi_\tau(\varphi)) \, d\tau. \tag{2.11.5}$$

By the Theorem 2.11.1 the smoothness of the invariant torus (2.11.1) of the system (2.11.2) depends essentially on only the properties of the Green function $G_0(\tau, \varphi)$ and the solution $\varphi_t(\varphi)$ of the system prescribing the flow of trajectories (2.11.2) on the torus (2.11.1). The functions a, A and f are considered to belong to the space $C^l(\mathcal{T}_m)$ with finite $l \geq 1$. We establish conditions ensuring the inequality (2.11.3).

Consider the case when the torus (2.11.1) is exponentially stable. By the results of Section 2.3 this case takes place when

$$\inf_{\varphi \in \mathcal{T}_m} \beta(\varphi) = \beta_0 > 0, \tag{2.11.6}$$

where $\beta(\varphi)$ is determined by

$$\inf_{S \in \mathfrak{N}_0} \max_{\|x\|=1} \frac{\langle [S(\varphi) A(\varphi) + \frac{1}{2}\dot{S}(\varphi)] x, x \rangle}{\langle S(\varphi) x, x \rangle} \leq -\beta(\varphi), \tag{2.11.7}$$

$\mathfrak{N}_0$ is a set of $n \times n$-dimensional positive definite symmetric matrices belonging to $C^1(\mathcal{T}_m; a)$.

We define the value of $a(\varphi)$ setting

$$\sup_{S_1 \in \mathfrak{N}_1} \min_{\|\psi\|=1} \frac{\langle [S_1(\varphi) \frac{\partial a(\varphi)}{\partial \varphi} + \frac{1}{2}\dot{S}_1(\varphi)] \psi, \psi \rangle}{\langle S_1(\varphi) \psi, \psi \rangle} \geq a(\varphi), \tag{2.11.8}$$

where $\mathfrak{N}_1$ is a set of $m \times m$-dimensional positive definite symmetric matrices belonging to $C^1(\mathcal{T}_m)$.

Theorem 2.11.2 *Let the inequality (2.11.6) be satisfied and*

$$\inf_{\varphi \in \mathcal{T}_m} [\beta(\varphi) + l\alpha(\varphi)] > 0. \tag{2.11.9}$$

Then the invariant torus (2.11.1) of the system (2.11.2) belongs to the space $C^l(\mathcal{T}_m)$.

Proof We first incorporate the roughness of the conditions of the theorem due to which these conditions ensure the existence of the matrices $S(\varphi) \in \mathfrak{N}_0$, $S_1(\varphi) \in \mathfrak{N}_1$ and the functions $\beta(\varphi)$, $\alpha(\varphi)$ which satisfy inequalities of the type (2.11.7), (2.11.8)

$$\max_{\|x\|=1} \frac{\langle [S(\varphi)A(\varphi) + \frac{1}{2}\dot{S}(\varphi)]x, x \rangle}{\langle S(\varphi)x, x \rangle} \leq -\beta(\varphi),$$

$$\min_{\|\psi\|=1} \frac{\langle [S_1(\varphi)\frac{\partial a(\varphi)}{\partial \varphi} + \frac{1}{2}\dot{S}_1(\varphi)]\psi, \psi \rangle}{\langle S_1(\varphi)\psi, \psi \rangle} \geq a(\varphi) \tag{2.11.10}$$

and conditions of the type (2.11.6), (2.11.9)

$$\beta(\varphi) \geq \beta_0, \quad \beta(\varphi) + l\alpha(\varphi) \geq \gamma, \tag{2.11.11}$$

where β_0 and γ are positive constants

$$\beta_0 = \inf_{\varphi \in \mathcal{T}_m} \beta(\varphi), \quad \gamma = \inf_{\varphi \in \mathcal{T}_m} [\beta(\varphi) + l\alpha(\varphi)].$$

When the first inequality (2.11.11) is satisfied, the second one is of independent value only due to the values of φ for which $\alpha(\varphi) < 0$. Therefore the function $\alpha(\varphi)$ can be considered to satisfy the condition

$$\alpha(\varphi) \leq 0, \quad \varphi \in \mathcal{T}_m. \tag{2.11.12}$$

Assuming that the inequality (2.11.13) is satisfied, we prove the estimate

$$\|D_\varphi^s \varphi_t(\varphi)\| \leq K \exp\left\{-\int_t^0 s\alpha(\varphi_\tau(\varphi))\,d\tau - \varepsilon t\right\}, \quad t \leq 0, \tag{2.11.13}$$

for any $1 \leq s \leq l$, arbitrarily small $\varepsilon > 0$, some K dependent on s, ε and independent of t, φ. Here $D_\varphi^s \varphi_t(\varphi)$ is any derivative of the order of s of the function $\varphi_t(\varphi)$ in the variables $\varphi = (\varphi_1, \ldots, \varphi_m)$. We apply the induction method for the proof. Let the system of equations

$$\frac{d\psi}{dt} = \frac{\partial a(\varphi_t(\varphi))}{\partial \varphi_t} \psi \tag{2.11.14}$$

which the matrix of partial derivatives $\frac{\partial \varphi_t(\varphi)}{\partial \varphi}$ of the function $\varphi_t(\varphi)$ satisfies. Setting $\psi_t^i = \frac{\partial \varphi_t(\varphi)}{\partial \varphi_i}$ and considering the function $V = \langle S_1(\varphi_t(\varphi))\psi_t^i, \psi_t^i \rangle$, in view of (2.11.10) for $S_1(\varphi)$ we find

$$\dot{V} \geq 2\alpha(\varphi_t(\varphi))V. \qquad (2.11.15)$$

Integrating the latter inequality and performing standard calculations we get from (2.11.15)

$$\|\psi_t^i\| \leq K \exp\left\{-\int_t^0 \alpha(\varphi_\tau(\varphi))\,d\tau\right\}, \quad t \leq 0,$$

thereby proving the estimate (2.11.13) for $s = 1$.

Assume that the inequality (2.11.13) is valid for all $s \leq l_1$. We prove that it is then valid for $s = l_1 + 1$ as well. We differentiate the identity obtained by the replacement of ψ in (2.11.14) by ψ_t^i, l_1 times and represent the result as

$$\frac{d}{dt}\left(D_\varphi^{l_1}\psi_t^i\right) = \frac{\partial a(\varphi_t(\varphi))}{\partial \varphi_t}\left(D_\varphi^{l_1}\psi_t^i\right) + R(\varphi_t(\varphi)),$$

where

$$R(\varphi_t(\varphi)) = D_\varphi^{l_1}\left[\frac{\partial a(\varphi_t(\varphi))}{\partial \varphi_t}\psi_t^i\right] - \frac{\partial a(\varphi_t(\varphi))}{\partial \varphi_t}D_\varphi^{l_1}\psi_t^i.$$

Since $R(\varphi_t(\varphi))$ has the form of a differential expression containing the products

$$D_\varphi^{l_1-j}\left(\frac{\partial a(\varphi_t(\varphi))}{\partial \varphi_t}\right)D_\varphi^j\left(\frac{\partial \varphi_t(\varphi)}{\partial \varphi_\nu}\right), \quad j = 0, 1, \ldots, l_1 - 1,$$

$$\nu = 1, \ldots, m,$$

with constant coefficients, then for $s \leq l_1$ (2.11.13) enables us to estimate $R(\varphi_t(\varphi))$ as follows. We have

$$D_\varphi^{l_1-j}\left(\frac{\partial a(\varphi_t(\varphi))}{\partial \varphi_t}\right)$$

$$= \sum_{\sigma=1}^{l_1-j} D_{\varphi_t}^\sigma\left(\frac{\partial a(\varphi_t(\varphi))}{\partial \varphi_t}\right)\sum_\alpha c_{\sigma\alpha}(D_\varphi\varphi_t(\varphi))^{\alpha_1}(D_\varphi^2\varphi_t(\varphi))^{\alpha_2}\ldots(D_\varphi^{l_1-j}\varphi_t(\varphi))^{\alpha_{l_1-j}},$$

where

$$\alpha_1 + \alpha_2 + \cdots + \alpha_{l_1-j} = \sigma,$$
$$\alpha_1 + 2\alpha_2 + \cdots + (l_1-j)\alpha_{l_1-j} = l_1 - j.$$

Therefore

$$\left\| D_\varphi^{l_1-j} \frac{\partial a(\varphi_t(\varphi))}{\partial \varphi_t} \right\|$$

$$\leq K_1 \exp\left\{ -\int_t^0 (l_1-j)\alpha(\varphi_\tau(\varphi))\, d\tau - (l_1-j)\varepsilon t \right\}, \quad t \leq 0,$$

from which

$$\left\| D_\varphi^{l_1-j} \frac{\partial a(\varphi_t(\varphi))}{\partial \varphi_t} D_\varphi^j \frac{\partial \varphi_t(\varphi)}{\partial \varphi_\nu} \right\|$$

$$\leq K_1 K \exp\left\{ -\int_t^0 [(l_1-j)+j+1]\alpha(\varphi_\tau(\varphi))\, d\tau \right\} \exp\left\{ -(l_1-j+1)\varepsilon t \right\}$$

$$\leq K_1 K \exp\left\{ -\int_t^0 (l_1+1)\alpha(\varphi_\tau(\varphi))\, d\tau - (l_1+1)\varepsilon t \right\}, \quad t \leq 0.$$

The estimate for $R(\varphi_t(\varphi))$ this proves:

$$\|R(\varphi_t(\varphi))\|$$

$$\leq K_2 \exp\left\{ -\int_t^0 (l_1+1)\alpha(\varphi_\tau(\varphi))\, d\tau - (l_1+1)\varepsilon t \right\}, \quad t \leq 0, \tag{2.11.16}$$

where K_2 is a positive constant independent of φ.

Since

$$\psi_0^i = (\underbrace{0,\ldots,0}_{i-1},\, 1,\, \underbrace{0,\ldots,0}_{m-i}),$$

then $[D_\varphi^{l_1}\psi_t^i]|_{t=0} = 0$ for $l_1 \geq 1$. Hence the derivative $D_\varphi^{l_1}\psi_t^i$ satisfies the relation

$$D_\varphi^{l_1}\psi_t^i = -\int_t^0 \Omega_\tau^t\left(\frac{\partial a}{\partial \varphi}\right) R(\varphi_\tau(\varphi))\, d\tau, \tag{2.11.17}$$

where $\Omega_\tau^t\,(\partial a/\partial \varphi_t)$ denotes a fundamental matrix of the solutions of the system (2.11.14) taking $t = \tau$ for the value of the identity matrix.

We estimate the integral (2.11.17). The columns of the matrix $\Omega_\tau^t\,(\partial a/\partial \varphi_t)$ are evidently, the solutions $\psi_t^i(\tau,\varphi)$ of the system (2.11.14) taking $t = \tau$ for the values of the unique basis vectors e_i of the spaces R^m. Standard arguments incorporating the inequalities (2.11.10) for $S_1(\varphi)$ allow us to obtain the estimate for $\psi(\tau, \varphi)$

$$\|\psi_t^i(\tau, \varphi)\| \leq K \exp\left\{-\int_t^\tau \alpha(\varphi_\tau(\varphi))\, d\tau\right\}, \quad t \leq \tau,$$

ensuring the inequalities for $\Omega_\tau^t (\partial a/\partial \varphi_t)$:

$$\left\|\Omega_\tau^t \left(\frac{\partial a}{\partial \varphi_t}\right)\right\| \leq K \exp\left\{-\int_t^\tau \alpha(\varphi_\tau(\varphi))\, d\tau\right\}, \quad t \leq \tau.$$

We estimate the integral (2.11.17) by mean of this inequality and get the estimate for $D_\varphi^{l_1} \psi_t^i$:

$$\|D_\varphi^{l_1} \psi_t^i\| \leq K K_2 \int_t^0 \exp\left\{-\int_t^\tau \alpha(\varphi_\tau(\varphi))\, d\tau - \int_\tau^0 (l_1 + 1)\alpha(\varphi_\tau(\varphi))\, d\tau\right.$$

$$\left. - (l_1 + 1)\varepsilon t\right\} d\tau \leq K_3 \exp\left\{-\int_t^0 \alpha(\varphi_\tau(\varphi))\, d\tau - (l_1 + 1)\varepsilon t\right\} \quad (2.11.18)$$

$$\times \int_t^0 \exp\left\{-\int_\tau^0 l_1 \alpha(\varphi_\tau(\varphi))\right\} d\tau, \quad t \leq 0.$$

Estimating the integral

$$I_1 = \int_t^0 \exp\left\{-\int_\tau^0 l_1 \alpha(\varphi_\tau(\varphi))\right\} d\tau, \quad t \leq 0,$$

we find that

$$I_1 = -t \exp\left\{-\int_t^0 l_1 \alpha(\varphi_\tau(\varphi))\right\} d\tau$$

$$- \int_t^0 l_1 \tau \alpha(\varphi_\tau(\varphi)) \exp\left\{-\int_\tau^0 l_1 \alpha(\varphi_\tau(\varphi))\, d\tau\right\} d\tau \quad (2.11.19)$$

$$\leq -t \exp\left\{-\int_t^0 l_1 \alpha(\varphi_\tau(\varphi))\right\} d\tau$$

$$\leq K(\varepsilon) \exp\left\{-\int_t^0 l_1 \alpha(\varphi_\tau(\varphi))\, d\tau - \varepsilon t\right\},$$

where $K(\varepsilon)$ is a positive constant which depends on ε.

The inequalities (2.11.18), (2.11.19) allow the estimate of function $D_\varphi^{l_1}\psi_t^i$ to have the form

$$\|D_\varphi^{l_1}\psi_t^i\| \leq K_3 K(\varepsilon) \exp\left\{-\int_t^0 (l_1+1)\alpha(\varphi_\tau(\varphi))\,d\tau - (l_1+2)\varepsilon t\right\}, \quad t \leq 0,$$

from which it follows

$$\|D_\varphi^{l_1+1}\varphi_t(\varphi)\| \leq K \exp\left\{-\int_t^0 (l_1+1)\alpha(\varphi_\tau(\varphi))\,d\tau - \varepsilon_1 t\right\}, \quad t \leq 0, \quad (2.11.20)$$

for some positive $K = K(\varepsilon_1)$ and arbitrary fixed $\varepsilon_1 > 0$. The inequality (2.11.20) proves the estimate (2.11.13) for $s = l_1 + 1$, and hence the inequality (2.11.13) is satisfied.

Consider the function $x_t(\tau,\varphi,x_0) = \Omega_\tau^t(\varphi)x_0$, where $\Omega_\tau^t(\varphi)$ is a fundamental matrix of solutions to the system

$$\frac{dx}{dt} = A(\varphi_t(\varphi))x, \quad (2.11.21)$$

taking $t = \tau$ for the value of the identity matrix. We prove the estimate

$$\|D_\varphi^s x_t(\tau,\varphi,x_0)\|$$
$$\leq K \exp\left\{-\int_\tau^t \beta(\varphi_\tau(\varphi))\,d\tau - \int_\tau^0 s\alpha(\varphi_\tau(\varphi))\,d\tau - \varepsilon\tau\right\}\|x_0\|, \quad (2.11.22)$$

for arbitrary $\tau \leq t \leq 0$, $0 \leq s \leq l$, arbitrarily small $\varepsilon > 0$ and some K, which is dependent on s, ε and independent of t, τ, φ. The proof is performed by induction. For $s = 0$ the inequality (2.11.22) follows from (2.11.10) for the matrix $S(\varphi)$ due to the standard estimates associated with the consideration of $V = \langle S(\varphi_t(\varphi))x_t(\tau,\varphi,x_0), x_t(\tau,\varphi,x_0)\rangle$ and its derivative in t. We assume that the estimate (2.11.22) is valid for all $0 \leq s \leq l_1 - 1$ and prove its validity for $s = l_1$. We differentiate the identity obtained by the replacement of x by $x_t(\tau,\varphi,x_0)$ in (2.11.21) l_1 times and represent the result in the form

$$\frac{d}{dt}\left[D_\varphi^{l_1}x_t(\tau,\varphi,x_0)\right]$$
$$= A(\varphi_t(\varphi))\left[D_\varphi^{l_1}x_t(\tau,\varphi,x_0)\right] + R_2(\varphi_t(\varphi), x_t(\tau,\varphi,x_0)),$$

where

$$R_2(\varphi_t(\varphi), x_t(\tau, \varphi, x_0))$$
$$= D_\varphi^{l_1}\left[A(\varphi_t(\varphi))x_t(\tau, \varphi, x_0)\right] - A(\varphi_t(\varphi))D_\varphi^{l_1}x_t(\tau, \varphi, x_0).$$

The function $R_2(\varphi_t(\varphi), x_t(\tau, \varphi, x_0))$ has the form of a differential expression containing the products $D_\varphi^{l_1-j}A(\varphi_t(\varphi))D_\varphi^j x_t(\tau, \varphi, x_0)$, $j = 0, \ldots, l_1 - 1$, with constant coefficients. The inequalities (2.11.13) and (2.11.22) allow an estimation of $R_2(\varphi_t(\varphi), x_t(\tau, \varphi, x_0))$ as follows

$$D_\varphi^{l_1-j}A(\varphi_t(\varphi)) = \sum_{\sigma=1}^{l_1-j} D_{\varphi_t}^\sigma A(\varphi_t(\varphi)) \sum_\alpha c_{\sigma\alpha}(D_\varphi \varphi_t(\varphi))^{\alpha_1}$$
$$\times (D_\varphi^2 \varphi_t(\varphi))^{\alpha_2} \ldots (D_\varphi^{l_1-j}\varphi_t(\varphi))^{\alpha_{l_1-j}},$$
$$\alpha_1 + \alpha_2 + \cdots + \alpha_{l_1-j} = \sigma,$$
$$\alpha_1 + 2\alpha_2 + \cdots + (l_1-j)\alpha_{l_1-j} = l_1 - j,$$

from which

$$\left\|D_\varphi^{l_1-j}A(\varphi_t(\varphi))\right\| \leq K \exp\left\{-(l_1-j)\int_t^0 \alpha(\varphi_\tau(\varphi))\,d\tau - \varepsilon t\right\}, \quad t \leq 0.$$

Then for any $0 \leq j \leq l_1 - 1$ and $\tau \leq t \leq 0$

$$\left\|D_\varphi^{l_1-j}A(\varphi_t(\varphi))D_\varphi^j x_t(\tau, \varphi, x_0)\right\| \leq K_4 K \exp\left\{-(l_1-j)\int_t^0 \alpha(\varphi_\tau(\varphi))\,d\tau\right\}$$
$$\times \exp\left\{-\int_\tau^t \beta(\varphi_\tau(\varphi))\,d\tau - j\int_\tau^0 \alpha(\varphi_\tau(\varphi))\,d\tau - \varepsilon(t+\tau)\right\}\|x_0\|$$
$$\leq K_4 K \exp\left\{-\int_t^\tau \beta(\varphi_\tau(\varphi))\,d\tau\right\} \exp\left\{-l_1\int_\tau^0 \alpha(\varphi_\tau(\varphi))\,d\tau - 2\varepsilon\tau\right\}\|x_0\|,$$

which results in the estimate

$$\|R_2(\varphi_t(\varphi)), x_t(\tau, \varphi, x_0))\|$$
$$\leq K_5 \exp\left\{-\int_\tau^t \beta(\varphi_\tau(\varphi))\,d\tau - l_1\int_\tau^0 \alpha(\varphi_\tau(\varphi))\,d\tau - 2\varepsilon\tau\right\}\|x_0\| \qquad (2.11.23)$$

for all $\tau \leq t \leq 0$ and some positive K_5 independent of t, φ, τ.

Since $[D_\varphi^s x_t(\tau, \varphi, x_0)]|_{t=\tau} = 0$ for any $s \geq 1$, then

$$D_\varphi^{l_1} x_t(\tau, \varphi, x_0) = \int_\tau^t \Omega_{\tau_1}^t R_2(\varphi_{\tau_1}(\varphi)) x_{\tau_1}(\tau, \varphi, x_0) \, d\tau_1. \qquad (2.11.24)$$

The inequality (2.11.23) yields the estimate for integral (2.11.24):

$$\|D_\varphi^{l_1} x_t(\tau, \varphi, x_0)\| \leq K_6 \int_\tau^t \exp\left\{-\int_{\tau_1}^t \beta(\varphi_\tau(\varphi)) \, d\tau - \int_\tau^{\tau_1} \beta(\varphi_\tau(\varphi)) \, d\tau\right\}$$

$$\times \exp\left\{-l_1 \int_\tau^0 \alpha(\varphi_\tau(\varphi)) \, d\tau - 2\varepsilon\tau\right\} d\tau_1 \, \|x_0\| \qquad (2.11.25)$$

$$= K_6(t-\tau) \exp\left\{-\int_\tau^t \beta(\varphi_\tau(\varphi)) \, d\tau\right\}$$

$$\times \exp\left\{-l_1 \int_\tau^0 \alpha(\varphi_\tau(\varphi)) \, d\tau - 2\varepsilon\tau\right\} \|x_0\|, \quad \tau \leq t \leq 0.$$

Hence

$$\|D_\varphi^{l_1} x_t(\tau, \varphi, x_0)\|$$

$$\leq K(\varepsilon_1) \exp\left\{-\int_\tau^t \beta(\varphi_\tau(\varphi)) \, d\tau - \int_\tau^0 l_1 \alpha(\varphi_\tau(\varphi)) \, d\tau - \varepsilon_1 \tau\right\} \|x_0\|,$$

for all $\tau \leq t \leq 0$, arbitrarily small $\varepsilon_1 > 0$ and some $K(\varepsilon_1) > 0$. The latter inequality proves the estimate (2.11.22) for $s = l_1$ and hence, for any $0 \leq s \leq l$.

We let $t = 0$ in (2.11.22). Then

$$\|D_\varphi^s x_0(\tau, \varphi, x_0)\|$$
$$\leq K \exp\left\{-\int_\tau^0 [\beta(\varphi_\tau(\varphi)) + s\alpha(\varphi_\tau(\varphi))] \, d\tau - \varepsilon\tau\right\} \|x_0\|, \qquad (2.11.26)$$

for all $\tau \leq 0$. Since the exponentially stable invariant torus of system (2.11.2) is under consideration, then the homogeneous system of equations corresponding to (2.11.2) has the Green function $G_0(\tau, \varphi)$ determined by the matrix $\Omega_\tau^0(\varphi)$ for

$\tau \leq 0$ and the zero matrix for $\tau > 0$. The inequality (2.11.3) is therefore satisfied for $\tau > 0$, and for $\tau \leq 0$

$$x_0(\tau, \varphi, f(\varphi_\tau(\varphi))|_l \leq K\, e^{\gamma \tau}\, |f|_l. \tag{2.11.27}$$

We prove (2.11.27). We have

$$D_\varphi^s x_0(\tau, \varphi, f(\varphi_t(\varphi))) = \sum_{p+\sigma \leq s} D_\varphi^\sigma \frac{\partial^p x_0(\tau, \varphi, f)}{\partial f^p}$$
$$\times \sum_\alpha c_{\sigma p \alpha} (D_\varphi f(\varphi_\tau(\varphi)))^{\alpha_1} \ldots (D_\varphi^l f(\varphi_\tau(\varphi)))^{\alpha_l}, \tag{2.11.28}$$

$$\alpha_1 + \alpha_2 + \cdots + \alpha_l = p,$$
$$\alpha_1 + 2\alpha_2 + \cdots + l\alpha_l + \sigma = s.$$

Since further

$$D_\varphi^j f(\varphi_\tau(\varphi)) = \sum_{\sigma_1 \leq j} D_{\varphi_\tau}^{\sigma_1} f(\varphi_\tau(\varphi)) \sum_\alpha c_{\sigma_1 \alpha} (D_\varphi \varphi_\tau(\varphi))^{\alpha_1} \ldots (D_\varphi^j \varphi_\tau(\varphi))^{\alpha_j},$$
$$\alpha_1 + \alpha_2 + \cdots + \alpha_j = \sigma_1, \quad \alpha_1 + 2\alpha_2 + \cdots + j\alpha_j = j,$$

then for all $\tau \leq 0$

$$\|D_\varphi^j f(\varphi_\tau(\varphi))\| \leq K \exp\left\{-j \int_\tau^0 \alpha(\varphi_\tau(\varphi))\, d\tau - \varepsilon \tau\right\} |f|_j. \tag{2.11.29}$$

The function $x_0(\tau, \varphi, f)$ is linear relative to f; consequently

$$\left\| D_\varphi^\sigma \frac{\partial^p x_0(\tau, \varphi, f)}{\partial f^p} \right\|$$
$$\leq \text{const} \left[\|D_\varphi^\sigma x_0(\tau, \varphi, f)\| \delta_{0p} + \sum_{j=1}^n \|D_\varphi^\sigma x_0(\tau, \varphi, e_j)\| \delta_{1p} \right],$$

where e_j is a basis vector, $\delta_{\alpha p}$ equal zero for $\alpha \neq p$ and to one for $\alpha = p$.

In view of (2.11.26) we have

$$\left\| D_\varphi^\sigma \frac{\partial^p x_0(\tau, \varphi, f)}{\partial f^p} \right\| \leq K[\|f\|\delta_{0p} + \delta_{1p}]$$
$$\times \exp\left\{-\int_\tau^0 \beta(\varphi_\tau(\varphi))\, d\tau\right\} \exp\left\{-\sigma \int_\tau^0 \alpha(\varphi_\tau(\varphi))\, d\tau - \varepsilon \tau\right\}, \quad \tau \leq 0. \tag{2.11.30}$$

From (2.11.29) and (2.11.30) we get

$$\|D^s_\varphi x_0(\tau, \varphi, f(\varphi_\tau(\varphi)))\|$$
$$\leq K \exp\left\{-\int_\tau^0 [\beta(\varphi_\tau(\varphi)) + s\alpha(\varphi_\tau(\varphi))]\, d\tau - \varepsilon\tau\right\} |f|_s,$$

for any $s \leq l$ and $\tau \leq 0$. The second inequality (2.11.11) ensures the estimate

$$|x_0(\tau, \varphi, f(\varphi_\tau(\varphi)))|_l \leq K e^{(\gamma-\varepsilon)\tau} |f|_l, \quad \tau \leq 0,$$

which is sufficient for the inequality (2.11.3) to be satisfied for $\varepsilon < \gamma$.

This completes the proof of Theorem 2.11.2.

Analysis of the last proof shows that it remains valid if the fundamental matrices of the solutions $\Omega^t_\tau(\varphi)$ and $\Omega^t_\tau(\partial a(\varphi)/\partial\varphi)$ of the systems (2.11.21) and (2.11.14) respectively satisfy the inequalities

$$\|\Omega^t_\tau(\varphi)\| \leq K \exp\left\{-\int_\tau^t \beta(\varphi_\tau(\varphi))\, d\tau\right\}, \quad \tau \leq t,$$

$$\left\|\Omega^t_\tau\left(\frac{\partial a(\varphi)}{\partial\varphi}\right)\right\| \leq K \exp\left\{\int_t^\tau \alpha(\varphi_\tau(\varphi))\, d\tau\right\}, \quad \tau \geq t,$$
(2.11.31)

where $\beta(\varphi)$ and $\alpha(\varphi)$ obey the conditions (2.11.6) and (2.11.9).

Since the matrices $\Omega^t_0(\varphi)$, $\Omega^t_\tau(\partial a(\varphi)/\partial\varphi)$ satisfy identities of the form

$$\Omega^t_0(\varphi_z(\varphi)) \equiv \Omega^{t+z}_z(\varphi),$$
$$\Omega^t_0\left(\frac{\partial a(\varphi_z(\varphi))}{\partial\varphi}\right) \equiv \Omega^{t+z}_z\left(\frac{\partial a(\varphi)}{\partial\varphi}\right),$$

then the inequalities

$$\|\Omega^t_0(\varphi)\| \leq K \exp\left\{-\int_0^t \beta(\varphi_\tau(\varphi))\, d\tau\right\}, \quad 0 \leq t,$$

$$\left\|\Omega^t_0\left(\frac{\partial a(\varphi)}{\partial\varphi}\right)\right\| \leq K \exp\left\{\int_t^0 \alpha(\varphi_\tau(\varphi))\, d\tau\right\}, \quad t \leq 0,$$
(2.11.32)

follow from (2.11.31). In fact, by the cited relations (2.11.32) we have the estimate for $\Omega_\tau^t(\varphi)$:

$$\|\Omega_\tau^t(\varphi)\| = \|\Omega_\tau^{t-\tau+\tau}(\varphi)\|$$

$$= \|\Omega_0^{t-\tau}(\varphi_\tau(\varphi))\| \leq K \exp\left\{-\int_0^{t-\tau} \beta(\varphi_{\tau_1}(\varphi_\tau(\varphi)))\, d\tau_1\right\}$$

$$= K \exp\left\{-\int_0^{t-\tau} \beta(\varphi_{\tau+\tau_1}(\varphi))\, d\tau_1\right\} = K \exp\left\{-\int_\tau^t \beta(\varphi_\tau(\varphi))\, d\tau\right\}, \quad t - \tau \geq 0,$$

which coincides with the estimate (2.11.31) for $\Omega_\tau^t(\varphi)$.

We have the following assertion.

Corollary 2.11.1 *The assertions of Theorem 2.11.2 remain valid, if $\Omega_0^t(\varphi)$ and $\Omega_0^t(\partial a/\partial \varphi)$ satisfy (2.11.32), the values $\beta(\varphi)$ and $\alpha(\varphi)$ of which obey (2.11.6) and (2.11.9).*

In particular, if

$$\|\Omega_0^t(\varphi)\| \leq K \exp\{-\beta_0 t\}, \quad t \geq 0,$$

$$\|\Omega_0^t(\partial a/\partial \varphi)\| \leq K \exp\{-\alpha_0 t\}, \quad t \leq 0,$$

where $\beta_0 = \text{const} > 0$, $\alpha_0 = \text{const} > 0$, then the conditions (2.11.6), (2.11.9) are satisfied for l satisfying the inequality $l < \beta_0/\alpha_0$. The invariant torus (2.11.1) of system (2.11.2) then belongs to $C^l(\mathcal{T}_m)$.

It should be noted that l which is determined by the conditions of Theorem 2.11.2 is reachable as the minimal possible smoothness of the torus (2.11.1). This is shown by the example of the system

$$\frac{d\varphi}{dt} = -\sin \varphi,$$

$$\frac{dx}{dt} = -bx + \sin^b \varphi, \tag{2.11.33}$$

consisting of two scalar equations. The invariant torus (2.11.1) of this system is determined by

$$u(\varphi) = -\tan \frac{\varphi}{2} \int_{\pi/2}^{\varphi/2} \frac{\cos^{2b-1} t}{\sin t}\, dt, \quad 0 < \varphi < 2\pi,$$

which behaves as the function $-c\varphi^b \log \varphi$ for $\varphi \to 0$. Assuming that b is an integer and computing the b-th derivative of function $u(\varphi)$ we make sure that this derivative

LINEAR EXTENSION OF DYNAMICAL SYSTEMS ON A TORUS 203

is not bounded. The inequality (2.11.9) for system (2.11.33) is $\inf_{\varphi \in T_1}[b-l\cos\varphi] > 0$ and is satisfied for $l = b - 1$. Hence, the smoothness of the torus (2.11.1) of the system (2.11.33) determined by (2.11.9) coincides with its truth smoothness.

2.12 On the Dependence of Green Functions on Parameters

Consider the system of differential equations with a parameter

$$\frac{d\varphi}{dt} = a(\varphi, \mu),$$
$$\frac{dx}{dt} = A(\varphi, \mu)x, \tag{2.12.1}$$

where $\varphi \in R^m$, $x \in R^n$, $\mu \in [0, \mu_0]$, $a(\varphi, \mu)$, $A(\varphi, \mu)$ are vector and matrix functions respectively, 2π-periodic in every variable φ_j, $j = 1, \ldots, m$. Assume that for every value of the parameter $\mu \in [0, \mu_0]$ the system (2.12.1) has a Green function of the problem on invariant tori $G_0(\tau, \varphi, \mu)$ satisfying the estimate

$$\|G_0(\tau, \varphi, \mu)\| \le K \exp\{-\gamma|\tau|\}, \tag{2.12.2}$$

where K, γ are positive constants independent of τ, φ, $\mu \in [0, \mu_0]$. We designate by $\omega(z)$ the continuous non-decreasing function defined on $[0, \mu_0]$ for which $\omega(0) = 0$. We impose the following smoothness conditions on the functions $a(\varphi, \mu)$, $A(\varphi, \mu)$

$$\|a(\varphi, \mu) - a(\bar\varphi, \bar\mu)\| \le \alpha_1\|\varphi - \bar\varphi\| + \alpha_2\omega(|\mu - \bar\mu|),$$
$$\|A(\varphi, \mu) - A(\bar\varphi, \bar\mu)\| \le \beta_1\|\varphi - \bar\varphi\| + \beta_2\omega(|\mu - \bar\mu|), \tag{2.12.3}$$

for all $\varphi, \bar\varphi \in R^m$, $\mu \in [0, \mu_0]$.

We find the estimates for the difference $\varphi_t^\mu(\varphi) - \varphi_t^{\bar\mu}(\bar\varphi)$, where $\varphi_t^\mu(\varphi)$ is a solution of the first system (2.12.1) $\left(\varphi_t^\mu(\varphi)\big|_{t=0} = \varphi\right)$:

$$\frac{d}{dt}\left(\varphi_t^\mu(\varphi) - \varphi_t^{\bar\mu}(\bar\varphi)\right) = a(\varphi_t^\mu(\varphi), \mu) - a(\varphi_t^{\bar\mu}(\bar\varphi), \bar\mu),$$

$$\varphi_t^\mu(\varphi) - \varphi_t^{\bar\mu}(\bar\varphi) = \varphi - \bar\varphi + \int_0^t [a(\varphi_\sigma^\mu(\varphi), \mu) - a(\varphi_\sigma^{\bar\mu}(\bar\varphi), \bar\mu)]\, d\sigma.$$

Analysing the latter equality for $t \ge 0$ we get

$$\|\varphi_t^\mu(\varphi) - \varphi_t^{\bar\mu}(\bar\varphi)\| \le \|\varphi - \bar\varphi\| + \alpha_1\int_0^t \|\varphi_\sigma^\mu(\varphi) - \varphi_\sigma^{\bar\mu}(\bar\varphi)\|\, d\sigma + \alpha_2\omega(|\mu - \bar\mu|)t.$$

Designate the right-hand side of the inequality obtained by $F(t)$ and differentiate it

$$\frac{dF(t)}{dt} \leq \alpha_1 F(t) + \alpha_2 \omega(|\mu - \bar{\mu}|).$$

Hence it follows that

$$F(t) \leq \left(\|\varphi - \bar{\varphi}\| + \frac{\alpha_2}{\alpha_1} \omega(|\mu - \bar{\mu}|) \right) \exp\{\alpha_1 t\}.$$

Similarly, when the case of $t < 0$ is treated we verify that

$$\|\varphi_t^\mu(\varphi) - \varphi_t^{\bar{\mu}}(\bar{\varphi})\| \leq \left(\|\varphi - \bar{\varphi}\| + \frac{\alpha_2}{\alpha_1} \omega(|\mu - \bar{\mu}|) \right) \exp\{\alpha_1 |t|\}. \qquad (2.12.4)$$

Applying (2.6.4) and in view of (2.6.3) we find

$$\|A(\varphi_t^\mu(\varphi), \mu) - A(\varphi_t^{\bar{\mu}}(\bar{\varphi}), \bar{\mu})\| \leq \beta_1 \|\varphi_t^\mu(\varphi) - \varphi_t^{\bar{\mu}}(\bar{\varphi})\| + \beta_2 \omega(|\mu - \bar{\mu}|)$$

$$\leq \left(\beta_1 \|\varphi - \bar{\varphi}\| + \frac{\beta_1 \alpha_2}{\alpha_1} \omega(|\mu - \bar{\mu}|) \right) \exp\{\alpha_1 |t|\} + \beta_2 \omega(|\mu - \bar{\mu}|) \qquad (2.12.5)$$

$$\leq \left(\beta_1 \|\varphi - \bar{\varphi}\| + \begin{vmatrix} \beta_1 & \beta_2 \\ -\alpha_1 & \alpha_2 \end{vmatrix} \alpha_1^{-1} \omega(|\mu - \bar{\mu}|) \right) \exp\{\alpha_1 |t|\}.$$

On the other hand

$$\|A(\varphi_t^\mu(\varphi), \mu) - A(\varphi_t^{\bar{\mu}}(\bar{\varphi}), \bar{\mu})\| \leq 2\|A\|_0, \qquad (2.12.6)$$

where $\|A\|_0 = \max\limits_{\substack{\varphi \in R^m \\ \mu \in [0, \mu_0]}} \|A(\varphi, \mu)\|$. In view of the estimates (2.12.5), (2.12.6) we get $\|A(\varphi_t^\mu(\varphi), \mu) - A(\varphi_t^{\bar{\mu}}(\bar{\varphi}), \bar{\mu})\|$

$$\leq (2\|A\|_0)^{\frac{\nu}{\nu+1}} \left(\beta_1 \|\varphi - \bar{\varphi}\| + \begin{vmatrix} \beta_1 & \beta_2 \\ -\alpha_1 & \alpha_2 \end{vmatrix} \alpha_1^{-1} \omega(|\mu - \bar{\mu}|) \right)^{\frac{1}{\nu+1}} \qquad (2.12.7)$$

$$\times \exp\left\{ \frac{\alpha_1}{\nu+1} |t| \right\}, \quad \nu \geq 0.$$

An integral representation is sought for the difference of Green functions $G_0(\tau, \varphi, \mu) - G_0(\tau, \bar{\varphi}, \bar{\mu})$. Consider the difference of the corresponding Green functions of the problem on bounded solutions $G_t(\tau, \varphi, \mu) - G_t(\tau, \bar{\varphi}, \bar{\mu})$ and differentiate it in t ($t \neq \tau$)

$$\frac{d}{dt}(G_t(\tau, \varphi, \mu) - G_t(\tau, \bar{\varphi}, \bar{\mu}))$$

$$= A(\varphi_t^\mu(\varphi), \mu) G_t(\tau, \varphi, \mu) - A(\varphi_t^{\bar{\mu}}(\bar{\varphi}), \bar{\mu}) G_t(\tau, \bar{\varphi}, \bar{\mu}) \qquad (2.12.8)$$

$$= A(\varphi_t^\mu(\varphi), \mu)(G_t(\tau, \varphi, \mu) - G_t(\tau, \bar{\varphi}, \bar{\mu}))$$

$$+ (A(\varphi_t^\mu(\varphi), \mu) - A(\varphi_t^{\bar{\mu}}(\bar{\varphi}), \bar{\mu})) G_t(\tau, \bar{\varphi}, \bar{\mu}).$$

Treating this equality as an inhomogeneous system and taking into account the continuity of its solution

$$(G_{\tau+0}(\tau,\varphi,\mu) - G_{\tau+0}(\tau,\bar{\varphi},\bar{\mu})) - (G_{\tau-0}(\tau,\varphi,\mu) - G_{\tau-0}(\tau,\bar{\varphi},\bar{\mu}))$$
$$= (G_{\tau+0}(\tau,\varphi,\mu) - G_{\tau-0}(\tau,\varphi,\mu)) - (G_{\tau+0}(\tau,\bar{\varphi},\bar{\mu}) - G_{\tau-0}(\tau,\bar{\varphi},\bar{\mu}))$$
$$= I_n - I_n = 0,$$

and its boundedness for all $t \in R$, we get

$$G_t(\tau,\varphi,\mu) - G_t(\tau,\bar{\varphi},\bar{\mu})$$
$$= \int_{-\infty}^{\infty} G_t(\sigma,\varphi,\mu) [A(\varphi_\sigma^\mu(\varphi),\mu) - A(\varphi_\sigma^{\bar{\mu}}(\bar{\varphi}),\bar{\mu})] G_\sigma(\tau,\bar{\varphi},\bar{\mu}) \, d\sigma.$$

Thus,

$$G_0(\tau,\varphi,\mu) - G_0(\tau,\bar{\varphi},\bar{\mu})$$
$$= \int_{-\infty}^{\infty} G_0(\sigma,\varphi,\mu) [A(\varphi_\sigma^\mu(\varphi),\mu) - A(\varphi_\sigma^{\bar{\mu}}(\bar{\varphi}),\bar{\mu})] G_\sigma(\tau,\bar{\varphi},\bar{\mu}) \, d\sigma. \qquad (2.12.9)$$

In view of (2.12.2), (2.12.7) and (2.12.9) we find

$$\|G_0(\tau,\varphi,\mu) - G_0(\tau,\bar{\varphi},\bar{\mu})\|$$
$$\leq \int_{-\infty}^{\infty} \|G_0(\sigma,\varphi,\mu)\| \, \|A(\varphi_\sigma^\mu(\varphi),\mu) - A(\varphi_\sigma^{\bar{\mu}}(\bar{\varphi}),\bar{\mu})\| \, \|G_\sigma(\tau,\bar{\varphi},\bar{\mu})\| \, d\sigma$$
$$\leq K^2 (2\|A\|_0)^{\frac{\nu}{\nu+1}} \left(\beta_1 \|\varphi - \bar{\varphi}\| + \left| \begin{matrix} \beta_1 & \beta_2 \\ -\alpha_1 & \alpha_2 \end{matrix} \right| \alpha_1^{-1} \omega(|\mu - \bar{\mu}|) \right)^{\frac{1}{\nu+1}}$$
$$\times \int_{-\infty}^{\infty} \exp\left\{ -\gamma(|\sigma| + |\sigma - \tau|) + \frac{\alpha_1}{\nu+1}|\sigma| \right\} d\sigma.$$

Moreover, $\nu \geq 0$ is taken so that

$$\gamma - \frac{\alpha_1}{\nu+1} > 0. \qquad (2.12.10)$$

Inequality (2.12.10) ensures the convergence of the last integral.

Summing up the above presentation gives:

Theorem 2.12.1 *Let*

(1) *the system (2.12.1) have a unique Green function of the problem on invariant tori, satisfying estimates (2.12.2) for every fixed value of the parameter* $\mu \in [0, \mu_0]$;

(2) *the smoothness conditions (2.12.3) be satisfied for functions* $a(\varphi, \mu)$, $A(\varphi, \mu)$.

Then the Green function $G_0(\nu, \varphi, \mu)$ *is continuous with respect to the totality of variables* φ, μ *and, moreover, the estimate*

$$\|G_0(\tau, \varphi, \mu) - G_0(\tau, \bar{\varphi}, \bar{\mu})\|$$
$$\leq \frac{2(\nu+1)}{\alpha_1} K^2 (2\|A\|_0)^{\frac{\nu}{\nu+1}} \left(\beta_1 \|\varphi - \bar{\varphi}\| + \left| \begin{array}{cc} \beta_1 & \beta_2 \\ -\alpha_1 & \alpha_2 \end{array} \right| \alpha_1^{-1} \omega(|\mu - \bar{\mu}|) \right)^{\frac{1}{\nu+1}}$$
$$\times \exp\left\{ -\left(\gamma - \frac{\alpha_1}{\nu+1} \right) |\tau| \right\}$$

is satisfied. In addition, the constant $\nu \geq 0$ *can be taken to be arbitrary provided that (2.12.10) holds true.*

Remark 2.12.1 If the parameter μ does not enter the first equation of (2.12.1), then under the conditions of Theorem 2.12.1 for the difference of the Green functions the estimate

$$\|G_0(\tau, \varphi, \mu) - G_0(\tau, \bar{\varphi}, \bar{\mu})\| \leq K_\varepsilon \omega(|\mu - \bar{\mu}|) \exp\{-(\gamma - \varepsilon)|\tau|\},$$

is satisfied, i.e. the constant $\nu \geq 0$ is not important.

Remark 2.12.2 In the case when the system (2.12.1) has a set of Green functions for some $\mu \in [0, \mu_0]$, we consider an extended system of equations

$$\frac{d\varphi}{dt} = a(\varphi),$$
$$\frac{dx}{dt} = A(\varphi)x,$$
$$\frac{dy}{dt} = -x - A^*(\varphi)y,$$

which has only one Green function.

Corollary 2.12.1 *For the system of equations*

$$\frac{d\varphi}{dt} = a(\varphi) + \mu^k a(\varphi), \quad \frac{dx}{dt} = A(\varphi)x$$

the norm of the difference of the Green functions $\|G_0(\tau, \varphi, \mu) - G_0(\tau, \varphi, 0)\|$ *has the order of smallness in μ for $\mu \to 0$ which is not lower than $\mu k/\nu + 1$, where ν is determined by (2.12.10).*

Now abandoning the conditions (2.12.3) we assume that the functions $a(\varphi, \mu)$, $A(\varphi, \mu)$ are continuous with respect to the totality of variables φ, μ, $\varphi \in R^m$, $\mu \in [0, \mu_0]$ and 2π-periodic in φ_j, $j = 1, \ldots, m$. Then the following result is valid.

Theorem 2.12.2 *Let the following conditions be satisfied*

(1) *for every value of the parameter $\mu \in [0, \mu_0]$ the system of equations (2.6.1) has a unique Green function $G_0(\tau, \varphi, \mu)$ satisfying the estimate (2.12.2) with constants K, γ independent of the parameter $\mu \in [0, \mu_0]$;*

(2) *the integral $\int_{+0} \frac{d\sigma}{\omega_1(\sigma)}$, where the function $\omega_1(\sigma)$ is determined by $\|a(\varphi) - a(\bar{\varphi})\| \leq \omega_1(\|\varphi - \bar{\varphi}\|)$ is divergent.*

Then the Green function $G_0(\tau, \varphi, \mu)$ is continuous in the parameter $\mu \in [0, \mu_0]$.

Proof For the difference of the solutions $\varphi_t^\mu(\varphi) - \varphi_t^{\bar{\mu}}(\bar{\varphi})$ for all $t \in R$

$$\|\varphi_t^\mu(\varphi) - \varphi_t^{\bar{\mu}}(\bar{\varphi})\| \leq F^{-1}(F(0; \omega_2(|\mu - \bar{\mu}|)); \omega_2(|\mu - \bar{\mu}|)),$$

where

$$F(u; v) = \int_d^u \frac{d\sigma}{\omega_1(\sigma) - v},$$

and the function ω_2 is determined by

$$\max_{|\mu - \bar{\mu}| \leq \sigma} \|a(\varphi, \mu) - a(\varphi, \bar{\mu})\| \leq \omega_2(\sigma).$$

Estimation of the difference of the Green functions $G_0(\tau, \varphi, \mu) - G_0(\tau, \varphi, \bar{\mu})$ yields

$$\|G_0(\tau, \varphi, \mu) - G_0(\tau, \varphi, \bar{\mu})\|$$

$$\leq \int_{-\infty}^{\infty} \|G_0(\sigma, \varphi, \mu)\| \, \|A(\varphi_\sigma^\mu(\varphi), \mu) - A(\varphi_\sigma^{\bar{\mu}}(\varphi), \mu)\| \, \|G_\sigma(\tau, \varphi, \bar{\mu})\| \, d\sigma \qquad (2.12.11)$$

$$\leq K^2 \exp\{-(\gamma - \nu)|\tau|\} \int_{-\infty}^{\infty} \exp\{-\nu|\sigma|\} \, [\omega(A; F^{-1}(F(0; \omega_2(|\mu - \bar{\mu}|))$$

$$+ |\sigma|; \omega_2(|\mu - \bar{\mu}|))) + \omega_2(A; |\mu - \bar{\mu}|)] \, d\sigma.$$

From which it follows that

$$\|G_0(\tau, \varphi, \mu) - G_0(\tau, \varphi, \bar{\mu})\|$$

$$\leq 2K \exp\{-(\gamma - \nu)|\tau|\} \left[\int_0^{-F(0; \omega_2(|\mu - \bar{\mu}|))} \exp\{-\nu\sigma\} \right.$$

$$\times \omega(A; F^{-1}(F(0; \omega_2(|\mu - \bar{\mu}|)) + |\sigma|; \omega_2(|\mu - \bar{\mu}|))) \, d\sigma$$

$$\left. + (c) \exp\{\nu F(0; \omega_2(A; |\mu - \bar{\mu}|))\} + \tfrac{1}{\nu} \omega_2(A; |\mu - \bar{\mu}|) \right]$$

$$= 2K^2 \exp\{-(\gamma-\nu)|\tau|\} \Bigg[\int_0^d \exp\{-\nu F(u;\omega_2(|\mu-\bar\mu|))$$

$$+ \nu F(0;\omega_2(|\mu-\bar\mu|))\} \frac{\omega(A;u)}{\omega_1(u)+\omega_2(|\mu-\bar\mu|)} du$$

$$+ (c)\exp\left\{\nu F(0;\omega_2(A;|\mu-\bar\mu|) + \tfrac{1}{\nu}\omega_2(A;|\mu-\bar\mu|)\right\} \Bigg].$$

Now, it is clear that in order to prove the vanishing of the right-hand side as $\bar\mu \to \mu$, it is sufficient to show that the first additive in square brackets also vanishes. This is implied by the following inequality

$$\int_0^d \exp\{-\nu F(u;\omega_2) + \nu F(0;\omega_2)\} \frac{\omega(A;u)}{\omega_1(u)+\omega_2} du \le \frac{\omega(A;u_1)}{\nu}$$

$$+ \exp\{-\nu F(u_1;0)\} \frac{\omega(A;u_1)}{\omega_1(d)} (d-u_1) \exp\{\nu F(0;\omega_2(|\mu-\bar\mu|))\},$$

for all $u_1 \in (0,d)$.

Remark 2.12.3 In the case when $\omega(A;u) \le c\omega_1(u)$ and $u^{-1}\omega_1(u) \xrightarrow[u \to +0]{} \infty$, for the difference of the Green functions $G_0(\tau,\varphi,\mu) - G_0(\tau,\varphi,\bar\mu)$ the estimate

$$\|G_0(\tau,\varphi,\mu) - G_0(\tau,\varphi,\bar\mu)\| \le (c)\exp\{-(\gamma-\nu)|\tau|\}$$

$$\times \left[\frac{1}{\nu}\omega_2(A;|\mu-\bar\mu|) + \exp\{-\nu F(0;\omega_2(|\mu-\bar\mu|))\}\right],$$

is valid, where (c) is a constant.

Now consider the system (2.12.1) under the assumption that the functions $a(\varphi,\mu)$ and $A(\varphi,\mu)$ are continuously differentiable in φ_j, $j=1,\ldots,m$, and the parameter μ. A question arises: will the Green function $G_0(\tau,\varphi,\mu)$ be continuously differentiable in μ? It turns out that under some additional assumptions the answer is "yes".

For $\bar\varphi = \varphi$ we divide the both parts of (2.12.9) into the difference $\mu - \bar\mu$ and pass to the limit as $\bar\mu \to \mu$ in view of the continuity of the Green function $G_0(\tau,\varphi,\mu)$ in μ. We get the formal equality

$$\frac{d}{d\mu} G_0(\tau,\varphi,\mu)$$

$$= \int_{-\infty}^{\infty} G_0(\sigma,\varphi,\mu) \left[\sum_{j=1}^m \frac{\partial A}{\partial \varphi_{\sigma j}^\mu} \frac{\partial \varphi_{\sigma j}^\mu}{\partial \mu} + \frac{\partial A}{\partial \mu}\right] G_\sigma(\tau,\varphi,\mu)\, d\sigma. \qquad (2.12.12)$$

Hence, taking into account the zero initial values $\frac{\partial}{\partial \mu} \varphi_t^\mu(\varphi)\big|_{t=0} = 0$ we find that

$$\frac{\partial}{\partial \mu} \varphi_t^\mu(\varphi) = \int_0^t \Omega_\sigma^t \left(\frac{\partial a}{\partial \varphi}\right) \frac{\partial a(\varphi, \mu)}{\partial \mu}\bigg|_{\varphi=\varphi_\sigma^\mu(\varphi)} d\sigma, \qquad (2.12.13)$$

where $\Omega_\sigma^t \left(\frac{\partial a}{\partial \varphi}\right)$ denotes a normal fundamental matrix of solutions to the linear system of the differential equations

$$\frac{dy}{dt} = \left(\frac{\partial a(\varphi, \mu)}{\partial \mu}\bigg|_{\varphi=\varphi_t^\mu(\varphi)}\right) y. \qquad (2.12.14)$$

We determine the positive number a by

$$\left|\left\langle \frac{\partial a(\varphi, \mu)}{\partial \mu} \eta, \eta \right\rangle\right| \leq \alpha \|\eta\|^2, \qquad (2.12.15)$$

where η is an arbitrary m-dimensional vector from R^m. Inequality (2.12.15) provides the possibility of obtaining the estimate of the norm of the fundamental matrix of solutions

$$\left\|\Omega_\sigma^t \left(\frac{\partial a}{\partial \varphi}\right)\right\| \leq M \exp\{\alpha |t - \sigma|\}. \qquad (2.12.16)$$

Making the change of variable $t - \sigma \to \sigma$ in (2.12.13) and in view of (2.12.16) we find

$$\left\|\frac{\partial}{\partial \mu} \varphi_t^\mu(\varphi)\right\| \leq \frac{M}{\alpha} \left\|\frac{\partial a}{\partial \mu}\right\|_0 \exp\{\alpha |t|\}. \qquad (2.12.17)$$

Thus the expression under the integral sign in the right-hand side of (2.12.11) can be estimated as

$$\|N(\sigma, \tau, \varphi, \mu)\| \leq (c) \exp\{-\gamma|\sigma| + \alpha|\sigma| - \gamma|\sigma - \tau|\}, \qquad (2.12.18)$$

where c is a constant. Assuming that the inequality

$$\alpha < \gamma \qquad (2.12.19)$$

is satisfied and using the estimate (2.12.18) we get

$$\left\|\frac{\partial}{\partial \mu} G_0(\tau, \varphi, \mu)\right\| \leq (c) \exp\{-(\gamma - \alpha)|\tau|\}. \qquad (2.12.20)$$

Thus we come to the following conclusion.

Theorem 2.12.3 *Let*

(1) *the first condition of the Theorem 2.12.1 be satisfied;*
(2) *the functions $a(\varphi, \mu)$ and $A(\varphi, \mu)$ be continuously differentiable in φ_j, $j = 1, \ldots, m$, and μ;*
(3) *the inequality (2.12.19) be satisfied, where α is taken from (2.12.15).*

Then the Green function $G_0(\tau, \varphi, \mu)$, $\tau \neq 0$, is continuously differentiable in μ and for its derivatives the representation (2.12.11) and estimate (2.12.20) are valid.

Our aim is to establish conditions for l-multiple differentiability of the Green function in the parameter μ.

Assume that the functions $a(\varphi, \mu)$, $A(\varphi, \mu)$ are continuously differentiable in φ_j, $j = 1, \ldots, m$, and μ up to the order l ($l \geq 2$) including. We differentiate (2.12.12) in μ

$$\frac{d}{dt}\left(\frac{\partial^2 \varphi_t^\mu(\varphi)}{\partial \mu^2}\right) = \frac{\partial a(\varphi,\mu)}{\partial \mu}\bigg|_{\varphi=\varphi_t^\mu(\varphi)}\left(\frac{\partial^2 \varphi_t^\mu(\varphi)}{\partial \mu^2}\right)$$

$$+ \left[\sum_{j=1}^m \frac{\partial}{\partial \varphi_j}\left(\frac{\partial a(\varphi,\mu)}{\partial \varphi}\right)\bigg|_{\varphi=\varphi_t^\mu(\varphi)} \frac{\partial \varphi_t^\mu(\varphi)}{\partial \mu} + \frac{\partial}{\partial \mu}\left(\frac{\partial a(\varphi,\mu)}{\partial \varphi}\right)\bigg|_{\varphi=\varphi_t^\mu(\varphi)}\right]$$

$$\times \frac{\partial}{\partial \mu}\varphi_t^\mu(\varphi) + \frac{\partial}{\partial \varphi}\left(\frac{\partial a(\varphi,\mu)}{\partial \mu}\right)\bigg|_{\varphi=\varphi_t^\mu(\varphi)}\left(\frac{\partial}{\partial \mu}\varphi_t^\mu(\varphi)\right) \qquad (2.12.21)$$

$$+ \frac{\partial^2 a(\varphi,\mu)}{\partial \mu^2}\bigg|_{\varphi=\varphi_t^\mu(\varphi)}.$$

Treating (2.12.21) as an inhomogeneous system of equations, we get the equality for the partial derivative $\frac{\partial^2}{\partial \mu^2}(\varphi_t^\mu(\varphi))$:

$$\frac{\partial^2}{\partial \mu^2}(\varphi_t^\mu(\varphi)) = \int_0^t \Omega_\sigma^t\left(\frac{\partial a}{\partial \varphi}\right) f_\sigma \, d\sigma, \qquad (2.12.22)$$

where f_σ designates the inhomogeneous part in (2.12.21) for which the estimate

$$\|f_\sigma\| \leq (c) \exp\{2\alpha|\sigma|\} \qquad (2.12.23)$$

holds true. We consider (2.12.22) for $t \leq 0$. An estimate of it gives

$$\left\|\frac{\partial^2}{\partial \mu^2}\varphi_t^\mu(\varphi)\right\| \leq (c) \int_t^0 \exp\{\alpha\sigma - \alpha t - 2\alpha\sigma\} \, d\sigma \leq (c) \exp\{-2\alpha t\}.$$

An estimate for $t > 0$ is made similarly. Proceeding with the higher-order derivatives we verify that

$$\left\|\frac{\partial^\rho}{\partial \mu^\rho} \varphi_t^\mu(\varphi)\right\| \leq (c) \exp\{\rho\alpha|t|\}, \tag{2.12.24}$$

where ρ is an integer which is not larger than l.

The expression $N(\sigma, \tau, \varphi, \mu)$ under the integral sign in the right-hand side of (2.12.11) is differentiated in the parameter μ and estimated in view of (2.12.20)

$$\left\|\frac{\partial}{\partial \mu} N(\sigma, \tau, \varphi, \mu)\right\|$$

$$\leq \left\|\frac{\partial}{\partial \mu} G_0(\tau, \varphi, \mu)\right\| \left(\sum_{j=1}^m \left\|\frac{\partial}{\partial \varphi_j} A\right\|_0 \left\|\frac{\partial}{\partial \mu} \varphi_{\sigma j}^\mu(\varphi)\right\| + \left\|\frac{\partial}{\partial \mu} A\right\|_0\right) \|G_\sigma(\tau, \varphi, \mu)\|$$

$$+ \|G_0(\sigma, \varphi, \mu)\| \left[\sum_{j=1}^m \left(\sum_{k=1}^m \left\|\frac{\partial^2 A}{\partial \varphi_k \partial \varphi_j}\right\|_0 \left\|\frac{\partial}{\partial \mu} \varphi_{\sigma k}^\mu(\varphi)\right\| \left\|\frac{\partial}{\partial \mu} \varphi_{\sigma j}^\mu(\varphi)\right\|\right.\right.$$

$$+ \left\|\frac{\partial A}{\partial \varphi_j}\right\|_0 \left\|\frac{\partial^2}{\partial \mu^2} \varphi_{\sigma j}^\mu(\varphi)\right\| + \left\|\frac{\partial^2 A}{\partial \varphi_j \partial \mu}\right\|_0 \left\|\frac{\partial}{\partial \mu} \varphi_{\sigma j}^\mu(\varphi)\right\|\right)$$

$$+ \sum_{k=1}^m \left\|\frac{\partial^2 A}{\partial \varphi_k \partial \mu}\right\|_0 \left\|\frac{\partial}{\partial \mu} \varphi_{\sigma k}^\mu(\varphi)\right\| + \left\|\frac{\partial^2 A}{\partial \mu^2}\right\|_0 \right] \|G_\sigma(\tau, \varphi, \mu)\|$$

$$+ \|G_0(\sigma, \varphi, \mu)\| \left(\sum_{j=1}^m \left\|\frac{\partial A}{\partial \varphi_j}\right\|_0 \left\|\frac{\partial}{\partial \mu} \varphi_{\sigma j}^\mu(\varphi)\right\| + \left\|\frac{\partial A}{\partial \mu}\right\|_0\right) \left\|\frac{\partial}{\partial \mu} G_\sigma(\tau, \varphi, \mu)\right\|$$

$$\leq (c)_1 \exp\{-\gamma(|\sigma| + |\sigma - \tau| + 2\alpha|\sigma|\}$$

$$+ (c)_2 \exp\{-(\gamma - 2\alpha)(|\sigma - \tau| + |\sigma|) - 2\alpha|\sigma - \tau|\}.$$

Hence it is clear that the inequality $\gamma > 2\alpha$ ensures the existence of the second variable with respect to the parameter μ of the Green function.

Estimating the higher-order derivatives of the function $N(\sigma, \tau, \varphi, \mu)$ in the same way we conclude that the inequality

$$\gamma - l\alpha > 0, \tag{2.12.25}$$

ensures the existence of continuous derivatives of the Green function $G_0(\tau, \varphi, \mu)$ in the parameter μ up to the order l. In addition the estimate

$$\left\|\frac{\partial^\rho}{\partial \mu^\rho} G_0(\tau, \varphi, \mu)\right\| \leq (C)_\rho \exp\{-(\gamma - \rho\alpha)|\tau|\} \tag{2.12.26}$$

is valid.

Condition (2.12.25) is sufficient for the Green function $G_0(\tau, \varphi, \mu)$ to be continuously differentiable in φ_j, $j = 1, \ldots, m$, up to the order l included. Making an estimate similar to the above we make sure that the partial derivatives $D_\varphi^\rho G_0(\tau, \varphi, \mu)$, $1 \leq \rho < l$, are also continuously differentiable in the parameter μ up to the order $l - \rho$.

Then we have the following result.

Theorem 2.12.4 *Let the first condition of Theorem 2.12.1 be satisfied, and moreover,*

(1) *the functions $a(\varphi, \mu)$ and $A(\varphi, \mu)$ are continuously differentiable in φ, μ up to the order l;*

(2) *inequality (2.12.25) is satisfied.*

Then the partial derivatives of the Green function in the variables φ of the order ρ $D_\varphi^\rho G_0(\tau, \varphi, \mu)$ are continuously differentiable in the parameter μ up to the order $l - \rho$ included. Moreover, the estimate

$$\left\| \frac{\partial^k}{\partial \mu^k} (D_\varphi^\rho) G_0(\tau, \varphi, \mu)) \right\| \leq (C)_{k,\rho} \exp\left\{ -[\gamma - \alpha(k + \rho)]|\tau| \right\},$$

is valid, where $1 \leq \rho < l$, $0 \leq k \leq l - \rho$, $(C)_{k,\rho}$ is a constant independent of τ, φ and $\mu \in [0, m_0]$.

2.13 Continuity and Differentiability of the Green Function

Let $\mu(\sigma)$ be a continuous non-decreasing scalar function defined on some interval $[0, d]$, $d > 0$, which satisfies the conditions

$$\mu(0) = 0, \quad \inf_{\sigma \in (0, d]} \frac{\mu(\sigma)}{\sigma} > 0,$$

$$F(u) = \int_d^u \frac{d\sigma}{\mu(\sigma)} \xrightarrow[u \to +0]{} -\infty. \quad (2.13.1)$$

The following functions σ, $\sigma|\ln \sigma|$, $\sigma|\ln \sigma| \ln |\ln \sigma|$ are of this type, supplied with a zero value at point $\sigma = 0$.

We designate by H^μ the class of functions $\Phi(\varphi)$, 2π-periodic in φ_j, $j = 1, \ldots, m$, for which the estimate

$$\|\Phi(\varphi) - \Phi(\bar\varphi)\| \leq \mu(\|\varphi - \bar\varphi\|),$$

is satisfied for all $\varphi, \bar\varphi \in R^m$, $\|\varphi - \bar\varphi\| \leq d$.

Theorem 2.13.1 *Let $a(\varphi)$, $A(\varphi)$ be the functions of class H^μ and for system (2.2.4) there exists the unique Green function of the problem on invariant tori which satisfies the estimate (2.6.11). Then for every fixed $\nu \in (0, d_0)$, where*

$$d_0 = \min\left\{\gamma, \inf_{\sigma \in (0,d]} \frac{\mu(\sigma)}{\sigma}\right\},$$

the following estimate for the difference $G_0(\tau, \varphi) - G_0(\tau, \bar{\varphi})$ is valid:

$$\|G_0(\tau, \varphi) - G_0(\tau, \bar{\varphi})\| \leq M_\nu \exp\{-(\gamma - \nu)|\tau| + \nu F(\|\varphi - \bar{\varphi}\|)\}, \quad (2.13.2)$$

for all $\varphi, \bar{\varphi} \in R^m$, $\|\varphi - \bar{\varphi}\| \leq d$, $M_\nu \to \infty$, provided that $\nu \to \inf_{\sigma \in (0,d]} \frac{\mu(\sigma)}{\sigma}$, $d \to +0$.

Proof For the Green functions difference the representation (2.7.8) yields

$$\|G_0(\tau, \varphi) - G_0(\tau, \bar{\varphi})\|$$

$$\leq K^2 \int_{-\infty}^{\infty} \exp\{-\gamma(|\sigma| + |\sigma - \tau|)\} \tilde{\mu}(A; \|\varphi_\sigma(\varphi) - \varphi_\sigma(\bar{\varphi})\|) \, d\sigma, \quad (2.13.3)$$

where $\tilde{\mu}(A; \sigma) = \mu(\sigma)$, $\sigma \in [0, d]$, $\max_{\|\varphi - \bar{\varphi}\| \leq \sigma} \|A(\varphi) - A(\bar{\varphi})\| \leq \tilde{\mu}(A; \sigma) \leq \tilde{\mu}_0 < \infty$, $\sigma \in [d, \infty)$. We estimate the norm of the difference in solutions $\varphi_t(\varphi) - \varphi_t(\bar{\varphi})$ for $t \geq 0$

$$\|\varphi_t(\varphi) - \varphi_t(\bar{\varphi})\| \leq \|\varphi - \bar{\varphi}\| + \int_0^t \tilde{\mu}(a; \|\varphi_\sigma(\varphi) - \varphi_\sigma(\bar{\varphi})\|) \, d\sigma.$$

Hence it follows that

$$\|\varphi_t(\varphi) - \varphi_t(\bar{\varphi})\| \leq \tilde{F}^{-1}(a; F(\|\varphi - \bar{\varphi}\|) + |t|), \quad (2.13.4)$$

where

$$\tilde{F}(a; u) = \int_d^u \frac{d\sigma}{\tilde{\mu}(a; \sigma)}, \quad \|\varphi - \bar{\varphi}\| \leq d.$$

In view of (2.13.4) we proceed with the estimate (2.13.3) as

$$\|G_0(\tau, \varphi) - G_0(\tau, \bar{\varphi})\| \leq K^2 \exp\{-(\gamma - \nu)|\tau|\} \int_{-\infty}^{\infty} \exp\{-\gamma|\sigma|\}$$

$$\times \tilde{\mu}(A; \tilde{F}^{-1}(a; F(\|\varphi - \bar{\varphi}\|) + |\sigma|)) \, d\sigma \leq 2K^2 \exp\{-(\gamma - \nu)|\tau|\}$$

$$\times \left[\int_0^{-F(\|\varphi - \bar{\varphi}\|)} \exp\{-\nu\sigma\} \mu(F^{-1}(F(\|\varphi - \bar{\varphi}\|) + \sigma)) \, d\sigma \right.$$

$$\left. + \tilde{\mu}_0 \int_{-F(\|\varphi - \bar{\varphi}\|)}^{\infty} \exp\{-\nu\sigma\} \, d\sigma\right].$$

Consider the integral

$$\int_0^{-F(\|\varphi-\bar\varphi\|)} \exp\{-\nu\sigma\}\mu(F^{-1}(F(\|\varphi-\bar\varphi\|)+\sigma))\,d\sigma$$

$$= \int_{F(\|\varphi-\bar\varphi\|)}^0 \exp\{-\nu\sigma\}\mu(F^{-1}(\sigma))\,d\sigma\,\exp\{\nu F(\|\varphi-\bar\varphi\|)\}$$

$$= \left\{ F^{-1}(\sigma)\exp\{-\nu\sigma\}\Big|_{F(\|\varphi-\bar\varphi\|)}^0 \right.$$

$$\left. + \int_{F(\|\varphi-\bar\varphi\|)}^0 F^{-1}(\sigma)\exp\{-\nu\sigma\}\,d\sigma \right\}\exp\{\nu F(\|\varphi-\bar\varphi\|)\}.$$

Hence we get

$$\int_{F(\|\varphi-\bar\varphi\|)}^0 [\mu(F^{-1}(\sigma)) - \nu F^{-1}(\sigma)]\exp\{-\nu\sigma\}\,d\sigma \qquad (2.13.5)$$

$$= d - \|\varphi - \bar\varphi\|\exp\{-\nu F(\|\varphi-\bar\varphi\|)\}.$$

Let us now show that a constant $c > 0$ exists such that for all $\sigma \leq 0$ the inequality

$$c\mu(F^{-1}(\sigma)) \leq \mu(F^{-1}(\sigma)) - \nu F^{-1}(\sigma), \qquad (2.13.6)$$

is satisfied. In fact, (2.13.6) holds, provided that

$$\frac{\mu(F^{-1}(\sigma))}{F^{-1}(\sigma)} \geq \frac{\nu}{1-c},$$

which follows due to the choice of constant

$$\nu \in \left(0, \min\left\{\gamma, \inf_{\sigma\in(0,d]}\frac{\mu(\sigma)}{\sigma}\right\}\right],$$

and the fact that $0 < F^{-1}(\sigma) \leq d$, $\sigma \leq 0$. Evidently, the constant $c > 0$ is to be taken close to zero, if the constant ν is close to the value $\inf_{\sigma\in(0,d]}\frac{\mu(\sigma)}{\sigma}$, $d \to +0$.

The relations (2.13.5) and (2.13.6) imply the estimate

$$\int_{F(\|\varphi-\bar\varphi\|)}^0 \mu(F^{-1}(\sigma))\exp\{-\nu\sigma\}\,d\sigma \leq \frac{d}{c},$$

which together with the equality

$$\int_{-F(\|\varphi-\bar\varphi\|)}^{0} \exp\{-\nu t_1\}\,dt_1 = \frac{1}{\nu}\exp\{\nu F(\|\varphi-\bar\varphi\|)\},$$

complete the proof of the Theorem 2.13.1.

Remark 2.13.1 We note that under the condition $\frac{\mu(\sigma)}{\sigma} \underset{\sigma\to+0}{\longrightarrow} \infty$ the function $\exp\{\nu F(\sigma)\}$ tends to zero sufficiently weakly as $\sigma \to +0$; $\exp\{\nu F(\sigma)\} \geq c_\lambda \sigma^\lambda$, $0 < \lambda \leq 1$, where c_λ is a positive constant which is independent of σ. In the case when $\mu(\sigma) = \sigma|\ln\sigma|$ the function $\exp\{\nu F(\sigma)\}$ is $|\ln\sigma|^{-\nu}$.

Remark 2.13.2 The above result is immediately applicable to state the continuity of the invariant torus of the system of equations (2.1.9) for $F(\varphi) \in C^0(\mathcal{T}_m)$.

Remark 2.13.3 The conditions of Theorem 2.13.1 are sufficient for the projection matrix $C(\varphi)$ found in the structure of the Green function to be continuous, since the matrix has the representation

$$C(\varphi) = G_0(\tau,\varphi)\Omega_0^\tau(\varphi;A), \quad \tau < 0.$$

Remark 2.13.4 The conditions of Theorem 2.13.1 allow conclusions to be made on the continuity of the invariant toroidal manifold of the system of Riccati type equations

$$\frac{d\varphi}{dt} = a(\varphi),$$

$$\frac{dX}{dt} = -XA_{21}(\varphi)X + A_{11}(\varphi)X - XA_{22}(\varphi) + A_{12}(\varphi),$$

where X is a quadratic or rectangular matrix, provided that $A_{ij}(\varphi) \in H^\mu$, $i,j = 1,2$. Moreover, the estimate

$$\langle A_{11}(\varphi)x_1, x_1\rangle + \langle A_{12}(\varphi)x_2, x_1\rangle - \langle A_{21}(\varphi)x_1, x_2\rangle$$
$$- \langle A_{22}(\varphi)x_2, x_2\rangle \leq -\gamma\|x\|^2, \quad \gamma > 0$$

is satisfied, ensuring the existence of the toroidal manifold.

Let the functions $a(\varphi)$ and $A(\varphi)$ be continuous with respect to the totality of variables φ. Then the question arises: is the Green function $G_0(\tau,\varphi)$ continuous in the variables φ? To answer the question we consider the difference of Green functions $G_0(\tau,\varphi) - G_0(\tau,\bar\varphi)$ which is estimated as

$$\|G_0(\tau,\varphi) - G_0(\tau,\bar\varphi)\| \leq K^2 \exp\{-(\gamma-\nu)|\tau|\} \int_{-\infty}^{\infty} \exp\{-\gamma|\sigma|\}$$

$$\times \mu(A; F^{-1}(F(\|\varphi-\bar\varphi\|) + |\sigma|))\,d\sigma = K^2 \exp\{-(\gamma-\nu)|\tau|\}\mathcal{I}_\nu(\|\varphi-\bar\varphi\|),$$

where

$$F(u) = \int_d^u \frac{d\sigma}{\mu(a;\sigma)},$$

$$\mu(a;\sigma) \geq \max_{\|\varphi-\bar{\varphi}\|\leq\sigma} \|a(\varphi) - a(\bar{\varphi})\|.$$

This shows that the integral $\mathcal{I}_\nu(\|\varphi - \bar{\varphi}\|)$ tending to zero as $\bar{\varphi} \to \varphi$ and some $\nu \in (0,\gamma)$ is a sufficient condition for the Green function $G_0(\tau,\varphi)$ being continuous in φ.

Theorem 2.13.2 *The necessary and sufficient condition for the integral $\mathcal{I}_\nu(\|\varphi - \bar{\varphi}\|)$ to tend to zero as $\bar{\varphi} \to \varphi$ and some arbitrary fixed $\nu \in (0,\gamma)$ is the divergence of the integral $\int_{+0} \frac{d\sigma}{\mu(a;\sigma)}$, i.e. the condition*

$$\lim_{u \to +0} F(u) = -\infty. \tag{2.13.7}$$

Proof We transform the integral $\mathcal{I}_\nu(\|\varphi - \bar{\varphi}\|)$ as

$$\mathcal{I}_\nu = 2 \int_{-\infty}^\infty \exp\{-\gamma\sigma\} \mu(A; F^{-1}(F(\|\varphi - \bar{\varphi}\|) + \sigma))\, d\sigma$$

$$= 2 \int_{F(\|\varphi-\bar{\varphi}\|)}^\infty \exp\{-\nu\sigma\} \mu(A; F^{-1}(\sigma))\, d\sigma\, \exp\{\nu F(\|\varphi - \bar{\varphi}\|)\}$$

$$= 2 \exp\{\nu F(\|\varphi - \bar{\varphi}\|)\} \left[\int_{F(\|\varphi-\bar{\varphi}\|)}^0 \exp\{-\nu\sigma\} \mu(A; F^{-1}(\sigma))\, d\sigma \right.$$

$$\left. + \int_0^\infty \exp\{-\nu\sigma\} \mu(A; F^{-1}(\sigma))\, d\sigma \right] \leq 2 \exp\{\nu F(\|\varphi - \bar{\varphi}\|)\}$$

$$\times \left[\int_{\|\varphi-\bar{\varphi}\|}^d \exp\{-\nu F(\sigma)\} \frac{\mu(A;\sigma)}{\mu(a;\sigma)}\, d\sigma + \frac{\sup_{\sigma \geq d} \mu(A;\sigma)}{\nu} \right].$$

Now let us prove that

$$\int_{\|\varphi-\bar{\varphi}\|}^d \frac{\mu(A;\sigma)}{\mu(a;\sigma)} \exp\{\nu[F(\|\varphi - \bar{\varphi}\|) - F(\sigma)]\}\, d\sigma \xrightarrow[\bar{\varphi}\to\varphi]{} 0.$$

To this end it is sufficient to show that for every prescribed $\varepsilon > 0$ there exists $u_0 > 0$ such that for all $u \in (0, u_0)$ the inequality

$$\int_u^d \frac{\mu(A;\sigma)}{\mu(a;\sigma)} \exp\{-\nu F(\sigma) + \nu F(u)\} \, d\sigma < \varepsilon,$$

holds true.

We take $u_1 > 0$ from the condition $\mu(A; u_1) < \frac{\varepsilon}{2}\nu$ and consider the integral

$$\int_u^{u_1} \frac{\mu(A;\sigma)}{\mu(a;\sigma)} \exp\{-\nu F(\sigma) + \nu F(u)\} \, d\sigma$$

$$= -\frac{1}{\nu} \mu(A;\sigma) \exp\{-\nu F(\sigma) + \nu F(u)\}\bigg|_u^{u_1}$$

$$+ \frac{1}{\nu} \int_u^{u_1} \exp\{-\nu F(\sigma) + \nu F(u)\} \mu'(A;\sigma) \, d\sigma$$

$$\leq \frac{1}{\nu} \mu(A;u) - \frac{1}{\nu} \mu(A;u_1) \exp\{-\nu F(u_1) + \nu F(u)\} + \frac{1}{\nu} \mu(A;\sigma)\bigg|_u^{u_1}$$

$$\leq \frac{1}{\nu} \mu(A;u_1) < \frac{\varepsilon}{2}.$$

Then we take u_2 to be dependent on u_1 from the inequality

$$\int_{u_1}^d \frac{\mu(A;\sigma)}{\mu(a;\sigma)} \exp\{-\nu F(\sigma)\} \, d\sigma \, \exp\{\nu F(u_2)\} < \frac{\varepsilon}{2}.$$

This choice is possible due to the condition (2.13.7). Setting $u_0 = \min\{u_1, u_2\}$ we obtain the desired inequality.

Now we demonstrate that the integral $\mathcal{I}_\nu(\|\varphi - \bar\varphi\|)$ tending to zero givs the condition (2.13.7) being satisfied. Suppose on the contrary, i.e. the integral $\mathcal{I}_\nu$ tending to zero yields the estimate $\int_u^d \frac{d\sigma}{\mu(a;\sigma)} < c < \infty$, then $F(\|\varphi - \bar\varphi\|) > -c$. Hence

$$\mathcal{I}_\nu(\|\varphi - \bar\varphi\|) = 2 \int_{F(\|\varphi-\bar\varphi\|)}^\infty \exp\{-\nu\sigma\} \mu(A; F^{-1}(\sigma)) \, d\sigma \, \exp\{\nu F(\|\varphi - \bar\varphi\|)\}$$

$$\geq 2 \exp\{-\nu c\} \int_0^\infty \exp\{-\nu\sigma\} \mu(A; F^{-1}(\sigma)) \, d\sigma.$$

The latter contradicts the integral $\mathcal{I}_\nu$ tending to zero.

This completes the proof of the Theorem 2.13.2.

Assume that the functions $a(\varphi)$ and $A(\varphi)$ of the system (2.2.4) are continuously differentiable once in the variable φ, and that the continuity moduli in the partial derivatives $\partial A(\varphi)/\partial \varphi_k$, $\partial a(\varphi)/\partial \varphi_k$ are bounded

$$\max\left\{\mu\left(\frac{\partial A}{\partial \varphi_k}; \sigma\right), \mu\left(\frac{\partial a}{\partial \varphi_k}; \sigma\right)\right\} \leq c\mu(\sigma), \qquad (2.13.8)$$

where $\mu(\sigma)$ is a fixed monotone increasing function $\mu(0) = 0$, c is a positive constant.

We specify the constant $\alpha > 0$ as

$$\max_{\|\eta\|=1}\left\|\frac{\partial a}{\partial \varphi}\eta\right\| \leq \alpha.$$

It is known that the inequality $\alpha < \gamma$ ensures the existence of the first-order partial derivatives of the Green function $G_0(\tau,\varphi)$ in the variables φ. We characterize the continuity moduli of these derivatives. To this end the partial derivatives of the Green function are represented as

$$\frac{\partial}{\partial \varphi_i} G_0(\sigma,\varphi) = \int_{-\infty}^{\infty} G_0(\sigma,\varphi) \sum_{k=1}^{m} \frac{\partial A}{\partial \varphi_{\sigma k}} \frac{\partial \varphi_{\sigma k}}{\partial \varphi_i} G_\sigma(\tau,\varphi)\,d\sigma.$$

In view of the Green function boundedness we get for the difference $\frac{\partial}{\partial \varphi_i}G_0(\tau,\varphi) - \frac{\partial}{\partial \bar\varphi_i}G_0(\sigma,\bar\varphi)$

$$\left\|\frac{\partial}{\partial \varphi_i}G_0(\tau,\varphi) - \frac{\partial}{\partial \bar\varphi_i}G_0(\tau,\bar\varphi)\right\|$$

$$\leq K_1 \int_{-\infty}^{\infty} \exp\{-\gamma_0|\sigma-\tau| + \alpha|\sigma|\} \|G_0(\sigma,\varphi) - G_0(\sigma,\bar\varphi)\|\,d\sigma$$

$$+ K_2 \int_{-\infty}^{\infty} \exp\{-\gamma_0|\sigma| - \gamma_0|\sigma-\tau| + \alpha|\sigma|\}$$

$$\times \left\|\left.\frac{\partial}{\partial \varphi_k}A(\varphi)\right|_{\varphi=\varphi_\sigma(\varphi)} - \left.\frac{\partial}{\partial \varphi_k}A(\varphi)\right|_{\varphi=\varphi_\sigma(\bar\varphi)}\right\|\,d\sigma \qquad (2.13.9)$$

$$+ K_3 \int_{-\infty}^{\infty} \exp\{-\gamma_0(|\sigma| + |\sigma-\tau|)\} \left\|\frac{\partial}{\partial \varphi_i}\varphi_\sigma(\varphi) - \frac{\partial}{\partial \bar\varphi_i}\varphi_\sigma(\bar\varphi)\right\|\,d\sigma$$

$$+ K_4 \int_{-\infty}^{\infty} \exp\{-\gamma_0|\sigma| + \alpha|\sigma|\} \|G_\sigma(\tau,\varphi) - G_\sigma(\tau,\bar\varphi)\|\,d\sigma.$$

The simultaneous satisfaction of the inequalities

$$\|G_t(\tau,\varphi) - G_t(\tau,\bar{\varphi})\| = \|G_0(\tau - t, \varphi_t(\varphi)) - G_0(\tau - t, \varphi_t(\bar{\varphi}))\|$$
$$\leq (K) \exp\{-(\gamma_0 - \alpha)|t - \tau|\} \|\varphi_t(\varphi) - \varphi_t(\bar{\varphi})\|$$
$$\leq (K) \exp\{-(\gamma_0 - \alpha)|t - \tau| + \alpha|t|\} \|\varphi - \bar{\varphi}\|,$$
$$\|G_t(\tau,\varphi) - G_t(\tau,\bar{\varphi})\| \leq \|G_t(\tau,\varphi)\| + \|G_t(\tau,\bar{\varphi})\|$$
$$\leq 2K_0 \exp\{-\gamma_0|t - \tau|\},$$

for all $z \geq 0$ implies

$$\|G_t(\tau,\varphi) - G_t(\tau,\bar{\varphi})\| \quad (2.13.10)$$
$$\leq (K) \exp\left\{-\gamma_0|t - \tau| + \frac{\alpha}{z+1}(|t| + |t - \tau|)\right\} \|\varphi - \bar{\varphi}\|^{\frac{1}{z+1}}.$$

Further we estimate the norm of the difference

$$\frac{\partial}{\partial \varphi_i} \varphi_t(\varphi) - \frac{\partial}{\partial \bar{\varphi}_i} \varphi_t(\bar{\varphi}).$$

From equation (2.13.1) we get for this difference, for positive t,

$$\left\|\frac{\partial}{\partial \varphi_i} \varphi_t(\varphi) - \frac{\partial}{\partial \bar{\varphi}_i} \varphi_t(\bar{\varphi})\right\|$$
$$\leq \alpha \int_0^t \left\|\frac{\partial}{\partial \varphi_i} \varphi_\sigma(\varphi) - \frac{\partial}{\partial \bar{\varphi}_i} \varphi_\sigma(\bar{\varphi})\right\| d\sigma + c \int_0^t \exp\{\alpha\sigma\} \mu \|\varphi_\sigma(\varphi) - \varphi_\sigma(\bar{\varphi})\| d\sigma,$$

or

$$\left\|\frac{\partial}{\partial \varphi_i} \varphi_t(\varphi) - \frac{\partial}{\partial \bar{\varphi}_i} \varphi_t(\bar{\varphi})\right\|$$
$$\leq (c) \exp\{\alpha\sigma\} \int_0^t \mu \|\varphi_\sigma(\varphi) - \varphi_\sigma(\bar{\varphi})\| d\sigma, \quad t > 0.$$

Taking into account the exponential estimate for the partial derivative

$$\left\|\frac{\partial}{\partial \varphi_i} \varphi_t(\varphi)\right\| \leq (K) \exp\{\alpha|t|\},$$

we find for all $z \geq 0$

$$\left\|\frac{\partial}{\partial \varphi_i} \varphi_t(\varphi) - \frac{\partial}{\partial \bar{\varphi}_i} \varphi_t(\bar{\varphi})\right\| \quad (2.13.11)$$
$$\leq (c) \exp\left\{\left(\alpha + \frac{\alpha}{z+1}\right)|t|\right\} \mu^{1/(z+1)} (\|\varphi - \bar{\varphi}\|), \quad t > 0.$$

The estimate (2.13.11) is proved in the same way for all $t < 0$. Finally, estimating the norm of the difference

$$\frac{\partial}{\partial \varphi_k} A(\varphi)\bigg|_{\varphi=\varphi_t(\varphi)} - \frac{\partial}{\partial \varphi_k} A(\varphi)\bigg|_{\varphi=\varphi_t(\bar{\varphi})}$$

for all $z \geq 0$, $t \in R$ yields

$$\left\|\frac{\partial}{\partial \varphi_k} A(\varphi)\bigg|_{\varphi=\varphi_t(\varphi)} - \frac{\partial}{\partial \varphi_k} A(\varphi)\bigg|_{\varphi=\varphi_t(\bar{\varphi})}\right\| \\ \leq (K) \exp\left\{\frac{\alpha}{z+1}|t|\right\} \mu^{1/(z+1)} (\|\varphi - \bar{\varphi}\|). \quad (2.13.12)$$

In view of (2.13.9) estimates (2.13.11) and (2.13.12) make it possible to proceed with the inequalities for the difference of the Green function partial derivatives as follows

$$\left\|\frac{\partial}{\partial \varphi_i} G_0(\tau, \varphi) - \frac{\partial}{\partial \bar{\varphi}_i} G_0(\tau, \bar{\varphi})\right\|$$

$$\leq (K)\bigg[\int_{-\infty}^{\infty} \exp\left\{-\gamma|\sigma - \tau| + \alpha|\sigma| - \gamma|\sigma| + \frac{\alpha}{z+1}|\sigma|\right\} d\sigma$$

$$\times \|\varphi - \bar{\varphi}\|^{1/(z+1)} + \int_{-\infty}^{\infty} \exp\left\{-\gamma(|\sigma| + |\sigma - \tau|) + \alpha|\sigma|\right.$$

$$\left. + \frac{\alpha}{z+1}|\sigma|\right\} d\sigma\, \mu^{1/(z+1)} (\|\varphi - \bar{\varphi}\|)$$

$$+ \int_{-\infty}^{\infty} \exp\left\{-\gamma(|\sigma| + |\sigma - \tau|) + \left(\alpha + \frac{\alpha}{z+1}\right)|\sigma|\right\} d\sigma\, \mu^{1/(z+1)} (\|\varphi - \bar{\varphi}\|)$$

$$+ \int_{-\infty}^{\infty} \exp\left\{-\gamma|\sigma| + \alpha|\sigma| - \gamma|\sigma - \tau| + \frac{\alpha}{z+1}(|\sigma| + |\sigma - \tau|)\right\} d\sigma$$

$$\times \|\varphi - \bar{\varphi}\|^{1/(z+1)}\bigg] \leq (K) \exp\left\{-\left(\gamma - \alpha - \frac{\alpha}{z+1}\right)|\tau|\right\} \mu^{1/(z+1)} (\|\varphi - \bar{\varphi}\|),$$

where the positive constant z is taken from the condition $\gamma - \alpha - \alpha/(z+1) > 0$. Thus, if $\gamma > 2\alpha$, then the constant $z \geq 0$ can be taken to be equal to zero.

Assume that the functions $a(\varphi)$ and $A(\varphi)$ are continuously differentiable in φ up to the order $|l| = \sum_{i=1}^{m} l_i$, and the continuity moduli of its partial derivatives of the order $|l|$ are bounded from above by a continuity modulus $\mu(\sigma)$

$$\max\left[\mu(D_\varphi^l A; \sigma), \mu(D_\varphi^l a; \sigma)\right] \leq (K)\mu(\sigma). \tag{2.13.13}$$

It is known that a sufficient condition for the existence of continuous derivatives up to the order $|l|$ in φ of the Green function $G_0(\tau, \varphi)$, $\tau \neq 0$, is the inequality (2.12.25) being satisfied, where it is necessary to set $l = |l|$. The derivatives of the $|l|$-th order are represented as

$$D_\varphi^l G_0(\tau, \varphi) = \int_{-\infty}^{\infty} \sum_{|\lambda_1|+|\lambda_2|+|\lambda_3|=|l|+1} c_{\lambda_1,\lambda_2,\lambda_3} D_\varphi^{\lambda_1} G_0(\sigma, \varphi)$$
$$\times D_\varphi^{\lambda_2} \left[\sum_{k=1}^m \frac{\partial A}{\partial \varphi_{\sigma k}} \frac{\partial \varphi_{\sigma k}}{\partial \varphi_i}\right] D_\varphi^{\lambda_3} G_0(\tau, \varphi) \, d\sigma, \tag{2.13.14}$$

where $c_{\lambda_1,\lambda_2,\lambda_3}$ are some constants, $\lambda_i = (\lambda_{i1}, \ldots, \lambda_{im})$, $\lambda_{ij} \geq 0$, $|\lambda_i| = \sum_{j=1}^m \lambda_{ij}$. The following estimates are valid

$$\left\|D_\varphi^{\lambda_3} G_\sigma(\tau, \varphi)\right\| \leq (K) \exp\left\{-(\gamma - |\lambda_3|\alpha)|\sigma - \tau| + \alpha|\lambda_3||\sigma|\right\},$$
$$|\lambda_3| \leq |l| - 1; \tag{2.13.15}$$

$$\left\|D_\varphi^{\lambda_2}\left[\sum_{k=1}^m \frac{\partial A}{\partial \varphi_{tk}} \frac{\partial \varphi_{tk}}{\partial \varphi_i}\right]\right\| \leq (K) \exp\left\{(|\lambda_2| + 1)\alpha|t|\right\},$$
$$|\lambda_2| \leq |l| - 1; \tag{2.13.16}$$

$$\left\|D_\varphi^{\lambda_2}\left[\frac{\partial A(\varphi_t(\varphi))}{\partial \varphi_{tk}} \frac{\partial \varphi_{tk}(\varphi)}{\partial \varphi_i}\right] - D_{\bar\varphi}^{\lambda_2}\left[\frac{\partial A(\varphi_t(\bar\varphi))}{\partial \varphi_{tk}} \frac{\partial \varphi_{tk}(\bar\varphi)}{\partial \bar\varphi_i}\right]\right\|$$
$$\leq (K) \exp\left\{(|\lambda_2| + 1)\alpha|t|\right\} \|\varphi - \bar\varphi\|, \quad |\lambda_2| \leq |l| - 2; \tag{2.13.17}$$

$$\left\|D_\varphi^{\lambda_3} G_\sigma(\tau, \varphi) - D_{\bar\varphi}^{\lambda_3} G_\sigma(\tau, \varphi)\right\|$$
$$\leq (K) \exp\left\{-(\gamma - |\lambda_3|\alpha - \alpha)|\sigma - \tau| + (|\lambda_3| + 1)\alpha|\sigma|\right\} \|\varphi - \bar\varphi\|, \tag{2.13.18}$$
$$|\lambda_3| \leq |l| - 1.$$

Estimating the derivatives of the order of $|l| - 1$ in the variables φ of the functions $\frac{\partial}{\partial \varphi_i} A(\varphi)$, $\frac{\partial}{\partial \varphi_i} a(\varphi)$ and analysing its differences at various points $\varphi, \bar\varphi \in R^m$, we find that

$$\left\| D_{\varphi_t}^\lambda \frac{\partial A(\varphi_t(\varphi))}{\partial \varphi_{tk}} - D_{\bar\varphi_t}^\lambda \frac{\partial A(\varphi_t(\bar\varphi))}{\partial \varphi_{tk}} \right\|$$
$$\leq (K) \exp\left\{ \frac{\alpha}{z+1} |t| \right\} \mu^{1/(z+1)} (\|\varphi - \bar\varphi\|);$$
(2.13.19)

$$\left\| D_\varphi^\lambda \frac{\partial \varphi_t(\varphi)}{\partial \varphi_i} - D_{\bar\varphi}^\lambda \frac{\partial \varphi_t(\bar\varphi)}{\partial \bar\varphi_i} \right\|$$
$$\leq (K) \exp\left\{ \left(|\lambda|\alpha + \frac{\alpha}{z+1} \right) |t| \right\} \mu^{1/(z+1)} (\|\varphi - \bar\varphi\|).$$
(2.13.20)

With regard to (2.13.14)–(2.13.20), we arrive at the estimate

$$\|D_\varphi^l G_0(\tau, \varphi) - D_{\bar\varphi}^l G_0(\tau, \varphi)\|$$
$$\leq (K) \exp\left\{ -\left(\gamma - |l|\alpha - \frac{\alpha}{z+1} \right) |t| \right\} \mu^{1/(z+1)} (\|\varphi - \bar\varphi\|), \quad z \geq 0.$$
(2.13.21)

Summing up the above yields the following result.

Theorem 2.13.3 *Let the following conditions be satisfied:*
(1) *the functions $a(\varphi)$ and $A(\varphi)$ are continuously differentiable in the variables φ_j, $j = 1, \ldots, m$, up to the order of $|l|$ and the continuity moduli of its partial derivatives of the order of $|l|$ satisfy the inequalities (2.13.13);*
(2) *there exists a unique Green function (2.6.10) with the estimate (2.6.11);*
(3) *the inequality $\gamma - |l|\alpha > 0$ is satisfied, where*

$$\alpha \geq \max_{\|\eta\|=1} \left\| \frac{\partial a}{\partial \varphi} \eta \right\|.$$

Then for the partial derivatives of the Green function $G_0(\tau, \varphi)$ in the variables φ of the order of $|l|$ the estimate (2.13.21) is valid.

2.14 Invariant Tori of Linear Extensions with a Degenerate Matrix at the Derivatives

Consider the system of differential equations of the type

$$\dot\varphi = a(\varphi), \quad B(\varphi)\dot x = A(\varphi)x + f(\varphi), \tag{2.14.1}$$

where $\varphi \in R^m$, $x \in R^n$, $\dot\varphi = \frac{d\varphi}{dt}$, $a(\varphi), B(\varphi), A(\varphi) \in C^r(R^m, 2\pi)$. In order to indicate the value of the period of the functions $F(\varphi)$ the space $C^r(\mathcal{T}_m)$ is designated by $C^r(R^m, 2\pi)$.

Definition 2.14.1 The equality $x = u(\varphi)$ specifies the *smooth invariant torus* of the system (2.8.1), if $u(\varphi) \in C^l(R^m, 2\pi)$, $1 \leq l \leq r$, and the identity

$$B(\varphi) \frac{du(\varphi)}{d\varphi} a(\varphi) \equiv A(\varphi)u(\varphi) + f(\varphi)$$

is satisfied.

Lemma 2.14.1 *Let the scalar function $\rho(\varphi)$ belong to the space $C^r(R^m, 2\pi)$. Then the vector-function $x = u(\varphi)$ determining the invariant torus of one of the systems*

$$\dot{\varphi} = \rho(\varphi)a(\varphi), \quad B(\varphi)\dot{x} = A(\varphi)x + f(\varphi); \qquad (2.14.2)$$

$$\dot{\varphi} = a(\varphi), \quad \rho(\varphi)B(\varphi)\dot{x} = A(\varphi)x + f(\varphi), \qquad (2.14.3)$$

also specifies the smooth invariant torus of the other system.

Proof In fact, in view of the commutativity of the scalar function $\rho(\varphi)$ with any matrix function we get one and the same partial equation for the vector-functions which determine the smooth invariant torus of the systems (2.14.2) and (2.14.3)

$$\rho(\varphi)B(\varphi) \frac{du(\varphi)}{d\varphi} a(\varphi) \equiv A(\varphi)u(\varphi) + f(\varphi).$$

Hence the validity of the Lemma 2.14.1 follows immediately.

Lemma 2.14.2 *With respect to the system of equations*

$$\dot{\varphi} = a(\varphi), \quad \rho(\varphi)\dot{x} = b(\varphi)x + c(\varphi), \qquad (2.14.4)$$

where the vector-function $a(\varphi)$ and the scalar functions $\rho(\varphi)$, $b(\varphi)$, $c(\varphi)$ belong to the space $C^r(R^m, 2\pi)$, $r \geq 1$, let the following conditions be satisfied:

(1) *there exists a scalar function $s(\varphi) \in C^1(R^m, 2\pi)$ such that for all $\varphi \in R^m$ and $s(\varphi) \neq 0$*

$$\frac{ds(\varphi)}{d\varphi} a(\varphi)\rho(\varphi) + 2b(\varphi)s(\varphi) \leq -2\gamma < 0; \qquad (2.14.5)$$

(2) *the inequality*

$$\gamma_1 - l\alpha > 0, \qquad (2.14.6)$$

is satisfied, where

$$\gamma_1 = \frac{\gamma}{|s|_0}, \quad \alpha \geq \max_{\|\eta\|=1} \left|\left\langle \frac{\partial[a(\varphi)\rho(\varphi)]}{\partial\varphi} \eta, \eta \right\rangle\right|,$$

l is a positive integer.

Then the system of equations (2.14.4) has a unique smooth invariant torus $x = u(\varphi) \in C^l(R^m, 2\pi)$ *represented by*

$$u(\varphi) = \begin{cases} \int\limits_{-\infty}^{0} c(\varphi_\tau(\varphi)) \exp\left\{\int\limits_{\tau}^{0} b(\varphi_\sigma(\varphi))\, d\sigma\right\} d\tau, & s(\varphi) > 0, \\ -\int\limits_{0}^{\infty} c(\varphi_\tau(\varphi)) \exp\left\{\int\limits_{\tau}^{0} b(\varphi_\sigma(\varphi))\, d\sigma\right\} d\tau, & s(\varphi) < 0, \end{cases} \qquad (2.14.7)$$

where φ_t *is a solution of system*

$$\dot{\varphi} = \rho(\varphi)a(\varphi), \quad \varphi_0(\varphi) = \varphi.$$

Proof Let $s(\varphi) > 0$. We designate $m_1 = \min\limits_{\varphi \in R^m} s(\varphi)$, $M = \max\limits_{\varphi \in R^m} s(\varphi)$. In view of (2.8.5), we get for the solution $x(t; \varphi)$ of equation $\dot{x} = b(\varphi_t(\varphi))x$, where $\varphi \in R^m$,

$$\frac{d}{dt}[s(\varphi_t(\varphi))x^2(t;\varphi)] \leq -2\gamma x^2(x;\varphi) \leq -\frac{2\gamma}{M} s(\varphi_t(\varphi))x^2(t;\varphi),$$

$$s(\varphi_t(\varphi))x^2(t;\varphi) \leq s(\varphi_\tau(\varphi))x^2(\tau;\varphi) \exp\left\{-\frac{2\gamma}{M}(t-\tau)\right\}, \quad \tau \leq t.$$

Hence

$$|x(t;\varphi)| \leq \left(\frac{M}{m_1}\right)^{1/2} |x(\tau;\varphi)| \exp\left\{-\frac{\gamma}{M}(t-\tau)\right\}, \quad \tau \leq t. \qquad (2.14.8)$$

Thus, the function $x(t, \varphi) = \int\limits_{-\infty}^{t} c(\varphi_\tau(\varphi)) \exp\left\{\int\limits_{\tau}^{t} b(\varphi_\sigma(\varphi))\, d\sigma\right\} dt$, is the only solution which is bounded on the whole R axis of the equation

$$\dot{x} = b(\varphi_t(\varphi))x + c(\varphi_t(\varphi)),$$

which possesses the property $x(0; \varphi_t(\varphi)) \equiv x(t; \varphi)$. Similar arguments for the case of $s(\varphi) < 0$ result in the second equality (2.14.7). By the immediate differentiation under the integral sign in (2.14.7) we make sure that the inequality (2.14.6) means that $u(\varphi) \in C^l(R^m, 2\pi)$.

Let us prove the uniqueness of the invariant torus. Assume that the system (2.14.4) has one more smooth invariant toroidal manifold $x = v(\varphi) \not\equiv u(\varphi)$. Then, the function $z(\varphi) = u(\varphi) - v(\varphi)$ which is bounded for all $\varphi \in R^m$ has to satisfy on the trajectories of solutions $\varphi_t(\varphi)$ of the system $\dot{\varphi} = \rho(\varphi)a(\varphi)$ for the equation $\dot{z} = b(\varphi_t(\varphi))z$. By virtue of inequality (2.14.8) the unique solution of this equation bounded on R for $s(\varphi) \neq 0$ is the trivial one. Therefore $v(\varphi) \equiv u(\varphi)$, $\forall \varphi \in R^m$.

The Lemma 2.14.2 is proved.

Consider the following system of equations
$$\dot{\varphi} = a(\varphi), \qquad (2.14.9)$$
$$\langle g(\varphi), \dot{x}\rangle = \langle p_1(\varphi), x\rangle + f_1(\varphi), \quad 0 = \langle p_i(\varphi), x\rangle + f_i(\varphi),$$
$$i = 2,\ldots,n,$$
where $\langle p_i(\varphi), x\rangle = \sum_{j=1}^{n} p_{ij}(\varphi) x_j$, $g(\varphi) = (g_1(\varphi),\ldots,g_n(\varphi))$, $a(\varphi)$, $g(\varphi)$, $p_i(\varphi)$, $f_i(\varphi) \in C^{r+1}(R^m, 2\pi)$, $r \geq 1$. It is of interest to establish conditions under which the system of equations (2.14.9) has a unique smooth invariant torus for each function $f(\varphi) \in C^{r+1}(R^m, 2\pi)$.

We introduce the designations
$$P_2 = \begin{pmatrix} p_2 \\ \vdots \\ p_n \end{pmatrix}, \quad F_2 = \begin{pmatrix} f_2 \\ \vdots \\ f_n \end{pmatrix},$$
$$\Delta = \begin{vmatrix} p_1 \\ P_2 \end{vmatrix}, \quad \Delta_1 = \begin{vmatrix} g \\ P_2 \end{vmatrix}, \quad \Delta_2 = \begin{vmatrix} \dot{g} \\ P_2 \end{vmatrix},$$
where $|\mathcal{D}| = \det \mathcal{D}$ ($\mathcal{D}$ is an arbitrary matrix) and prove the following assertion.

Theorem 2.14.1 *Let* rank $P_2(\varphi) \equiv n - 1$ *and all conditions of the Lemma 2.14.2 be satisfied, where* $\rho(\varphi) = \Delta_1(\varphi)$, $b(\varphi) = \Delta(\varphi) + \Delta_2(\varphi) - \dot{\Delta}_1(\varphi)$. *Then the system of equations (2.14.9) has a unique smooth invariant torus* $x = u(\varphi) \in C^l(R^m, 2\pi)$, $1 \leq l \leq r$ *for every vector-function* $f(\varphi) \in C^{r+1}(R^m, 2\pi)$.

Proof We represent the general solution of the algebraic system as
$$x = v(\varphi) y + q(\varphi), \qquad (2.14.10)$$
where
$$v(\varphi) = \begin{pmatrix} v_1 \\ \vdots \\ v_n \end{pmatrix}, \quad v_i = (-1)^{i+1} \begin{vmatrix} p_{21} & \cdots & p_{2i-1} & p_{2i+1} & \cdots & p_{2n} \\ \vdots & \ddots & \vdots & \vdots & \ddots & \vdots \\ p_{n1} & \cdots & p_{ni-1} & p_{ni+1} & \cdots & p_{nn} \end{vmatrix},$$
$$q = -P_2^*(P_2 P_2^*)^{-1} F_2,$$
y is an arbitrary scalar function. Substituting (2.14.10) into the first equation (2.14.9) yields
$$\langle g(\varphi), v(\varphi)\rangle \dot{y} = (\langle p_1(\varphi), v(\varphi)\rangle - \langle g(\varphi), \dot{v}(\varphi)\rangle) y$$
$$+ \langle p_1(\varphi), q(\varphi)\rangle - \langle g(\varphi), \dot{q}(\varphi)\rangle + f_1(\varphi).$$
Seeing that $\langle p_1(\varphi), v(\varphi)\rangle = \Delta(\varphi)$, $\langle g(\varphi), v(\varphi)\rangle = \Delta_1(\varphi)$, $\langle g(\varphi), \dot{v}(\varphi)\rangle = \dot{\Delta}_1(\varphi) - \Delta_2(\varphi)$ and taking the assertions of Lemma 2.14.2 into account we verify the validity of Theorem 2.14.1.

Theorem 2.14.2 *Let the system of equations (2.14.1), all coefficients of which belong to the space $C^r(R^m, 2\pi)$, $r \geq 2$, satisfy the conditions:*

(1) *the matrix $A(\varphi)$ is non-degenerate for all $\varphi \in R^m$;*
(2) *there exists $n \times k$-matrix $\hat{B}(\varphi) \in C^r(R^m, 2\pi)$, $r \geq 2$, such that*

$$\operatorname{rank}[B(\varphi), \hat{B}(\varphi)] \equiv 1.$$

Then a fixed number $\delta > 0$ can be indicated such that the inequality

$$\|a\|\,\|B\|\,\|A^{-1}\| + \left\|\frac{\partial}{\partial \varphi_i}a\right\|\|B\|\,\|A^{-1}\| + \|a\|\,\|B\|\left\|\frac{\partial}{\partial \varphi_i}A^{-1}\right\|$$
$$+ \|a\|\left\|\frac{\partial}{\partial \varphi_i}B\right\|\|A^{-1}\| \leq \delta, \quad i = 1,\ldots, m, \tag{2.14.11}$$

ensures the existence of a unique smooth invariant torus $x = u(\varphi) \in C^{r-1}(R^m, 2\pi)$, $r \geq 2$, of the system of equations (2.14.1) for any function $f(\varphi) \in C^r(R^m, 2\pi)$.

Proof We first transform the system of the equations (2.14.1) to the form of the system (2.14.9). Based on the properties of the matrix $[B(\varphi), \hat{B}(\varphi)]$ one can state the existence of an n-dimensional vector $\mu(\varphi) = (\mu_1(\varphi), \ldots, \mu_n(\varphi))$ such that $\mu(\varphi) \in C^r(R^m, 4\pi)$, $\|\mu\| \equiv 1$, and $\mu(\varphi)[B(\varphi), \hat{B}(\varphi)] = \lambda(\varphi)$, where $\|\lambda(\varphi)\| \neq 0$, $\forall \varphi \in R^m$. Then for every vector $\nu(\varphi)$ orthogonal to $\mu(\varphi)$ the equality $\nu(\varphi)[B(\varphi), \hat{B}(\varphi)] \equiv 0$ is valid for all $\varphi \in R^m$.

It is known (see Bylov et al. [1]) that the vector $\mu(\varphi)$ can not be always added by the vectors from $C^r(R^m, 4\pi)$ up to a complete basis in R^n. Therefore we construct the matrix of the dimensions $(n+1) \times n$

$$\begin{bmatrix} \mu(\varphi) \\ I_n - \mu^*(\varphi)\mu(\varphi) \end{bmatrix} = \begin{bmatrix} \mu(\varphi) \\ Q_1(\varphi) \end{bmatrix} = Q(\varphi),$$

which possesses the following properties

$$Q(\varphi) \in C^r(R^m, 4\pi), \quad \operatorname{rank} Q(\varphi) \equiv n, \quad Q_1(\varphi)\mu^*(\varphi) \equiv 0. \tag{2.14.12}$$

Since the lines of the matrix $Q_1(\varphi)$ are orthogonal to the vector $\mu^*(\varphi)$, then $Q_1(\varphi)[B(\varphi), \hat{B}(\varphi)] = 0$. Multiplying the second system of equations (2.14.1) by the matrix $Q(\varphi)$ on the left, we come to the equivalent system

$$\dot{\varphi} = a(\varphi), \quad \mu(\varphi)B(\varphi)\dot{x} = \mu(\varphi)A(\varphi)x + \mu(\varphi)f(\varphi); \tag{2.14.13}$$

$$0 = Q_1(\varphi)[A(\varphi)x + f(\varphi)]. \tag{2.14.14}$$

With regard to non-degeneracy of the matrix $A(\varphi)$ the general solution of the algebraic system (2.14.14) can be represented as

$$x = A^{-1}(\varphi)\mu^*(\varphi)y - A^{-1}(\varphi)f(\varphi), \tag{2.14.15}$$

where y is an arbitrary scalar function. Substituting (2.14.15) into the second differential equation (2.14.13) we get the system of the form (2.14.4), where

$$\rho(\varphi) = \mu B A^{-1} \mu^*,$$

$$b(\varphi) = 1 - D(\varphi) = 1 - \mu B \frac{d}{dt}(A^{-1}\mu^*), \qquad (2.14.16)$$

$$C(\varphi) = \mu B \frac{d}{dt}(A^{-1}f).$$

The result of Lemma 2.14.2 is applied to the given system. To this end, function $b(\varphi)$ is required to be close to 1, which allows us to take $s(\varphi) \equiv -1$ and the smallness of function $\rho(\varphi)$ which provides the required smoothness of the torus. In the case under consideration inequality (2.14.6) becomes

$$1 - |D(\varphi)| - (r-1)\left\|\frac{\partial}{\partial \varphi}[a(\varphi)\rho(\varphi)]\right\| > 0. \qquad (2.14.17)$$

Estimation of $|D(\varphi)|$ and $\left\|\frac{\partial}{\partial \varphi}[a(\varphi)\rho(\varphi)]\right\|$ yields

$$|D(\varphi)| \le \|B(\varphi)\| \, \|a(\varphi)\| \left[\left(\sum_{i=1}^{m}\left\|\frac{\partial A^{-1}(\varphi)}{\partial \varphi_i}\right\|^2\right)^{1/2}\right.$$

$$\left. + \|A^{-1}\|\left(\sum_{i=1}^{m}\left\|\frac{\partial \mu(\varphi)}{\partial \varphi_i}\right\|^2\right)^{1/2}\right],$$

$$\left\|\frac{\partial}{\partial \varphi}[a(\varphi)\rho(\varphi)]\right\| \le \|B(\varphi)A^{-1}(\varphi)\|\left\|\frac{\partial}{\partial \varphi}a(\varphi)\right\|$$

$$+ \sqrt{m}\,\|a\|\max_{\substack{1\le i\le m \\ \varphi\in R^m}}\left[2\|B(\varphi)A^{-1}(\varphi)\|\left\|\frac{\partial}{\partial \varphi_i}\mu(\varphi)\right\|\right.$$

$$\left. + \left\|\left(\frac{\partial}{\partial \varphi_i}B(\varphi)\right)A^{-1}(\varphi)\right\| + \left\|B(\varphi)\frac{\partial}{\partial \varphi_i}A^{-1}(\varphi)\right\|\right].$$

Hence it is clear that the choice of a sufficiently small $\delta > 0$ in the inequality (2.14.11) ensures the that estimate (2.14.17) is satisfied and this means that the system of equations (2.14.1) has a unique $r - 1$ times continuously differentiable invariant torus $x = u(\varphi)$. Moreover, due to the uniqueness $u(\varphi)$ is a 2π-periodic function in φ_j, $j = 1, \ldots, m$.

Thus, the Theorem 2.14.2 is proved.

Remark 2.14.1 Applying estimate

$$\left\|\frac{\partial}{\partial \varphi}\mu^*(\varphi)\right\| \le 2\sqrt{n+k}\left\|\frac{\partial}{\partial \varphi}[B(\varphi),\hat{B}(\varphi)]\right\|\|[B(\varphi),\hat{B}(\varphi)]\|^{-1},$$

228   DICHOTOMIES AND STABILITY

where the matrix norms are Euclidean, one can show a precise value of the constant δ in (2.14.11).

We cite an example illustrating Theorem 2.14.2. Consider the system of differential equations with quasiperiodic coefficients and the degenerate matrix at the derivatives

$$p \begin{pmatrix} \sin t \sin(\sqrt{2}t) & 1+\cos t \\ (1-\cos t)\sin(\sqrt{2}t) & \sin t \end{pmatrix} \begin{pmatrix} \dot{x}_1 \\ \dot{x}_2 \end{pmatrix} = \begin{pmatrix} 0 & 1 \\ 1 & 0 \end{pmatrix} \begin{pmatrix} x_1 \\ x_2 \end{pmatrix} \\ + \begin{pmatrix} f_1(t,\sqrt{2}t) \\ f_2(t,\sqrt{2}t) \end{pmatrix}, \qquad (2.14.18)$$

where p is a fixed parameter from the interval $(0,1]$, $f_i(t,\sqrt{2}t)$, $i=1,2$, are some fixed quasiperiodic functions $f_i(\varphi_1,\varphi_2) \in C^r(R^2, 2\pi)$. The problem is to find the values of the parameter $p \in (0,1]$ for which the system of equations (2.14.18) has a unique quasiperiodic solution for every inhomogeneity $f_i(t,\sqrt{2}t)$, $i=1,2$.

Consider the extended system of differential equations

$$\dot{\varphi}_1 = 1, \quad \dot{\varphi}_2 = \sqrt{2}, \qquad (2.14.19)$$

$$p \begin{pmatrix} \sin \varphi_1 \sin \varphi_2 & 1+\cos \varphi_1 \\ (1-\cos \varphi_1)\sin \varphi_1 & \sin \varphi_1 \end{pmatrix} \begin{pmatrix} \dot{x}_1 \\ \dot{x}_2 \end{pmatrix} = \begin{pmatrix} 0 & 1 \\ 1 & 0 \end{pmatrix} \begin{pmatrix} x_1 \\ x_2 \end{pmatrix} \\ + \begin{pmatrix} f_1(\varphi_1,\varphi_2) \\ f_2(\varphi_1,\varphi_2) \end{pmatrix}.$$

In this case rank $B(\varphi) \in [0,1]$. However the matrix $B(\varphi)$ can be completed up to the matrix $[B(\varphi), \hat{B}(\varphi)]$ of the first rank. Therefore, it is sufficient to take the vector-column $\begin{pmatrix} \sin \varphi_1 \\ 1 - \cos \varphi_1 \end{pmatrix}$ for $\hat{B}(\varphi_1)$. Note that the unit vector-function

$$\mu(\varphi_1) = \left(\cos \frac{\varphi_1}{2}, \sin \frac{\varphi_1}{2} \right)$$

is proportional to every column of the matrix $[B(\varphi_1,\varphi_2), \hat{B}(\varphi_1)]$. The multiplication of the second system of equation (2.14.19) by the matrix

$$\begin{bmatrix} \mu(\varphi) \\ I_2 - \mu^*(\varphi)\mu(\varphi) \end{bmatrix} = \frac{1}{2} \begin{bmatrix} 2\cos\frac{\varphi_1}{2} & 2\sin\frac{\varphi_1}{2} \\ 1-\cos \varphi_1 & -\sin \varphi_1 \\ -\sin \varphi_1 & 1+\cos \varphi_1 \end{bmatrix}$$

yields
$$\dot\varphi_1 = 1, \quad \dot\varphi_2 = \sqrt{2},$$

$$2p\left(\sin\varphi_2 \sin\frac{\varphi_1}{2}, \cos\frac{\varphi_1}{2}\right)\begin{pmatrix}\dot x_1\\\dot x_2\end{pmatrix} = \left(\cos\frac{\varphi_1}{2}, \sin\frac{\varphi_1}{2}\right)$$

$$\times\left[\begin{pmatrix}x_2\\x_1\end{pmatrix} + \begin{pmatrix}f_1(\varphi_1,\varphi_2)\\f_2(\varphi_1,\varphi_2)\end{pmatrix}\right],$$

$$0 = \begin{bmatrix}1-\cos\varphi_1 & -\sin\varphi_1\\-\sin\varphi_1 & 1+\cos\varphi_1\end{bmatrix}\left[\begin{pmatrix}x_2\\x_1\end{pmatrix} + \begin{pmatrix}f_1(\varphi_1,\varphi_2)\\f_2(\varphi_1,\varphi_2)\end{pmatrix}\right].$$

The general solution of the algebraic system of equations is

$$x_1 = \left(\sin\frac{\varphi_1}{2}\right)y - f_2(\varphi_1,\varphi_2), \quad x_2 = \left(\cos\frac{\varphi_1}{2}\right)y - f_1(\varphi_1,\varphi_2).$$

Substituting it into the differential equation, we find

$$\rho(\varphi)\dot y = b(\varphi)y + c(\varphi),$$

where

$$\rho(\varphi) = p(1 + \sin\varphi_2 + \cos\varphi_1 - \cos\varphi_1\sin\varphi_2),$$

$$b(\varphi) = 1 - \frac{p}{2}(\sin\varphi_2\sin\varphi_1 - \sin\varphi_1),$$

$$c(\varphi) = 2p\left[\sin\varphi_2\sin\frac{\varphi_1}{2}\left(\frac{\partial f_2}{\partial\varphi_1} + \sqrt{2}\frac{\partial f_2}{\partial\varphi_2}\right) + \cos\frac{\varphi_1}{2}\left(\frac{\partial f_1}{\partial\varphi_1} + \sqrt{2}\frac{\partial f_1}{\partial\varphi_2}\right)\right].$$

The values $|D(\varphi)|$ and $\left\|\frac{\partial}{\partial\varphi}[a(\varphi)\rho(\varphi)]\right\|$ are estimated so as to find the values of p for which inequality (2.14.17) holds

$$|D(\varphi)| \le p, \quad \left\|\frac{\partial}{\partial\varphi}[a(\varphi)\rho(\varphi)]\right\| \le p\sqrt{24}.$$

It is clear that inequality (2.14.17) is satisfied whenever $p \in (0,(5r-4)^{-1})$. Therefore, the system of equations (2.14.18) has a unique quasiperiodic solution $x_i(t,\sqrt{2}t)$, $x_i(\varphi_1,\varphi_2) \in C^{r-1}(R^2,2\pi)$ for every inhomogeneity $f_i(t,\sqrt{2}t)$, $f_i(\varphi_1,\varphi_2) \in C^r(R^2,2\pi)$ as soon as $p \in (0,(5r-4)^{-1})$. This solution can be represented as

$$\begin{pmatrix}x_1\\x_2\end{pmatrix} = -\begin{pmatrix}f_2(t,\sqrt{2}t)\\f_1(t,\sqrt{2}t)\end{pmatrix}$$

$$-\begin{pmatrix}\sin\frac{t}{2}\\\cos\frac{t}{2}\end{pmatrix}\int_0^\infty c(\varphi_\tau(t,\sqrt{2}t))\exp\left\{\int_\tau^0 b(\varphi_\sigma(t,\sqrt{2}t))\,d\sigma\right\}d\tau,$$

where $\varphi_t(\varphi_1, \varphi_2) = (\varphi_{t1}(\varphi_1, \varphi_2), \varphi_{t2}(\varphi_1, \varphi_2))$ is a solution of the system of the equations

$$\dot{\varphi}_1 = p(1 + \sin \varphi_2 + \cos \varphi_1 - \cos \varphi_1 \sin \varphi_2),$$
$$\dot{\varphi}_2 = p\sqrt{2}(1 + \sin \varphi_2 + \cos \varphi_1 - \cos \varphi_1 \sin \varphi_2).$$

The second condition is of essential importance for the proof of Theorem 2.14.2. We note in this regard that there exist matrix functions $B(\varphi) \in C^r(R^m, 2\pi)$ which cannot be complemented by the matrix function $\hat{B}(\varphi) \in C^r(R^m, 2\pi)$ so that $rank\,[B(\varphi), \hat{B}(\varphi)] \equiv 1$. For example,

$$B(\varphi) = \begin{pmatrix} \sin \varphi_1 & 0 \\ \sin \varphi_2 & 0 \end{pmatrix}.$$

For the system of equations

$$\dot{\varphi} = a(\varphi), \qquad (2.14.20)$$

$$p \begin{pmatrix} b_{11}(\varphi) & b_{12}(\varphi) \\ b_{21}(\varphi) & b_{22}(\varphi) \end{pmatrix} \begin{pmatrix} \dot{x}_1 \\ \dot{x}_2 \end{pmatrix} = \begin{pmatrix} a_{11}(\varphi) & a_{12}(\varphi) \\ a_{21}(\varphi) & a_{22}(\varphi) \end{pmatrix} \begin{pmatrix} x_1 \\ x_2 \end{pmatrix} + \begin{pmatrix} f_1(\varphi) \\ f_2(\varphi) \end{pmatrix},$$

where $a, b_{ij}, a_{ij}, f_i \in C^r(R^m, 2\pi)$, $i, j = 1, 2$, $p \in [0, p_0]$ is a parameter, the following result can be stated.

Theorem 2.14.3 *Let it be known for the system of equations (2.14.20) that*

(1) $\quad \det \begin{pmatrix} a_{11}(\varphi) & a_{12}(\varphi) \\ a_{21}(\varphi) & a_{22}(\varphi) \end{pmatrix} \neq 0, \quad \forall \varphi \in R^m,$

(2.14.21)

(2) $\quad rank \begin{pmatrix} b_{11}(\varphi) & a_{11}(\varphi) \\ b_{21}(\varphi) & a_{21}(\varphi) \end{pmatrix} \equiv 1, \quad rank \begin{pmatrix} b_{12}(\varphi) & a_{12}(\varphi) \\ b_{22}(\varphi) & a_{22}(\varphi) \end{pmatrix} \equiv 1.$

Then a value of $p_1 \in (0, p_0]$ can be found such that for all $p \in (0, p_1]$ the system of equations (2.14.20) has a unique smooth invariant torus $x_i = u_i(\varphi) \in C^{r-1}(R^m, 2\pi)$, $i = 1, 2$, for every inhomogeneity $f_i(\varphi) \in C^r(R^m, 2\pi)$, $i = 1, 2$.

Proof Let, for example, the second equality (2.14.21) be satisfied. We designate by $(\mu_1(\varphi), \mu_2(\varphi))$ the solution of the algebraic system of equations

$$b_{12}(\varphi)\mu_1 + b_{22}(\varphi)\mu_2 = 0,$$
$$a_{12}(\varphi)\mu_1 + a_{22}(\varphi)\mu_2 = 0,$$

where $\mu_i \in C^r(R^m, 4\pi)$. Multiplying the second system from the left (2.14.20) by the non-degenerate matrix $\begin{pmatrix} \mu_1(\varphi) & \mu_2(\varphi) \\ -\mu_2(\varphi) & -\mu_1(\varphi) \end{pmatrix}$ we get

$$\dot\varphi = a(\varphi),$$

$$p \begin{pmatrix} b'_{11}(\varphi) & 0 \\ b'_{21}(\varphi) & b'_{22}(\varphi) \end{pmatrix} \begin{pmatrix} \dot x_1 \\ \dot x_2 \end{pmatrix} = \begin{pmatrix} a'_{11}(\varphi) & 0 \\ a'_{21}(\varphi) & a'_{22}(\varphi) \end{pmatrix} \begin{pmatrix} x_1 \\ x_2 \end{pmatrix} \quad (2.14.22)$$
$$+ \begin{pmatrix} f'_1(\varphi) \\ f'_2(\varphi) \end{pmatrix}.$$

The first condition of Theorem 2.14.3 and Lemma 2.14.2 enable a statement to be made on the existence of a unique smooth invariant torus $x_1 = u_1(\varphi) \in C^r(R^m, 4\pi)$ of the system of equations

$$\dot\varphi = a(\varphi), \quad pb'_{11}(\varphi)\dot x_1 = a'_{11}(\varphi)x_1 + f_1(\varphi),$$

for sufficiently small $p > 0$. Substituting $x_1 = u_1(\varphi)$ into the second equation of system (2.14.22) we arrive at the system

$$\dot\varphi = a(\varphi),$$
$$pb'_{22}(\varphi)\dot x_2 = a'_{22}(\varphi)x_2 + (f'_2(\varphi) - pb'_{21}(\varphi)\dot u_1(\varphi) + a'_{21}(\varphi)u_1(\varphi)),$$

for which, in view of Lemma 2.14.2, $x_2 = u_2(\varphi) \in C^{r-1}(R^m, 4\pi)$. Since the torus $x_i = u_i(\varphi)$, $i = 1, 2$, of the system (2.14.20) is unique and the coefficients are 2π-periodic, then $u_i(\varphi) \in C^{r-1}(R^m, 2\pi)$.

2.15 Bounded Invariant Manifolds of Dynamical Systems and their Smoothness

The analysis of the system (2.1.1) shows that some results presented above are also valid if the vector function $a(\varphi)$ and the matrix function $A(\varphi)$ are not periodic in φ. Moreover, the vector function $a(\varphi)$ can be assumed to not even be bounded on R^m. Therefore the results of this section which immediately apply to these systems are presented without proof while essentially different ones are proved.

Similarly to the system (2.1.1) we consider the system of differential equations of the type

$$\frac{dg}{dt} = \omega(g), \quad \frac{dh}{dt} = H(g)h, \quad (2.15.1)$$

where $g \in R^m$, $h \in R^n$, the vector-function $\omega(g)$ and the matrix function $H(g)$ are definite for all $g \in R^m$ and continuous with respect to the totality of variables

$g_1, \ldots, g_m$. Additionally, it is always assumed in the vector function $\omega(g)$ that the Cauchy problem

$$\frac{dg}{dt} = \omega(g), \quad g|_{t=0} = g_0$$

has a unique solution $g_t(g_0)$ determined for all $t \in R$ and continuously dependent on g_0 for every fixed $g_0 \in R^m$. We designate a space of the functions $F(g)$ which is continuous with respect to the totality of the variables $g_1, \ldots, g_m$ and bounded on R^m by $C^0(R^m)$, where $C'(R^m; \omega)$ is a subspace of $C^0(R^m)$ of functions $F(g)$ such that the function $F(g_t(g_0))$ is continuously differentiable in t for all $t \in R$, $g_0 \in R^m$. Besides,

$$\left.\frac{d}{dt} F(g_t(g))\right|_{t=0} \stackrel{\text{def}}{=} \dot{F}(g) \in C^0(R^m).$$

Definition 2.15.1 We claim that the system (2.15.1) has the *Green function of the problem on bounded invariant manifolds*, provided that there exists an $n \times n$-dimensional matrix $C(g) \in C^0(R^m)$ satisfying the estimates

$$\begin{aligned}\|\Omega_0^t(g) C(g)\| &\le K \exp\{-\gamma t\}, \quad t > 0, \\ \|\Omega_0^t(g)[C(g) - I_n]\| &\le K \exp\{\gamma t\}, \quad t < 0,\end{aligned} \tag{2.15.2}$$

where the positive constants K and γ do not depend on $g \in R^m$, $t \in R$, and $\Omega_0^t(g)$ is the matriciant of the linear system

$$\frac{dh}{dt} = H(g_t(g))h, \quad \Omega_0^t(g)|_{t=0} = I_n.$$

In addition, a function of the type

$$G_0(\tau, \varphi) = \begin{cases} \Omega_\tau^0(g) C(g_\tau(g)), & \tau \le 0, \\ \Omega_\tau^0(g)[C(g_\tau(g)) - I_n], & \tau > 0, \end{cases} \tag{2.15.3}$$

is called the *Green function of the problem on bounded invariant manifolds* for the system (2.15.1).

Evidently, estimates (2.15.2) are equivalent to the estimate

$$\|G_0(\tau, g)\| \le K \exp\{-\gamma |\tau|\}, \tag{2.15.4}$$

with one and the same positive constants K and γ.

Definition 2.15.2 The system of equations

$$\frac{dg}{dt} = \omega(g),$$

$$\frac{dh}{dt} = H(g)h + f(g), \quad f(g) \in C^0(R^m), \tag{2.15.5}$$

has a *bounded invariant manifold*, providing that there exists a function $h = u(g) \in C'(R^m; \omega)$ such that for all $g \in R^m$ the identity

$$\frac{du}{dt} \equiv H(g)u(g) + f(g)$$

is satisfied. Besides, the equality $h = u(g)$ is said to specify a bounded invariant manifold of the system (2.15.5).

Further, the matrix function $H(g)$ is assumed to belong to the space $C^0(R^m)$. We formulate the basic results.

Theorem 2.15.1 *Let there exist an $n \times n$-dimensional symmetric matrix function $S(g) \in C'(R^m; \omega)$ satisfying the conditions*

$$\det S(g) \neq 0, \quad \forall g \in R; \tag{2.15.6}$$

$$\langle [\dot{S}(g) + S(g)H(g) + H^*(g)S(g)]h, h\rangle \leq -\beta \|h\|^2, \tag{2.15.7}$$

where $\beta = \text{const} > 0$, $h \in R^n$. Then the system of the equations (2.15.1) has a unique Green function (2.15.3) with the estimate (2.15.4). Also, the matrix $C(g) \in C'(R^m; \omega)$ possesses the properties

$$C^2(g) \equiv C(g), \quad C(g_\tau(g)) \equiv \Omega_0^\tau(g)C(g)\Omega_\tau^0(g), \tag{2.15.8}$$

and the constants K and γ in (2.15.4) can be represented as

$$K = (2 + \sqrt{2})\left(\frac{\|H\|_0 \|S\|_0}{\beta}\right)^{3/2}, \quad \gamma = \frac{\beta}{2\|S\|_0}.$$

Theorem 2.15.2 *Let there exist the Green function (2.15.3) with the estimate (2.15.4) and let one of the identities (2.15.8) be satisfied. Then*

(1) *the other identity (2.15.8) is satisfied;*
(2) *the Green function (2.15.3) is unique;*
(3) *there exist the symmetric matrices $S(g) \in C'(R^m; \omega)$ satisfying the conditions (2.15.6) and (2.15.7) and one of the matrices is*

$$S(g) = \int_0^\infty [\Omega_0^\sigma(g)C(g)]^* \Omega_0^\sigma(g)C(g)\, d\sigma$$

$$- \int_{-\infty}^0 [\Omega_0^\sigma(g)(C(g) - I_n)]^* \Omega_0^\sigma(g)(C(g) - I_n)\, d\sigma; \tag{2.15.9}$$

(4) each matrix $S(g) \equiv S^*(g) \in C'(R^m;\omega)$ which satisfies inequality (2.15.7) will also satisfy the condition (2.15.6);

(5) the conjugated system of equations

$$\frac{dg}{dt} = \omega(g), \quad \frac{dh}{dt} = -H^*(g)h, \qquad (2.15.10)$$

also has a unique Green function.

Theorem 2.15.3 *Let there exist an n-dimensional symmetric matrix function $S(g) \in C'(R^m;\omega)$ satisfying the condition*

$$\langle [\dot{S}(g) - S(g)H^*(g) - H(g)S(g)]h, h\rangle \leq -\beta\|h\|^2, \qquad (2.15.11)$$

where $\beta = \text{const} > 0$ and for some $g = g_0 \in R^m$,

$$\det S(g_0) = 0. \qquad (2.15.12)$$

Then the following assertion is valid.

1. The system of equations (2.15.1) has many Green functions (2.15.3) with the estimate (2.15.4), and the system (2.15.10) has no Green function.

2. There exists an n-dimensional symmetric matrix function $W(g) \not\equiv 0$, $W(g) \in C'(R^m;\omega)$ possessing the following properties

(a)

$$h = \int_{-\infty}^{\infty} W(g)(\Omega_0^\tau(g))^* \varphi(g_\tau(g))\, d\tau + \int_{-\infty}^{\infty} G_0(\tau, g) f(g_\tau(g))\, d\tau, \qquad (2.15.13)$$

where $\varphi(g)$ is an arbitrary vector function from the space $C^0(R^m)$ specifying all bounded invariant manifolds of the system (2.15.5) for any fixed vector function $f(g) \in C^0(R^m)$;

(b) *for all $t, \tau \in R$ and $g \in R^m$*

$$\|\Omega_0^t(g)W(g)(\Omega_0^\tau(g))^*\| \leq K \exp\{-\gamma|t-\tau|\}; \qquad (2.15.14)$$

(c)

$$W(g_t(g)) \equiv \Omega_0^t(g) W(g)(\Omega_0^t(g))^*; \qquad (2.15.15)$$

where

$$W(g) \equiv \int_{-\infty}^{\infty} W(g)(\Omega_0^\tau(g))^* \Omega_0^\tau(g) W(g)\, d\tau. \qquad (2.15.16)$$

3. For the system of equations

$$\frac{dg}{dt} = \omega(g),$$

$$\frac{dh}{dt} = -H^*(g)h + f(g), \quad f(g) \in C^0(R^m),$$

to have a bounded invariant manifold, it is necessary and sufficient that the vector function $f(g)$ satisfies the identity

$$\int_{-\infty}^{\infty} \langle f(g_\tau(g)), u(g_\tau(g)) \rangle \equiv 0, \qquad (2.15.17)$$

for every function $u = u(g)$ specifying a bounded invariant manifold of the system (2.15.1). Besides, (2.15.17) is equivalent to

$$\int_{-\infty}^{\infty} \left(W(g)(\Omega_0^\tau(g))^*, f(g_\tau(g)) \right) d\tau \equiv 0. \qquad (2.15.18)$$

4. For every $n \times n$ sign-definite matrix $B(g) \in C^0(R^m)$ ($|\langle B(g)h, h\rangle| \geq \beta_0 \|h\|^2$, $\beta_0 = \text{const} > 0$) there exist unique $n \times n$-dimensional matrices $C(g)$, $W(g) \in C'(R^m; \omega)$ satisfying the identity

$$\Omega_t^0(g) C(g_t(g)) \Omega_0^t(g) \equiv C(g) + \int_t^0 W^*(g)(\Omega_0^\sigma(g))^* B(g_\sigma(g)) \Omega_0^\sigma(g) \, d\sigma, \qquad (2.15.19)$$

and the estimates (2.15.2) and (2.15.14). Moreover, if $B(g) \equiv B^*(g)$ then also $W(g) \equiv W^*(g)$, $\forall g \in R^m$.

5. The extended system

$$\frac{dg}{dt} = \omega(g), \quad \frac{dh}{dt} = H(g)h,$$

$$\frac{dy}{dt} = -h - H^*(g)y, \quad y \in R^n,$$

has a unique Green function.

It should also be noted that the theorems similar to Theorems 2.7.4, 2.7.7, 2.8.2 and 2.8.3 are also valid for this case.

For example, consider the equation

$$\left\langle \omega, \frac{\partial h}{\partial g} \right\rangle = (\tanh \langle \lambda, g \rangle)h + \varphi(g), \qquad (2.15.20)$$

where $\omega = (\omega_1, \ldots, \omega_m)$, $\omega_i \neq 0$, $\lambda = (\lambda_1, \ldots, \lambda_m)$,

$$\left\langle \omega, \frac{\partial h}{\partial g} \right\rangle = \sum_{i=1}^{m} \omega_i \frac{\partial h}{\partial g}, \qquad \langle \lambda, g \rangle = \sum_{i=1}^{m} \lambda_i g_i,$$

$\varphi \in C^1(R^m)$ and $C^1(R^m)$ is a subspace of $C^0(R^m)$ of the function $F(g)$ which has continuous derivatives of the first order in every variable g_i, $i = 1, \ldots, m$.

Let us show that the inequality

$$\langle \lambda, \omega \rangle < 0. \qquad (2.15.21)$$

is a necessary and sufficient condition for the existence of solutions $h = h(g) \in C^1(R^m)$ of the equation (2.15.20) for every function $\varphi(g) \in C^1(R^m)$. Let the inequality (2.15.21) be violated $\langle \lambda_0, \omega_0 \rangle \geq 0$ for some numerical vectors $\lambda_0, \omega \in R^m$. Then the equation

$$\left\langle \omega_0, \frac{\partial h}{\partial g} \right\rangle = (\tanh \langle \lambda_0, g \rangle)h + 1, \qquad (2.15.22)$$

has no solutions bounded on R^m. Actually, if there is such a solution $h = h(g) \in C^1(R^m)$, then $h = h(\omega_0 t)$ is a solution bounded on R of the equation

$$\frac{dh}{dt} = (\tanh (\langle \lambda, \omega_0 \rangle t))h + 1.$$

However this equation has no solutions bounded on R. Now let condition (2.15.21) be satisfied. Then we take a scalar function $S(g) = \tanh \langle \lambda, g \rangle$ for $S(g)$ and check condition (2.15.11)

$$\dot{S}(g) - 2S(g) \tanh \langle \lambda, g \rangle = (\langle \lambda, g \rangle - 2 \sinh^2 \langle \lambda, g \rangle) \cosh^{-2} \langle \lambda, g \rangle \leq -\delta < 0.$$

So, condition (2.15.21) is necessary and sufficient for equation (2.15.20) to possess a solution $h = h(g) \in C^1(R^m)$ for every function $\varphi(g) \in C^1(R^m)$.

We treat the problem of the representation of all solutions of the equation (2.15.20) bounded on R^m under the condition (2.15.21). Consider the system

$$\frac{dg}{dt} = \omega, \qquad \frac{dh}{dt} = (\tanh \langle \lambda, g \rangle)h.$$

Substituting the solution of the first system $g_t(g) = \omega t + g$ into the last equation yields

$$\frac{dh}{dt} = (\tanh(-\mu t + \nu))h, \qquad (2.15.23)$$

where $\mu = -\langle \lambda, \omega \rangle$, $\nu = \langle \lambda, g \rangle$. For (2.15.21) the matriciant is of the form

$$\Omega_0^t(\nu) = (\cosh \nu)^{1/\mu}(\cosh(\mu t - \nu))^{-1/\mu}. \qquad (2.15.24)$$

The equality (see (2.15.16))

$$W = W \int_{-\infty}^{\infty} (\cosh \nu)^{2/\mu} [\cosh(\mu t - \nu)]^{-2\mu} \, dt \, W$$

unambiguously yields

$$W(\nu) = (\cosh \nu)^{-2\mu} \mu \left(\int_{-\infty}^{\infty} (\cosh \sigma)^{-2/\mu} \, d\sigma \right)^{-1}$$

The general solution bounded on R^m of the equation

$$\left\langle \omega, \frac{\partial h}{\partial g} \right\rangle = (\tanh \langle \lambda, g \rangle)h,$$

is described by

$$h = (\cosh \langle \lambda, g \rangle)^{\langle \lambda, \omega \rangle^{-1}} \int_{-\infty}^{\infty} [\cosh(\langle \lambda, \omega \rangle \tau + \langle \lambda, g \rangle)]^{\langle \lambda, \omega \rangle^{-1}} f(\omega \tau + g) \, d\tau, \quad (2.15.25)$$

where $f(g)$ is an arbitrary function from the space $C^1(R^m)$. Obviously the second term in (2.15.25) is a first integral of the system

$$\frac{dg}{dt} = \omega,$$

so it equals some function of $m - 1$ first integrals

$$F(\bar{g}_1, \ldots, \bar{g}_{m-1})\bar{g}_i = \omega_i g_{i+1} - \omega_{i+1} g_i, \quad i = 1, \ldots, (m-1).$$

Thus it is possible to replace (2.15.25) by

$$h = (\cosh \langle \lambda, g \rangle)^{\langle \lambda, \omega \rangle^{-1}} F\left(\begin{vmatrix} \omega_1 & \omega_2 \\ g_1 & g_2 \end{vmatrix}, \ldots, \begin{vmatrix} \omega_{m-1} & \omega_m \\ g_{m-1} & g_m \end{vmatrix} \right), \qquad (2.15.26)$$

where $F(\bar{g}_1, \ldots, \bar{g}_{m-1})$ is an arbitrary function from the space $C^1(R^{m-1})$. To write down a solution to equation (2.15.20) which is partially bounded on (R^m) it is sufficient to find a function $C(\nu) \in C^1(R)$ such that the estimate $\|G_t(0,\nu)\| \leq K \exp\{-|t|\}$ holds true, where

$$G_t(0,\nu) = \begin{cases} (\cosh \nu)^{1/\mu} [\cosh(\mu t - \nu)]^{-1/\mu} c(\nu), & t \geq 0, \\ (\cosh \nu)^{1/\mu} [\cosh(\mu t - \nu)]^{-1/\mu} [c(\nu) - 1], & t < 0. \end{cases} \quad (2.15.27)$$

The existence of this function is equivalent to the existence of the function $c(\nu) \in C^1(R)$ satisfying the inequalities

$$|c(\nu)| \leq K (\cosh \nu)^{-1/\mu} \exp\{-\nu/\mu\},$$
$$|c(\nu) - 1| \leq K (\cosh \nu)^{-1/\mu} \exp\{\nu/\mu\}, \quad (2.15.28)$$

where the constant K does not depend on ν.

Consider two cases: 1) $m \geq 1$ and 2) $0 < \mu < 1$. In the first case it is possible to take the function $c(\nu) = (2\cosh \nu)^{-1} \exp\{-\nu\}$ for the function $c(\nu)$. In the second case we first consider the domain determined by the inequalities $-y_1(\nu) \leq y \leq y_1(\nu)$, $2 - y_2(\nu) \leq y \leq y_2(\nu)$ in the plane of the variables ν and y, where

$$y_1(\nu) = K(\cosh \nu)^{-1/\mu} \exp\{-\nu/\mu\},$$
$$y_2(\nu) = 1 + K(\cosh \nu)^{-1/\mu} \exp\{\nu/\mu\}. \quad (2.15.29)$$

For the function $y = c(\nu)$ satisfying (2.15.28) to exist, it is necessary and sufficient that the graphs of the functions $y = y_1(\nu)$ and $y = 2 - y_2(\nu)$ do not intersect, i.e. $y_1(\nu) \geq 2 - y_2(\nu)$. We show that for all $\nu \in R$ and $\mu \in (0,1)$

$$[(\cosh \nu)^{-1} \exp\{-\nu\}]^{1/\mu} + [(\cosh \nu)^{-1} \exp\{\nu\}]^{1/\mu} \geq 2. \quad (2.15.30)$$

Designate $x = (\cosh \nu)^{-1} \exp\{-\nu\}$ and find the least value of the function $f(x) = x^{1/\mu} + (2-x)^{1/\mu}$ on the interval $(0,2)$; $\min_{x \in (0,2)} f(x) = f(1) = 2$. In view of (2.15.29) we get $y_1(\nu) + y_2(\nu) - 2 \geq 2K - 1$ from (2.15.30). It is then clear that the choice of the constant $K \geq 1/2$ ensures that the graphs of the functions $y = y_1(\nu)$ and $y = 2 - y_2(\nu)$ are non-intersecting, and hence, the possibility of choosing the function $c(\nu)$ which satisfies the estimate (2.15.28).

In order to find the function $c(\nu)$ the matriciant of the extended system

$$\frac{dh}{dt} = [\tanh(-\mu t + \nu)]h,$$
$$\frac{dy}{dt} = -h - [\tanh(-\mu t + \nu)]y, \quad (2.15.31)$$

is represented as

$$\tilde{\Omega}_0^t(\nu) = \begin{bmatrix} \dfrac{(\cosh \nu)^{1/\mu}}{[\cosh (\mu t - \nu)]^{-1/\mu}} & 0 \\ \omega(t,\nu,\mu) & \dfrac{[\cosh (\mu t - \nu)]^{1/\mu}}{(\cosh \nu)^{1/\mu}} \end{bmatrix}, \qquad (2.15.32)$$

where

$$\omega(t,\nu,\mu) = -[\cosh (\mu t - \nu)]^{1/\mu} (\cosh \nu)^{1/\mu} \int_0^t [\cosh (\mu t - \nu)]^{-2/\mu} \, d\tau. \qquad (2.15.33)$$

Since it is known in advance that system (2.15.31) is exponentially e-dichotomous on the whole R axis, there exists a two-dimensional numerical vector $\tilde{c} = \begin{pmatrix} c \\ c_2 \end{pmatrix}$ such that

$$\tilde{\Omega}_0^t(\nu)\tilde{c} \underset{t \to +\infty}{\longrightarrow} 0, \quad \tilde{\Omega}_0^t(\nu)\begin{pmatrix} c-1 \\ c_2 \end{pmatrix} \underset{t \to +\infty}{\longrightarrow} 0. \qquad (2.15.34)$$

In view of the structure of the matriciant (2.15.32), (2.15.33) and condition (2.15.34) we conclude

$$[\cosh (\mu t - \nu)]^{1/\mu}$$
$$\times \left[-(\cosh \nu)^{1/\mu} \int_0^t [\cosh (\mu\tau - \nu)]^{-2/\mu} d\tau \, c + (\cosh \nu)^{-1/\mu} c_2 \right] \underset{t \to +\infty}{\longrightarrow} 0,$$

$$[\cosh (\mu t - \nu)]^{1/\mu} \qquad (2.15.35)$$

$$\times \left[-(\cosh \nu)^{1/\mu} \int_0^t [\cosh (\mu\tau - \nu)]^{-2/\mu} d\tau \, (c-1) + (\cosh \nu)^{-1/\mu} c_2 \right] \underset{t \to -\infty}{\longrightarrow} 0.$$

Since $\cosh (\mu t - \nu) \underset{|t| \to +\infty}{\longrightarrow} \infty$ then (2.15.35) implies

$$-(\cosh \nu)^{1/\mu} \int_0^\infty [\cosh (\mu\tau - \nu)]^{-2/\mu} d\tau \, c + (\cosh \nu)^{-1/\mu} c_2 = 0,$$

$$(\cosh \nu)^{1/\mu} \int_{-\infty}^0 [\cosh (\mu\tau - \nu)]^{-2/\mu} d\tau \, (c-1) + (\cosh \nu)^{-1/\mu} c_2 = 0.$$

Hence we determine $c(\nu)$

$$c(\nu) = \int_{\nu}^{\infty} (\cosh \sigma)^{-2/\mu} \, d\sigma \left(\int_{-\infty}^{\infty} (\cosh \sigma)^{-2/\mu} \, d\sigma \right)^{-1}. \qquad (2.15.36)$$

Thus, in view of (2.15.26), (2.15.36), all solutions of (2.15.20) bounded on R^m can be represented for any $\langle \lambda, \omega \rangle < 0$ as

$$h = (\cosh \langle \lambda, g \rangle)^{\langle \lambda, \omega \rangle^{-1}} F \left(\begin{vmatrix} \omega_1 & \omega_2 \\ g_1 & g_2 \end{vmatrix}, \ldots, \begin{vmatrix} \omega_{m-1} & \omega_m \\ g_{m-1} & g_m \end{vmatrix} \right)$$

$$+ [\cosh \langle \lambda, g \rangle]^{\langle \lambda, \omega \rangle^{-1}} \left\{ \int_{-\infty}^{\infty} [\cosh (\langle \lambda, \omega \rangle \tau + \langle \lambda, g \rangle)]^{-\langle \lambda, \omega \rangle^{-1}} \right.$$

$$\times \int_{\langle \lambda, \omega \rangle \tau + \langle \lambda, g \rangle}^{\infty} (\cosh \sigma)^{-2/\langle \lambda, \omega \rangle} \, d\sigma \, \varphi(\omega \tau + g) \, d\tau$$

$$- \int_0^{\infty} [\cosh (\langle \lambda, \omega \rangle \tau + \langle \lambda, g \rangle)]^{-\langle \lambda, \omega \rangle^{-1}} \int_{-\infty}^{\langle \lambda, \omega \rangle \tau + \langle \lambda, g \rangle} (\cosh \sigma)^{2/\langle \lambda, \omega \rangle} \, d\sigma \, \varphi(\omega \tau + g) \, d\tau \right\}$$

$$\times \left\{ \int_{-\infty}^{\infty} (\cosh \sigma)^{2/\langle \lambda, \omega \rangle} \, d\sigma \right\}^{-1}.$$

Essential differences are discovered between the systems (2.15.1) and (2.2.4) when the smoothness is analysed. In the case when, for example, the function $A(\varphi)$ is 2π-periodic in φ_j, $j = 1, \ldots, m$, and continuously differentiable, the derivative $\frac{\partial}{\partial \varphi_j} A(\varphi)$ is bounded by virtue of periodicity. Besides, the function $H(g)$ is not given on the compact manifold, therefore, the derivative $\frac{\partial}{\partial g_j} H(g)$ cannot be bounded on R^m any more.

In the system (2.15.1) we assume that the vector-function $\omega(g) \in C^1(R^{m-1})$ satisfies the estimates

$$\|\omega(g)\| \leq \alpha_1 \|g\| + \alpha_2, \quad \sup_{\varphi \in R^m} \left\| \frac{\partial}{\partial g_i} \omega(g) \right\| < \infty, \qquad (2.15.37)$$

where $\alpha_1, \alpha_2 = \text{const} > 0$. The function $g \sin \ln(1 + g^2)$ is the example of such a function. The matrix function $H(g)$ is definite on the whole space R^m, continuous with respect to the whole totality of variables and has continuous partial derivatives of the first order $\frac{\partial}{\partial g_i} H(g)$, $i = 1, \ldots, m$. It is also assumed that

$$\left\| \frac{\partial}{\partial g_i} H(g) \right\| \leq \alpha_3 \|g\|^{\nu} + \alpha_4, \quad \|H(g)\| \leq \alpha_5, \qquad (2.15.38)$$

where α_3, α_4, α_5, $\nu = \text{const} > 0$. The space of such functions $H(g)$, which are not necessarily matrix, is designated by $C_\nu^1(R^m)$. We prove the following result.

Theorem 2.15.4 *Let the system of equations (2.15.1) have a unique Green function (2.15.3) satisfying the estimate (2.15.4) and the estimates (2.15.37), (2.15.38) be satisfied. Then if the inequality*

$$\alpha_1 \nu + \alpha_0 < \gamma, \qquad (2.15.39)$$

is satisfied, where

$$\alpha_0 = \sup_{\varphi \in R^m} \left(\max_{\|\eta\|=1} \left| \left\langle \frac{\partial \omega(g)}{\partial g} \eta, \eta \right\rangle \right| \right), \qquad (2.15.40)$$

the system (2.15.5) has a unique smooth bounded invariant manifold for every fixed vector-function $f(g) \in C_\nu^1(R^m)$ determined by

$$h = \int_{-\infty}^{\infty} G_0(\tau, g) f(g_\tau(g)) \, d\tau = u_0(g) \in C_\nu^1(R^m). \qquad (2.15.41)$$

Proof In the conditions of Theorem 2.15.4 we state that the Green function $G_0(\tau, g)$ not only depends continuously on g, but also has continuous partial derivatives $\frac{\partial}{\partial g_i} G_0(\tau, g)$ in every variable g_i, $i = 1, \ldots, m$ ($\tau \neq 0$), and, moreover, the estimate

$$\left\| \frac{\partial}{\partial g_i} G_0(\tau, g) \right\| \leq (K_1 \|g\|^\nu + \bar{K}_1 |\tau|^\nu + K_2) \exp\{-(\gamma - \alpha_0 - \alpha_1 \nu)|\tau|\}, \qquad (2.15.42)$$

holds for all $\tau \in R$. For the difference of the Green functions $G_0(\tau, g) - G_0(\tau, \bar{g})$ it is easy to get the representation

$$G_0(\tau, g) - G_0(\tau, \bar{g}) = \int_{-\infty}^{\infty} G_0(\sigma, g)[H(g_\sigma(g)) - H(g_\sigma(\bar{g}))] G_\sigma(\tau, g) \, d\sigma. \qquad (2.15.43)$$

As in the estimate (2.15.4) the constants K and γ do not depend on g and τ. The boundedness of the matrix function $H(g)$ on the whole space R^m ensures uniform convergence with respect to parameters g, $\bar{g}$ of the integral (2.15.43). Therefore, the right-hand side of the inequality

$$\|G_0(\tau, g) - G_0(\tau, \bar{g})\| \leq K^2 \int_{-\infty}^{\infty} \exp\{-\gamma|\sigma|\} \|H(g_\sigma(g)) - H(g_\sigma(\bar{g}))\| \, d\sigma$$

is a continuous function $Z(g,\bar{g})$ with respect to the totality of variables g, $\bar{g}$ and for $g = \bar{g}$, the result $Z(g,\bar{g}) = 0$ is valid. This allows the conclusion on continuity of the Green function $G_0(\tau, g)$ with respect to variables g to be made.

Assuming the variable $\bar{g}$ is only different from g in the first coordinate $g_1 \neq \bar{g}_1$, $g_i = \bar{g}_i$, $i = 2, \ldots, m$ in (2.15.43), we divide (2.15.43) by the difference $g_1 - \bar{g}_1$

$$\frac{G_0(\tau, g) - G_0(\tau, \bar{g})}{g_1 - \bar{g}_1} = \int_{-\infty}^{\infty} G_0(\tau, g) \frac{H(g_\sigma(g)) - H(g_\sigma(\bar{g}))}{g_1 - \bar{g}_1} G_\sigma(\tau, \bar{g}) \, d\sigma, \quad (2.15.44)$$

and formally pass to the limit as $\bar{g}_1 \to g_1$. As a result we have

$$\frac{\partial G_0(\tau, g)}{\partial g_1} = \int_{-\infty}^{\infty} N_1(\sigma, g, \tau) \, d\sigma$$

$$= \int_{-\infty}^{\infty} G_0(\tau, g) \left[\sum_{j=1}^{m} \frac{\partial H(g_\sigma(g))}{\partial g_{\sigma j}} \frac{\partial g_{\sigma j}}{\partial g_1} \right] G_\sigma(\tau, g) \, d\sigma. \quad (2.15.45)$$

In view of (2.15.37) we derive an estimate for the solutions $g_t(g)$ of the system

$$\frac{dg}{dt} = \omega(g).$$

We represent $g_t(g)$ in the integral form

$$g_t(g) = g + \int_0^t \omega(g_\tau(g)) \, d\tau$$

and assume that $t \geq 0$, then

$$\|g_t(g)\| \leq \|g\| \exp\{\alpha_1 |t|\} + \alpha_2 \alpha_1^{-1}(\exp\{\alpha_1 |t|\} - 1). \quad (2.15.46)$$

The second term is estimated as follows

$$\alpha_2 \alpha_1^{-1}(\exp\{\alpha_1 |t|\} - 1) = \alpha_2 \exp\{\theta(t)|t|\} |t| \leq \alpha_2 \exp\{\alpha_1 |t|\}|t|.$$

In a similar manner we verify that the estimate (2.15.46) is also valid for $t < 0$. Therefore, for all $t \in R$ the solution $g_t(g)$ is estimated as

$$\|g_t(g)\| \leq (\|g\| + \alpha_2 |t|) \exp\{\alpha_1 |t|\}. \quad (2.15.47)$$

The estimates (2.15.38) and (2.15.47) allows us to obtain

$$\left\| \frac{\partial}{\partial g_{\sigma j}} H(g_\sigma(g)) \right\| \leq \alpha_3 \|g_\sigma(g)\|^\nu + \alpha_4$$

$$\leq \alpha_3 \exp\{\alpha_1 \nu |\sigma|\} (\|g\| + \alpha_2 |\sigma|)^\nu + \alpha_4.$$

$$(2.15.48)$$

Treating the function $f(x) = (x+d)^\nu(x^\nu + d^\nu)^{-1}$ as a function of x with a parameter we make sure that for all $d > 0$ the estimate

$$f(x) \le K(\nu) = \begin{cases} 2^{\nu-1}, & \nu > 1, \\ 1, & \nu \in [0,1], \end{cases} \qquad (2.15.49)$$

is satisfied. Then (2.15.48) can be represented as

$$\left\| \frac{\partial}{\partial g_{\sigma j}} H(g_\sigma(g)) \right\| \le \alpha_3 K(\nu) \exp\{\alpha_1 \nu |\sigma|\} (\|g\|^\nu + \alpha_2^\nu |\sigma|^\nu) + \alpha_4. \qquad (2.15.50)$$

It is easy to obtain the estimate for the derivative $\partial g_t(g)/\partial g_i$:

$$\left\| \frac{\partial}{\partial g_i} g_t(g) \right\| \le K_0 \exp\{\alpha_0 |t|\}, \qquad (2.15.51)$$

where the constant α_0 is specified by (2.15.40). In view of (2.15.50) and (2.15.51) we estimate (2.15.45)

$$\|N_1(\sigma, g, \tau)\| \le K^2 \exp\{-\gamma(|\sigma| + |\sigma - \tau|)\} \sum_{j=1}^m \left\| \frac{\partial}{\partial g_{\sigma j}} H(g_\sigma(g)) \right\|$$

$$\times \left\| \frac{\partial}{\partial g_1} g_{\sigma j}(g) \right\| \le K_3 \exp\{-\gamma(|\sigma| + |\sigma - \tau|) + (\alpha_0 + \alpha_1 \nu)|\sigma|\} \qquad (2.15.52)$$

$$\times (\|g\|^\nu + \alpha_2^\nu |\sigma|^\nu) + K_4 \exp\{-\gamma(|\sigma| + |\sigma - \tau|) + \alpha_0 |\sigma|\},$$

where $K_3 = K^3 K_0 K(\nu) \alpha_3 \sqrt{m}$, $K_4 = K^2 K_0 \alpha_4$. From here it is clear that the inequality (2.15.39) ensures uniform convergence of the integral in the right-hand side of (2.15.43).

We show for any positive numbers γ, μ and ν, providing $\gamma > \mu$, that

$$\mathcal{I} = \int_{-\infty}^{\infty} |\sigma|^\nu \exp\{-\gamma(|\sigma| + |\sigma - \tau|) + \mu|\sigma|\} d\sigma$$

$$\le [2\gamma(2\gamma - \mu)^{-1} \mu^{-1} |\tau|^\nu + \Gamma(\nu+1)(1 + K(\nu))(2\gamma - \mu)^{-\nu-1}] \qquad (2.15.53)$$

$$\times \exp\{-(\gamma - \mu)|\tau|\},$$

where $\Gamma(\nu+1)$ is a gamma function, and the constant $K(\nu)$ is determined by (2.15.49). The integral on the left is represented in the form of three integrals

$$\mathcal{I} = \int_{-\infty}^0 + \int_0^\tau + \int_\tau^\infty = \mathcal{I}_1 + \mathcal{I}_2 + \mathcal{I}_3.$$

Letting $\tau \geq 0$ we get

$$\mathcal{I}_1 = \exp\{-\gamma\tau\} \int_{-\infty}^{0} |\sigma|^{\nu} \exp\{(2\gamma - \mu)\sigma\} d\sigma = \exp\{-\gamma\tau\} \quad (2.15.54)$$

$$\times \int_0^{\infty} \delta^{\nu} \exp\{-(2\gamma - \mu)\sigma\} d\sigma = \Gamma(\nu + 1)(2\gamma - \mu)^{-\nu-1} \exp\{-\gamma\tau\}.$$

As a result of integration by parts the integral $\mathcal{I}_2$ becomes

$$\mathcal{I}_2 = \mu^{-1}\left[\tau^{\nu} \exp\{\mu\tau\} - \nu \int_0^{\tau} \sigma^{\nu-1} \exp\{\mu\sigma\} d\sigma\right] \exp\{-\gamma\tau\},$$

hence

$$\mathcal{I}_2 \leq \mu^{-1}\tau^{\nu} \exp\{-(\gamma - \mu)\tau\}. \quad (2.15.55)$$

We make the change of variable $\sigma - \tau = \sigma_1$ in the integral $\mathcal{I}_3$ and in view of (2.15.49) find

$$\mathcal{I}_3 = \exp\{\gamma\tau\} \int_0^{\infty} (\sigma_1 + \tau)^{\nu} \exp\{-(2\gamma - \mu)(\sigma_1 + \tau)\} d\sigma_1$$

$$\leq K(\nu) \exp\{-(\gamma - \mu)\tau\} \left[\int_0^{\infty} \sigma_1^{\nu} \exp\{-(2\gamma - \mu)\sigma_1\} d\sigma_1 \right.$$

$$\left. + \tau^{\nu} \int_0^{\infty} \exp\{-(2\gamma - \mu)\sigma_1\} d\sigma_1\right] = K(\nu) \exp\{-(\gamma - \mu)\tau\} \quad (2.15.56)$$

$$\times \left[\Gamma(\nu + 1)(2\gamma - \mu)^{-\nu-1} + \tau^{\nu}(2\gamma - \mu)^{-1}\right].$$

Summing up the estimates (2.15.54)–(2.15.56) and taking into account the evenness with respect to the parameter τ of the integral $\mathcal{I}$ we arrive at estimate (2.15.53).

The estimates (2.15.52) and (2.15.53) and the equality (2.15.44) gives (2.15.42), where the constants K_1, $\bar{K}_1$ and K_2 are determined as

$$K_1 = 2K_3(\alpha_0 + \alpha_1\nu)^{-1}, \quad \bar{K}_1 = 2K_3\alpha_2^{\nu}(2\gamma - \alpha_0 - \alpha_1\nu)^{-1}(\alpha_0 + \alpha_1\nu)^{-1},$$

$$K_2 = 2K_4\alpha_0^{-1} + K_3\alpha_2^{\nu}\Gamma(\nu + 1)(1 + K(\nu))(2\gamma - \alpha_0 - \alpha_1\nu)^{-\nu-1}.$$

We show that the function $u_0(g)$ determined by (2.15.33) belongs to the space $C_{\nu}^1(R^m)$. Differentiating the integrand expression in (2.15.43) with respect to the parameter g_i we get

$$\frac{\partial G_0(\tau, g)}{\partial g_i} f(g_{\tau}(g)) + G_0(\tau, g) \sum_{j=1}^{m} \frac{\partial f(g_{\tau}(g))}{\partial g_{\tau j}} \frac{\partial g_{\tau j}}{\partial g_i} = N_2(\tau, g). \quad (2.15.57)$$

We estimate the expression (2.15.57) in view of (2.15.4), (2.15.42), (2.15.47), (2.15.51) and the function $f(g)$ belonging to the space $C^1_\nu(R^m)$

$$\|N_2(\tau,g)\| \leq \|f\|_0 (K_1 \|g\|^\nu + \bar{K}_1 |\tau|^\nu + K_2) \exp\{-(\gamma - \alpha_0 - \alpha_1 \nu)|\tau|\}$$
$$+ KK_0 \exp\{-\gamma|\tau| + \alpha_0|t|\} [M_1(\nu)(\|g\|^\nu + |\tau|^\nu)] \exp\{\nu\alpha_1|\tau|\} + M_2. \qquad (2.15.58)$$

It becomes clear that if inequality (2.15.39) is satisfied, the integral $\int_{-\infty}^{\infty} N_2(\tau, g)\, d\tau$ converges uniformly in $g \in D$ on every bounded domain $D \subset R^m$. The estimate (2.15.58) indicates that the derivative $\partial u_0(g)/\partial g_i$ is estimated as

$$\left\|\frac{\partial u_0(g)}{\partial g_i}\right\| \leq \bar{\alpha}_3 \|g\|^\nu + \bar{\alpha}_4, \quad \text{i.e.} \quad u_0(g) \in C^1_\nu(R^m).$$

This completes the proof of Theorem 2.15.4.

Theorem 2.15.5 *For some non-degenerate symmetric matrix $S(g) \in C^1_\nu(R^m)$ let the inequality*

$$\left\langle \left[\sum_{i=1}^m \frac{\partial S(g)}{\partial g_i} \omega_i(g) - S(g)H^*(g) - H(g)S(g)\right]h, h\right\rangle \leq -\beta(g)\|h\|^2, \qquad (2.15.59)$$

be satisfied, where $\beta(g) \geq \beta_0 = \text{const} > 0$. Then there exists a unique Green function (2.15.3) with the estimate (2.15.4), where the constant γ is determined by

$$\gamma = \frac{1}{2} \inf_{g \in R^m} \frac{\beta(g)}{\max_{\|h\|=1} |\langle S(g)h, h\rangle|}. \qquad (2.15.60)$$

Proof System (2.15.1) is assumed to have a unique Green function (2.15.3) with the estimate (2.15.4). Let us show the existence of a unique Green function $\tilde{G}_0(\tau, g)$ for the conjugated system (2.15.10) and establish the relationship between $G_0(\tau, g)$ and $\tilde{G}_0(\tau, g)$. To this end the matrix Green function $G_t(\tau, g)$ of the problem of solutions bounded on R is transposed and multiplied by -1

$$-G_t^*(\tau, g) = -\begin{cases} \Omega_\tau^{0*}(g) C^*(g) \Omega_0^{t*}(g), & \tau \leq t, \\ \Omega_\tau^{0*}(g) [C^*(g) - I_n] \Omega_0^{t*}(g), & \tau > t. \end{cases}$$

Designating $P(g) = I_n - C^*(g)$ and replacing t and τ gives

$$-G_\tau^*(t, g) = -\begin{cases} \Omega_t^{0*}(g) P^*(g) \Omega_0^{\tau*}(g), & \tau < t, \\ \Omega_t^{0*}(g) [P^*(g) - I_n] \Omega_0^{\tau*}(g), & \tau \geq t. \end{cases} \qquad (2.15.61)$$

Since $\Omega_t^{0*}(g)$ is a matriciant of the conjugated system

$$\frac{dh}{dt} = -H^*(g_t(g))h, \qquad (2.15.62)$$

and $P(g)$ is a projection matrix $P^2(g) \equiv P(g)$, then the equality (2.15.61) determines the Green function of the problem of solutions bounded on R for the system (2.15.62). Therefore

$$\tilde{G}_0(\tau, g) \equiv -G_\tau^*(0, g), \qquad (2.15.63)$$

and hence, the Green function $\tilde{G}_0(\tau, g)$ of the conjugated system (2.15.10) satisfies the same estimate (2.15.4) as $G_0(\tau, g)$ with the same constants K, γ.

The inequality (2.15.59) with a continuously differentiable and non-degenerate matrix function $S(g)$ bounded on R^m and a positive function $\beta(g)$ ($\beta(g) \geq \beta_0 > 0$) provides the existence of a unique Green function $\tilde{G}_0(\tau, g)$ satisfying the estimate (2.15.4). We show that the constant γ in this estimate can be represented by (2.15.60). Designate

$$\mu(g) = \max_{\|h\|=1} \|S(g)h\|, \quad \mu_0 = \sup_{g \in R^m} \mu(g),$$
$$\|H\|_0 = \sup_{g \in R^m} \left(\max_{\|h\|=1} \|H(g)h\| \right), \quad \varepsilon_0 = \beta_0(3\|H\|_0)^{-1}. \qquad (2.15.64)$$

Note that for any solutions $h(t; g)$ of the system (2.15.62) the inequality

$$\frac{d}{dt}[\langle S(g_t(g))h(t;g), h(t;g)\rangle \pm \varepsilon_0\|h(t;g)\|^2] \leq -\frac{\beta_0}{3}\|h(t;g)\|^2 \qquad (2.15.65)$$

is satisfied. Therefore for all non-trivial damping on $+\infty$ solutions $h^+(t;g)$ we have

$$\langle S(g_t(g))h^+(t;g), h^+(t;g)\rangle - \varepsilon_0\|h^+(t;g)\|^2 > 0. \qquad (2.15.66)$$

If the inequality (2.15.66) is violated at some point $t = t^0$:

$$\langle S(g_{t^0}(g))h^+(t^0;g), h^+(t^0;g)\rangle - \varepsilon_0\|h^+(t^0;g)\|^2 < 0,$$

then by virtue of (2.15.65)

$$\langle S(g_t(g))h^+(t;g), h^+(t;g)\rangle - \varepsilon_0\|h^+(t;g)\|^2$$
$$\leq \langle S(g_{t^0}(g))h^+(t^0;g), h^+(t^0;g)\rangle - \varepsilon_0\|h^+(t^0;g)\|^2,$$

for all $t \geq t^0$. This contradicts the tending to zero on $+\infty$ of the solution $h^+(t;g)$. Together with (2.15.65) the estimate

$$\frac{d}{dt}\langle S(g_t(g))h^+(t;g), h^+(t;g)\rangle \leq -\beta(g_t(g))\|h^+(t;g)\|^2, \qquad (2.15.67)$$

is satisfied. Hence for $\tau \leq t$ we get

$$\langle S(g_t(g))h^+(t;g), h^+(t;g)\rangle \leq \langle S(g_\tau(g))h^+(\tau;g), h^+(\tau;g)\rangle$$
$$\times \exp\left\{-\int_\tau^t \beta(g_\sigma(g))\mu^{-1}(g_\sigma(g))\,d\sigma\right\}. \qquad (2.15.68)$$

Inequalities (2.15.66) and (2.15.68) imply

$$\|h^+(t;g)\| \leq (3\|H\|_0\mu_0\beta_0^{-1})^{1/2}\|h^+(\tau;g)\|$$
$$\times \exp\left\{-2^{-1}\int_\tau^t \beta(g_\sigma(g))\mu^{-1}(g_\sigma(g))\,d\sigma\right\}, \quad \tau \leq t. \qquad (2.15.69)$$

A similar estimate is satisfied for the solutions $h^-(t;g)$ of the system (2.15.62) damping on $-\infty$

$$\|h^-(t;g)\| \leq (3\|H\|_0\mu_0\beta_0^{-1})^{1/2}\|h^-(\tau;g)\|$$
$$\times \exp\left\{2^{-1}\int_\tau^t \beta(g_\sigma(g))\mu^{-1}(g_\sigma(g))\,d\sigma\right\}, \qquad (2.15.70)$$

for all $t \leq \tau$. Using the estimates (2.15.69) and (2.15.70) we estimate the Green function (2.15.61) as

$$\|G_\tau^*(t,g)\| \leq K \exp\left\{-2^{-1}\inf_{g\in R^m}\frac{\beta(g)}{\mu(g)}|t-\tau|\right\},$$

where $K = \text{const} > 0$. Hence it is clear that the constant γ can be taken in the form of (2.15.60).

This completes the proof of Theorem 2.15.5.

In the case when the matrix $S(g)$ degenerates for some $g = g_0 \in R^m$, the extended system

$$\frac{dg}{dt} = \omega(g),$$
$$\frac{dh}{dt} = H(g)h, \qquad (2.15.71)$$
$$\frac{dx}{dt} = -h - H^*(g)x,$$

is considered which already has a unique Green function $\bar{G}_0(\tau, g)$ which satisfies the estimate $\|\bar{G}_0(\tau, g)\| \leq K \exp\{-\gamma|\tau|\}$.

Also, it is of interest to determine the constant γ via the matrix $S(g)$. The investigations in this direction result in the following.

Theorem 2.15.6 *Let the inequality (2.15.59) be satisfied for some symmetric matrix* $S(g) \in C_\nu^1(R^m)$ *and let there exist at least one point* $g = g_0 \in R^m$ *such that* $\det S(g_0) = 0$. *Then for every* $\varepsilon > 0$ *there exists a constant* $K(\varepsilon)$ *such that for the Green function* $\tilde{G}_0(\tau, g)$ *of the system (2.15.71) the estimate*

$$\|\tilde{G}_0(\tau, g)\| \leq K(\varepsilon) \exp\{-(\Theta - \varepsilon)|\tau|\}, \qquad (2.15.72)$$

is satisfied, where

$$\Theta = \frac{\beta_0}{6\mu_0} = \frac{\inf \beta(g)}{6 \sup_{g \in R^m} \left(\max_{\|h\|=1} |\langle S(g)h, h \rangle|\right)}. \qquad (2.15.73)$$

Proof We fix a number $\varepsilon^0 \in (0, \theta)$ and consider the system of differential equations for all values of the parameter $\delta \in [-\varepsilon^0, \varepsilon^0]$

$$\frac{dh}{dt} = H(g)h + \delta h,$$

$$\frac{dx}{dt} = -h - H^*(g)x + \delta x, \qquad (2.15.74)$$

$$\frac{dg}{dt} = \omega(g).$$

We demonstrate the existence of positive numbers λ such that the derivative of the quadratic form $V(g, h, x) = \lambda \langle h, x \rangle + \langle S(g)x, x \rangle$ computed along the solutions of the system (2.15.74) is negative definite. Let us compute this derivative:

$$\dot{V}(g, h, x) = \lambda \langle Hh + \delta h, x \rangle + \lambda \langle h, -h - H^*x + \delta x \rangle$$
$$+ \langle \dot{S}x, x \rangle + \langle S(-h - H^*x + \delta x), x \rangle + \langle Sx, -h - H^*x + \delta x \rangle.$$

Estimate $\dot{V}$:

$$\dot{V}(g, h, x) \leq -\lambda \|h\|^2 + 2(|\delta|\lambda + \mu_0)\|h\|\|x\| - (\beta_0 - 2|\delta|\mu_0)\|x\|^2.$$

For the negative definiteness of V it is sufficient that

$$\lambda > 0, \quad \begin{vmatrix} -\lambda & |\delta|\lambda + \mu_0 \\ |\delta|\lambda + \mu_0 & -(\beta_0 - 2|\delta|\mu_0) \end{vmatrix} > 0.$$

Hence we find

$$\lambda^2 \delta^2 + 4|\delta|\mu_0 \lambda - \beta_0 \lambda + \mu_0^2 < 0. \qquad (2.15.75)$$

For the solutions of this inequality to exist, it is necessary that the discriminant be positive

$$D = (4|\delta|\mu_0 - \beta_0)^2 - 4\mu_0^2 \delta^2 = (\beta_0 - 6|\delta|\mu_0)(\beta_0 - 2|\delta|\mu_0) > 0. \qquad (2.15.76)$$

Since $\delta \in [-\varepsilon^0, \varepsilon^0]$, $0 < \varepsilon^0 < \theta = \beta_0(6\mu_0)^{-1}$, then the inequality (2.15.76) is satisfied. In order to find the solution to inequality (2.15.75) we take into account that

$$\lambda_1 = \frac{\beta_0 - 4|\delta|\mu_0 + D^{1/2}}{2\delta^2} > \frac{2\mu_0^2}{\beta_0 - 4\mu_0|\delta|},$$

$$\lambda_2 = \frac{\beta_0 - 4|\delta|\mu_0 - D^{1/2}}{2\delta^2} < \frac{2\mu_0^2}{\beta_0 - 4\mu_0|\delta|}.$$

Therefore, for

$$\lambda = 2\mu_0^2(\beta_0 - 4\mu_0|\delta|)^{-1} = \lambda_0, \qquad (2.15.77)$$

the inequality (2.15.75) is satisfied.

Thus, by virtue of the non-degeneracy of the quadratic form $V(g, h, x)$ the system (2.15.71) has a unique Green function $\bar{G}_0(\tau, g; \delta)$ for every fixed $\delta \in [-\varepsilon^0, \varepsilon^0]$ satisfying the estimate

$$\|\bar{G}_0(\tau, g; \delta)\| \leq K(\delta) \exp\{-\gamma(\delta)|\tau|\}. \qquad (2.15.78)$$

Since λ_0 in (2.15.77) has the largest value for $\delta = \pm\varepsilon^0$, then the estimate (2.15.78) can obviously be represented as

$$\|\bar{G}_0(\tau, g; \delta)\| \leq K(\varepsilon^0) \exp\{-\gamma(\delta)|\tau|\}, \qquad (2.15.79)$$

where the constant $K(\varepsilon^0)$ only depends on $\varepsilon^0 \in (0, \theta)$. In view of the structure of the Green function only

$$\bar{G}_0(\tau, g; \delta) = \begin{cases} \bar{C}(g)\bar{\Omega}_\tau^0(g; \delta), & \tau \leq 0, \\ [\bar{C}(g) - I_n]\bar{\Omega}_\tau^0(g; \delta), & \tau > 0, \end{cases}$$

where $\bar{C}(g)$ is a $2n \times 2n$-dimensional projection matrix, $\bar{\Omega}_\tau^0(g; \delta) = \bar{\Omega}_\tau^t(g; \delta)|_{t=0}$, $\bar{\Omega}_\tau^t(g; \delta)$ is a matriciant of the system

$$\frac{dh}{dt} = H(g_t(g))h + \delta h,$$

$$\frac{dx}{dt} = -h - H^*(g_t(g))x + \delta x,$$

we get

$$\bar{G}_0(\tau, g; \delta) \equiv \bar{G}_0(\tau, g; 0) \exp\{-\delta\tau\}, \qquad (2.15.80)$$

for all $\delta \in [-\varepsilon^0, \varepsilon^0]$, $\tau \in R$. The estimate (2.15.79) together with (2.15.80) imply

$$\|\bar{G}_0(\tau, g; 0)\| \leq K(\varepsilon^0) \exp\{-\varepsilon^0|\tau|\}. \qquad (2.15.81)$$

Actually, if this inequality is violated, there exists a sequence $\tau^{(n)}$, $g^{(n)}$ such that $\|\bar{G}_0(\tau^{(n)}, g^{(n)}; 0)\| \exp\{-\varepsilon^0 |\tau^{(n)}|\} \xrightarrow[n\to\infty]{} \infty$. Evidently, a subsequence $\tau^{(n_k)}$ can be taken such that either $|\tau^{(n_k)}| = \tau^{(n_k)}$ or $|\tau^{(n_k)}| = -\tau^{(n_k)}$ for all $k = 1, 2, \ldots$. In the first case

$$\|\bar{G}_0(\tau^{(n_k)}, g^{(n_k)}; 0)\| \exp\{\varepsilon^0 \tau^{(n_k)}\} \xrightarrow[k\to\infty]{} \infty,$$

which contradicts the estimate (2.15.79) when $\delta = -\varepsilon^0$. Now the validity of the inequality (2.15.81) and the estimate (2.15.72) becomes obvious.

Remark 2.15.1 Since the value $\varepsilon > 0$ in (2.15.72) can be taken arbitrarily small, then under the conditions of Theorem 2.15.6 the inequality $\alpha_1 \nu + \alpha_0 < \beta_0 (6\mu_0)^{-1}$ ensures the existence of a set of smooth bounded invariant manifolds of the system (2.15.5) for every fixed vector function $f(g) \in C^1_\nu(R^m)$.

2.16 Comments and References

Section 2.0 The investigation of the stability of nonlinear multifrequency oscillation processes is closely related to the perturbation theory of invariant toroidal manifolds of dynamical systems. The Second Chapter deals with this theory. The investigations in this direction are based on outstanding works by Bogoliubov [1-3], Bogoliubov and Mitropolsky [1, 2], Bogoliubov, Mitropolsky and Samoilenko [1], Grebenikov and Riabov [1], Griebienikow, J.A. and Riabov, J.A. [1], Samoilenko [1, 2], Kryloff and Bogolyubov [1,2]. Attention should be paid to works related to the investigations of Chapter 2 by Akhiezer [1], Andronov, Vitt and Khaikin [1], Arnol'd [1], Bakaj and Stepanovsky [1], Barbashin and Krasovsky [1], Bibikov [1], Bibikov and Pliss [1], Birkhoff [1], Bohl [1], Bohr [1], Bronstein [1], Cesari [1], Erugin [1, 2], Filatov and Sharova [1], Hale [1], Hayashi [1], Kurzweil [1], LaSalle and Lefschetz [1], Letov [1], Levitan [1], Lomov [1], Lyapunov [1, 2], Lyashchenko [1], Malkin [1], Mandel'shtam [1], Martynyuk [1], Martynyuk and Gutowski [1], Martynyuk and Obolensky [1], Martynyuk and Tsyganovsky [1], Mel'nikov [1], Mitropolsky [1-3], Mitropolsky and Lykova [1], Mitropolsky, Samoilenko and Martynyuk [1], Mishchenko and Rozov [1], Mozer [1], Mukhamadiev [1], Myshkis [1], Neimark [1], Nemytzky [1], Nemytzky and Stepanov [1], Osipenko [1], Perestyuk [1], Poincaré [1, 2], Povzner [1], Rozov [1], Rumjantsev [1], Rouch, Habets and Laloy [1], Sobolev [1], Sternberg [1], Tikhnov [1], Trofimchuk [1], Vasil'eva and Butuzov [2], Vishik and Ljusternik [1], Zadiraka [1], Zhautykov [1], Zhikov and Levitan [1], Zubov [1].

Sections 2.1 – 2.6 Theorems 2.1.1, 2.1.2, 2.2.1, 2.3.1, 2.4.1, 2.6.1, 2.6.2 were proved by Samoilenko [5]. The proof of Theorem 2.5.1 is due to Samoilenko and Kulik [1].

Section 2.7 The Theorems 2.7.1 – 2.7.7 are due to Kulik [5, 8, 11].

Section 2.8 The problems of the non-uniqueness of the Green function were discussed by Mitropolsky and Kulik [2], Mitropolsky, Samoilenko and Kulik [2], Samoilenko and Kulik [7].

Section 2.9 Theorems 2.9.1 – 2.9.4 are due to Ordynskaya and Kulik [1].

Section 2.10 The results of this section are new.

Sections 2.11 – 2.13 Theorems 2.11.1, 2.11.2 are due to Samoilenko [5]. The dependence of the Green function on parameters (Theorems 2.12.1 – 2.12.4, 2.13.1 – 2.13.3) were discussed by Kulik [2], Samoilenko and Kulik [5].

Section 2.14 Contains results by Kulik and Eremenko [1].

Section 2.15 The ideas of the results are due to Mitropolsky and Kulik [2, 3].

3 Splittability of Linear Extensions of Dynamical Systems on a Torus

3.0 Introduction

It is known that every linear system which is e-dichotomous on the whole R axis

$$\frac{dx}{dt} = A(t)x, \quad A(t) \in C^0(R),$$

can always be transformed by a change of Lyapunov variables

$$x = L(t)y, \tag{3.0.1}$$

to the split form

$$\frac{dy_1}{dt} = B^+(t)y_1, \quad \frac{dy_2}{dt} = B^-(t)y_2,$$

coordinated with the e-dichotomy of the initial system. In this regard a difficult and important problem arises: whether the matrix $L(t)$ in the change of variables (3.0.1) can always be taken to be quasiperiodic, if the matrix $A(t)$ is quasiperiodic in t in the initial system? It turned out that the answer is negative. It should also be noted that this problem is closely connected with the possibility of introducing local coordinates x, φ in the neighbourhood of the torus $\mathcal{T}_m$ and to complement the r-frame which is 2π-periodic in many variables up to a complete basis in R^n.

Investigating the property of e-dichotomy of the torus $x = 0$ of the system

$$\frac{d\varphi}{dt} = a(\varphi), \quad \frac{dx}{dt} = A(\varphi)x \qquad (3.0.2)$$

by means of the Lyapunov functions of variable sign

$$V(\varphi, x) = \langle S(\varphi)x, x \rangle \qquad (3.0.3)$$

we noticed that the possibility of simplifying the quadratic form (3.0.3) influences the possibility of separating the variables x in the system (3.0.2) by means of the change of variables $x = L(\varphi)y$.

This interconnection is studied in Chapter 3 of the present book. It also treats interesting problems on the reduction of the system (3.0.2) to the diagonal form, the connection of the splittability of the system of type (3.0.2) with the solvability of algebraic systems, on the possibility of three-block splitting of the system (3.0.2). Here the algebraic problems of k-block splittability are distinguished.

3.1 Sufficient Conditions for Splittability of Linear Extensions of Dynamical Systems on a Torus

Consider a linear system of differential equations which is exponentially dichotomous on R

$$\frac{dx}{dt} = A(\omega t)x, \quad x \in R^n, \qquad (3.1.1)$$

with a continuous quasiperiodic coefficient matrix $A(\omega t) = A(\omega_1 t, \ldots, \omega_m t)$, where $\omega = (\omega_1, \ldots, \omega_m)$ is a frequency basis, $t \in R$, $A(\varphi) \in C^0(\mathcal{T}_m)$. It is well known that the system (3.1.1) can be transformed by means of the change of Lyapunov variables

$$x = L(t)y \qquad (3.1.2)$$

to the split form

$$\frac{dy_1}{dt} = B^+(t)y_1, \quad \frac{dy_2}{dt} = B^-(t)y_2, \qquad (3.1.3)$$

which is coordinated with the exponential dichotomy of the system (3.1.1) on the whole R axis, i.e. all non-trivial solutions of the system

$$\frac{dy_i}{dt} = B^{\pm}(t)y_i, \quad i = 1, 2,$$

damp on $\pm\infty$ and increase on $\mp\infty$. There arises a question: whether it is always possible to take the Lyapunov matrix $L(t)$ in (3.1.2) to be quasiperiodic in t? This question was answered by Bylov et al. [1]. It turned out that such a quasiperiodic matrix $L(t)$ does not always exist. Therefore additional conditions are to be established under which such a quasiperiodic matrix $L(t)$ does exist (see Anosov [1], Bylov [1], Bylov et al. [1], Kulik [5, 12], Kulik and Eremenko [1], Lyashchenko [1], Samoilenko [3, 5] and Trofimchuk [1]).

Note that the system (3.1.1) can be replaced by

$$\frac{d\varphi}{dt} = \omega, \quad \frac{dx}{dt} = A(\varphi)x.$$

Therefore a more general system (2.2.4) from Chapter 2 is considered.

Theorem 3.1.1 *Let there exist a non-degenerate symmetric matrix $S(\varphi) \in C'(\mathcal{T}_m; a)$ of the block diagonal form*

$$S(\varphi) = \mathrm{diag}\{S_1(\varphi), -S_2(\varphi)\}, \tag{3.1.4}$$

where the matrices $S_1(\varphi)$ and $S_2(\varphi)$ are positive definite

$$\langle S_i(\varphi)\eta_i, \eta_i \rangle \geq \beta_i \|\eta_i\|^2, \quad \beta_i = \mathrm{const} > 0, \quad i = 1, 2, \tag{3.1.5}$$

$\eta_1 \in R^r$, $\eta_2 \in R^{n-r}$, *satisfying the condition*

$$\langle [\dot{S}(\varphi) + S(\varphi)A(\varphi) + A^*(\varphi)S(\varphi)]x, x \rangle \leq -2\gamma \|x\|^2, \quad \gamma > 0. \tag{3.1.6}$$

Then there exists a non-degenerate $n \times n$-dimensional matrix

$$L(\varphi) = \begin{pmatrix} I_r & L_1(\varphi) \\ L_2(\varphi) & I_{n-r} \end{pmatrix} \in C'(\mathcal{T}_m; a) \tag{3.1.7}$$

which satisfies the splittability condition

$$L^{-1}(\varphi)[A(\varphi)L(\varphi) - \dot{L}(\varphi)] = \begin{pmatrix} I_r & L_1(\varphi) \\ L_2(\varphi) & I_{n-r} \end{pmatrix}^{-1}$$

$$\times \left[\begin{pmatrix} A_{11}(\varphi) & A_{12}(\varphi) \\ A_{21}(\varphi) & A_{22}(\varphi) \end{pmatrix} \begin{pmatrix} I_r & L_1(\varphi) \\ L_2(\varphi) & I_{n-r} \end{pmatrix} - \begin{pmatrix} 0 & \dot{L}_1(\varphi) \\ \dot{L}_2(\varphi) & 0 \end{pmatrix} \right] \tag{3.1.8}$$

$$= \begin{pmatrix} A_{11}(\varphi) + A_{12}(\varphi)L_2(\varphi) & 0 \\ 0 & A_{22}(\varphi) + A_{21}(\varphi)L_1(\varphi) \end{pmatrix}.$$

Also, the rectangular matrix functions $L_i(\varphi) \in C'(\mathcal{T}_m; a)$ satisfy the identities

$$\dot{L}_1(\varphi) \equiv -L_1(\varphi)A_{21}(\varphi)L_1(\varphi) + A_{11}(\varphi)L_1(\varphi) - L_1(\varphi)A_{22}(\varphi) + A_{12}(\varphi),$$

$$\dot{L}_2(\varphi) \equiv -L_2(\varphi)A_{12}(\varphi)L_2(\varphi) + A_{22}(\varphi)L_2(\varphi) - L_2(\varphi)A_{11}(\varphi) + A_{21}(\varphi)$$

and the estimates

$$\max\{\|S_1^{1/2}(\varphi)L_1(\varphi)S_2^{-1/2}(\varphi)\|,\ \|S_2^{1/2}(\varphi)L_2(\varphi)S_1^{-1/2}(\varphi)\|\} \le 1 - \varepsilon_0, \qquad (3.1.9)$$

where ε_0 is a fixed positive number which is sufficiently small.

Proof Alongside the function $V(\varphi, x) = \langle S(\varphi)x, x\rangle$ we consider the function

$$V_\varepsilon(\varphi, x) = V(\varphi, x) - \varepsilon[\langle S_1(\varphi)x_1, x_1\rangle + \langle S_2(\varphi)x_2, x_2\rangle] = V(\varphi, x) - \varepsilon V_1(\varphi, x).$$

Obviously, for the derivative $\dot{V}_\varepsilon(\varphi, x)$ computed along solutions of the system

$$\frac{d\varphi}{dt} = a(\varphi), \qquad \frac{dx}{dt} = A(\varphi)x$$

the estimate

$$\dot{V}_\varepsilon(\varphi, x) = \dot{V}(\varphi, x) - \varepsilon \dot{V}_1(\varphi, x) \le -2(\gamma - \varepsilon M)\|x\|^2 \qquad (3.1.10)$$

is valid, where M is a positive constant determined by

$$M \ge \max_{\varphi \in \mathcal{T}_m}\left(\|S(\varphi)A(\varphi)\| + \frac{1}{2}\|\dot{S}(\varphi)\|\right).$$

Consider the properties of the solutions $x(t, \varphi)$ of the linear system

$$\frac{dx}{dt} = A(\varphi_t(\varphi))x, \qquad (3.1.11)$$

which satisfies the inequality

$$V_\varepsilon(\varphi_t(\varphi),\ x(t, \varphi)) > 0, \qquad (3.1.12)$$

for all $t \in R$, $\varphi \in \mathcal{T}_m$ and some $\varepsilon > 0$. If such solutions exist, then the inequality

$$\frac{d}{dt} V(\varphi_t(\varphi),\ x(t,\varphi)) \le -2\gamma \|x(t,\varphi)\|^2 \le -2\frac{\gamma}{\beta_1^0} V(\varphi_t(\varphi),\ x(t,\varphi)),$$

where $\beta_1^0 \ge \max_{\|x_1\|=1} \langle S_1(\varphi)x_1, x_1\rangle$, holds true along these solutions, solving which one gets

$$V(\varphi_t(\varphi),\ x(t,\varphi)) \le \exp\left[-\frac{2\gamma}{\beta_1^0}(t - \tau)\right] V(\varphi_\tau(\varphi),\ x(t,\varphi)), \qquad \tau \le t.$$

Consequently,

$$\varepsilon \min\{\beta_1, \beta_2\} \|x(t,\varphi)\|^2 \le \varepsilon[\langle S_1(\varphi_t(\varphi))x_1, x_1\rangle + \langle S_2(\varphi_t(\varphi))x_2, x_2\rangle]$$
$$= V(\varphi_t(\varphi),\ x(t,\varphi)) - V_\varepsilon(\varphi_t(\varphi),\ x(t,\varphi)) \le V(\varphi_t(\varphi),\ x(t,\varphi))$$
$$\le \beta_1^0 \exp\left[-\frac{2\gamma}{\beta_1^0}(t-\tau)\right]\|x(\tau,\varphi)\|^2, \qquad \tau \le t.$$

Thus we have that the solutions $x(t,\varphi)$ of the system (3.1.1), for which the inequality (3.1.12) is satisfied for all $t \in R$ and some $\varepsilon > 0$, satisfy the estimate

$$\|x(t,\varphi)\| \leq \left(\frac{\beta_1^0}{\varepsilon \min(\beta_1, \beta_2)}\right)^{1/2} \exp\left[-\gamma_1(t-\tau)\right] \|x(\tau,\varphi)\|, \quad \tau \leq t, \qquad (3.1.13)$$

where $\gamma_1 = \gamma/\beta_1^0$.

Let us prove that for every φ and a fixed, sufficiently small $\varepsilon > 0$ ($\gamma > \varepsilon M$) there are precisely r linearly independent solutions of the system (3.1.11) which satisfy the condition (3.1.12). To this end we fix the vector $x_1^0 = (x^{10}, \ldots, x^{r0})$, $\|x_1^0\| = 1$, and consider the solutions $x = (x_1(t,\varphi), x_2(t,\varphi))$ of the system (3.1.11), taking the value $x_1(0,\varphi) = x_1^0$ for $t = 0$. By virtue of the positiveness of the quadratic form $\langle S_1(\varphi)x_1, x_1\rangle$ the inequality (3.1.12) is equivalent to the condition

$$\frac{\langle S_2(\varphi_t(\varphi))x_2(t,\varphi), x_2(t,\varphi)\rangle}{\langle S_1(\varphi_t(\varphi))x_1(t,\varphi), x_1(t,\varphi)\rangle} \leq \frac{1-\varepsilon}{1+\varepsilon}. \qquad (3.1.14)$$

We fix a point $\varphi = \varphi_0 \in \mathcal{T}_m$ and consider the function

$$W(t, W_0) = x_2(t,\varphi_0) \langle S_1(\varphi_t(\varphi_0))x_1(t,\varphi_0), x_1(t,\varphi_0)\rangle^{-1/2}. \qquad (3.1.15)$$

The solution $(x_1(t,\varphi_0), x_2(t,\varphi_0))$ of the system (3.1.18) takes the value $(x_1^0, W_0\langle S_1(\varphi_0)x_1^0, x_1^0\rangle^{1/2})$ for $t = 0$. Now, for a fixed $\varphi = \varphi_0 \in \mathcal{T}_m$ the condition (3.1.14) becomes

$$\langle S_2(\varphi_t(\varphi))W(t, W_0), W(t, W_0)\rangle \leq \frac{1-\varepsilon}{1+\varepsilon}. \qquad (3.1.16)$$

In the space of variables W and t consider the surface $\mathfrak{N}$ formed by the points for which

$$\langle S_2(\varphi_t(\varphi))W, W\rangle = \frac{1-\varepsilon}{1+\varepsilon}$$

and designate by $\mathfrak{M}$ a set of points W, $t = 0$, determined by

$$\langle S_2(\varphi_0)W, W\rangle = \frac{1-\varepsilon}{1+\varepsilon}.$$

Evidently, the set $\mathfrak{M}$ is a ball (ellipsoid) in r-dimensional space R^r. It is easy to notice that if the function $W(t, W_0)$ takes the value belonging to the surface $\mathfrak{N}$ at some time $t = t_1$ the values of this function at later times $t > t_1$ will be out of the surface $\mathfrak{N}$, i.e. every point from $\mathfrak{N}$ is a strict output point of the family of the functions $W(t, W_0)$.

Let us prove the existence of the point $W_0 \in \mathfrak{M}$ such that the inequality (3.1.16) equivalent to (3.1.13) for all $t \in R$ is valid. Let there exist no such point $W_0 \in \mathfrak{M}$.

Then, whatever the point $W_0 \in \mathfrak{M}$ is, a value $t = t_1(W_0) > 0$ is found such that the curve determined by the function $W(t, W_0)$ gets onto the surface $\mathfrak{N}$ at time $t = t_1$. Let Π_0 be a mapping of $\mathfrak{M}$ into $\mathfrak{N}$ obeying the law

$$\Pi_0 = \begin{cases} (W_0, 0), & \text{if } W_0 \in \mathfrak{M} \cap \mathfrak{N}, \\ (W(t_1(W_0), W_0), t_1(W_0)), & \text{if } W_0 \in \mathfrak{M} \setminus \mathfrak{N}. \end{cases}$$

The continuity of the functions $W(t, W_0)$ with respect to W_0 which follows from the continuous dependence of the solutions to system (3.1.11) on the initial conditions, implies the continuity of the mapping Π_0. We designate the mapping of $\mathfrak{N}$ into the hyperplane $(W, 0)$ projecting every point of $\mathfrak{N}$ into the space W by Π_1, and the mapping of point $(W_0, 0)$ ($\|W_0\| \neq 0$) on the boundary $\mathfrak{M} \cap \mathfrak{N}$ acting as follows by Π_2. We draw an axis in the space of variables W via the points $(0, 0)$ and $(W_0, 0)$. The intersection of this axis with the boundary $\mathfrak{M} \cap \mathfrak{N}$ yields the result of the mapping Π_2 action. The mappings Π_1 and Π_2 are evidently continuous. The composition of the three mappings $\Pi_2 \Pi_1 \Pi_0$ is a retraction of the ball $\mathfrak{M}$ on its boundary $\mathfrak{M} \cap \mathfrak{N}$. By the theory of continuous mappings the boundary of the ball $\mathfrak{M}$ is not a retractor of the ball $\mathfrak{M}$ itself. The contradiction obtained shows the existence of the point $W_0 \in \mathfrak{M}$, such that for all $t \in R$ and a fixed vector $\varphi_0 \in \mathcal{T}_m$ the condition (3.1.16) holds. Thus, there exist solutions $x(t, \varphi_0)$ of the system (3.1.11) for which (3.1.12) holds true. Changing the numerical vector x_1^0, we get precisely r linearly independent initial values determining the solutions of the system (3.1.11) subjected to the condition (3.1.12) with $\varphi = \varphi_0$ and for all $t \in R$. The arbitrariness of the choice of φ_0 demonstrates the existence of r linearly independent solutions to the system (3.1.11) satisfying (3.1.13) for any $\varphi \in \mathcal{T}_m$. Now we can designate the subspace stretched on the vectors determining the initial values of these solutions by E^+.

By arguments similar to the above we establish the existence of an $n - r$-dimensional subspace E^- for the function $-V(\varphi, x)$ possessing the property that the solutions $x(t, \varphi)$ starting at $t = 0$ for points $x(0, \varphi) \in E^-$ are subject to the condition

$$\|x(t, \varphi)\| \leq \left(\frac{\beta_2^0}{\varepsilon \min(\beta_1, \beta_2)} \right)^{1/2} e^{\gamma_2(t-\tau)} \|x(\tau, \varphi)\|, \quad t \leq \tau, \tag{3.1.17}$$

for all $\varphi \in \mathcal{T}_m$, where $\gamma_2 = \gamma/\beta_2^0$, $\beta_2^0 \geq \max_{\|x_2\|=1} \langle S_2 x_2, x_2 \rangle$. The arbitrariness of the choice of values x_1^0 when the subspace E^+ is constructed and that of values x_2^0 for E^- enables us to take the vectors (e_j^1, x_2^j), $j = \overline{1, r}$, and (x_1^i, e_i^2), $i = \overline{r+1, r}$ for the basis vectors of E^+ and E^- respectively. Let $\begin{pmatrix} I_r \\ L_2 \end{pmatrix}$ be the matrix formed by the vectors (e_j^1, x_2^j), and $\begin{pmatrix} L_1 \\ I_{n-r} \end{pmatrix}$ be the matrix formed by the vectors (x_1^i, e_i^2).

We will prove that the determinant of the matrix

$$L(\varphi) = \begin{pmatrix} I_r & L_1(\varphi) \\ L_2(\varphi) & I_{n-r} \end{pmatrix} \quad (3.1.18)$$

differs from zero for all $\varphi \in \mathcal{T}_m$. This will be sufficient for the claim that R^n is a direct sum of the subspaces E^+ and E^-.

We verify that

$$\|S_2^{1/2}(\varphi) L(\varphi) S_1^{-1/2}(\varphi)\| \leq \left(\frac{1-\varepsilon}{1+\varepsilon}\right)^{1/2}, \quad (3.1.19)$$

for all $\varphi \in \mathcal{T}_m$, where the operator norm is $\|L\| = \max_{\|x\|=1} \|Lx\|$. Assume that the inequality (3.1.19) is violated for some $\varphi = \varphi_0 \in \mathcal{T}_m$. This means that a numerical vector x_1^0, $\|x_1^0\| = 1$, is found such that

$$\|S_1^{1/2}(\varphi_0) L_2(\varphi_0) S_1^{-1/2}(\varphi_0) x_1^0\| > \left(\frac{1-\varepsilon}{1+\varepsilon}\right)^{1/2}.$$

For the solution $x = (x_1(t, \varphi_0), x_2(t, \varphi_0))$ of the system (3.1.11) for $\varphi = \varphi_0$ taking the value $x(0, \varphi_0) = (S_1^{-1/2}(\varphi_0) x_1^0, L_2(\varphi_0) S_1^{-1/2}(\varphi_0) x_1^0)$ for $t = 0$ the estimate (3.1.13) is valid, since the initial value of this solution belongs to the space E^+. On the other hand, for the same solution $x(t, \varphi_0)$ at point $t = 0$

$$V_\varepsilon(\varphi_0, x(0, \varphi_0)) < 0. \quad (3.1.20)$$

In fact,

$$\frac{\langle S_2(\varphi_0) x_2(0, \varphi_0), x_2(0, \varphi_0) \rangle}{\langle S_1(\varphi_0) x_1(0, \varphi_0), x_1(0, \varphi_0) \rangle} = \frac{\langle S_2^{1/2} x_2, S_2^{1/2} x_2 \rangle}{\langle S_1^{1/2} x_1, S_1^{1/2} x_1 \rangle}$$

$$= \frac{\|S_2^{1/2} L_2 S_1^{-1/2} x_1^0\|^2}{\|x_1^0\|^2} > \frac{1-\varepsilon}{1+\varepsilon}.$$

The inequality (3.1.20) implies the estimate $V_\varepsilon(\varphi_t(\varphi_0), x(t, \varphi_0)) < 0$ for all $t \geq 0$; the application of which yields

$$-V_\varepsilon(\varphi_t(\varphi_0), x(t, \varphi_0)) < e^{2\gamma_2 t} [-V_\varepsilon(\varphi_0, x(0, \varphi_0))], \quad t \geq 0, \quad \gamma_2 > 0. \quad (3.1.21)$$

The inequality (3.1.21) contradicts (3.1.13). Thus, the estimate (3.1.19) is proved.

The estimation of the norm of the matrix $S_1^{1/2} L_1 S_2^{-1/2}$ is proved in a similar way

$$\|S_1^{1/2}(\varphi) L_1(\varphi) S_2^{-1/2}(\varphi)\| \leq \left(\frac{1-\varepsilon}{1+\varepsilon}\right)^{1/2}. \quad (3.1.22)$$

From (3.1.19) and (3.1.22) we get

$$\left\| \begin{pmatrix} 0 & S_1^{1/2} L_1 S_2^{-1/2} \\ S_2^{1/2} L_2 S_1^{-1/2} & 0 \end{pmatrix} \right\| = \sup_{\|x\|=1} (\|S_1^{1/2} L_1 S_2^{-1/2} x_2\|^2$$

$$+ \|S_2^{1/2} L_2 S_1^{-1/2} x_1\|^2)^{1/2} \leq \left(\frac{1-\varepsilon}{1+\varepsilon} \right)^{1/2}.$$

Now the non-degeneracy of the matrix $L(\varphi)$ becomes obvious due to the following representation

$$L(\varphi) = \begin{pmatrix} I_r & 0 \\ 0 & I_{n-r} \end{pmatrix} + \begin{pmatrix} 0 & L_1 \\ L_2 & 0 \end{pmatrix} = \begin{pmatrix} 0 & S_2^{-1/2} \\ S_1^{-1/2} & 0 \end{pmatrix}$$

$$\times \left[\begin{pmatrix} I_r & 0 \\ 0 & I_{n-r} \end{pmatrix} + \begin{pmatrix} 0 & S_1^{1/2} L_1 S_2^{-1/2} \\ S_2^{1/2} L_2 S_1^{-1/2} & 0 \end{pmatrix} \right] \begin{pmatrix} 0 & S_1^{1/2} \\ S_2^{1/2} & 0 \end{pmatrix}.$$

The above arguments demonstrate that the system of vectors forming the subspaces E^+ and E^- is linearly independent in totality and, therefore, forms a basis in R^n. Consequently, the space R^n is a direct sum of the subspaces E^+ and E^-.

It is clear that the vector $x_2^0 = L_2(\varphi) x_1^0$ solely corresponds to every numerical vector x_1^0 ($\|x_1^0\| \neq 0$), so that $x^0 = (x_1^0, x_2^0) \in E^+$. Similarly, the vector x_2^0 ($\|x_2^0\| \neq 0$) corresponds to $x_1^0 = L_1(\varphi) x_2^0$ and $(x_1^0, x_2^0) \in E^-$. We designate by

$$X^+(t, \varphi) = \begin{pmatrix} X_1^+(t, \varphi) \\ X_2^+(t, \varphi) \end{pmatrix} = \begin{pmatrix} x_1^{(1)+}(t, \varphi) & \cdots & x_1^{(r)+}(t, \varphi) \\ x_2^{(1)+}(t, \varphi) & \cdots & x_2^{(r)+}(t, \varphi) \end{pmatrix}$$

the matrix composed of r linearly independent solutions of the system (3.1.11), vanishing exponentially as $t \to \infty$; $X_1^+(t, \varphi)$ is an $r \times r$-dimensional matrix composed of the first r lines of the matrix X^+. Correspondingly,

$$X^-(t, \varphi) = \begin{pmatrix} X_1^-(t, \varphi) \\ X_2^-(t, \varphi) \end{pmatrix}$$

denotes the matrix composed of $n - r$ linearly independent solutions vanishing as $t \to -\infty$; $X_2^-(t, \varphi)$ is an $(n-r) \times (n-r)$-dimensional matrix, composed of the last $n - r$ lines of the matrix $X^-(t, \varphi)$. It is easy to see that $\det X_1^+(t, \varphi) \neq 0$ and $\det X_2^-(t, \varphi) \neq 0$ for all $t \in R$ and $\varphi \in \mathcal{T}_m$. Moreover, the matrices $X_1^+(t, \varphi)$ and $X_2^-(t, \varphi)$ are fundamental matrices of solutions of the systems of equations

$$\frac{dx_1}{dt} = [A_{11}(\varphi_t(\varphi)) + A_{12}(\varphi_t(\varphi)) L_2(t, \varphi)] x_1,$$

$$\frac{dx_2}{dt} = [A_{21}(\varphi_t(\varphi)) L_1(t, \varphi) + A_{22}(\varphi_t(\varphi))] x_2,$$

respectively, where the matrix functions L_1 and L_2 are determined by

$$L_1(t,\varphi) = X_1^-(t,\varphi)(X_2^-(t,\varphi))^{-1},$$
$$L_2(t,\varphi) = X_2^+(t,\varphi)(X_1^+(t,\varphi))^{-1}. \qquad (3.1.23)$$

Here the functions $L_2(t,\varphi)$ and $L_1(t,\varphi)$ do not depend on the choice of $X^+(t,\varphi)$ and $X^-(t,\varphi)$. Thus, if $x^+(t,\varphi)$ is a solution of the system (3.1.11) vanishing as $t \to +\infty$, the correlation $x_2^+(t,\varphi) \equiv L_2(t,\varphi)x_1^+(t,\varphi)$ is valid for its coordinates. Similarly, for the solutions $x^-(t,\varphi)$ decreasing on $-\infty$, $x_1^-(t,\varphi) \equiv L_1(t,\varphi)x_2^-(t,\varphi)$.

For the functions (3.1.20)

$$\max_{\|x_1\|=1} \|S_2^{1/2}(\varphi_t(\varphi))L_2(t,\varphi)S_1^{-1/2}(\varphi_t(\varphi))x_1\| \leq \left(\frac{1-\varepsilon}{1+\varepsilon}\right)^{1/2},$$
$$\max_{\|x_2\|=1} \|S_1^{1/2}(\varphi_t(\varphi))L_1(t,\varphi)S_2^{-1/2}(\varphi_t(\varphi))x_2\| \leq \left(\frac{1-\varepsilon}{1+\varepsilon}\right)^{1/2}, \qquad (3.1.24)$$

for all $t \in R$, $\varphi \in \mathcal{T}_m$ and some $\varepsilon > 0$ independent of t, φ. Prove, for example, the first inequality (3.1.24). Let for some $t = t_1 \in R$, φ^*, x_1^*, $\|x_1^*\| = 1$, the opposite inequality

$$\|S_2^{1/2}(\varphi_{t_1}(\varphi^*))L_2(t_1,\varphi^*)S_1^{-1/2}(\varphi_{t_1}(\varphi^*))x_1^*\| > \left(\frac{1-\varepsilon}{1+\varepsilon}\right)^{1/2} \qquad (3.1.25)$$

hold. We decompose $S_1^{-1/2}(\varphi_{t_1}(\varphi^*))x_1^*$ into the vectors $\{x_1^{(i)+}(t_1,\varphi^*)\}_{i=1}^r$:

$$S_1^{-1/2}(\varphi_{t_1}(\varphi^*))x_1^* = \sum_{i=1}^r \mu_i x_1^{(i)+}, \quad \sum_{i=1}^r \mu_i^2 \neq 0.$$

Now consider the solution $\bar{x}^*$ of the system (3.1.11) for $\varphi = \varphi^*$ of the form

$$\bar{x}^*(t,\varphi^*) = \sum_{i=1}^r \mu_i x_1^{(i)+}(t,\varphi^*).$$

On the one hand this solution vanishes as $t \to +\infty$. On the other hand, in view of (3.1.25) we find

$$V_\varepsilon(\varphi_{t_1}(\varphi^*), \bar{x}^*(t_1,\varphi^*)) = (1+\varepsilon)\langle S_1(\varphi_{t_1}(\varphi^*))\bar{x}_1^*(t_1,\varphi^*), \bar{x}_1^*(t_1,\varphi^*)\rangle$$
$$\times \left[\frac{1-\varepsilon}{1+\varepsilon} - \frac{\langle S_2^{1/2}(\varphi_{t_1}(\varphi^*))\bar{x}_2^*(t_1,\varphi^*), S_2^{1/2}(\varphi_{t_1}(\varphi^*))\bar{x}_2^*(t_1,\varphi^*)\rangle}{\langle S_1^{1/2}(\varphi_{t_1}(\varphi^*))\bar{x}_1^*(t_1,\varphi^*), S_1^{1/2}(\varphi_{t_1}(\varphi^*))\bar{x}_1^*(t_1,\varphi^*)\rangle}\right]$$
$$= (1+\varepsilon)\langle S_1\bar{x}_1^*, \bar{x}_1^*\rangle\left[\frac{1-\varepsilon}{1+\varepsilon} - \frac{\|S_2^{1/2}L_2S_1^{-1/2}x_1^*\|^2}{\|x_1^*\|^2}\right] < 0.$$

The contradiction obtained proves the first inequality (3.1.24). The second inequality (3.1.24) is proved similarly.

Differentiating the equalities (3.1.23), we verify that the functions $L_1(t,\varphi)$ and $L_2(t,\varphi)$ satisfy the equations

$$\frac{d}{dt}L_1 = - L_1 A_{21}(\varphi_t(\varphi))L_1 + A_{11}(\varphi_t(\varphi))L_1 \\ - L_1 A_{22}(\varphi_t(\varphi)) + A_{12}(\varphi_t(\varphi)), \\ \frac{d}{dt}L_2 = - L_2 A_{12}(\varphi_t(\varphi))L_2 + A_{22}(\varphi_t(\varphi))L_2 \\ - L_2 A_{11}(\varphi_t(\varphi)) + A_{21}(\varphi_t(\varphi)). \quad (3.1.26)$$

It is easy to show that the unique solutions of the system (3.1.26) which satisfy the estimates (3.1.24) are the functions determined by (3.1.23). The uniqueness of the solutions $L_1(t,\varphi)$ and $L_2(t,\varphi)$ implies $L_i(t,\varphi) \equiv L_i(0,\varphi_t(\varphi))$, $i = 1,2$. In (3.1.11) we make the change of variables $L(\varphi_t(\varphi))y = x$, where the matrix $L(\varphi)$ is determined by (3.1.18). The matrix $L(\varphi_t(\varphi))$ is represented as

$$L(\varphi_t(\varphi)) = \bar{\bar{L}}_t(\varphi)\bar{L}_t(\varphi)$$

$$= \begin{pmatrix} X_1^+(t,\varphi) & L_1(\varphi_t(\varphi))X_2^-(t,\varphi) \\ L_2(\varphi_t(\varphi))X_1^+(t,\varphi) & X_2^-(t,\varphi) \end{pmatrix} \begin{pmatrix} (X_1^+(t,\varphi))^{-1} & 0 \\ 0 & (X_2^-(t,\varphi))^{-1} \end{pmatrix}.$$

Since $\bar{\bar{L}}_t(\varphi)$ is a fundamental matrix of the solutions to (3.1.11), then after the change of variables $x = \bar{\bar{L}}_t(\varphi)z$ the system (3.1.11) becomes $dz/dt = 0$. A further change of variables $z = \bar{L}_t(\varphi)y$ yields

$$\frac{dy}{dt} = -\bar{L}_t^{-1}(\varphi)\frac{d\bar{L}_t(\varphi)}{dt}y = -\begin{pmatrix} X_1^+(t,\varphi) & 0 \\ 0 & X_2^-(t,\varphi) \end{pmatrix}$$

$$\times \begin{pmatrix} -(X_1^+)^{-1}(A_{11} + A_{12}L_2) & 0 \\ 0 & -(X_2^-)^{-1}(A_{21}L_1 + A_{22}) \end{pmatrix} \begin{pmatrix} y_1 \\ y_2 \end{pmatrix}$$

$$= \begin{pmatrix} [A_{11}(\varphi_t(\varphi)) + A_{12}(\varphi_t(\varphi))L_2(\varphi_t(\varphi))]y_1 \\ [A_{21}(\varphi_t(\varphi))L_1(\varphi_t(\varphi)) + A_{22}(\varphi_t(\varphi))]y_2 \end{pmatrix}.$$

This completes the proof of the Theorem 3.1.1.

Remark 3.1.1 One can assume the existence of the matrix $Q(\varphi) \in C'(\mathcal{T}_m;a)$ such that $Q^*(\varphi)S(\varphi)Q(\varphi) = \text{diag}\{S_1(\varphi), -S_2(\varphi)\}$ in the conditions of Theorem 3.1.1. This is sufficient for the conclusion to be made on the existence of the change of variables $x = L(\varphi_t(\varphi)y$ transforming the system (1.3.11) to the split form.

3.2 Reversibility of the Theorem on Splittability

Lemma 3.2.1 *Let the n-dimensional symmetric matrix function $S(\varphi) \in C^q(\mathcal{T}_m)$, $q \geq 0$, satisfy the condition on positive definiteness, i.e.*

$$\min_{\|x\|=1} \langle S(\varphi)x, x \rangle \geq \gamma = \text{const} > 0, \qquad (3.2.1)$$

for all $\varphi \in \mathcal{T}_m$. Then there exists a non-degenerate matrix function $T(\varphi) \in C^q(\mathcal{T}_m)$ such that

$$T^*(\varphi)S(\varphi)T(\varphi) \equiv I_n, \qquad (3.2.2)$$

for all $\varphi \in \mathcal{T}_m$.

Proof This result is well known for the case when the matrix S is constant. One should follow the proof from Kurosh [1] for the case of a variable matrix function $S(\varphi) \in C^q(\mathcal{T}_m)$, $q \geq 0$. To this end, the quadratic form $\langle S(\varphi)x, x\rangle$ is given in the form

$$\langle S(\varphi)x, x\rangle = s_{11}(\varphi)x_1^2 + \left(\sum_{i=2}^n s_{1i}(\varphi)x_i\right)x_1 + \cdots + s_{nn}(\varphi)x_n^2.$$

The estimate (3.2.1) implies that $s_{11}(\varphi) \geq \gamma > 0$. Then

$$\langle S(\varphi)x, x\rangle = s_{11}(\varphi)\left[x_1 + \frac{1}{2}\left(\sum_{i=2}^n \frac{s_{1i}(\varphi)}{s_{11}(\varphi)}x_i\right)\right]^2 + \bar{s}_{22}(\varphi)x_2^2$$

$$+ \left(\sum_{i=3}^n \bar{s}_{2i}(\varphi)x_i\right)x_2 + \cdots + \bar{s}_{nn}(\varphi)x_n^2.$$

Thus, after the non-degenerate change of variables

$$z_1 = \sqrt{s_{11}(\varphi)}\, x_1 + \frac{1}{2}\sum_{i=2}^n \frac{s_{1i}(\varphi)}{\sqrt{s_{11}(\varphi)}}x_i, \qquad z_i = x_i, \quad i = \overline{2, n},$$

the coefficients of which belong to the space $C^q(\mathcal{T}_m)$, the quadratic form $\langle S(\varphi)x, x\rangle$ becomes

$$\langle ST_1z, T_1z\rangle = z_1^2 + \bar{s}_{22}(\varphi)z_2^2 + \left(\sum_{i=3}^n \bar{s}_{2i}(\varphi)z_i\right)z_2 + \cdots + \bar{s}_{nn}(\varphi)z_n^2.$$

Since the obtained quadratic form is still positive definite, then $\bar{s}_{22}(\varphi) > 0$.

Proceeding in the same way with the transformation of the quadratic form, we verify that under the non-degenerate change of variables $x = T(\varphi)y = T_1(\varphi) \cdots \times T_{n-1}(\varphi)y$ the quadratic form $\langle S(\varphi)x, x\rangle$ is transformed to the sum of squares, i.e. (3.2.2) is valid.

Lemma 3.2.2 *Let an n-dimensional symmetric matrix function*

$$S(\varphi) = \begin{pmatrix} S_{11}(\varphi) & S_{12}(\varphi) \\ S_{21}(\varphi) & S_{22}(\varphi) \end{pmatrix}, \quad S_{12}^*(\varphi) \equiv S_{21}(\varphi), \quad S_{ii}^*(\varphi) \equiv S_{ii}(\varphi),$$

belong to the space $C^q(\mathcal{T}_m)$, $q \geq 0$, and correspondingly for the quadratic r- and $n-r$-dimensional blocks $S_{11}(\varphi)$ and $S_{22}(\varphi)$ the conditions

$$\langle S_{11}(\varphi)x_1, x_1 \rangle \leq -\gamma \|x_1\|^2,$$
$$\langle S_{22}(\varphi)x_2, x_2 \rangle \geq \gamma \|x_2\|^2, \quad \gamma = \text{const} > 0,$$

are satisfied for all $\varphi \in \mathcal{T}_m$, $x_1 \in R^r$, $x_2 \in R^{n-r}$. Then there exists a non-degenerate $n \times n$-dimensional matrix function $T(\varphi) \in C^q(\mathcal{T}_m)$ satisfying the identity

$$T^*(\varphi)S(\varphi)T(\varphi) = \text{diag}\{-I_r, I_{n-r}\}, \tag{3.2.3}$$

for all $\varphi \in \mathcal{T}_m$.

Proof Consider the quadratic form

$$\langle S(\varphi)x, x \rangle = \langle S_{11}(\varphi)x_1, x_1 \rangle + 2\langle S_{12}(\varphi)x_2, x_1 \rangle + \langle S_{22}(\varphi)x_2, x_2 \rangle,$$

and in view of the Lemma 3.2.1 reduce it by the non-degenerate change of variables $x_1 = \bar{T}(\varphi)y_1$, $x_2 = \tilde{\bar{T}}(\varphi)y_2$, $\bar{T}(\varphi), \tilde{\bar{T}}(\varphi) \in C^q(\mathcal{T}_m)$, to the form

$$-\langle y_1, y_1 \rangle + 2\langle \bar{T}^*(\varphi)S_{12}(\varphi)\tilde{\bar{T}}(\varphi)y_2, y_1 \rangle + \langle y_2, y_2 \rangle$$
$$= -\langle y_1, y_1 \rangle - \langle M_{12}^*(\varphi)y_1, M_{12}^*(\varphi)y_1 \rangle + \langle y_2 + M_{12}^*(\varphi)y_1, y_2 + M_{12}^*(\varphi)y_1 \rangle$$
$$= -\langle (I_r + M_{12}(\varphi)M_{12}^*(\varphi))y_1, y_1 \rangle + \langle z_2, z_2 \rangle,$$

where $M_{12} = \bar{T}^* S_{12} \tilde{\bar{T}}$, $z_2 = y_2 + M_{12}^* y_1$. Since

$$\langle [I_r + M_{12}(\varphi)M_{12}^*(\varphi)]y_1, y_1 \rangle = \|y_1\|^2 + \|M_{12}^*(\varphi)y_1\|^2 \geq \|y_1\|^2,$$

the result of the Lemma 3.2.1 ensures the reduction of the quadratic form $\langle [I_r + M_{12}(\varphi)M_{12}^*(\varphi)]y_1, y_1 \rangle$ to the sum of squares by the non-degenerate change of variables $y_1 = \tilde{T}(\varphi)z_1$. Thus, taking for the matrix $T(\varphi) \in C^q(\mathcal{T}_m)$ the matrix

$$T(\varphi) = \begin{pmatrix} \bar{T}(\varphi)\tilde{T}(\varphi) & 0 \\ -\tilde{\bar{T}}(\varphi)M_{12}(\varphi)\tilde{T}(\varphi) & \tilde{\bar{T}}(\varphi) \end{pmatrix},$$

we verify that (3.2.3) is satisfied. This completes the proof of Lemma 3.2.2.

Assume that the system (3.1.8) is exponentially dichotomous on the whole R axis uniformly in φ, i.e. for every fixed $\varphi \in \mathcal{T}_m$ the space R^n is solely representable in the form of a sum of two subspaces $E^+(\varphi)$ and $E^-(\varphi)$ of the dimensions r and $n - r$ correspondingly, so that the estimates

$$\|\Omega_0^t(\varphi;A)x^+\| \leq K\|\Omega_0^\tau(\varphi;A)x^+\| \exp\{-\beta(t-\tau)\}, \quad \tau \leq t, \quad x^+ \in E^+,$$
$$\|\Omega_0^t(\varphi;A)x^-\| \leq K\|\Omega_0^\tau(\varphi;A)x^-\| \exp\{\beta(t-\tau)\}, \quad t \leq \tau, \quad x^- \in E^- \quad (3.2.4)$$

are satisfied, where the positive constants K and β do not depend on $\varphi \in \mathcal{T}_m$. Also, we assume the existence of a non-degenerate $n \times n$-dimensional matrix function $L(\varphi) \in C^s(\mathcal{T}_m)$, $s \geq 1$, such that

$$L^{-1}(\varphi)A(\varphi)L(\varphi) - L^{-1}(\varphi)\frac{\partial L(\varphi)}{\partial \varphi}a(\varphi) = \text{diag}\{A^+(\varphi), A^-(\varphi)\} \equiv \bar{A}(\varphi), \quad (3.2.5)$$

where $A^+(\varphi)$ and $A^-(\varphi)$ are correspondingly $r \times r$- and $(n-r) \times (n-r)$-dimensional square matrix functions from the space $C^{\bar{s}}(\mathcal{T}_m)$, $\bar{s} = \min\{s, q\}$, and for the matriciants $\Omega_\tau^t(\varphi; A^+)$ and $\Omega_\tau^t(\varphi; A^-)$

$$\|\Omega_\tau^t(\varphi; A^+)\| \leq K \exp\{-\beta(t-\tau)\}, \quad \tau \leq t,$$
$$\|\Omega_\tau^t(\varphi; A^-)\| \leq K \exp\{\beta(t-\tau)\}, \quad t \leq \tau. \quad (3.2.6)$$

Theorem 3.2.1 *Let the system (3.1.8) be exponentially dichotomous on the whole R axis uniformly in φ, and there exists a non-degenerate matrix function $L(\varphi) \in C^s(\mathcal{T}_m)$, $s \geq 1$, which is coordinated with the dichotomy block splitting (3.2.5). Then for any symmetric matrix function $S(\varphi) \in C^1(\mathcal{T}_m)$ which satisfies the estimate*

$$\left\langle \left(\frac{\partial S(\varphi)}{\partial \varphi}a(\varphi) + S(\varphi)A(\varphi) + A^*(\varphi)S(\varphi)\right)x, x \right\rangle \leq -\gamma \|x\|^2, \quad (3.2.7)$$

where $\gamma = \text{const} > 0$, $x \in R^n$, there exists a matrix function $T(\varphi) \in C^{\min\{q,s\}}(\mathcal{T}_m)$ satisfying (3.2.3).

Proof Since the matrix $L(\varphi) \in C^s(\mathcal{T}_m)$ makes the block splitting (3.2.5), and $S(\varphi) \in C^q(\mathcal{T}_m)$ satisfies the estimate (3.2.7), then for all $y \in R^n$

$$\left\langle \left(\frac{\partial \bar{S}(\varphi)}{\partial \varphi}a(\varphi) + \bar{S}(\varphi)\bar{A}(\varphi) + \bar{A}^*(\varphi)\bar{S}(\varphi)\right)y, y \right\rangle \leq -\gamma_1 \|y\|^2, \quad (3.2.8)$$

where $\bar{S} = L^*SL$, $\gamma = \text{const} > 0$. In fact, designating $\dot{S} = \frac{\partial S(\varphi)}{\partial \varphi}a(\varphi)$ we have for the estimate (3.2.7)

$$\langle L^*(\dot{S} + SA + A^*S)Ly, y \rangle \leq -\gamma \|Ly\|^2,$$

or

$$\langle (L^*\dot{S}L + \dot{L}^*SL + L^*S\dot{L} - \dot{L}^*SL - L^*SL + \dot{L}^*SLL^{-1}AL + L^*A^*L^{*-1}L^*SL)y, y\rangle$$
$$\leq -\gamma \|L^{-1}\|^{-2}\|y\|^2.$$

Hence it follows that (3.2.8) is satisfied, where $\gamma_1 = \min_\varphi \|L^{-1}(\varphi)\|^{-2}$.

Let us now verify that the matrix function $S(\varphi)$ satisfies the conditions of Lemma 3.2.2. We present it in the block form

$$\bar{S}(\varphi) = \begin{pmatrix} \bar{S}_{11}(\varphi) & \bar{S}_{12}(\varphi) \\ \bar{S}_{21}(\varphi) & \bar{S}_{22}(\varphi) \end{pmatrix},$$

and show that (3.2.8) ensures (3.2.1) for the matrices $\bar{S}_{11}(\varphi)$ and $\bar{S}_{22}(\varphi)$. The inequality (3.2.8) can obviously be reduced to the form

$$\frac{d}{dt}\langle \bar{S}(\varphi_t(\varphi))\Omega_\tau^t(\varphi;\bar{A})y, \Omega_\tau^t(\varphi;\bar{A})y\rangle \leq -\gamma_1\|\Omega_\tau^t(\varphi;\bar{A})y\|^2.$$

Integrating both parts of the inequality from τ to t, $\tau \leq t$, yields

$$\langle \bar{S}(\varphi_t(\varphi))\Omega_\tau^t(\varphi;\bar{A})y, \Omega_\tau^t(\varphi;\bar{A})y\rangle - \langle \bar{S}(\varphi_\tau(\varphi))y, y\rangle$$
$$\leq -\gamma_1 \int_\tau^t \|\Omega_\tau^\sigma(\varphi;\bar{A})y\|^2 \, d\sigma. \quad (3.2.9)$$

Now we set $y = \operatorname{col}\{y_1, 0\}$ in (3.2.9), where y_1 is a non zero r-dimensional vector. Then

$$\langle \bar{S}_{11}(\varphi_t(\varphi))\Omega_\tau^t(\varphi;A^+)y, \Omega_\tau^t(\varphi;A^+)y_1\rangle - \langle \bar{S}_{11}(\varphi_\tau(\varphi))y_1, y_1\rangle$$
$$\leq -\gamma_1 \int_\tau^t \|\Omega_\tau^\sigma(\varphi;A^+)y_1\|^2 \, d\sigma.$$

We pass to the limit as $t \to \infty$ in the inequality, taking into account the estimates (3.2.6) and the boundedness of the function $\bar{S}_{11}(\varphi_t(\varphi))$

$$-\langle \bar{S}_{11}(\varphi_\tau(\varphi))y_1, y_1\rangle \leq -\gamma_1 \int_t^\infty \|\Omega_\tau^\sigma(\varphi;A^+)y_1\|^2 \, d\sigma,$$

i.e.

$$\langle \bar{S}_{11}(\varphi)y_1, y_1\rangle \geq \gamma_1 \int_0^\infty \|\Omega_0^\sigma(\varphi;A^+)y_1\|^2 \, d\sigma.$$

Hence it follows that $\langle \bar{S}_{11}(\varphi)y_1, y_1\rangle \geq \bar{\gamma}_1\|y_1\|^2$, $\bar{\gamma}_1 = \text{const} > 0$. We similarly verify the negative definiteness of the quadratic form $\langle \bar{S}_{22}(\varphi)y_2, y_2\rangle$. Thus, the result of Lemma 3.2.2 completes the proof of the Theorem 3.2.1.

Remark 3.2.1 A similar result is valid, if less smoothness is required of the functions $a(\varphi)$, $A(\varphi)$ and $S(\varphi)$. One can assume that $a(\varphi) \in C_{\text{Lip}}(\mathcal{T}_m)$, $A(\varphi) \in C^0(\mathcal{T}_m)$ and $S(\varphi) \in C'(\mathcal{T}_m; a)$.

Remark 3.2.2 Since there exist exponentially dichotomous linear systems of differential equations with quasiperiodic coefficients which are not split by the quasiperiodic change of variables, there are non-degenerate symmetric matrix functions $S(\varphi) \in C^q(\mathcal{T}_m)$, $q \geq 1$ (even among the trigonometrical polynomials) having r negative and $n-r$ positive eigenvalues, for which no continuous matrix function $T(\varphi) \in C^0(\mathcal{T}_m)$ exists which realizes the identity (3.2.3).

The above results together with those from Samoilenko [3] give the following theorem.

Theorem 3.2.2 *Let a n-dimensional symmetric matrix function $S(\varphi) \in C^q(\mathcal{T}_m)$ have one negative eigenvalue for all $\varphi \in \mathcal{T}_m$, while the rest of the eigenvalues are positive, and $m < n - 1$, where m is a number of variables $\varphi_1, \ldots, \varphi_m$, n is the dimension of the matrix $S(\varphi)$. Then there exists an $n \times n$-dimensional matrix function $T(\varphi)$ which is continuously differentiable in φ up to the order q, is 4π-periodic in φ and is such that $T^*(\varphi)S(\varphi)T(\varphi) = \text{diag}\{-1, I_{n-1}\}$.*

Consider the case when $n = 2$ and prove an auxiliary assertion.

Lemma 3.2.3 *Let $S(\varphi)$ be a two-dimensional matrix belonging to the space $C'(\mathcal{T}_m; a)$ and having eigenvalues $\lambda_1(\varphi)$ and $\lambda_2(\varphi)$ of opposite signs. Then there exists a non-degenerate continuous and 4π-periodic in φ matrix $Q(\varphi)$ which is continuously differentiable in t for $\varphi = \varphi_t(\varphi)$ and such that*

$$S(\varphi) = Q(\varphi) \text{diag}\{d_1(\varphi), -d_2(\varphi)\} Q^*(\varphi), \tag{3.2.10}$$

where $d_1(\varphi)$ and $d_2(\varphi)$ are positive functions for all $\varphi \in \mathcal{T}_m$.

Proof First we note that if (3.2.10) is possible with the matrix $Q(\varphi)$ satisfying the conditions of Lemma 3.2.3, it is possible with $d_1(\varphi) = \lambda_1(\varphi)$, $d_2(\varphi) = -\lambda_2(\varphi)$ if $\lambda_1(\varphi)$ is considered to be a positive eigenvalue and λ_2 is considered to be a negative eigenvalue of the matrix $S(\varphi)$. Then however

$$\det S(\varphi) = [\det Q(\varphi)]^2 d_1(\varphi)(-d_2(\varphi)) = [\det Q(\varphi)]^2 \lambda_1(\varphi)\lambda_2(\varphi)$$
$$= [\det Q(\varphi)]^2 \det S(\varphi),$$

and consequently

$$[\det Q(\varphi)]^2 = 1.$$

If $\det Q(\varphi) = -1$, then the matrix

$$Q(\varphi) \, \text{diag}\{1, -1\}$$

realizes the representation (3.2.10) and has the determinant equal to 1. Thus, if the representation (3.2.10) is possible, then the representation

$$S(\varphi) = Q(\varphi) \, \text{diag}\{\lambda_1(\varphi), \lambda_2(\varphi)\} Q^*(\varphi) \qquad (3.2.11)$$

is also possible, with a 4π-periodic matrix $Q(\varphi)$ satisfying the condition of Lemma 3.2.3 and such that

$$\det Q(\varphi) = 1. \qquad (3.2.12)$$

That is why we search for a matrix Q realizing the representation (3.2.11) and subject to (3.2.12).

Let

$$S(\varphi) = \begin{pmatrix} a & b \\ b & d \end{pmatrix}, \quad Q = \begin{pmatrix} x & y \\ x_1 & y_1 \end{pmatrix}.$$

Condition (3.2.12) becomes the equation

$$xy_1 - yx_1 = 1, \qquad (3.2.13)$$

and (3.2.11) becomes the system of equations

$$\begin{aligned} ay_1 - by &= \lambda_1 x, & -ax_1 + bx &= \lambda_2 y, \\ by_1 - dy &= \lambda_1 x, & -bx_1 + dx &= \lambda_2 y_1. \end{aligned} \qquad (3.2.14)$$

Substituting x and x_1 found by the first two equations of (3.2.14) into the other two equations we get

$$(ad - b^2)y = \lambda_1 \lambda_2 y, \quad (ad - b^2)y_1 = \lambda_1 \lambda_2 y_1.$$

Therefore, in order to find $Q(\varphi)$ it is sufficient to take x and x_1 relative to y and y_1 in accordance with the first two equalities of (3.2.14) and satisfy (3.2.13). In view of the choice of x and x_1 the latter becomes

$$ay_1^2 - 2byy_1 + dy^2 = \lambda_1. \qquad (3.2.15)$$

Thus, the problem is reduced to the determination of two functions $y(\varphi)$ and $y_1(\varphi)$ which are continuously periodic in φ reducing (3.2.15) to the identity for all $\varphi \in \mathcal{T}_m$. Since any two functions $y(\varphi)$ and $y_1(\varphi)$ satisfying (3.2.15) do not take zero values simultaneously, then one can set

$$y = R \cos \psi, \quad y_1 = R \sin \psi, \qquad (3.2.16)$$

and reduce (3.2.15) to the form

$$R^2[a \sin^2 \psi - 2b \sin \psi \cos \psi + d \cos^2 \psi] = \lambda_1,$$

or, by obvious transformations, to the form

$$(d - a) \cos 2\psi - ab \sin 2\psi = \frac{2\lambda_1}{R^2} - (a + d). \qquad (3.2.17)$$

The eigenvalues of the matrix $S(\varphi)$ have opposite signs, and hence

$$(d - a)^2 + 4b^2 \neq 0,$$

which allows (3.2.17) to be transformed to

$$\cos(2\psi + \alpha) = \frac{2\lambda_1 - (a + d)R^2}{R^2[(d - a)^2 + 4b^2]^{1/2}}, \qquad (3.2.18)$$

where α is determined as

$$\cos \alpha = \frac{d - a}{[(d - a)^2 + 4b^2]^{1/2}}, \quad \sin \alpha = \frac{2b}{[(d - a)^2 + 4b^2]^{1/2}}.$$

Since $\lambda_1(\varphi) > 0$, $\lambda_2(\varphi) < 0$, then $(a+d)^2 < (d-a)^2 + 4b^2$. Therefore, taking $R(\varphi)$ large enough, one can make the absolute value of the right-hand side of (3.2.18) smaller than 1. For such a choice of $R(\varphi)$ the equation (3.2.18) has the solution in the form of the function

$$\psi = \frac{1}{2} \arccos \frac{2\lambda_1 - (a + d)R^2}{R^2[(d - a)^2 + 4b^2]^{1/2}} - \frac{\alpha}{2}.$$

We determine the properties of the solution as a function of the variable φ. The choice of a sufficiently large $R = R(\varphi)$ from the subspace $C^1(\mathcal{T}_m; a)$ ensures the function

$$\arccos \frac{2\lambda_1 - (a + d)R^2}{R^2[(d - a)^2 + 4b^2]^{1/2}} = 2\Phi(\varphi),$$

belonging to the space $C^1(\mathcal{T}_m; a)$. Applying the theorem on the argument to the pair of the functions

$$\frac{d(\varphi) - a(\varphi)}{[(d(\varphi) - a(\varphi))^2 + 4b^2(\varphi)]^{1/2}}, \quad \frac{2b(\varphi)}{[(d(\varphi) - a(\varphi))^2 + 4b^2(\varphi)]^{1/2}},$$

we verify that

$$\alpha = \alpha(\varphi) = \langle k, \varphi \rangle + 2\Phi_1(\varphi),$$

where $k = (k_1, \ldots, k_m)$ is an integer-valued vector, Φ_1 is a function from $C'(\mathcal{T}_m; a)$. Then however

$$\psi = \Phi(\varphi) - \Phi_1(\varphi) - \frac{\langle k, \varphi \rangle}{2} = \Phi_0(\varphi) - \frac{\langle k, \varphi \rangle}{2},$$

where $\Phi_0 \in C'(\mathcal{T}_m; a)$.

The change of variables (3.2.16) implies that the functions

$$y = y(\varphi) = R(\varphi) \cos \left(\Phi_0(\varphi) - \frac{\langle k, \varphi \rangle}{2} \right),$$

$$y_1 = y_1(\varphi) = R(\varphi) \sin \left(\Phi_0(\varphi) - \frac{\langle k, \varphi \rangle}{2} \right),$$

satisfy the equality (3.2.15) for all $\varphi \in \mathcal{T}_m$. These functions have the period 4π in φ_ν, $\nu = 1, \ldots, m$, are continuous in φ, and have continuous derivatives in t for $\varphi = \varphi_t(\varphi)$. Then the matrix $Q(\varphi)$ possesses similar properties as well, the elements of which are the functions $y(\varphi)$ and $y_1(\varphi)$ and the functions $x(\varphi)$ and $x_1(\varphi)$ are expressed via the former.

In order to prove that the doubling of the period sometimes actually takes place when the matrix $S(\varphi)$ is represented in the form of (3.2.10), we consider the representation mentioned above for the matrix

$$S(\varphi) = \begin{pmatrix} -\cos \varphi & \sin \varphi \\ \sin \varphi & \cos \varphi \end{pmatrix}, \qquad (3.2.19)$$

where φ is a scalar. The numbers $\lambda_1 = 1$ and $\lambda_2 = -1$ are the eigenvalues of this matrix. The eigenvectors of the matrix can be taken as $\left(\sin \frac{\varphi}{2}, \cos \frac{\varphi}{2} \right)$ for $\lambda_1 = 1$ and $\left(\cos \frac{\varphi}{2}, -\sin \frac{\varphi}{2} \right)$ for $\lambda_2 = -1$. Hence the eigenvectors of the matrix $S(\varphi)$ cannot be periodically dependent on φ with the period 2π.

We call the system (3.2.4) two-dimensional, if $x = (x_1, x_2)$ is a two-dimensional vector, and C'-*block-split* with the period doubling, when the splitting is made by the matrix $L(\varphi)$ which is continuous in φ, has a continuous derivative in t for $\varphi = \varphi_t(\varphi)$ and is 4π-periodic in φ.

The following result is a corollary of Lemma 3.2.3 and Theorem 3.2.1.

Theorem 3.2.3 *Let $a \in C_{\mathrm{Lip}}(\mathcal{T}_m)$, $A \in C^0(\mathcal{T}_m)$ and the system (1.2.4) be two-dimensional and exponentially dichotomous. Then it is C'-block-split with period doubling.*

For $n \geq 3$ the period doubling is not sufficient for the C'-block-splittability of the exponentially dichotomous system.

3.3 On Triangulation and the Relationship of C'-block Splittability of a Linear System with the Problem on r-frame Complementability up to the Periodic Basis in R^n

We proceed with the system of equations

$$\frac{d\varphi}{dt} = a(\varphi), \quad \frac{dx}{dt} = A(\varphi)x, \tag{3.3.1}$$

viewing a and A as functions from $C_{\text{Lip}}(\mathcal{T}_m)$ and $C^0(\mathcal{T}_m)$ respectively. We set

$$Lu = \dot{u} - A(\varphi)u,$$

which prescribes the operator L on the space of functions $C'(\mathcal{T}_m; a)$. The scalar function $\lambda = \lambda(\varphi)$ from $C^0(\mathcal{T}_m)$ is called an *eigenvalue of the operator L* providing that

$$Lu = \lambda u \tag{3.3.2}$$

has a non-trivial solution $u = u(\varphi) \not\equiv 0$, $\varphi \in \mathcal{T}_m$ in $C^1(\mathcal{T}_m; a)$. The eigenvalue of the operator L is called the *eigennumber*, if it does not depend on φ. The non-trivial solution $C'(\mathcal{T}_m; a)$ of (3.3.2) is called the *eigenfunction of the operator L corresponding to the eigenvalue λ.*

Theorem 3.3.1 *Let the operator L have eigenvalues $\lambda_1, \lambda_2, \ldots, \lambda_p$ and*

$$u_1(\varphi), \ u_2(\varphi), \ldots, \ u_p(\varphi) \tag{3.3.3}$$

be eigenfunctions of the operator L corresponding to these values. Then the subspace

$$x = u_1(\varphi)y_1 + \cdots + u_p(\varphi)y_p, \quad y_i \in R, \quad \varphi \in \mathcal{T}_m, \tag{3.3.4}$$

is as invariant set of the system (3.3.1) and on this surface the system (3.3.1) is equivalent to the system

$$\frac{d\varphi}{dt} = a(\varphi), \quad \frac{dy_i}{dt} = -\lambda_i(\varphi)y_i, \quad i = 1, \ldots, p. \tag{3.3.5}$$

Proof We fixed the point (φ_0, x_0) on the surface (3.3.4) setting

$$x_0 = \sum_{j=1}^{p} u_j(\varphi_0)y_j^0, \quad y_j^0 \in R,$$

and consider the motion $\varphi_t(\varphi_0)$, $x_t(\varphi_0, x_0)$ of the system (3.3.1) beginning at this point. Let

$$x(t, \varphi_0, x) = \sum_{j=1}^{p} u_j(\varphi_t(\varphi_0))y_j(t, \varphi_0, y_j^0),$$

where $\varphi_t(\varphi_0)$, $y_i(t,\varphi_0,y_i^0)$ is a solution of the system (3.3.5) with the initial conditions φ_0, y_i^0. Differentiating the function $x(t,\varphi_0,x_0)$ in t we verify that

$$\frac{dx(t,\varphi_0,x_0)}{dt} = \sum_{j=1}^{p}[\dot{u}_j(\varphi_t(\varphi_0))y_j(t,\varphi_0,y_j^0) - \lambda_j(\varphi_t(\varphi_0))u_j(\varphi_t(\varphi_0))y_j(t,\varphi_0,y_j^0)]$$

$$= A(\varphi_t(\varphi_0))\sum_{j=1}^{p}u_j(\varphi_t(\varphi_0))y_j(t,\varphi_0,y_j^0) = A(\varphi_t(\varphi_0))x(t,\varphi_0,x_0).$$

Hence it follows that $\varphi_t(\varphi_0)$, $x(t,\varphi_0,x_0)$ is a solution of the system (3.3.1) determined by the initial conditions φ_0, x_0. Then however $x(t,\varphi_0,x_0) = x_t(\varphi_0,x_0)$, $t \in R$. The latter means that any motion $\varphi_t(\varphi_0)$, $x_t(\varphi_0,x_0)$ where (φ_0,x_0) belongs to the surface (3.3.4) is determined by (3.3.4) for any $t \in R$ providing φ, y_j in (3.3.4 is replaced by the solution of the system (3.3.5). This is sufficient to prove Theorem 3.3.1.

Note that the surface (3.3.4) is a manifold which is homeomorphic to the direct product of the torus $\mathcal{T}_m$ by the plane R^p whenever vectors (3.3.3) form a p-frame in R^n. In particular, it can be the separatrix manifold of the exponentially dichotomous system.

Theorem 3.3.2 *System (3.3.1) is reduced by the change of variables*

$$x = U(\varphi)y + V(\varphi)z, \quad \det[U(\varphi), V(\varphi)] \neq 0, \tag{3.3.6}$$

to the block-triangular form

$$\frac{d\varphi}{dt} = a(\varphi), \quad \frac{dz}{dt} = Q(\varphi)z, \quad \frac{dy}{dt} = -D(\varphi)y + Q_1(\varphi)z, \tag{3.3.7}$$

where $U(\varphi) = (u_1(\varphi),\ldots,u_p(\varphi))$, $V(\varphi) = (v_1(\varphi),\ldots,v_{n-p}(\varphi))$ are the matrices from $C'(\mathcal{T}_m;a)$, $D(\varphi) = \text{diag}\{\lambda_1(\varphi),\ldots,\lambda_p(\varphi)\} \in C^0(\mathcal{T}_m)$ if and only if the functions $\lambda_j(\varphi)$, $j = 1,\ldots,p$, are eigenvalues of the operator L, and $u_j(\varphi)$, $j = 1,\ldots,p$, are the eigenfunctions of the operator L corresponding to them.

Proof First we assume that $\lambda_j(\varphi)$, $j = 1,\ldots,p$, are eigenvalues of the operator L, and $u_j(\varphi)$, $j = 1,\ldots,p$, are the eigenfunctions of the operator L corresponding to them, so that

$$Lu_j(\varphi) = \lambda_j(\varphi)u_j(\varphi), \quad j = \overline{1,p}, \quad \varphi \in \mathcal{T}_m. \tag{3.3.8}$$

The equalities (3.3.8) are equivalent to a matrix equality

$$LU(\varphi) = U(\varphi)D(\varphi), \quad \varphi \in \mathcal{T}_m. \tag{3.3.9}$$

Let $V(\varphi)$ be the matrix possessing the properties mentioned in Theorem 3.3.2. We shall verify that the change of variables (3.3.6) reduces the system (3.3.1) to (3.3.7). The motion $\varphi = \varphi_t(\varphi_0)$, $x = x_t(\varphi_0, x_0)$ of the system (3.3.1) is shown to be defined by (3.3.6) for any $\varphi_0 \in \mathcal{T}_m$, $x_0 \in R^n$, $t \in R$, if in (3.3.6) $\varphi = \varphi_t(\varphi_0)$, $y = y_t$, $z = z_t$ and y_t, z_t are taken for the solution of the system (3.3.7). With this in mind we set

$$\dot{U}(\varphi_t(\varphi))y_t + U(\varphi_t(\varphi))\frac{dy_t}{dt} + \dot{V}(\varphi_t(\varphi))z_t + V(\varphi_t(\varphi))\frac{dz_t}{dt}$$
$$= A(\varphi_t(\varphi))U(\varphi_t(\varphi))y_t + A(\varphi_t(\varphi))V(\varphi_t(\varphi))z_t,$$

for all $t \in R$, $\varphi \in \mathcal{T}_m$. In view of (3.3.9) we get for dy_t/dt and dz_t/dt

$$U(\varphi_t(\varphi))\left(\frac{dy_t}{dt} + D(\varphi_t(\varphi))y_t\right) + V(\varphi_t(\varphi))\frac{dz_t}{dt} \qquad (3.3.10)$$
$$= [A(\varphi_t(\varphi))V(\varphi_t(\varphi)) - \dot{V}(\varphi_t(\varphi))]z_t.$$

We designate the matrix inverse to the matrix $\Phi(\varphi) = [U(\varphi), V(\varphi)]$ by

$$\begin{pmatrix} U_1(\varphi) \\ V_1(\varphi) \end{pmatrix} = \Phi^{-1}(\varphi).$$

Then the identity $\Phi^{-1}(\varphi)\Phi(\varphi) = I_n$ can be represented as

$$U_1(\varphi)U(\varphi) = I_n, \quad U_1(\varphi)V(\varphi) = 0,$$
$$V_1(\varphi)U(\varphi) = 0, \quad V_1(\varphi)V(\varphi) = I_n,$$

where I_n and 0 are the identity and zero matrices of the corresponding dimensions.

Multiplying on the left (3.3.10) by $\Phi^{-1}(\varphi_t(\varphi))$, we get for dy_t/dt and dz_t/dt the system

$$\frac{dy_t}{dt} = -D(\varphi_t(\varphi))y_t + U_1(\varphi_t(\varphi))[A(\varphi_t(\varphi))V(\varphi_t(\varphi)) - \dot{V}(\varphi_t(\varphi))]z_t,$$

$$\frac{dz_t}{dt} = V_1(\varphi_t(\varphi))[A(\varphi_t(\varphi))V(\varphi_t(\varphi)) - \dot{V}(\varphi_t(\varphi))]z_t,$$

proving that the change of variables (3.3.6) reduces the initial system (3.3.1) to (3.3.7)

$$\frac{d\varphi}{dt} = a(\varphi), \quad \frac{dz}{dt} = V_1(\varphi)[A(\varphi)V(\varphi) - \dot{V}(\varphi)]z,$$
$$\frac{dy}{dt} = -D(\varphi)y + U_1(\varphi)[A(\varphi)V(\varphi) - \dot{V}(\varphi)]z.$$

Conversely, let the system (3.3.1) be reduced to the form (3.3.7) by the change of variables (3.3.6). The motion $\varphi_t(\varphi)$, $x_t(\varphi, x_0)$ is then related to the solution $\varphi_t(\varphi)$, y_t, z_t of the system (3.3.7) by

$$x_t(\varphi, x_0) = U(\varphi_t(\varphi))y_t + V(\varphi_t(\varphi))z_t.$$

Differentiating the correlation in t yields

$$\dot{U}(\varphi_t(\varphi))y_t + U(\varphi_t(\varphi))\left[-D(\varphi_t(\varphi))y_t + Q_1(\varphi_t(\varphi))z_t\right] + \dot{V}(\varphi_t(\varphi))z_t$$
$$+ V(\varphi_t(\varphi))Q(\varphi_t(\varphi))z_t = A(\varphi_t(\varphi))U(\varphi_t(\varphi))y_t + A(\varphi_t(\varphi))V(\varphi_t(\varphi))z_t.$$

Considering this identity for the solution y_t, z_t, $z_t \equiv 0$, we come to a new identity

$$\left[\dot{U}(\varphi_t(\varphi)) - A(\varphi_t(\varphi))U(\varphi_t(\varphi)) - U(\varphi_t(\varphi))D(\varphi_t(\varphi))\right]y_t \equiv 0,$$

which due to the arbitrary choice of the initial value y_0 becomes for $t = 0$

$$\dot{U}(\varphi) - A(\varphi)U(\varphi) = U(\varphi)D(\varphi), \quad \varphi \in \mathcal{T}_m.$$

However, $U(\varphi)$ then satisfies the matrix equation (3.3.9) and hence, λ_j, $j = 1, \ldots, p$, are the eigenvalues and $u_j(\varphi)$, $j = 1, \ldots, p$, are the corresponding eigenfunctions of the operator L.

Theorem 3.3.2 obviously yields the following result which shows the relationship of the C'-block splittability of the linear system (3.3.1) to the problem of r-frame complementability up to the periodic basis in R^n.

Theorem 3.3.3 *The change of variables (3.3.6) with the matrix $[U(\varphi), V(\varphi)] \in C'(\mathcal{T}_m; a)$ reduces the system (3.3.1) to the block-triangular form (3.3.7) if and only if the p-frame (3.3.3) is complemented up to the periodic basis in R^n.*

By Theorem 3.3.3 the system (3.3.1), the eigenfunctions of which form a frame which is not complementable up to a periodic basis in R^n, is not reduced by (3.3.6) to the block-triangular form (3.3.7). System (3.3.7) has the general form of the block-triangular system when y is a one-dimensional vector. Therefore, if the operator L only has the eigenvalues such that the eigenfunctions corresponding to them cannot be complemented up to the periodic basis in R^n, then the system (3.3.1) cannot be reduced by the transformation with the coefficients from the space $C'(\mathcal{T}_m; a)$ to the block-triangular form with one- and $n-1$-dimensional diagonal blocks. This fact can be employed to construct exponentially dichotomous systems which are not C'-block-splittable.

We claim that the exponentially dichotomous system (3.3.1) is C'-*block-splittable with the rectification of separatrix manifolds*, if the separatrix manifolds of the split system are its coordinate planes, and the motions on them are determined by the block subsystems of the split system.

Theorem 3.3.4 *Let the system (3.3.1) be exponentially dichotomous. Assume that there are r positive eigenvalues among those of the operator L and the eigenfunctions corresponding to them form an r-frame in R^n. Then the system (3.3.1) is C'-block splittable with the rectification of the separatrix manifolds, if the mentioned frame is complemented up to the periodic basis in R^n, and is not if this is not the case.*

Proof Designate the positive eigenvalues of the operator L by

$$\lambda_1(\varphi), \ldots, \lambda_r(\varphi) \tag{3.3.11}$$

and by

$$u_1(\varphi), \ldots, u_r(\varphi) \tag{3.3.12}$$

the eigenfunctions of the operator corresponding to them. Assume that the functions (3.3.12) form a r-frame complemented up to the periodic basis in R^n. Then the change (3.3.6) with the matrix $V(\varphi)$ from $C'(\mathcal{T}_m; a)$ complementing $U(\varphi) = (u_1(\varphi), \ldots, u_r(\varphi))$ up to the periodic basis in R^n reduces the system (3.3.1) to (3.3.7). For the change (3.3.6) the property of exponential dichotomy of the system (3.3.1) is presented; therefore system (3.3.7) is exponentially dichotomous.

The number r determines, as has been stated before, the dimension of the separatrix manifold M^+ of the exponentially dichotomous system. The positiveness of the eigenvalues $\lambda_j(\varphi)$, $j = 1, \ldots, r$, and the type of system (3.3.7) yield that its separatrix manifold M^+ is prescribed by the equation

$$z = 0, \quad \varphi \in \mathcal{T}_m. \tag{3.3.13}$$

In terms of the variables φ, x the manifold M^+ is described by

$$C(\varphi)x = x, \quad \varphi \in \mathcal{T}_m, \tag{3.3.14}$$

where $C(\varphi)$ is a projector from $C'(\mathcal{T}_m; a)$. Substituting the expression (3.3.6) into equation (3.3.14) for the variables x expressed via y and z we get the following equation for the manifold M^+ of the system (3.3.7)

$$C(\varphi)U(\varphi)y + C(\varphi)V(\varphi)z = U(\varphi)y + V(\varphi)z. \tag{3.3.15}$$

The points of the manifold (3.3.13) satisfy the equation (3.3.15). This is possible due to the arbitrariness of the choice of y only if

$$C(\varphi)U(\varphi) = U(\varphi), \quad \varphi \in \mathcal{T}_m. \tag{3.3.16}$$

Consider the equation of the separatrix manifold M^-

$$C(\varphi)x = 0, \quad \varphi \in \mathcal{T}_m.$$

When the change (3.3.6) is made, it becomes a separatrix manifold of the system (3.3.7) with the equation

$$C(\varphi)[U(\varphi)y + V(\varphi)z] = 0, \quad \varphi \in \mathcal{T}_m,$$

which in view of (3.3.16) becomes

$$U(\varphi)y + C(\varphi)V(\varphi)x = 0. \tag{3.3.17}$$

The matrix $U^*(\varphi)U(\varphi)$ is the Gramm matrix of linearly independent vectors, thus; it is non-degenerate for all $\varphi \in \mathcal{T}_m$. Multiplying equation (3.3.17) by the matrix $U^*(\varphi)$ first, and then by the matrix $[U^*(\varphi)U(\varphi)]^{-1}$ we arrive at

$$y = -[U^*(\varphi)U(\varphi)]^{-1}U^*(\varphi)C(\varphi)V(\varphi)z, \quad \varphi \in \mathcal{T}_m, \tag{3.3.18}$$

which any solution of (3.3.17) satisfies. For a fixed $\varphi \in \mathcal{T}_m$ equality (3.3.18) prescribes an $n - r$-dimensional plane in the space of variables y, z. Since the manifold M^- is an $n - r$-dimensional plane of the space of variables y, z for any fixed $\varphi \in t$, then M^- coincides with (3.3.18). This proves that equality (3.3.18) is the equation of the separatrix manifold M^- of the system (3.3.7).

Passing from the variables y to y_1 in (3.3.7) according to

$$y = y_1 - [U^*(\varphi)U(\varphi)]^{-1}U^*(\varphi)C(\varphi)V(\varphi)z,$$

we transform the system (3.3.7) into a system of the same type but possessing a separatrix manifold M^- with the equation

$$y_1 = 0, \quad \varphi \in \mathcal{T}_m.$$

The latter is possible only if the system of the equations for φ, z, y_1 has the form of (3.3.7) with $y = y_1$ and $Q \equiv 0$, and hence, is a block-diagonal system with the diagonal blocks $Q(\varphi)$ and $-D(\varphi) = -\operatorname{diag}\{\lambda_1(\varphi), \ldots, \lambda_r(\varphi)\}$. The C'-block splittability of the system (3.3.1) is proved. The form of the split system implies that, while splitting, the separatrix manifolds M^+ and M^- are rectified.

Let us now examine the eigenfunctions (3.3.12) of the operator L which form a r-frame which cannot be complemented up to the periodic basis in R^n. Assume that, in spite of this fact, the system (3.3.1) is reduced by the change of variables $x = U_1(\varphi)y_1 + V_1(\varphi)z_1$ with the matrix

$$\Phi(\varphi) = [U_1(\varphi), V_1(\varphi)] \in C'(\mathcal{T}_m), \quad \det \Phi \neq 0, \quad \varphi \in \mathcal{T}_m,$$

to the block-diagonal form

$$\frac{d\varphi}{dt} = a(\varphi), \quad \frac{dy_1}{dt} = Q_1(\varphi)y_1, \quad \frac{dz_1}{dt} = Q_2(\varphi)z_1, \tag{3.3.19}$$

with rectification of the separatrix manifolds. The equation

$$z_1 = 0, \quad \varphi \in \mathcal{T}_m,$$

describes the separatrix manifold M^+ of the system (3.3.19). Then the surface

$$x = U_1(\varphi)y_1, \quad y_1 \in R^r, \quad \varphi \in \mathcal{T}_m, \qquad (3.3.20)$$

is an invariant set of the system (3.3.1) the motions of which are specified by the first two equations of the system (3.3.19) and therefore, damps by the exponential law as $t \to +\infty$. By Theorem 3.2.1 the surface

$$x = U(\varphi)y, \quad y \in R^r, \quad \varphi \in \mathcal{T}_m, \qquad (3.3.21)$$

is also the invariant set of the system (3.3.1). The motions on it are determined by the system (3.3.5) and by virtue of the positiveness of the values λ_j, $j = 1, \ldots, r$, also damps by the exponential law as $t \to +\infty$.

For any fixed $\varphi \in \mathcal{T}_m$ both surfaces (3.3.20) and (3.3.21) determine r-dimensional planes in the space x. The characteristics of motion of the system (3.3.1) on surfaces (3.3.20) and (3.3.21) prove that these surfaces are one and the same separatrix manifold M^+ of the system (3.3.1) expressed in the systems of coordinates (φ, y_1) and (φ, y) respectively. For a fixed $\varphi \in \mathcal{T}_m$ are two Cartesian systems of coordinates y_1 and y superposed at the origin there determined in one and the same Euclidean space $R^r(\varphi)$. Therefore a non-degenerate $r \times r$-dimensional matrix $R(\varphi)$ is found, determined for every $\varphi \in \mathcal{T}_m$ and such that the passage from the system of coordinates y_1 to the system of coordinates y is described by $y_1 = R(\varphi)y$.

In order that (3.3.20) and (3.3.21) determine one and the same plane for any $\varphi \in \mathcal{T}_m$, it is necessary that the matrix $R(\varphi)$ satisfies the identity

$$U_1(\varphi)R(\varphi) = U(\varphi), \quad \varphi \in \mathcal{T}_m. \qquad (3.3.22)$$

By means of the matrix $\Phi(\varphi)$ identity (3.3.22) becomes

$$\Phi(\varphi)\begin{pmatrix} R(\varphi) \\ 0 \end{pmatrix} = U(\varphi), \quad \varphi \in \mathcal{T}_m,$$

and $R(\varphi) \in C'(\mathcal{T}_m; a)$ can be obtained by inverting the matrix $\Phi(\varphi)$.

This proves that the matrix $R(\varphi)$ in (3.3.22) belongs to the space $C'(\mathcal{T}_m; a)$. Then the representation

$$[U(\varphi), V_1(\varphi)] = [U_1(\varphi)R(\varphi), V_1(\varphi)] = [U_1(\varphi), V_1(\varphi)] \begin{pmatrix} R(\varphi) & 0 \\ 0 & I_{n-r} \end{pmatrix}$$

yields by virtue of the non-degeneracy of the matrices $[U_1(\varphi), V_1(\varphi)] = \Phi(\varphi)$ and $R(\varphi)$ that

$$\det [U(\varphi), V_1(\varphi)] \neq 0, \quad \varphi \in \mathcal{T}_m. \tag{3.3.23}$$

Since $V_1(\varphi) \in C'(\mathcal{T}_m; a)$, the inequality (3.3.23) ensures the complementability of the r-frame $U(r)$ up to the periodic basis in R^n. We get a contradiction which completes the proof of Theorem 3.3.4.

The assertions of Theorem 3.3.4 can be applied to construct exponentially dichotomous systems (3.3.1) which are not C'-block splittable for an arbitrary size of the diagonal blocks.

An obvious corollary from Theorem 3.3.2 is the result below on C'-diagonalization of the system (3.3.1).

Corollary 3.3.1 *In order to reduce the system (3.3.1) by the transformation*

$$x = U(\varphi)y, \quad U \in C'(\mathcal{T}_m : a), \quad \det U(\varphi) \neq 0, \quad \varphi \in \mathcal{T}_m,$$

to the diagonal form

$$\frac{dy_j}{dt} = -\lambda_j(\varphi)y_j, \quad j = 1, \ldots, n,$$

it is necessary and sufficient that the operator L of the system has n eigenvalues $\lambda_1(\varphi), \ldots, \lambda_n(\varphi)$, the eigenfunctions of which form a basis in R^n.

3.4 Reducing of Linearized Systems to a Diagonal Form

Alongside the system (3.1.8) we consider the family of linear systems of differential equations of the type

$$\frac{dx}{dt} = [A(\varphi_t(\varphi)) - \lambda I_n]x, \tag{3.4.1}$$

where λ is a scalar parameter, $\lambda \in R$, $A(\varphi) \in C^r(R^m, 2\pi)$, $r \geq 1$.

Definition 3.4.1 We claim that system (3.1.8) has n *one-dimensional exponentially separated solution subspaces*, if there exist $n-1$ non-intersecting intervals $(\bar{l}_i, l_{i+1})$, $i = 1, \ldots, n-1$, such that the system (3.4.1) is exponentially dichotomous on the whole R axis for $\lambda \in (\bar{l}_i, l_{i+1})$, and moreover, $\dim E_i^+(\varphi) \equiv i$, where $E_i^+(\varphi)$ is a linear subspace of the initial points of solutions of the corresponding system vanishing on $+\infty$.

We designate by $d = \min_i |l_{i+1} - \bar{l}_i|$, and $\alpha(\varphi)$ the largest eigenvalue of the matrix $\frac{1}{2}\left[\frac{\partial a(\varphi)}{\partial \varphi} + \left(\frac{\partial a(\varphi)}{\partial \varphi}\right)^*\right]$, $\max_\varphi \alpha(\varphi) = \alpha$.

Theorem 3.4.1 *Let the system of the equations (3.1.8) have n one-dimensional exponentially separated solution subspaces and*

$$\alpha < \frac{1}{2} dr^{-1}, \quad r \geq 1. \tag{3.4.2}$$

Then

(1) *there exists a non-degenerate continuously differentiable up to the order of r orthogonal matrix $T(\varphi)$, 4π-periodic in φ_j, $j = 1, \ldots, m$, and such that the change of variables $T(\varphi_t(\varphi))y = x$ reduces the system (3.1.8) to the triangular form*

$$\frac{dx}{dt} = B(\varphi_t(\varphi)), \quad B = \{b_{ij}\}_{i,j=1}^n, \quad b \equiv 0, \quad i > j;$$

(2) *there exists a non-degenerate 4π-periodic matrix $L(\varphi) \in C^r(R^m, 4\pi)$ satisfying the equality*

$$L^{-1}(\varphi)[A(\varphi)L(\varphi) - \dot{L}(\varphi)] = \mathrm{diag}\,\{p_1(\varphi), \ldots, p_n(\varphi)\},$$

where $p_i(\varphi)$ are scalar functions from the space $C^r(R^m, 4\pi)$.

Proof Together with subspaces $E_i^+(\varphi)$ we consider the projection matrices on these subspaces along the corresponding subspaces $E_i^-(\varphi)$ which are designated by $C_i(\varphi)$. For every such matrix

$$C_i(\varphi_t(\varphi))\Omega_0^t(\varphi) \equiv \Omega_0^t(\varphi)C_i(\varphi), \tag{3.4.3}$$

where $\Omega_0^t(\varphi) \equiv \Omega_0^t(\varphi; A)$. Let a non-zero vector x be taken from the subspace $E_i^+(\varphi_0)$, then $\Omega_0^t(\varphi_0)x \equiv x_t(\varphi_0) \in E_i^+(\varphi_t(\varphi_0))$ for all $t \in R$. Actually, in view of (3.4.3) we get

$$C_i(\varphi_t(\varphi_0))x_t(\varphi_0) = C_i(\varphi_t(\varphi_0))\Omega_0^t(\varphi_0)x = \Omega_0^t(\varphi_0)C_i(\varphi_0)x = \Omega_0^t(\varphi_0)x.$$

Note that the property of belonging to solutions $x_t(\varphi_0)$ to the family of the subspaces $E_i^+(\varphi_t(\varphi_0))$ is preserved for the system (3.4.1) as well as for every fixed $\lambda \in R$.

Consider the algebraic system of equations

$$[C_1(\varphi) - I_n]x = 0, \tag{3.4.4}$$

and show that it has 4π-periodic in φ_j, $j = 1, \ldots, m$, continuously differentiable up to the order of the $r \geq 1$ unit solution $x = e(\varphi)$, $\|e(\varphi)\| = 1$. To prove this we fix $\varepsilon > 0$ so that

$$\max\{\|C_1(\varphi) - C_1(\overline{\varphi})\|, \|C_1(\varphi) - C_1(\overline{\varphi})\|\,\|C_1\|_0\} < \frac{1}{2}, \tag{3.4.5}$$

for all $\varphi, \overline{\varphi} \in R^m$, $\|\varphi - \overline{\varphi}\| < \varepsilon$, $\|C_1\|_0 = \max_{\varphi} \|C_1(\varphi)\|$. Let x^0 ($\|x^0\| = 1$) be a constant vector, satisfying the system of equations $[C_1(0) - I_n]x = 0$. Then for every vector $\varphi \in R^m$ the vector $e(\varphi)$ is constructed as follows. Consider a set of points $\{\varphi^{(i)}\}_{i=1}^n$, $\varphi^{(0)} = 0$, $\varphi^{(n)} = \varphi$, for which $\|\varphi^{(i)} - \varphi^{(i+1)}\| < \varepsilon$, $i = 1, \ldots, n-1$, with the sequence

$$C_1(\varphi^{(i)})x^{(i-1)}\|C_1(\varphi^{(i)})x^{(i-1)}\|^{-1} = x^{(i)}, \quad i = 1, \ldots, n-1,$$
$$C_1(\varphi)x^{(n-1)}\|C_1(\varphi)x^{(n-1)}\|^{-1} = e(\varphi). \tag{3.4.6}$$

The vector $C_1(\varphi^{(i)})x^{(i-1)}$ is non-zero for all $i = 1, \ldots, n$,

$$\|C_1(\varphi^{(i)})x^{(i-1)}\| = \|C_1(\varphi^{(i-1)})x^{(i-2)}\|^{-1}\|C_1(\varphi^{(i)})C_1(\varphi^{(i-1)})x^{(i-2)}\|$$
$$= \|C_1(\varphi^{(i-1)})x^{(i-2)}\|^{-1}\|[C_1(\varphi^{(i)}) - C_1(\varphi^{(i-1)})]C_1(\varphi^{(i-1)})x^{(i-2)} \tag{3.4.7}$$
$$+ C_1(\varphi^{(i-1)})x^{(i-2)}\| \geq 1 - \|C_1(\varphi^{(i)}) - C_1(\varphi^{(i-1)})\| > \frac{1}{2}.$$

The value of the vector $e(\varphi)$ obtained by (3.4.6) is shown to be independent of the choice of the set of points $\{\varphi^{(i)}\}_{i=1}^n$. To prove this, it is sufficient to demonstrate that

$$\frac{C_1(\varphi^{(i+1)})x^{(i)}}{\|C_1(\varphi^{(i+1)})x^{(i)}\|} = \frac{C_1(\varphi^{(i+1)})C_1(\widetilde{\varphi}^{(i)})x^{(i)}}{\|C_1(\varphi^{(i+1)})C_1(\widetilde{\varphi}^{(i)})x^{(i)}\|}, \tag{3.4.8}$$

where $\widetilde{\varphi}^{(i)}$ is an arbitrary point for which $\|\widetilde{\varphi}^{(i)} - \varphi^{(i)}\| < \varepsilon$ and $\|\widetilde{\varphi}^{(i)} - \varphi^{(i+1)}\| < \varepsilon$. We make the following estimations

$$\|C_1(\varphi^{(i+1)})x^{(i)} - C_1(\varphi^{(i+1)})C_1(\widetilde{\varphi}^{(i)})x^{(i)}\| \leq \|C_1\|_0\|[I_n - C_1(\widetilde{\varphi}^{(i)})]x^{(i)}\|$$
$$= \|C_1\|_0\|[I_n - C_1(\widetilde{\varphi}^{(i)})]x^{(i)} - [I_n - C_1(\varphi^{(i)})]x^{(i)}\| \tag{3.4.9}$$
$$\leq \|C_1\|_0\|C_1(\widetilde{\varphi}^{(i)}) - C_1(\varphi^{(i)})\| < \frac{1}{2}.$$

Estimates (3.4.7) and (3.4.9) imply

$$C_1(\varphi^{(i+1)})x^{(i)} = \rho C_1(\varphi^{(i+1)})C_1(\widetilde{\varphi}^{(i)})x^{(i)}, \quad \rho > 0. \tag{3.4.10}$$

In view of (3.4.10) and the fact that the left and right-hand side vectors of (3.4.8) belong to the one-dimensional subspace $E_1^+(\varphi^{(i+1)})$ one sees the validity of equality (3.4.8).

Let us consider the smoothness of the vector function $e(\varphi)$ in the variables φ. For this we fix a point $\varphi = \varphi_0 \in R^m$ and consider the vectors $e(\varphi_0)$ and $e(\varphi)$, where φ varies within a sufficiently small neighbourhood of point φ_0. Therefore, the last equality (3.4.6) implies that the vector $x^{(n-1)}$ only depends on φ and hence, the smoothness of vector function $e(\varphi)$ coincides with the smoothness of the projection matrix $C_1(\varphi)$. We demonstrate that condition (3.4.2) ensures the continuous differentiability of the matrix function $C_1(\varphi)$ in the variables φ_j, $j = 1, \ldots, m$, up to the order of r. To this end, the Green function $G_t(\tau, \varphi, \lambda_1)$ of the problem of solutions for the system (3.4.1) bounded on R is considered, when $\lambda = \lambda_1 \in (\bar{l}_1, l_2)$. We have the estimate

$$\|G_t(\tau, \varphi, \lambda_1)\| \leq K_{\lambda_1} \exp\{-\gamma(\lambda_1)|t - \tau|\},$$

where $0 < \gamma(\lambda_1) < \min(|l_2 - \lambda_1|, |\bar{l}_1 - \lambda_1|)$; $\gamma(\lambda_1)$ takes the largest value for $\lambda_1 = \frac{1}{2}(l_2 + \bar{l}_1)$. Besides, the positive constant γ can be taken arbitrarily close to the value $\frac{1}{2}(l_2 - \bar{l}_1)$. The results presented in Chapter 2 give that the inequality $r\alpha < \frac{1}{2}(l_2 - \bar{l}_1)$ ensures continuous differentiability of the Green function $G_0(\tau, \varphi, \lambda_1)$ in variables φ up to the order of r. Now the identity $C_1(\varphi) = G_0(\tau, \varphi, \lambda_1)\Omega_0^\tau(\varphi, \lambda_1)$, $\tau < 0$, enables us to ascertain the continuous differentiability of the projection matrix $C_1(\varphi)$ in the variables φ up to the order of r as well.

We consider the unit vectors $e(\varphi_1, \ldots, \varphi_m)$ and $e(\varphi_1 + 2\pi, \ldots, \varphi_m)$. Both of the vectors satisfy one and the same equation $C_1(\varphi)e = e$, i.e. the vectors $e(\varphi_1, \ldots, \varphi_m)$ and $e(\varphi_1 + 2\pi, \ldots, \varphi_m)$ belong to the one-dimensional subspace $E_1^+(\varphi)$. Therefore, two cases are possible:

(a) $e(\varphi_1, \ldots, \varphi_m) = e(\varphi_1 + 2\pi, \ldots, \varphi_m)$;

(b) $e(\varphi_1, \ldots, \varphi_m) = -e(\varphi_1 + 2\pi, \ldots, \varphi_m)$.

In case (a) the vector $e(\varphi)$ is 2π-periodic in φ_1. In case (b) one can conclude the 4π-periodicity in φ_1 of the vector $e(\varphi)$. Thus, the existence of the smooth non-trivial solution of system (3.4.4) is proved.

Since the solutions of the system (3.1.8) belong to the space $E_1^+(\varphi_t(\varphi))$ under the initial condition $x|_{t=0} = x_0 \in E_1^+(\varphi)$, they can be represented as

$$x_t(\varphi) = e(\varphi_t(\varphi))\rho(t, \varphi), \qquad (3.4.11)$$

where $\rho(t, \varphi) \equiv \|x_t(\varphi)\|$ or $\rho(t, \varphi) \equiv -\|x_t(\varphi)\|$ for all $t \in R$. In fact, if, for example, for some $t = t_1$ the equality

$$x_{t_1}(\varphi)\|x_{t_1}(\varphi)\|^{-1} = -e(\varphi_{t_1}(\varphi)), \quad x_{t_1}(\varphi) \in E_1^+(\varphi_{t_1}(\varphi)),$$

holds, then it also holds for all $t = t_2$ close enough to t_1

$$x_{t_2}(\varphi)\|x_{t_2}(\varphi)\|^{-1} = -e(\varphi_{t_2}(\varphi)).$$

Suppose on the contrary, i.e. $x_{t_2}(\varphi)\|x_{t_2}(\varphi)\| = e(\varphi_{t_2}(\varphi))$. Then

$$\big\|x_{t_1}(\varphi)\|x_{t_1}(\varphi)\|^{-1} - x_{t_2}(\varphi)\|x_{t_2}(\varphi)\|^{-1}\big\|$$
$$= \|e(\varphi_{t_1}(\varphi)) + e(\varphi_{t_2}(\varphi))\| \geq 2 - \|e(\varphi_{t_1}(\varphi)) - e(\varphi_{t_2}(\varphi))\|.$$

In this inequality there is a small value on the left and a value close to two on the right. The contradiction obtained proves the representation of the solutions of (3.1.8) by the equality (3.4.11).

Now we prove the complementability of the vector $e(\varphi)$ up to the complete basis in R^n. Consider the system of equations (3.4.1) for $\lambda = \lambda_2 \in (\bar{l}_2, l_3)$. Let $C_2(\varphi)$ be the projection matrix onto the subspace $E_2^+(\varphi)$. Then $C_2(\varphi)C_1(\varphi) \equiv C_1(\varphi)C_2(\varphi) \equiv C_1(\varphi)$ since $E_1^+(\varphi) \subset E_2^+(\varphi)$. Let us discuss the algebraic system of equations

$$[C_2(\varphi) - C_1(\varphi)]x = x. \qquad (3.4.12)$$

We show that $[C_2(\varphi) - C_1(\varphi)]$ is the projection matrix of unit rank. In fact, squaring this matrix yields

$$[C_2(\varphi) - C_1(\varphi)]^2 = C_2^2(\varphi) - C_2(\varphi)C_1(\varphi) - C_1(\varphi)C_2(\varphi) + C_1^2(\varphi)$$
$$\equiv C_2(\varphi) - C_1(\varphi).$$

Now we fix $\varphi = \varphi_0$ and reduce the matrix $C_1(\varphi_0)$ to the diagonal form

$$T(\varphi_0)C_1(\varphi_0)T^{-1}(\varphi_0) = J = \text{diag}\{0, 1, \ldots, 0\}.$$

Hence we find

$$T(\varphi_0)[C_2(\varphi_0) - C_1(\varphi_0)]T^{-1}(\varphi_0) = \tilde{C}_2(\varphi_0) - J.$$

When the equality

$$[\tilde{C}_2(\varphi_0) - J]^2 = \tilde{C}_2(\varphi_0) - J \qquad (3.4.13)$$

is satisfied, we obtain that $\tilde{c}_{21} = \tilde{c}_{23} = \cdots = \tilde{c}_{2n} = 0$, $\tilde{c}_{12} = \tilde{c}_{32} = \cdots = \tilde{c}_{n2} = 0$ and $\tilde{c}_{22} = 1$, where $\tilde{c}_{ij}$ are the elements of the matrix $\tilde{C}_2(\varphi_0)$. Thus, the rank of the matrix $[C_2(\varphi) - C_1(\varphi)]$ is shown to be equal to 1 for every $\varphi \in R^m$.

We designate the smooth non-trivial 4π-periodic solution of the system (3.4.12) by $x^{(2)}(\varphi)$, $(\|x^{(2)}(\varphi)\| \equiv 1)$. Note that the vector $x^{(2)}(\varphi)$ is linearly independent with the vector $e(\varphi)$ for every $\varphi \in R^m$, since $C_1(\varphi)x^{(2)}(\varphi) \equiv 0$ and $C_1(\varphi)e(\varphi) \equiv e(\varphi)$. Designate by $e_2(\varphi)$ the vector orthogonal to $e(\varphi)$ determined by

$$e_2(\varphi) = [e(\varphi) - \langle e(\varphi), x^{(2)}(\varphi)\rangle x^{(2)}(\varphi)]\|e(\varphi) - \langle e(\varphi), x^{(2)}(\varphi)\rangle x^{(2)}(\varphi)\|^{-1}.$$

Proceeding in a similar way with the consideration of the systems of algebraic equations

$$[C_{i+1}(\varphi) - C_i(\varphi)]x = x,$$

where $[C_{i+1}(\varphi) - C_i(\varphi)]$ are the projection matrices with the unit rank and applying to the non-zero solutions of these equations the orthogonalization method each time, we get n linearly independent continuously differentiable 4π-periodic orthonormed vectors $e_i(\varphi)$, $i = 1,\ldots,n$, $e_1 = e$, which make up the non-degenerate orthogonal matrix $T(\varphi) = [e_1(\varphi),\ldots,e_n(\varphi)]$.

Now we make the change of variables in (3.1.8)

$$x = T(\varphi_t(\varphi))y = \rho^{-1}(t,\varphi)T(\varphi_t(\varphi))\rho(t,\varphi)y.$$

The first column of the matrix $T(\varphi_t(\varphi))\rho(t,\varphi)$ is the solution of the system (3.1.8). Therefore after the change of variables the system (3.1.8) becomes

$$\frac{dx}{dt} = B(\varphi_t(\varphi))y,$$

where $B = \{b_{ij}\}_{i,j=1}^n$, $b_{21} \equiv b_{31} \equiv \cdots \equiv b_{n1} \equiv 0$.

Continuing with the orthogonal change of variables in the obtained system we arrive at a system of triangular form.

Applying quadratic forms to the triangular system and employing the result of Theorem 3.1.1 we prove the existence of a smooth change of variables reducing the system (3.1.8) to the diagonal form.

Theorem 3.4.1 is proved.

It seems that the group exponential separatibility of the solutions of the system (3.1.8) provides a possibility of block diagonalization of this system; however this is not true. The problem of block diagonalization is still unsolved. Our aim is to formulate and prove this result.

Definition 3.4.2 We claim that the system (3.1.8) has s *one-dimensional and one $n - r$-dimensional exponentially separated solution subspaces*, if there exist s non-intersecting intervals $(\bar{l}_1, l_2), (\bar{l}_2, l_3), \ldots, (\bar{l}_{k_0+1}, l_{k_0+2}), \ldots, (\bar{l}_s, l_{s+1})$ such that for any $\lambda = l_i \in (\bar{l}_i, l_{i+1})$ the system (3.4.1) is exponentially dichotomous on R, and

$$\dim E_i^+(\varphi) = \begin{cases} i, & i = 1,\ldots,k_0, \\ i + n - s - 1, & i = k_0 + 1,\ldots,s. \end{cases}$$

Lemma 3.4.1 *For any continuous rectangular matrix function $D(\varphi) = \{d_{ij}(\varphi)\}_{i,j=1}^{n,r}$ 2π-periodic in φ_j, $j = 1,\ldots,m$, with rank $D(\varphi) \equiv 1$ there exist two non-zero vector functions $\lambda(\varphi) = (\lambda_1(\varphi),\ldots,\lambda_n(\varphi))$, $\mu(\varphi) = (\mu_1(\varphi),\ldots\mu_r(\varphi))$, $\|\lambda(\varphi)\| = 1$, $\|\mu(\varphi)\| \neq 0$ 4π-periodic in φ_j, $j = 1,\ldots,m$ and such that for all $\varphi \in R^m$ the identity $\lambda(\varphi) \equiv \mu(\varphi)D(\varphi)$ holds true.*

Proof Consider the system of algebraic equations

$$D(\varphi)x = 0, \tag{3.4.14}$$

and show that the $n-1$-dimensional hyperplane determined by (3.4.14) can be expressed by

$$\langle \lambda(\varphi), x \rangle = 0, \qquad (3.4.15)$$

where the non-zero vector function $\lambda(\varphi)$ is continuous and 4π-periodic in φ_j, $j = 1,\ldots,m$. For this purpose, to every non-zero vector $d_j(\varphi) = (d_{i1}(\varphi),\ldots,d_{in}(\varphi))$ for a fixed vector $\varphi \in R^m$ the orthogonal projection matrix

$$P_i(\varphi) = I_n - \|d_i(\varphi)\|^{-2} d_i^*(\varphi) d_i(\varphi)$$

is put into correspondence. If the vector $d_i(\varphi)$ vanishes at some point $\varphi = \varphi_0$, then the identity matrix is put into correspondence to the vector $d_i(\varphi_0)$. Thus, there is a set of matrices

$$\{P_i(\varphi)\}_{i=1}^r \qquad (3.4.16)$$

for every fixed $\varphi \in R^m$, where there is at least one non-identity matrix. If there are several non-identity matrices, they are necessarily equal due to the coupled linear dependence of vectors $\{d_i\}_{i=1}^r$. We designate the non-identity matrix taken for every fixed vector $\varphi \in \mathcal{T}_m$ from (3.4.16) by $P_0(\varphi)$. Let us show that the matrix $P_0(\varphi)$ is continuous for all $\varphi \in R^m$. We fix a point $\varphi = \varphi_0 \in R^m$. For this point a non-zero vector $d_{i_0}(\varphi_0)$ exists, which due to continuity is also non-zero for all φ from some neighbourhood of the point φ_0: $\varphi \in U_{\varphi_0}$. It is clear that for all $\varphi \in U_{\varphi_0}$ $P_0(\varphi) \equiv P_{i_0}(\varphi)$. Hence, the continuity of the matrix function $P_0(\varphi)$ follows; the 2π-periodicity of this function is obvious. The system (3.4.14) and the system

$$P_0(\varphi)x = x \qquad (3.4.17)$$

determine one and the same $n-1$-dimensional hyperspace in R^n for every fixed $\varphi \in R^m$. We designate the continuous 4π-periodic solution of the algebraic system of the equation $P_0(\varphi)x = 0$, by $\lambda(\varphi)$, $\|\lambda(\varphi)\| \equiv 1$, the existence of which is shown in the proof of Theorem 3.4.1 and demonstrate that the system (3.4.17) and the equation (3.4.15) are equivalent. Let $x(\varphi)$ be a solution of the system (3.4.17), then

$$\langle \lambda(\varphi), x(\varphi) \rangle = \langle \lambda(\varphi), P_0(\varphi)x(\varphi) \rangle = \langle P_0(\varphi)\lambda(\varphi), x(\varphi) \rangle \equiv 0.$$

Now le $x(\varphi)$ be a solution of the equation (3.4.15). Then, have for an arbitrary vector $\mu \in R^n$ we have

$$\langle [P_0(\varphi) - I_n]x(\varphi), \mu \rangle = \langle x(\varphi), [P_0(\varphi) - I_n]\mu \rangle = \langle x(\varphi), \rho(\varphi)\lambda(\varphi) \rangle \equiv 0,$$

where $\rho(\varphi)$ is a scalar function. Therefore, $P_0(\varphi)x(\varphi) \equiv x(\varphi)$.

Let us show that the vector function $\lambda(\varphi)$ can be represented in the form of a linear combination of the lines of the matrix $D(\varphi)$. Consider the system of algebraic equations

$$d_{11}(\varphi)a_1 + d_{21}(\varphi)a_2 + \cdots + d_{r1}(\varphi)a_r = \lambda_1(\varphi)a_{r+1},$$

$$\dots\dots\dots\dots\dots\dots\dots\dots\dots\dots\dots\dots\dots\dots\dots\dots\dots$$

$$d_{1n}(\varphi)a_1 + d_{2n}(\varphi)a_2 + \cdots + d_{rn}(\varphi)a_r = \lambda_n(\varphi)a_{r+1}.$$

Clearly, the rank of the matrix determining this system is equal to 1. Therefore it can be replaced by one equation

$$\sum_{i=1}^{r} \nu_i(\varphi)\alpha_i = \lambda_0(\varphi)\alpha_{r+1},$$

where $\sum_{i=1}^{r} \nu_i^2(\varphi) + \lambda_0^2(\varphi) \neq 0$ for all $\varphi \in R^m$. Moreover, let us show that

$$\lambda_0(\varphi) \neq 0, \quad \sum_{i=1}^{r} \nu_i^2(\varphi) \neq 0 \qquad (3.4.18)$$

for all $\varphi \in R^m$. Suppose on the contrary, then a point $\varphi = \varphi_0 \in R^m$ is found, for which, for instance, $\lambda_0(\varphi_0) = 0$. As the vector $(\nu_1(\varphi_0), \ldots, \nu_r(\varphi_0), -\lambda_0(\varphi_0))$ is proportional to every vector $(d_{1i}(\varphi_0), \ldots, d_{ri}(\varphi_)), -\lambda_i(\varphi_0))$, $i = 1, \ldots, n$, we get $\sum_{i=1}^{r} \lambda_i^2(\varphi_0) = 0$; however this is impossible. Now let $\sum_{i=1}^{r} \nu_i^2(\varphi_0) = 0$ at some point $\varphi_0 \in R^m$. Then obligatory $\lambda_0(\varphi_0) \neq 0$. Therefore, there exists a number i_0 such that $\sum_{j=1}^{r} d_{ji_0}^2(\varphi_0) = 0$, and $\lambda_{i_0}(\varphi_0) \neq 0$. This leads to the contradiction. Condition (3.4.18) is proved.

Now it is easy to verify that the vector function

$$\mu(\varphi) = \lambda_0(\varphi)\|\nu(\varphi)\|^{-2}(\nu_1(\varphi), \ldots, \nu_r(\varphi)),$$

which is at least 4π-periodic in φ_j, $j = 1, \ldots, m$, satisfies all the requirements of the lemma.

The algebraic system of equations $[C_1(\varphi) - I_n]x = 0$ has the non-trivial solution $e_1(\varphi)$. Let us show that under the additional conditions imposed on the number of variables φ, the vector $e_1(\varphi)$ is complemented up to the complete basis in R^n. Designate the unit vectors which are non-trivial solutions of equations $[C_i(\varphi) - C_{i-1}(\varphi)]x = x$, $i = 2, \ldots, k_0, k_0 + 2, \ldots, s$ by $e_i(\varphi)$. We prove that the equation

$$[C_{k_0+1}(\varphi) - C_{k_0}(\varphi)]x = x, \qquad (3.4.19)$$

where $[C_{k_0+1}(\varphi) - C_{k_0}(\varphi)]$ is a matrix of the rank $n - s$, has $n - s$ linearly independent solutions. Consider the obvious identity

$$I_n \equiv C_1 + (C_2 - C_1) + \cdots + (C_{k_0+1} - C_{k_0}) + \ldots$$
$$+ (C_s - C_{s-1}) + (I_n - C_s),$$

which yields

$$I_n - (C_{k_0+1} - C_{k_0}) \equiv C_1 + (C_2 - C_1) + \ldots$$
$$+ (C_{k_0} - C_{k_0-1}) + (C_{k_0+2} - C_{k_0+1}) + \cdots + (I_n - C_s).$$

Thus, system (3.4.19) is equivalent to the system

$$C_1 x + (C_2 - C_1)x + \cdots + (C_{k_0} - C_{k_0-1})x$$
$$+ (C_{k_0+2} - C_{k_0+1})x + \cdots + (I_n - C_s)x = 0. \quad (3.4.20)$$

If a non-zero vector-function $x(\varphi)$ satisfies the system of equations

$$[C_i(\varphi) - C_{i-1}(\varphi)]x = 0, \quad (3.4.21)$$

for all $i = 2, \ldots, k_0, k_0 + 2, \ldots, s$, then it is a solution of the system (3.4.20). Since the system (3.4.21) can be replaced by one equation $\langle \lambda_i(\varphi), x \rangle = 0$ then the system (3.4.19) is equivalent to the system

$$\langle \lambda_i(\varphi), x \rangle = 0, \quad i = 1, \ldots, k_0, k_0 + 2, \ldots, s. \quad (3.4.22)$$

Employing the results obtained in Samoilenko [3] we find that the condition

$$m + s < n, \quad (3.4.23)$$

is sufficient for the system (3.4.22) to have $n - s$ linearly independent solutions which are continuous and 4π-periodic in φ_j, $j = 1, \ldots, m$. Thus, the complementability of the vector $e_1(\varphi)$ to the complete basis in R^n is proved.

Summing up the above discussion yields the result.

Theorem 3.4.2 *Let the system of the equations (3.1.8) have s one-dimensional and one $n-s$-dimensional exponentially separated subspaces of solutions and let the inequalities (3.4.2) and (3.4.23) be satisfied. Then there exists a non-degenerate, continuously differentiable up to the order of $r \geq 1$ matrix $L(\varphi)$, 4π-periodic in φ_j, $j = 1, \ldots, m$, and such that the change of variables $x = L(\varphi_t(\varphi))y$ divides the*

system (3.1.8) into one $n - s$-dimensional block and s one-dimensional equations, i.e.
$$L^{-1}(\varphi)[A(\varphi)L(\varphi) - \dot{L}(\varphi)] = \text{diag}\,\{\mathcal{P}(\varphi), p_1(\varphi), \ldots, p_s(\varphi)\},$$
where $\mathcal{P}(\varphi)$ is an $(n-r) \times (n-r)$-dimensional matrix, $p_i(\varphi)$ are scalar functions.

3.5 On the Relationship of Exponentially Dichotomous Linear Expansions with the Algebraic System Solvability

Let the system (3.1.8) be exponentially dichotomous on the whole R axis uniformly in φ and let $C(\varphi)$ be the projection matrix on the subspace $E^+(\varphi)$ along $E^-(\varphi)$. In Samoilenko [5] it is shown that every non-degenerate matrix $T(\varphi) \in C^1(\mathcal{T}_m; a)$ reducing the projection matrix $C(\varphi)$ to the Jordan form $T_\varphi^{-1}(\varphi) C(\varphi) T(\varphi) = \text{diag}\,\{I_r, 0\}$ ensures the division of the system (3.1.8) coordinated with e-dichotomy, i.e. $T^{-1}(\varphi)[A(\varphi)T(\varphi) - \dot{T}(\varphi)] = \text{diag}\,\{A^+(\varphi), A^-(\varphi)\}$. Clearly, the columns of the matrix $T(\varphi)$ are non-trivial solutions of the algebraic systems of equations $C(\varphi)x = 0$ and $C(\varphi)x = x$. In this regard the following questions arise.

1. Let there be some rectangular matrix function $D(\varphi) \in C^0(\mathcal{T}_m)$ with the dimensions $l \times n$ and its rank, rank $D(\varphi) \equiv r$, does not depend on φ. How can the system of algebraic equations (3.4.14) be replaced by a simpler equivalent system?

2. Assume that there exists a non-degenerate symmetric matrix $S(\varphi) \in C'(\mathcal{T}_m; a)$ satisfying the condition (3.1.3). What is the relationship between the matrix $S(\varphi)$ and the projection matrix $C(\varphi)$?

Note that the system of algebraic equations (3.4.14) can always be replaced by the equivalent system
$$D^*(\varphi)D(\varphi)x = 0, \quad x \in R^n. \tag{3.5.1}$$

In fact, if $x = e(\varphi)$ is some solution of the algebraic system (3.5.1), then we have for every vector $\mu \in R^n$, $\langle D^*(\varphi)D(\varphi)e(\varphi), \mu \rangle \equiv 0$, and get, in particular, for $\mu = e(\varphi)$, $\langle D^*(\varphi)D(\varphi)e(\varphi), e(\varphi) \rangle \equiv 0$ or $\|D(\varphi)e(\varphi)\|^2 \equiv 0$, i.e. $x = e(\varphi)$ is a solution of the system (3.4.14). Clearly, the inverse is also valid, if $x = e(\varphi)$ is a solution of system (3.4.14), it is also a solution of the system (3.5.1).

Proposition 3.5.1 *The matrix $D(\varphi)$ in the system (3.4.14) can always be replaced by the $n \times n$-dimensional matrix of the orthogonal projection $P(\varphi) \in C^0(\mathcal{T}_m)$:*
$$P(\varphi)x = 0, \tag{3.5.2}$$
so that the systems (3.4.14) and (3.5.2) are equivalent. Moreover, $P(\varphi)$ is represented by the equality
$$P(\varphi) = \lim_{\varepsilon \to +0} [D^*(\varphi)D(\varphi) + \varepsilon I_n]^{-1} D^*(\varphi)D(\varphi). \tag{3.5.3}$$

Proof We fix a value $\varphi = \varphi_0 \in R^m$ and transform the symmetric matrix $B(\varphi_0) = D^*(\varphi_0)D(\varphi_0)$ to the Jordan form $Q^{-1}B(\varphi_0)Q = \text{diag}\{\lambda_1, \ldots, \lambda_r, 0, \ldots, 0\} = \text{diag}\{\Lambda, 0\}$, where Q is an orthogonal matrix. Also, for $\varepsilon > 0$

$$\begin{aligned}[B(\varphi_0) + \varepsilon I_n]^{-1}B(\varphi_0) &= \left[Q\begin{pmatrix}\Lambda & 0 \\ 0 & 0\end{pmatrix}Q^{-1} + \varepsilon I_n\right]^{-1} Q\begin{pmatrix}\Lambda & 0 \\ 0 & 0\end{pmatrix}Q^{-1} \\ &= Q\begin{pmatrix}(\Lambda+\varepsilon I_r)^{-1} & 0 \\ 0 & \frac{1}{\varepsilon}I_{n-r}\end{pmatrix}Q^{-1}Q\begin{pmatrix}\Lambda & 0 \\ 0 & 0\end{pmatrix} \\ &= Q\begin{pmatrix}(\Lambda+\varepsilon I_r)^{-1}\Lambda & 0 \\ 0 & 0\end{pmatrix}Q^{-1}.\end{aligned} \quad (3.5.4)$$

Hence it is clear

$$\lim_{\varepsilon \to +0}[B(\varphi_0) + \varepsilon I_n]^{-1}B(\varphi_0) = Q\begin{pmatrix}I_r & 0 \\ 0 & 0\end{pmatrix}Q^{-1} = P(\varphi_0).$$

Since Q is an orthogonal matrix, then obviously $P^*(\varphi_0) = P(\varphi_0) = P^2(\varphi_0)$, i.e. $P(\varphi_0)$ is the orthogonal projection matrix.

Let $x = x_0(\varphi)$ be a solution of the system (3.5.1), i.e. $B(\varphi)x_0(\varphi) \equiv 0$, then $[B(\varphi) + \varepsilon I_n]^{-1}B(\varphi)x_0(\varphi) \equiv 0$ for all $\varepsilon > 0$. Passing to the limit as $\varepsilon \to +0$ in the last identity, we get $P(\varphi)x_0(\varphi) \equiv 0$. Suppose on the contrary, $x = x^0(\varphi)$ is a solution of the system (3.5.2). Then we show that $B(\varphi)x^0(\varphi) \equiv 0$. Let this not be true. Then for some $\varphi = \varphi_0 \in R^m$

$$B(\varphi_0)x^0(\varphi_0) \neq 0. \quad (3.5.5)$$

Therefore, $[B(\varphi_0) + \varepsilon I_n]^{-1}B(\varphi_0)x^0(\varphi_0) \neq 0$ for all $\varepsilon > 0$. Estimating the difference $[B(\varphi_0) + \varepsilon I_n]^{-1}B(\varphi_0) - P(\varphi_0)$, we get in view of (3.5.4)

$$\begin{aligned}\|[B+\varepsilon I_n]^{-1}B - P\| &\leq \|Q\|^2 \left\|\begin{pmatrix}(\Lambda+\varepsilon I_r)^{-1}\Lambda & 0 \\ 0 & 0\end{pmatrix}\right\| \\ &= \|(\Lambda+\varepsilon I_r)^{-1}\Lambda - I_r\| = \|(\Lambda+\varepsilon I_r)^{-1}(\Lambda - \Lambda - \varepsilon I_r)\| \\ &= \varepsilon\|(\Lambda+\varepsilon I_r)^{-1}\| \leq \varepsilon\|\Lambda^{-1}\| \leq \varepsilon\lambda_0^{-1},\end{aligned} \quad (3.5.6)$$

where $\lambda_0 > 0$ is the smallest eigenvalue of the matrix Λ. In view of (3.5.6) we get

$$\begin{aligned}\|B(\varphi_0)x^0(\varphi_0)\| &\leq \|B(\varphi_0) + \varepsilon I_n\| \|[B(\varphi_0)+\varepsilon I_n]^{-1}B(\varphi_0)x^0(\varphi_0)\| \\ &\leq (\|B\|+\varepsilon)\varepsilon\lambda_0^{-1}\|x^0\|.\end{aligned}$$

Passing to the limit in the inequality obtained as $\varepsilon \to +0$, we get $\|B(\varphi)x^0(\varphi)\| = 0$, which contradicts (3.5.5).

Remark 3.5.1 Since the identity $B(\varphi)[B(\varphi) + \varepsilon I_n]^{-1} \equiv [B(\varphi) + \varepsilon I_n]^{-1}B(\varphi)$ is valid for the matrix $B(\varphi)$, then the projection matrix (3.5.3) can be represented as

$$P(\varphi) = \lim_{\varepsilon \to +0} D^*(\varphi)D(\varphi)[D^*(\varphi)D(\varphi) + \varepsilon I_n]^{-1}. \qquad (3.5.7)$$

Remark 3.5.2 The estimate (3.5.6) shows that a constant K exists independent of φ and $\varepsilon > 0$ and such that

$$\|[D^*(\varphi)D(\varphi) + \varepsilon I_n]^{-1}D^*(\varphi)D(\varphi) - P(\varphi)\| \leq K\varepsilon,$$

where $P(\varphi) \equiv P^*(\varphi) \equiv P^2(\varphi)$ is determined by (3.5.3) or (3.5.7).

Now we discuss a problem on the relationship of the non-degenerate symmetric matrix $S(\varphi) \in C'(\mathcal{T}_m; a)$ satisfying condition (3.1.3) and the projection matrix $C(\varphi)$ found in the structure of the Green function of the problem on the invariant tori.

Theorem 3.5.1 *Let there exist a non-degenerate matrix $S(\varphi) \equiv S^*(\varphi) \in C'(\mathcal{T}_m; a)$ satisfying (3.1.3). Then for all $\varphi \in R^m$, $x \in R^n$*

$$\langle [S(\varphi)C(\varphi) + C^*(\varphi)S(\varphi) - S(\varphi))]x, x\rangle \geq \delta\|x\|^2, \qquad (3.5.8)$$

where $\delta = \text{const} > 0$.

Proof We represent the inequality (3.1.3) in the form

$$\frac{d}{dt}\langle S(\varphi_t(\varphi))\Omega_0^t(\varphi)x, \Omega_0^t(\varphi)x\rangle \leq -\|\Omega_0^t(\varphi)x\|^2, \quad x \in R^n.$$

The small perturbations of the matrix $S(\varphi)$ do not change this inequality essentially. So, for a sufficiently small fixed $\varepsilon > 0$

$$\frac{d}{dt}\langle [S(\varphi_t(\varphi)) - \varepsilon I_n]\Omega_0^t(\varphi)x, \Omega_0^t(\varphi)x\rangle \leq -\gamma_\varepsilon\|\Omega_0^t(\varphi)x\|^2, \qquad (3.5.9)$$

where $\gamma_\varepsilon = \text{const} > 0$. Since $C(\varphi)$ is the projection matrix on the subspace $E^+(\varphi)$, then all solutions of the system (3.1.8) of the form $\Omega_0^t(\varphi)C(\varphi)x$, $x \in R^n$, vanish as $t \to +\infty$. Hence the estimate (3.5.9) implies

$$\langle [S(\varphi_t(\varphi)) - \varepsilon I_n]\Omega_0^t(\varphi)C(\varphi)x, \Omega_0^t(\varphi)C(\varphi)x\rangle \geq 0, \qquad (3.5.10)$$

for all $\varphi \in R^m$, $t \in R$, $x \in R^n$. Similarly, for all solutions of the system (3.1.8) of the form $\Omega_0^t(\varphi)[I_n - C(\varphi)]x$, $x \in R^n$, the inequality

$$\langle [S(\varphi_t(\varphi)) + \varepsilon I_n]\Omega_0^t(\varphi)[I_n - C(\varphi)]x, \Omega_0^t(\varphi)[I_n - C(\varphi)]x\rangle \leq 0 \qquad (3.5.11)$$

is valid for all $t \in R$, $\varphi \in R^m$, $x \in R^n$. Writing down the inequalities (3.5.10) and (3.5.11) for $t = 0$ we get

$$\langle S(\varphi)C(\varphi)x, C(\varphi)x \rangle \geq \varepsilon \|C(\varphi)x\|^2, \tag{3.5.12}$$

$$\langle S(\varphi)[I_n - C(\varphi)]x, [I_n - C(\varphi)]x \rangle \leq -\varepsilon \|[I_n - C(\varphi)]x\|^2. \tag{3.5.13}$$

Now we extract (3.5.13) from (3.5.12). Then

$$\langle S(\varphi)C(\varphi)x, x \rangle + \langle C^*(\varphi)S(\varphi)x, x \rangle - \langle S(\varphi)x, x \rangle$$
$$\geq \varepsilon[\|C(\varphi)x\|^2 + \|[I_n - C(\varphi)]x\|^2] \geq \frac{\varepsilon}{2}\|x\|^2.$$

Thus the theorem is proved.

Remark 3.5.3 The inequalities (3.5.8) imply (3.5.12) and (3.5.13). It is sufficient to replace x by $C(\varphi)x$ and x by $[I_n - C(\varphi)]x$ in (3.5.8).

Remark 3.5.4 Not every symmetric matrix $S(\varphi) \in C'(\mathcal{T}_m; a)$ satisfying the inequality (3.5.8) satisfies condition (3.1.3) (even if $S(\varphi)$ is continuously differentiable with respect to all variables φ_j, $j = 1, \ldots, m$). We can demonstrate this by a simple example:

$$\dot{x}_1 = x_2, \quad \dot{x}_2 = x_1 + 2x_2.$$

Since the eigennumbers of the matrix

$$A = \begin{pmatrix} 0 & 1 \\ 1 & 2 \end{pmatrix}$$

are equal to $1 + \sqrt{2}$ and $1 - \sqrt{2}$, then the system is exponentially dichotomous on R and the projection matrix on the subspace E^+ along E^- is of the form

$$C = \frac{1}{4}\begin{pmatrix} 2 + \sqrt{2} & -\sqrt{2} \\ -\sqrt{2} & 2 - \sqrt{2} \end{pmatrix}.$$

We take diag$\{1, -1\}$ for the matrix S. Obviously, condition (3.5.8) is satisfied, $SC + C^*S - S = \text{diag}\{\sqrt{2}/2, \sqrt{2}/2\}$. Also, condition (3.1.3) is not satisfied with the matrix $SA + A^*S = \text{diag}\{0, -4\}$.

Remark 3.5.5 For every projection matrix $P(\varphi) \equiv P^2(\varphi) \in C^0(\mathcal{T}_m)$ which is not associated in any way with the system (3.1.8) there exist matrices $S(\varphi) \in C^0(\mathcal{T}_m)$ satisfying the inequality

$$\langle [S(\varphi)P(\varphi) + P^*(\varphi)S(\varphi) - S(\varphi)]x, x \rangle \geq \|x\|^2, \tag{3.5.14}$$

for all $x \in R^n$, $\varphi \in R^m$.

Actually, we show that for the matrix $S(\varphi)$ the following matrix can be taken

$$S(\varphi) = 2[P(\varphi) + P^*(\varphi) - I_n]. \qquad (3.5.15)$$

Substituting (3.5.15) into the left-hand side of (3.5.14) yields

$$2\langle[(P + P^* - I_n)P + P^*(P + P^* - I_n) - (P + P^* - I_n)]x, x\rangle$$
$$= 2\langle[P^*P + P^*P - P - P^* + I_n]x, x\rangle = 2\langle P^*Px, x\rangle$$
$$+ 2\langle(P^* - I_n)(P - I_n)x, x\rangle = 2(\|Px\|^2 + \|(P - I_n)x\|^2) \geq \|x\|^2.$$

Theorem 3.5.2 *Let at least one of the matrices $S(\varphi) = S^*(\varphi) \in C^0(\mathcal{T}_m)$ satisfying the condition (3.5.14) be able to be represented as*

$$Q^*(\varphi)S(\varphi)Q(\varphi) = \operatorname{diag}\{S_1(\varphi), -S_2(\varphi)\}, \qquad (3.5.16)$$

where $S_1(\varphi)$, $S_2(\varphi)$ are positive definite matrices, $Q(\varphi) \in C^0(\mathcal{T}_m)$. Then rank $P(\varphi) \equiv r$ for all $\varphi \in R^m$, where r has the dimensions of the matrix $S_1(\varphi)$ and there exists a non-degenerate matrix $T(\varphi) \in C^0(\mathcal{T}_m)$ such that

$$T^{-1}(\varphi)P(\varphi)T(\varphi) = \operatorname{diag}\{I_r, 0\}. \qquad (3.5.17)$$

The inverse assertion takes place.

Theorem 3.5.3 *Let there exist a non-degenerate matrix $T(\varphi) \in C^0(\mathcal{T}_m)$ reducing the projection matrix $P(\varphi)$ to the Jordan form (3.5.17). Then for every symmetric matrix $S(\varphi) \in C^0(\mathcal{T}_m)$ satisfying condition (3.5.14) there exists the matrix $Q(\varphi) \in C^0(\mathcal{T}_m)$ ensuring the equality*

$$Q^*(\varphi)S(\varphi)Q(\varphi) = \operatorname{diag}\{I_r, -I_{n-r}\}. \qquad (3.5.18)$$

Proof Let the condition of Theorem 3.5.2 be satisfied, then after the change of variables $x = Q(\varphi)y$ in the inequality (3.5.14) we get

$$2\left\langle \begin{pmatrix} S_1(\varphi)\bar{P}_{11}(\varphi) & S_1(\varphi)\bar{P}_{12}(\varphi) \\ -S_2(\varphi)\bar{P}_{21}(\varphi) & -S_2(\varphi)\bar{P}_{22}(\varphi) \end{pmatrix} \begin{pmatrix} y_1 \\ y_2 \end{pmatrix}, \begin{pmatrix} y_1 \\ y_2 \end{pmatrix} \right\rangle \qquad (3.5.19)$$
$$-\langle S_1(\varphi)y_1, y_1\rangle + \langle S_2(\varphi)y_2, y_2\rangle \geq \|Q^{-1}(\varphi)\|^{-2}\|y\|^2.$$

Here $Q^{-1}(\varphi)P(\varphi)Q(\varphi) = \bar{P}(\varphi) = \{\bar{P}_{ij}(\varphi)\}_{i,j=1}^2$, $\bar{P}_{11}(\varphi)$ is an $r \times r$-dimensional block of the matrix $\bar{P}(\varphi)$, $y_1 \in R^r$, $y_2 \in R^{n-r}$. The estimate (3.5.19) implies the non-degeneracy of the matrices $\bar{P}_{11}(\varphi)$ and $I_{n-r} - \bar{P}_{22}(\varphi)$ for all $\varphi \in R^m$ in view of the positive definiteness of the matrices $S_1(\varphi)$, $S_2(\varphi)$. Since the matrix $\bar{P}_{11}(\varphi)$ has the dimensions $r \times r$, then rank $P(\varphi) \equiv \text{rank } \bar{P}(\varphi) \geq r$. On the other hand

$$\text{rank } \bar{P}(\varphi) + \text{rank}\,[I_n - \bar{P}(\varphi)] \equiv n,$$

and moreover, $\text{rank}\,[I_n - \bar{P}(\varphi)] \geq n - r$. Therefore, the only possibility which is left is rank $\bar{P}(\varphi) \equiv r$. Now, when discussing two systems of algebraic equations $[I_n - \bar{P}(\varphi)]x = 0$ and $\bar{P}(\varphi)x = 0$, we obtain the following expression for the matrix $T(\varphi)$

$$T(\varphi) = Q(\varphi) \begin{pmatrix} I_r & -\bar{P}_{11}^{-1}(\varphi)\bar{P}_{12}(\varphi) \\ [I_r - \bar{P}_{22}(\varphi)]^{-1}\bar{P}_{21}(\varphi) & I_{n-r} \end{pmatrix}. \tag{3.5.20}$$

Remark 3.5.6 If the projection matrix $P(\varphi)$ belongs to the space $C^q(\mathcal{T}_m)$, $q \geq 1$ (or $P(\varphi) \in C'(\mathcal{T}_m; a)$), then, in spite of the fact that the matrix $S(\varphi)$ is continuous only in the condition of Theorem 3.4.2, the matrix $T(\varphi)$ can always be taken from $C^q(\mathcal{T}_m)$ ($C'(\mathcal{T}_m; a)$), i.e. the smoothness of the matrix $T(\varphi)$ is not smaller than that of the projection matrix $P(\varphi)$.

In fact, the matrix $S(\varphi) \in C^0(\mathcal{T}_m)$ can always be made close by the matrices $S_n(\varphi) \in C^q(\mathcal{T}_m)$, $S_n(\varphi) \Rightarrow S(\varphi)$. Similarly, $Q_n(\varphi) \Rightarrow Q(\varphi)$, $Q_n(\varphi) \in C^q(\mathcal{T}_m)$. Therefore, there exists a symmetric matrix $S_{n_0}(\varphi) \in C^q(\mathcal{T}_m)$ satisfying the inequality

$$\langle [S_{n_0}(\varphi)P(\varphi) + P^*(\varphi)S_{n_0}(\varphi) - S_{n_0}(\varphi)]x, x \rangle \geq \frac{1}{2}\|x\|^2,$$

and, in addition, the diagonal blocks of the matrix $Q_{n_0}^*(\varphi)S_{n_0}(\varphi)Q_{n_0}(\varphi)$ corresponding to the blocks S_1 and S_2 are of a fixed sign. This is sufficient for the matrix $\bar{Q}(\varphi) \in C^q(\mathcal{T}_m)$ exist such that

$$\bar{Q}^*(\varphi)Q_{n_0}^*(\varphi)S_{n_0}(\varphi)Q_{n_0}(\varphi)\bar{Q}(\varphi) = \text{diag}\,\{I_r, -I_{n-r}\}.$$

Now assume that the condition of the Theorem 3.5.3 is satisfied. We make the change of variables $x = T(\varphi)y$ in (3.5.14) and designate $\bar{S}(\varphi) = T^*(\varphi)S(\varphi)T(\varphi) = \{\bar{S}_{ij}(\varphi)\}_{i,j=1}^2$; $S_{11}(\varphi)$ is an $r \times r$-dimensional matrix. Then

$$\langle \text{diag}\,\{\bar{S}_{11}(\varphi), -\bar{S}_{22}(\varphi)\}y, y\rangle \geq \|T(\varphi)y\|^2.$$

Hence it is clear that the matrices $\bar{S}_{11}(\varphi)$ and $-\bar{S}_{22}(\varphi)$ are positive definite. This is enough for the conclusion to be made on the existence of the matrix $\bar{Q}(\varphi) \in C^0(\mathcal{T}_m)$ such that $\bar{Q}^*\bar{S}\bar{Q} = \text{diag}\,\{I_r, -I_{n-r}\}$. Thus, for the matrix $Q(\varphi)$ ensuring (3.5.18) the matrix $Q(\varphi) = T(\varphi)\bar{Q}(\varphi)$ can be taken.

Remark 3.5.7 If we assume that there exists only a continuous matrix $T(\varphi) \in C^0(\mathcal{T}_m)$ accomplishing the division (3.1.17) for some projection matrix $P(\varphi) \in C^q(\mathcal{T}_m)$, $q \geq 1$, then the same smooth matrix $T(\varphi) \in C^q(\mathcal{T}_m)$ always exists.

3.6 Three Block Divisibility of Linear Extensions and Lyapunov Functions of Variable Sign

We return to the system (3.4.1) and assume that for the values λ_1 and λ_2 the system (3.4.1) is exponentially dichotomous on the whole R axis with the various dimensions of the subspaces $E_1^+(\varphi)$, $E_2^+(\varphi)$, dim $E_i^+(\varphi) \equiv r_i$, $i = 1, 2$, $r_1 < r_2$. Here we notice the separation of three subspaces $E_1^+(\varphi)$, $E_2^+(\varphi) \cap E_1^+(\varphi)$, $E_2^-(\varphi)$. There is a question of whether there exists a change of variables $x = L(\varphi_t(\varphi))y$, $L(\varphi) \in C'(\mathcal{T}_m; a)$, reducing the system to the corresponding three-block form.

Theorem 3.6.1 *For some scalar functions* $\lambda(\varphi), \bar{\lambda}(\varphi) \in C^0(\mathcal{T}_m)$ *let there exist non-degenerate symmetric matrices* $S(\varphi), \bar{S}(\varphi) \in C'(\mathcal{T}_m; a)$, *which satisfy the inequalities*

$$\langle [\dot{S}(\varphi) + S(\varphi)A(\varphi) + A^*(\varphi)S(\varphi) + \lambda(\varphi)S(\varphi)]x, x \rangle \leq -\|x\|^2; \quad (3.6.1)$$

$$\langle [\dot{\bar{S}}(\varphi) + \bar{S}(\varphi)A(\varphi) + A^*(\varphi)\bar{S}(\varphi) + \bar{\lambda}(\varphi)\bar{S}(\varphi)]x, x \rangle \leq -\|x\|^2, \quad (3.6.2)$$

and, moreover, both quadratic forms $\langle S(\varphi)x, x \rangle$, $\langle \bar{S}(\varphi)x, x \rangle$ *are reducible to the algebraic sum of squares, i.e. there exist non-degenerate matrices* $Q(\varphi), \bar{Q}(\varphi) \in C'(\mathcal{T}_m; a)$ *such that*

$$Q^*(\varphi)S(\varphi)Q(\varphi) = \text{diag}\{I_r, -I_{n-r}\}; \quad (3.6.3)$$

$$\bar{Q}^*(\varphi)\bar{S}(\varphi)\bar{Q}(\varphi) = \text{diag}\{I_{\bar{r}}, -I_{n-\bar{r}}\}. \quad (3.6.4)$$

In addition we assume that $\bar{r} < r$. *Then the inequality*

$$m < r - \bar{r}, \quad (3.6.5)$$

where m is the number of variables $\varphi_1, \ldots, \varphi_m$ *is sufficient for the non-degenerate matrix* $L(\varphi) \in C'(\mathcal{T}_m; a)$ *to exist such that*

$$L^{-1}(\varphi)[A(\varphi)L(\varphi) - \dot{L}(\varphi)] = \text{diag}\{B_1(\varphi), B_2(\varphi), B_3(\varphi)\}, \quad (3.6.6)$$

where the matrices B_1, B_2 *and* B_3 *have the dimensions* $\bar{r} \times \bar{r}$, $(r - \bar{r}) \times (r - \bar{r})$, $(n - r) \times (n - r)$ *respectively.*

Proof The inequality (3.6.1) ensures exponential dichotomy of the torus $x = 0$ of the system

$$\dot{\varphi} = a(\varphi), \quad \dot{x} = \left[A(\varphi) + \frac{1}{2}\lambda(\varphi)I_n\right]x.$$

In Chapter 2 it was proved that the projection matrix $C(\varphi)$ on the subspace $E^+(\varphi)$ along $E^-(\varphi)$ depends on the variables φ continuously, and its rank does not depend on φ: rank $C(\varphi) \equiv r$. Similarly, the condition (3.6.2) ensures exponential dichotomy of the torus $x = 0$ of the system

$$\dot\varphi = a(\varphi), \quad \dot x = \left[A(\varphi) + \frac{1}{2}\bar\lambda(\varphi)I_n\right]x.$$

The corresponding projection matrices $\bar C(\varphi)$ and $I_n - \bar C(\varphi)$ have the ranks $\bar r$ and $n - \bar r$.

Let us show that the difference $C(\varphi) - \bar C(\varphi)$ is a projection matrix. For this purpose we cite the estimates which follow from inequalities (3.6.1) and (3.6.2):

$$\exp\left\{\frac{1}{2}\int_\tau^t \lambda(\varphi_\sigma(\varphi))\,d\sigma\right\} \|\Omega_\tau^t(\varphi)C(\varphi_\tau(\varphi))\| \leq K\exp\{-\gamma(t-\tau)\},$$

$$\tau \leq t,$$

$$\exp\left\{\frac{1}{2}\int_\tau^t \lambda(\varphi_\sigma(\varphi))\,d\sigma\right\} \|\Omega_\tau^t(\varphi)[C(\varphi_\tau(\varphi)) - I_n]\|$$

$$\leq K\exp\{\gamma(t-\tau)\}, \quad t \leq \tau,$$

$$\exp\left\{\frac{1}{2}\int_\tau^t \bar\lambda(\varphi_\sigma(\varphi))\,d\sigma\right\} \|\Omega_\tau^t(\varphi)\bar C(\varphi_\tau(\varphi))\| \leq K\exp\{-\gamma(t-\tau)\}, \quad (3.6.7)$$

$$\tau \leq t,$$

$$\exp\left\{\frac{1}{2}\int_\tau^t \bar\lambda(\varphi_\sigma(\varphi))\,d\sigma\right\} \|\Omega_\tau^t(\varphi)[\bar C(\varphi_\tau(\varphi)) - I_n]\|$$

$$\leq K\exp\{\gamma(t-\tau)\}, \quad t \leq \tau.$$

We prove for all $\varphi \in R^m$ that

$$[C(\varphi) - \bar C(\varphi)]\bar C(\varphi) \equiv 0. \tag{3.6.8}$$

In view of the above proved invariance property

$$C(\varphi_t(\varphi))\Omega_0^t(\varphi) \equiv \Omega_0^t(\varphi)C(\varphi), \quad \bar C(\varphi_t(\varphi))\Omega_0^t(\varphi) \equiv \Omega_0^t(\varphi)\bar C(\varphi), \tag{3.6.9}$$

we transform the left-hand side of (3.6.8) as follows

$$[C(\varphi) - \bar{C}(\varphi)]\bar{C}(\varphi) = [C(\varphi) - I_n]\bar{C}(\varphi) = \Omega_0^t(\varphi)[C(\varphi_t(\varphi)) - I_n]$$

$$\times \Omega_0^t(\varphi)\bar{C}(\varphi) = \left\{\left[\Omega_0^t(\varphi) \exp\left\{\frac{1}{2}\int_t^0 \lambda(\varphi_\sigma(\varphi))\,d\sigma\right\}\right][C(\varphi_t(\varphi)) - I_n]\right\}$$

$$\times \left\{\Omega_0^t(\varphi) \exp\left\{\frac{1}{2}\int_0^t \bar{\lambda}(\varphi_\sigma(\varphi))\,d\sigma\right\}\bar{C}(\varphi)\right\} \qquad (3.6.10)$$

$$\times \exp\left\{\frac{1}{2}\int_0^t [\lambda(\varphi_\sigma(\varphi)) - \bar{\lambda}(\varphi_\sigma(\varphi))]\,d\sigma\right\}.$$

Because the dimensions r of the subspace E_1^+ of damping on $+\infty$ solutions of the system

$$\dot{x} = \left[A(\varphi_t(\varphi)) + \frac{1}{2}\lambda(\varphi_t(\varphi))I_n\right]x \qquad (3.6.11)$$

is larger than the dimensions $\bar{r}$ of the subspace E_2^+ of the same solutions of the system

$$\dot{x} = \left[A(\varphi_t(\varphi)) + \frac{1}{2}\bar{\lambda}(\varphi_t(\varphi))I_n\right]x, \qquad (3.6.12)$$

and every solution $x(t,\varphi)$ of the system (3.6.11) is related to the solution $\bar{x}(t,\varphi)$ of the system (3.6.12) by the equality

$$x(t,\varphi) = \bar{x}(t,\varphi) \exp\left\{\frac{1}{2}\int_0^t [\lambda(\varphi_\sigma(\varphi)) - \bar{\lambda}(\varphi_\sigma(\varphi))]\,d\sigma\right\},$$

we have

$$\lim_{t\to+\infty} \exp\left\{\frac{1}{2}\int_0^t [\lambda(\varphi_\sigma(\varphi)) - \bar{\lambda}(\varphi_\sigma(\varphi))]\,d\sigma\right\} = 0. \qquad (3.6.13)$$

Taking into account the second and the fourth estimates (3.6.7) and equality (3.6.13) we pass to the limit as $t \to +\infty$ in (3.6.10). Then we arrive at identity (3.6.8). The following identity is established in a similar way

$$\bar{C}(\varphi)[C(\varphi) - \bar{C}(\varphi)] \equiv 0. \qquad (3.6.14)$$

It is clear that (3.6.8) and (3.6.14) imply $[C(\varphi) - \bar{C}(\varphi)]^2 \equiv C(\varphi) - \bar{C}(\varphi)$.

Thus, the identity matrix I_n can be represented in the form of a sum of three projection matrices

$$I_n = \bar{C}(\varphi) + [C(\varphi) - \bar{C}(\varphi)] + [I_n - C(\varphi)]. \tag{3.6.15}$$

Each of these matrices has the invariance property (see also (3.6.9))

$$[C(\varphi_t(\varphi)) - \bar{C}(\varphi_t(\varphi))]\Omega_0^t(\varphi) \equiv \Omega_0^t(\varphi)[C(\varphi) - \bar{C}(\varphi)].$$

As the quadratic form $\langle \bar{S}(\varphi)x, x \rangle$ can be reduced to the algebraic sum of squares, then there exists a non-degenerate matrix $T(\varphi) \in C'(\mathcal{T}_m; a)$ such that $T^{-1}(\varphi)\bar{C}(\varphi)T(\varphi) = \text{diag}\{I_{\bar{r}}, 0\}$. Therefore after multiplication on the right by the matrix $T(\varphi)$ and on the left by $T^{-1}(\varphi)$ equality (3.6.15) becomes

$$I_n = \text{diag}\{I_{\bar{r}}, 0\} + \text{diag}\{0, C_1(\varphi)\} + \text{diag}\{0, I_{n-\bar{r}} - C_1(\varphi)\}, \tag{3.6.16}$$

where $C_1(\varphi)$ is a $(n - \bar{r}) \times (n - \bar{r})$-dimensional projection matrix.

We show that there exists a matrix $T_1(\varphi) \in C'(\mathcal{T}_m; a)$ such that

$$T_1^{-1}(\varphi)C_1(\varphi)T_1(\varphi) \equiv \text{diag}\{I_{r-\bar{r}}, 0\}. \tag{3.6.17}$$

Clearly, the existence of the matrix $Q(\varphi) \in C^0(\mathcal{T}_m)$ satisfying the identity (3.6.3) is equivalent to the solvability of two systems of algebraic equations $C(\varphi)x = 0$, $C(\varphi)x = x$, in $C'(\mathcal{T}_m; a)$. Therefore, the system

$$\begin{pmatrix} I_{\bar{r}} & 0 \\ 0 & C_1(\varphi) \end{pmatrix} \begin{pmatrix} y \\ z \end{pmatrix} = 0, \quad y \in R^{\bar{r}}, \quad z \in R^{n-\bar{r}},$$

has $n - r$ linearly independent solutions $\text{col}\{0, e_i\}$. The values of the coordinates e_i are the solutions of the system $C_1(\varphi)z = 0$.

Now consider the two systems of algebraic equations

$$\langle e_i(\varphi), z \rangle = 0, \quad i = 1, \ldots, n - r, \quad C_1^*(\varphi)z = z, \tag{3.6.18}$$

and show its equivalence. Let $z = u(\varphi)$ be the solution of the second system (3.6.18). Then it is easy to obtain the equality

$$\langle e_i(\varphi), u(\varphi) \rangle = \langle e_i(\varphi), C^*(\varphi)u(\varphi) \rangle = \langle C_1(\varphi)e_i(\varphi), u(\varphi) \rangle = 0.$$

Assume that $z = u(\varphi)$ is a solution of the first system (3.6.18). We show that for each of the values $\mu \in R^{n-\bar{r}}$ the scalar product $\langle [I_{n-\bar{r}} - C_1^*(\varphi)]u(\varphi), \mu \rangle$ equals to zero. Obviously, it follows from this that $C_1^*(\varphi)u(\varphi) \equiv u(\varphi)$. We fix some value $\varphi = \varphi_0 \in R^m$ and note that the vector

$$z_0 = [I_{n-\bar{r}} - C_1(\varphi)]\mu$$

is a solution of the algebraic system $C_1(\varphi_0)z = 0$. Then it can be decomposed into the vectors $e_i(\varphi_0)$

$$z_0 = \sum_{i=1}^{n-r} \nu_i e_i(\varphi_0).$$

Thus we get

$$\langle [I_{n-\bar{r}} - C_1^*(\varphi)]u(\varphi_0), \mu \rangle = \langle u(\varphi_0), [I_{n-\bar{r}} - C_1(\varphi)]\mu \rangle$$
$$= \left\langle u(\varphi_0), \sum_{i=1}^{n-r} \nu_i e_i(\varphi_0) \right\rangle = \sum_{i=1}^{n-r} \nu_i \langle u(\varphi_0), e_i(\varphi_0) \rangle = 0.$$

The arbitrariness of the choice of $\varphi_0 \in R^m$ proves the desired identity.

It is clear (see Samoilenko [3]) that the inequality (3.6.5) ensures the existence of $r - \bar{r}$ linearly independent solutions of the first of the systems (3.6.18).

Note that the solvability in $C'(\mathcal{T}_m; a)$ of the second system (3.6.18) involves the solvability of the conjugated system $C_1(\varphi, z) = z$. This follows from the identity $D(\varphi)C_1(\varphi) \equiv C_1^*(\varphi)D(\varphi)$, where the non-degenerate matrix $D(\varphi)$ is determined by

$$D(\varphi) = C_1^*(\varphi)C_1(\varphi) + [I_{n-\bar{r}} - C_1^*(\varphi)][I_{n-\bar{r}} - C_1(\varphi)].$$

The solvability in $C'(\mathcal{T}_m; a)$ of two algebraic systems $C_1(\varphi)z = 0$, $C_1(\varphi)z = z$ is equivalent to the existence of the non-degenerate matrix $T_1(\varphi) \in C'(\mathcal{T}_m; a)$ ensuring the division of the matrix $C_1(\varphi)$, i.e. the that (3.6.17) is satisfied. Thus, by means of a non-degenerate matrix of the form $L(\varphi) = T(\varphi) \operatorname{diag}\{I_{\bar{r}}, T_1(\varphi)\}$ the equality (3.6.15) is reduced to

$$I_n = L^{-1}(\varphi)\bar{C}(\varphi)L(\varphi) + L^{-1}(\varphi)[C(\varphi) - \bar{C}(\varphi)]L(\varphi)$$
$$+ L^{-1}(\varphi)[I_n - C(\varphi)]L(\varphi)$$
$$= \operatorname{diag}\{I_{\bar{r}}, 0, 0\} + \operatorname{diag}\{0, I_{r-\bar{r}}, 0\} + \operatorname{diag}\{0, 0, I_{n-\bar{r}}\}.$$

Taking into account the invariance conditions of each of the projection matrices $\bar{C}(\varphi)$, $C(\varphi) - \bar{C}(\varphi)$, $I_n - C(\varphi)$ we verify that the matrix $L(\varphi) \in C'(\mathcal{T}_m; a)$ satisfies condition (3.6.6). This completes the proof of Theorem 3.6.1.

Remark 3.6.1 Inequality (3.6.5) has a certain topological sense (see Bylov *et al.* [1]). Neglecting this in the conditions of Theorem 3.6.1 may result in the impossibility of making the three-blocked division (3.6.6).

Lemma 3.6.1 *Let the n-dimensional symmetric matrix*

$$S(\varphi) = \begin{pmatrix} S_{11}(\varphi) & S_{12}(\varphi) \\ S_{21}(\varphi) & -S_{22}(\varphi) \end{pmatrix} \in C'(\mathcal{T}_m; a),$$

be such that its $n-r$-dimensional rectangular block $-S_{22}(\varphi)$ is negative definite: $\langle -S_{22}(\varphi)\eta_2, \eta_2\rangle \leq -\gamma_2\|\eta_2\|^2$, $\eta_2 \in R^{n-r}$, $\gamma_2 > 0$. Then there exists an $n \times n$-dimensional matrix $Q(\varphi) \in C'(\mathcal{T}_m; a)$ such that

$$Q^*(\varphi)S(\varphi)Q(\varphi) = \operatorname{diag}\{\bar{S}_{11}(\varphi), -I_{n-r}\}. \tag{3.6.19}$$

Proof In the quadratic form

$$\langle S_{11}(\varphi)x_1, x_1\rangle + \langle S_{12}(\varphi)x_2, x_1\rangle + \langle S_{21}(\varphi)x_1, x_2\rangle - \langle S_{22}(\varphi)x_2, x_2\rangle$$

the change of a part of variables $x_2 = Q_2(\varphi)y_2$ is made so that the identity

$$Q_2^*(\varphi)S_{22}(\varphi)Q_2(\varphi) \equiv I_{n-r}, \quad Q_2(\varphi) \in C'(\mathcal{T}_m; a)$$

is satisfied. Then we get

$$-\langle y_2, y_2\rangle + \langle S_{21}(\varphi)x_1, Q_2(\varphi)y_2\rangle + \langle S_{12}(\varphi)Q_2(\varphi)y_2, x_1\rangle$$
$$+ \langle S_{11}(\varphi)x_1, x_1\rangle = -\|y_2 - Q_2^*(\varphi)S_{21}(\varphi)x_1\|^2$$
$$+ \langle [S_{21}(\varphi)Q_2(\varphi)Q_2^*(\varphi)S_{21}(\varphi) + S_{11}(\varphi)]x_1, x_1\rangle.$$

Hence it is clear that (3.6.19) is satisfied with a matrix $Q(\varphi) \in C'(\mathcal{T}_m; a)$ of the following type

$$Q(\varphi) = \begin{pmatrix} I_r & 0 \\ Q_2(\varphi)Q_2^*(\varphi)S_{21}(\varphi) & Q_2(\varphi) \end{pmatrix}.$$

Theorem 3.6.2 *For some scalar functions* $\lambda(\varphi), \bar{\lambda}(\varphi) \in C^0(\mathcal{T}_m)$ *let there exist n-dimensional symmetric matrices* $S(\varphi), \bar{S}(\varphi) \in C'(\mathcal{T}_m; a)$ *satisfying the conditions (3.6.1), (3.6.2) and having the block-diagonal form:*

$$S(\varphi) = \operatorname{diag}\{S_1(\varphi), -S_2(\varphi)\},$$
$$\bar{S}(\varphi) = \operatorname{diag}\{\bar{S}_1(\varphi), -\bar{S}_2(\varphi)\}, \tag{3.6.20}$$

where $S_1(\varphi)$ and $\bar{S}_1(\varphi)$ are r-dimensional, $S_1(\varphi), S_2(\varphi), \bar{S}_2(\varphi)$ are positive definite matrices, and $\bar{S}_1(\varphi)$ has $\bar{r}$ positive eigenvalues and $r - \bar{r}$ negative eigenvalues. Then, in order that the matrix $L(\varphi) \in C'(\mathcal{T}_m; a)$ exists which satisfies the condition (3.6.6) it is necessary and sufficient that the quadratic form $\langle\bar{S}_1(\varphi)x_1, x_1\rangle$ can be reduced to the algebraic sum of squares, i.e. the matrix $Q(\varphi) \in C'(\mathcal{T}_m; a)$ exists such that

$$Q^*(\varphi)\bar{S}_1(\varphi)Q(\varphi) = \operatorname{diag}\{I_{\bar{r}}, -I_{r-\bar{r}}\}. \tag{3.6.21}$$

Proof In the conditions of Theorem 3.6.1 the identity matrix I_n is represented in the form of a sum of three projection matrices, each of which has the invariance property. Moreover, the block diagonal form of the matrix $S(\varphi)$ with positive definite blocks $S_1(\varphi)$ and $S_2(\varphi)$ indicates the existence of the non-degenerate matrix

$$T(\varphi) = \begin{pmatrix} I_r & L_2(\varphi) \\ L_1(\varphi) & I_{n-r} \end{pmatrix} \in C'(\mathcal{T}_m; a), \qquad (3.6.22)$$

such that $T^{-1}(\varphi)C(\varphi)T(\varphi) = \operatorname{diag}\{I_r, 0\}$. Thus, the expansion of the identity matrix (3.6.15) by the matrix $T(\varphi)$ is reduced to

$$I_n = \operatorname{diag}\{C_1(\varphi), 0\} + \operatorname{diag}\{I_r - C_1(\varphi), 0\} + \operatorname{diag}\{0, I_{n-r}\}.$$

Hence it is clear that the division (3.6.6) is only possible only when the two systems of the algebraic equations in $C'(\mathcal{T}_m; a)$

$$C_1(\varphi)y_1 = 0, \quad C_1(\varphi)y_1 = y_1, \quad y_1 \in R' \qquad (3.6.23)$$

are solvable.

In the inequality $\langle [\bar{S}(\varphi)\bar{C}(\varphi) + \bar{C}^*(\varphi)\bar{S}(\varphi)]x, x \rangle \geq \delta \|x\|^2$ we make the change of variables $x = T(\varphi)y$. Then

$$\langle [\tilde{S}(\varphi)\tilde{C}(\varphi) + \tilde{C}^*(\varphi)\tilde{S}(\varphi) - \tilde{S}(\varphi)]y, y \rangle \geq \varepsilon \|y\|^2, \quad \varepsilon = \operatorname{const} > 0,$$

where

$$\tilde{S}(\varphi) = T^*(\varphi)\bar{S}(\varphi)T(\varphi), \quad \tilde{C}(\varphi) = T^{-1}(\varphi)\bar{C}(\varphi)T(\varphi) = \operatorname{diag}\{C_1(\varphi), 0\}.$$

Hence it follows that the r-dimensional block $\tilde{S}_{11}(\varphi)$ of the matrix $\tilde{S}(\varphi)$ ($\tilde{S}(\varphi) = \{S_{ij}(\varphi)\}_{i,j=1}^2$) satisfies the inequality

$$\langle [\tilde{S}_{11}(\varphi)C_1(\varphi) + C_1^*(\varphi)\tilde{S}_{11}(\varphi) - \tilde{S}_{11}(\varphi)]y_1, y_1 \rangle \geq \varepsilon \|y_1\|^2. \qquad (3.6.24)$$

Thus, the possibility of solving two systems of algebraic equations (3.6.23) in $C'(\mathcal{T}_m; a)$ is implied by the possibility of reducing the quadratic form

$$\langle \tilde{S}_{11}(\varphi)y_1, y_1 \rangle \qquad (3.6.25)$$

to the algebraic sum of squares. The representation of the matrix $T(\varphi)$ in the form of (3.6.22) enables us to establish the following relationship between the matrices $\tilde{S}_{11}(\varphi)$ and $\bar{S}_1(\varphi)$

$$\tilde{S}_{11}(\varphi) = \bar{S}_1(\varphi) - L_1^*(\varphi)\bar{S}_2(\varphi)L_1(\varphi). \qquad (3.6.26)$$

Since the inequality (3.6.24) holds for the matrix $\tilde{S}_{11}(\varphi)$, it has, similarly to $\bar{S}_1(\varphi)$, r positive and $r - \bar{r}$ negative eigenvalues.

Now we verify that the quadratic form (3.6.25) can be reduced to the algebraic sum of the squares if and only if the quadratic form $\langle \bar{S}_{11}(\varphi)y_1, y_1 \rangle$ is also reducible. Let there exist a non-degenerate matrix $\bar{Q}(\varphi) \in C'(\mathcal{T}_m; a)$ such that $\bar{Q}^*(\varphi)\bar{S}_1(\varphi)\bar{Q}(\varphi) = \text{diag}\{I_{\bar{r}}, -I_{r-\bar{r}}\}$, then (3.6.26) can be reduced to

$$\bar{Q}^*(\varphi)\tilde{S}_1(\varphi)\bar{Q}(\varphi) = \text{diag}\{I_{\bar{r}}, -I_{r-\bar{r}}\} - \bar{Q}^*(\varphi)L_1^*(\varphi)\bar{S}_2(\varphi)L_1(\varphi)\bar{Q}(\varphi)$$
$$= \begin{pmatrix} I_{\bar{r}} - \bar{\tilde{S}}_{11}(\varphi) & -\bar{\tilde{S}}_{12}(\varphi) \\ -\bar{\tilde{S}}_{21}(\varphi) & -[I_{r-\bar{r}} + \bar{\tilde{S}}_{22}(\varphi)] \end{pmatrix} = D(\varphi). \quad (3.6.27)$$

Since the matrix $\bar{S}_2(\varphi)$ is positive definite then for all $y_2 \in R^{r-\bar{r}}$ the inequality

$$\langle -[I_{r-\bar{r}} + \bar{S}_{22}(\varphi)]y_2, y_2 \rangle \leq -\|y_2\|^2$$

is satisfied. Therefore, due to the lemma proved above the matrix (3.6.27) can be reduced to the form

$$\bar{\bar{Q}}^*(\varphi)D(\varphi)Q(\varphi) = \text{diag}\{\bar{\tilde{S}}_{11}(\varphi), -I_{r-\bar{r}}\}, \quad \bar{\bar{Q}}(\varphi) \in C'(\mathcal{T}_m; a),$$

and, moreover, the matrix $D(\varphi)$ has r positive and $r - \bar{r}$ negative eigenvalues. Hence, it follows that the matrix $\tilde{S}_{11}(\varphi)$ is positive definite and then the quadratic form $\langle \tilde{S}_{11}(\varphi)y_1, y_1 \rangle$ can be reduced to the algebraic sum of squares by the change of variables $y_1 = M(\varphi)z_1$, $M \in C'(\mathcal{T}_m; a)$.

Similarly, we establish that the reducibility of the quadratic form (3.6.25) implies the reducibility of the quadratic form $\langle \bar{S}_1(\varphi)y_1, y_1 \rangle$. This completes the proof of Theorem 3.6.2.

Remark 3.6.2 In the above results the matrix $L(\varphi)$ which satisfies (3.6.6) belongs to the space $C'(\mathcal{T}_m; a)$. Sufficient conditions can be indicated for which this matrix belongs to the space $C^q(\mathcal{T}_m)$, $q = 1, 2, \ldots$. For this it is sufficient to assume that the functions $a(\varphi)$, $A(\varphi)$ belong to the space $C^q(\mathcal{T}_m)$ and the inequality

$$\max_{\|y\|=1} \left| \left\langle \frac{\partial a}{\partial \varphi} y, y \right\rangle \right| \max \left\{ \max_{\|x\|=1} |\langle S(\varphi)x, x \rangle|, \max_{\|x\|=1} |\langle \bar{S}(\varphi)x, x \rangle| \right\} < q$$

is satisfied.

3.7 Algebraic Problems of k-blocked Divisibility of Linear Extensions on a Torus

Definition 3.7.1 The equality

$$P(\varphi_0)x = 0, \quad x \in R^n, \quad \varphi \in R^m, \tag{3.7.1}$$

where $P(\varphi)$ is a rectangular matrix function, determines the *separatrix invariant manifold* of the system

$$\frac{d\varphi}{dt} = a(\varphi), \quad \frac{dx}{dt} = A(\varphi)x, \tag{3.7.2}$$

if $P(\varphi_0)x_0 = 0$ implies $P(\varphi_t(\varphi_0))\Omega_0^t(\varphi_0)x_0 \equiv 0$, for all $t \in R$.

Proposition 3.7.1 *If $P(\varphi)$ is an $n \times n$-dimensional projection matrix, then for the equality (3.7.1) to determine the separatrix invariant manifold of system (3.7.2) it is necessary and sufficient that*

$$P(\varphi_t(\varphi))\Omega_0^t(\varphi)[I_n - P(\varphi)] \equiv 0, \tag{3.7.3}$$

for all $t \in R$, $\varphi \in R^m$.

Proof Actually, for all $\eta \in R^n$

$$P(\varphi)[(I_n - P(\varphi))\eta] = 0.$$

Therefore,

$$P(\varphi_t(\varphi))\Omega_0^t(\varphi)[I_n - P(\varphi)]\eta = 0.$$

Hence the identity (3.7.3) follows. Since $P(\varphi_0)[I_n - P(\varphi_0)]x_0 = 0$ for every $x_0 \in R$, then (3.7.3) implies that (3.7.1) determines the separatrix invariant manifold of the system (3.7.2).

Corollary 3.7.1 *For the equalities $P(\varphi)x = 0$, $[I_n - P(\varphi)]x = 0$, where $P^2(\varphi) \equiv P(\varphi)$ to determine separatrix invariant manifolds of the system (3.7.2), it is necessary and sufficient that*

$$\Omega_0^t(\varphi)P(\varphi) = P(\varphi_t(\varphi))\Omega_0^t(\varphi). \tag{3.7.4}$$

Theorem 3.7.1 *Let the expansion of the identity matrix I_n into a sum of k projection matrices exist:*

$$I_n = P_1(\varphi) + P_2(\varphi) + \cdots + P_k(\varphi), \tag{3.7.5}$$

which satisfies the conditions

(1) $P_i^2(\varphi) \equiv P_i(\varphi) \in C'(\mathcal{T}_m; a)$, $\quad P_i(\varphi)P_j(\varphi) \equiv 0$, $\quad i \neq j$;

(2) *for every projection matrix $P_i(\varphi)$ the invariance condition (3.7.4) holds true;*
(3) *there exists a non-degenerate matrix $T(\varphi) \in C'(\mathcal{T}_m; a)$ which reduces all projection matrices $P_i(\varphi)$ to the Jordan form*

$$\begin{aligned}T^{-1}(\varphi)P_1(\varphi)T(\varphi) &= \operatorname{diag}\{I_{r_1}, 0, \ldots, 0\},\\ T^{-1}(\varphi)P_2(\varphi)T(\varphi) &= \operatorname{diag}\{0, I_{r_2}, \ldots, 0\},\\ &\cdots\cdots\cdots\cdots\cdots\cdots\cdots\cdots\cdots\cdots\\ T^{-1}(\varphi)P_k(\varphi)T(\varphi) &= \operatorname{diag}\{0, 0, \ldots, I_{r_k}\}.\end{aligned} \quad (3.7.6)$$

Then the change of variables $x = T(\varphi_t(\varphi))y$ divides the system (3.1.8) into k subsystems, i.e.

$$T^{-1}(\varphi)[A(\varphi)T(\varphi) - \dot{T}] = \operatorname{diag}\{B_1(\varphi), B_2(\varphi), \ldots, B_k(\varphi)\}, \quad (3.7.7)$$

where $B_i(\varphi)$ are $r_i \times r_i$-dimensional matrices.

Proof Consider (3.7.7) with the projection matrix $P(\varphi) = P_1(\varphi)$ and transform it as

$$\begin{aligned}&T^{-1}(\varphi_t(\varphi))\Omega_0^t(\varphi)T(\varphi)T^{-1}(\varphi)P_1(\varphi)T(\varphi)\\ &\equiv T^{-1}(\varphi_t(\varphi))P_1(\varphi_t(\varphi))T(\varphi_t(\varphi))T(\varphi_t(\varphi))\Omega_0^t(\varphi)T(\varphi).\end{aligned} \quad (3.7.8)$$

Note that $T^{-1}(\varphi_t(\varphi))\Omega_0^t(\varphi)T(\varphi) = \Omega_0^t(\varphi; B)$ is a matricant of the transformed system

$$\frac{dy}{dt} = T^{-1}(\varphi_t(\varphi))[A(\varphi_t(\varphi))T(\varphi_t(\varphi)) - \dot{T}(\varphi_t(\varphi))]y = B(\varphi_t(\varphi))y. \quad (3.7.9)$$

In view of the first equality (3.7.6) we represent (3.7.8) in the form

$$\Omega_0^t(\varphi; B)\operatorname{diag}\{I_{r_1}, 0, \ldots, 0\} \equiv \operatorname{diag}\{I_{r_1}, 0, \ldots, 0\}\Omega_0^t(\varphi; B).$$

Hence it is clear that the matricant $\Omega_0^t(\varphi; B)$ has the block-diagonal form $\operatorname{diag}\{\Omega_{01}^t(\varphi; B), \bar{\Omega}_0^t(\varphi; B)\}$. Therefore the matrix $B(\varphi)$ also has the block-diagonal form $\operatorname{diag}\{B_1(\varphi), \bar{B}(\varphi)\}$. Proceeding with the identities (3.7.4) with the matrices $P(\varphi) = P_1(\varphi), \ldots, P_k(\varphi)$ and the equalities (3.7.6) with one and the same matrix $T(\varphi)$ we conclude that the matrix $B(\varphi)$ is of the block-diagonal form (3.7.7).

Remark 3.7.1 The above assertion shows that the algebraic aspect of the problem on the possibility of dividing the system (3.7.2) with respect to the variables x into k subsystems consists in the possibility of transforming one expansion of the identity matrix in the form of a sum of variable projection matrices (3.7.5) into another expansion in the form of a sum of constant projection matrices.

Remark 3.7.2 The divisibility of the system (3.7.2) is not, generally speaking, a rough property relative to small perturbations of the functions $a(\varphi)$ and $A(\varphi)$, and its algebraic part in the transformation of one expansion of the identity matrix into the other is valid with respect to small perturbations of the projection matrix.

In fact, let there be one more expansion of the identity matrix $I_n = \bar{P}_1(\varphi) + \cdots + \bar{P}_k(\varphi)$ in addition to (3.7.5), where the projection matrices $\bar{P}_i(\varphi)$ are close enough to the matrices $P_i(\varphi)$. Then there is a non-degenerate matrix $T(\varphi) \in C'(\mathcal{T}_m; a)$ such that $T^{-1}(\varphi) P_i(\varphi) T(\varphi) \equiv \bar{P}_i(\varphi)$. We can take the following $T(\varphi) = \sum_{i=1}^{k} P_i(\varphi) \bar{P}_i(\varphi)$ for this matrix. The non-degeneracy of this matrix is implied by its representation

$$T(\varphi) = I_n + \sum_{i=1}^{k} P_i(\varphi)[\bar{P}_i(\varphi) - P_i(\varphi)].$$

For every matrix $P_i(\varphi)$ from (3.7.5) let there exist its own non-degenerate matrix $T_i(\varphi)$ reducing $P_i(\varphi)$ to the Jordan form. There arises a question whether there exist a common non-degenerate matrix $T(\varphi)$ in these conditions, which satisfy (3.7.6)? The following result is the answer.

Theorem 3.7.2 *Let there be an expansion of the identity matrix (3.7.5) with the projection matrices satisfying the first two conditions of Theorem 3.7.1. Then for the existence of a non-degenerate matrix $T(\varphi) \in C'(\mathcal{T}_m; a)$, which ensures the equalities (3.7.6), it is necessary and sufficient that each system of the algebraic equations*

$$[I_n - P_i(\varphi)]x = 0, \quad i = 1, \ldots, k, \qquad (3.7.10)$$

is solvable in $C'(\mathcal{T}_m; a)$.

Proof The necessity is obvious, since the representation (3.7.6) immediately implies the solvability of the systems (3.7.10) in $C'(\mathcal{T}_m; a)$. The unit solutions of (3.7.10) are designated by $e_{i1}(\varphi), \ldots, e_{ir_i}(\varphi)$. We now show that for the matrix $T(\varphi)$ the following matrix

$$T(\varphi) = [e_{11}(\varphi), \ldots, e_{1r_1}(\varphi), \ldots, e_{k1}(\varphi), \ldots, e_{rk_r}(\varphi)], \qquad (3.7.11)$$

can be taken. We first verify its non-degeneracy. For some $\varphi_0 \in R^m$ let $\det T(\varphi_0) = 0$, then the columns of the matrix $T(\varphi)$ must be linearly dependent

$$\sum_{i=1}^{r_1} \nu_{1i} e_{1i}(\varphi_0) + \cdots + \sum_{i=1}^{r_k} \nu_{ki} e_{ki}(\varphi_0) = 0,$$

and, also, $\nu_{11}^2 + \cdots + \nu_{kr_k}^2 \neq 0$. Without loss of generality one can assume $\nu_{11} = 1 \neq 0$. Thus, we have

$$e_{11}(\varphi_0) = -\sum_{i=2}^{r_1} \nu_{1i} e_{1i}(\varphi_0) - \cdots - \sum_{i=1}^{r_k} \nu_{ki} e_{ki}(\varphi_0). \tag{3.7.12}$$

The equality (3.7.12) is multiplied on the left by the matrix $P_1(\varphi_0)$ taking into account that

$$P_1(\varphi_0) P_j(\varphi_0) = 0, \quad j \neq 1, \quad P_1(\varphi_0) e_{1i}(\varphi_0) = e_{1i}(\varphi_0).$$

We get $e_{11}(\varphi_0) = -\sum_{i=2}^{r_1} \nu_{1i} e_{1i}(\varphi_0)$ which contradicts the linear independence of the vectors $e_{11}(\varphi_0), \ldots, e_{1r_1}(\varphi_0)$. Now we are to verify the equalities (3.7.6)

$$P_1(\varphi) T(\varphi) = [e_{11}(\varphi), \ldots, e_{1r_1}(\varphi), 0, \ldots, 0] = T(\varphi) \, \text{diag}\,\{I_{r_1}, 0, \ldots, 0\}.$$

The other equalities are verified in the same way.

Theorem 3.7.3 *In the conditions of Theorem 3.7.2 let it be known that all systems of the algebraic equations (3.7.10) but one*

$$[I_n - P_{j_0}(\varphi)] x = 0, \tag{3.7.13}$$

are solvable in $C'(\mathcal{T}_m; a)$. Then the inequality

$$m_{j_0} < r_{j_0} = \text{rank}\, P_{j_0}(\varphi), \tag{3.7.14}$$

where m_{j_0} is the number of variables φ of the matrix function $P_{j_0}(\varphi)$, ensures the solvability in $C'(\mathcal{T}_m; a)$ of the system (3.7.13).

Proof Note that every solution of the system $[I_n - P_i(\varphi)] x = 0$, $i \neq j_0$ is also a solution to the system

$$P_{j_0}(\varphi) x = 0, \tag{3.7.15}$$

and there are $n - r_{j_0}$ such linearly independent solutions. Therefore the system (3.7.15) is solvable in $C'(\mathcal{T}_m; a)$ and hence the conjugated system $P_{j_0}^*(\varphi) x = 0$ is also solvable. Its solutions are designated by $u_1(\varphi), \ldots, u_{n-r_{j_0}}(\varphi)$. We can easily verify that the system (3.7.13) is equivalent to the system consisting of $n - r_{j_0}$ equations

$$\langle u_i(\varphi), x \rangle = 0, \quad i = 1, \ldots, n - r_{j_0}. \tag{3.7.16}$$

Since the number of equations in (3.7.16) coincides with the rank of the matrix $P_{j_0}(\varphi)$, the inequality (3.7.14) ensures the solvability in $C'(\mathcal{T}_m; a)$ of the systems (3.7.16) and (3.7.13).

3.8 Comments and References

Section 3.0 The problem considered in Chapter 3 is associated with the transformation of variable matrices. Related problems were discussed by Blinov [1], Bylov [1], Bylov et al. [1], Gantmacher [1], Kolmogorov and Fomin [1], Kurosh [1], Lappo-Danilevsky [1], Samoilenko [3], Tkachenko [1].

Sections 3.1 – 3.2 Theorem 3.1.1 is due to Samoilenko and Kulik [1, 2]. Theorems 3.2.1 – 3.2.2 were cited by Kulik [5].

Section 3.3 The results are due to Samoilenko [5].

Section 3.4 The presentation is based on results by Samoilenko and Kulik [4].

Sections 3.5 – 3.7 The results are based on those by Kulik [1, 3], and Samoilenko and Kulik [6, 8, 9].

4 Problems of Perturbation Theory of Smooth Invariant Tori of Dynamical Systems

4.0 Introduction

The variations of the family of solutions to dynamical systems which start on the m-dimensional torus $\mathcal{T}_m$ are studied. Further investigations are made on linear extensions of dynamical systems on the torus which play a major part in perturbation theory.

New criteria of the exponential dichotomy of linear extension are set out and the conditions for roughness of the Green function are established with the smoothness indicator $r \geq 0$ for this extension.

Further, the previous results are employed to formulate theorems on the perturbation of a smooth invariant torus $\mathcal{T}_m$ under the conditions when $\mathcal{T}_m$ cannot be an e-dichotomous torus of the dynamical system. This problem has not been studied yet in perturbation theory.

Theorems are cited establishing the existence of Lyapunov functions of variable sign. Finally, interesting examples illustrate the application of the obtained results.

4.1 Solution Variations on the Manifold M

We consider a system of differential equations of the type

$$\frac{dx}{dt} = X(x) + \varepsilon X_1(x), \qquad (4.1.1)$$

where $X, X_1 \in C^r(R^n)$, $r \geq 2$, ε is a small parameter, $C^r(D)$ is a space of r times continuously differentiable functions in the domain $D \subseteq R^n$, $x \in R^n$. For $\varepsilon = 0$ let the system (4.1.1) (the unperturbed system) have an invariant manifold M of the type

$$M: x = f(\varphi). \qquad (4.1.2)$$

Here $f \in C^r(\mathcal{T}_m)$, $C^r(\mathcal{T}_m)$ is a space of r times continuously differentiable functions on the m-dimensional torus $\mathcal{T}_m$,

$$\text{rank } \frac{\partial f(\varphi)}{\partial \varphi} = m, \quad \forall \varphi \in \mathcal{T}_m. \qquad (4.1.3)$$

Manifold (4.1.2), (4.1.3) is an m-dimensional r times continuously differentiable (of smoothness class $C^r(\mathcal{T}_m)$) toroidal invariant manifold of the unperturbed system (4.1.1).

The perturbation theory for the manifold M is developed under the assumption that the small neighbourhood $V(M)$ of the manifold M is "well-organized" in R^n. The latter means that $V(M)$ is fibering into manifolds of the type of M

$$V(M) = \bigcup_{\|c\|<\delta} M_c, \quad M_c: x = f(\varphi) + B(\varphi)c,$$

where $B(\varphi)$ is an $(n \times (n-m))$-dimensional matrix from $C^r(\mathcal{T}_m)$ which together with the matrix $F(\varphi) = \partial f(\varphi)/\partial \varphi$ forms the 2π-periodic basis in R^n

$$\det [F(\varphi), B(\varphi)] \neq 0, \quad \forall \varphi \in \mathcal{T}_m,$$

c is an arbitrary value from the ball $\|c\|^2 = \sum_{\nu=1}^{n-m} c_\nu^2 \leq \delta$ of the space R^{n-m}, δ is a small positive number.

Such a structure of the neighbourhood M allows the system (4.1.1) to be written in coordinates φ, h (local coordinates in $V(M)$) related to the Euclidean ones by the formula

$$x = f(\varphi) + B(\varphi)h, \qquad (4.1.4)$$

in the form of the system of differential equations in

$$\begin{aligned}\frac{d\varphi}{dt} &= f(\varphi, h) + \varepsilon f_1(\varphi, h), \\ \frac{dh}{dt} &= P(\varphi, h)h + \varepsilon F(\varphi, h).\end{aligned} \qquad (4.1.5)$$

The designations used above are

$$f(\varphi, h) = a(\varphi) + L_1(\varphi, h)\bigg[X(f(\varphi) + B(\varphi)h) - X(f(\varphi))$$
$$- \left(\frac{\partial B(\varphi)}{\partial \varphi}\right) a(\varphi)h\bigg],$$

$$P(\varphi, h)h = L_2(\varphi, h)\bigg[X(f(\varphi) + B(\varphi)h) - X(f(\varphi)) \quad (4.1.6)$$
$$- \left(\frac{\partial B(\varphi)}{\partial \varphi}\right) a(\varphi)h\bigg],$$

$$f_1(\varphi, h) = L_1(\varphi, h)X_1(f(\varphi) + B(\varphi)h),$$
$$F(\varphi, h) = L_2(\varphi, h)X_1(f(\varphi) + B(\varphi)h),$$

$$a(\varphi) = \left[\left(\frac{\partial f(\varphi)}{\partial \varphi}\right)^* \left(\frac{\partial f(\varphi)}{\partial \varphi}\right)\right]^{-1} \left(\frac{\partial f(\varphi)}{\partial \varphi}\right)^* X(f(\varphi)), \quad (4.1.6a)$$

$(\partial f(\varphi)/\partial \varphi)^*$ is the matrix conjugated with $(\partial f(\varphi)/\partial \varphi)$, $L_1(\varphi, h)$ and $L_2(\varphi, h)$ are the matrix blocks with the dimensions $m \times n$ and $(n-m) \times n$ corresponding to the matrix inverse to $[F(\varphi) + (\partial B(\varphi)/\partial \varphi) h, B(\varphi)]$, and

$$\frac{\partial}{\partial \varphi} a(\varphi) = \sum_{\nu=1}^{m} \frac{\partial}{\partial \varphi_\nu} a_\nu(\varphi).$$

When the coordinates x are transformed into φ, h according to (4.1.4) the manifold M becomes the trivial invariant manifold of the system (4.1.5): $h = 0$ with a trajectory flow on it determined by the system of equations on $\mathcal{T}_m$

$$\frac{d\varphi}{dt} = a(\varphi), \quad (4.1.7)$$

where a is a function of (4.1.6a).

Let the neighbourhood M admit a fibering so that the change of variables (4.1.4) reduces system (4.1.1) to the system of equations (4.1.5) in $\mathcal{T}_m \times R^{n-m}$ in the neighbourhood M. By (4.1.6) the right-hand side of (4.1.5) belongs to the function space $C^r(\mathcal{T}_m \times K_\delta)$, where

$$K_\delta = \{h \in R^{n-m}: \|h\| \leq \delta\}, \quad (4.1.8)$$

and δ is a sufficiently small positive number.

The fact that the function belongs to the space $C^{r-1}(\mathcal{T}_m \times K_\delta)$ means its 2π-periodicity in φ_γ, $\forall \gamma = 1, \ldots, m$, and the existence of continuous derivatives of this function up to the order of $(r-1)$ included in the domain $\varphi \in \mathcal{T}_m$, $\|h\| \leq \delta_1$, where $\delta_1 - \delta > 0$ is a sufficiently small number. This implies the boundedness of derivatives of functions from the space $C^{r-1}(\mathcal{T}_m \times K_\delta)$ up to the order of $(r-1)$ included.

As before, we designate the solutions of equation (4.1.1) for $\varepsilon = 0$ by $x(t, N)$, the initial values of which belong to the set N, $N \in R^n$, and by $\Omega_0^t(A)$ the matriciant of the linear system

$$\frac{dx}{dt} = A(t)x, \quad x \in R^n.$$

For system (4.1.1) and (4.1.5) when $\varepsilon = 0$ we write the variation equations along the solutions

$$x = x(t, f(t)) = f(\varphi_t), \tag{4.1.9}$$

$$\varphi = \varphi_t = \varphi_t(\varphi), \quad h = h_t = 0, \tag{4.1.10}$$

where $\varphi_t(\varphi)$ is a solution of equation (4.1.7) taking the value $\varphi_0(\varphi) = \varphi$ for $t = 0$.

For solutions (4.1.9) we have

$$\frac{dz}{dt} = \frac{\partial X(f(\varphi_t))}{\partial x} z, \tag{4.1.11}$$

and for (4.1.10) in view of (4.1.6)

$$\frac{d\Theta}{dt} = \frac{\partial a(\varphi_t)}{\partial \varphi} \Theta + Q(\varphi_t)g,$$

$$\frac{dg}{dt} = P(\varphi_t)g, \tag{4.1.12}$$

where

$$Q(\varphi) = L_1(\varphi) \left[\frac{\partial X(f(\varphi_t))}{\partial x} B(\varphi) - \frac{\partial B(\varphi))}{\partial \varphi} a(\varphi) \right],$$

$$P(\varphi) = P(\varphi, 0), \quad L_1(\varphi) = L_1(\varphi, 0)$$

and $a(\varphi)$, $L_1(\varphi, h)$, $P(\varphi, h)$ are the functions of (4.1.6) and (4.1.6a).

Since the equations (4.1.1) and (4.1.5) are related by the change of variables formula (4.1.4) for $\varepsilon = 0$, its variations along the solutions (4.1.9) and (4.1.10) are related by the formula

$$z = \partial x = \left[\frac{\partial f(\varphi)}{\partial \varphi} + \frac{\partial B(\varphi)h}{\partial \varphi} \right]\bigg|_{\substack{\varphi=\varphi_t \\ h=0}} \partial\varphi + [B(\varphi)]\bigg|_{\substack{\varphi=\varphi_t \\ h=0}} \partial h$$

$$= \frac{\partial f(\varphi_t)}{\partial \varphi} \Theta + B(\varphi_t)g = [F(\varphi_t), B(\varphi_t)] \begin{pmatrix} \Theta \\ g \end{pmatrix}.$$

Therefore, the matricians of the systems (4.1.11) and (4.1.12) are correlated as

$$\Omega_0^t\left(\frac{\partial X(f)}{\partial x}\right)[F(\varphi),B(\varphi)] = [F(\varphi_t),B(\varphi_t)]\begin{pmatrix} \Omega_0^t\left(\frac{\partial a}{\partial \varphi}\right) & R_t \\ 0 & \Omega_0^t(P) \end{pmatrix},$$

$$R_t = \int_0^t \Omega_s^t\left(\frac{\partial a}{\partial \varphi}\right) Q(\varphi_s)\Omega_0^s(P)\,ds. \tag{4.1.13}$$

From (4.1.13) it follows that

$$\Omega_0^t\left(\frac{\partial X(f)}{\partial x}\right) F(\varphi) = F(\varphi_t)\Omega_0^t\left(\frac{\partial a}{\partial \varphi}\right). \tag{4.1.14}$$

By (4.1.14) the matricians $\Omega_0^t\left(\frac{\partial a}{\partial \varphi}\right)$ and $\Omega_0^t(P)$ determine for the behaviour of the solutions to systems of variation equations (4.1.10) and (4.1.11). The first one does not depend on B. Let us establish the dependence of the matricant $\Omega_0^t(P)$ on B.

We assume $C^l(\mathcal{T}_m)$-equivalence of the matricians $\Omega_0^t(P)$ and $\Omega_0^t(P_1)$ of the matrices $P = P(\varphi)$ and $P_1 = P_1(\varphi)$ from the space $C^l(\mathcal{T}_m)$ whenever a non-degenerate matrix $\Phi = \Phi(\varphi)$ from $C^l(\mathcal{T}_m)$ is found such that

$$\Omega_0^t(P) = \Phi(\varphi_t)\Omega_0^t(P_1)\Phi^{-1}(\varphi). \tag{4.1.15}$$

The dependence of $\Omega_0^t(P)$ on the matrix is stated by the following result.

Theorem 4.1.1 *When local coordinates are introduced from the space $C^r(\mathcal{T}_m)$ by means of the matrices $B = B(\varphi)$ and $B_1 = B_1(\varphi)$, the matricians $\Omega_0^t(P)$ and $\Omega_0^t(P_1)$ of the corresponding systems of equations with the matrices $P = P(\varphi)$ and $P_1 = P_1(\varphi)$ are $C^{r-1}(\mathcal{T}_m)$-equivalent.*

Proof Actually, since the matrices $[F,B]$ and $[F,B_1]$ are non-degenerate, then

$$[F,B] = [F,B_1]C, \tag{4.1.16}$$

where C is a non-degenerate matrix.

We shall specify this matrix. Let $\begin{pmatrix} L_1 \\ L_2 \end{pmatrix}$ be the matrix inverse to the matrix $[F,B]$, and hence such that

$$L_1 F = E_1, \quad L_1 B = O_1, \quad L_2 F = O_2, \quad L_2 B = E_2. \tag{4.1.17}$$

Here E_1, E_2, O_1, O_2 are identity and the zero matrices of appropriate dimensions. Then (4.1.16) implies

$$\begin{pmatrix} E_1 & 0 \\ 0 & E_2 \end{pmatrix} = \begin{pmatrix} E_1 & L_1 B_1 \\ 0 & L_2 B_1 \end{pmatrix} C, \tag{4.1.18}$$

and so the matrix $L_2 B_1$ is non-degenerate. In order that (4.1.18) is valid, it suffices to set C equal to the matrix

$$C = \begin{pmatrix} E_1 & -L_1 B_1 (L_2 B_1)^{-1} \\ 0 & (L_2 B_1)^{-1} \end{pmatrix} = \begin{pmatrix} E_1 & C_{12} \\ 0 & C_2 \end{pmatrix}, \qquad (4.1.19)$$

where C_2 is a non-degenerate matrix.

Equalities (4.1.16) and (4.1.19) yield the relation between the matrices B and B_1:

$$B = F C_{12} + B_1 C_2. \qquad (4.1.20)$$

Equalities (4.1.13) and (4.1.14) lead to the correlation

$$F(\varphi_t) R_t + B(\varphi_t) \Omega_0^t(P) = \Omega_0^t \left(\frac{\partial X(f)}{\partial x} \right) [F(\varphi) C_{12}(\varphi) + B_1(\varphi) C_2(\varphi)]$$

$$= F(\varphi_t) \Omega_0^t(P) \left(\frac{\partial a}{\partial \varphi} \right) C_{12}(\varphi) + F(\varphi_t) R_t C_2(\varphi) + B_1(\varphi_t) \Omega_0^t(P_1) C_2(\varphi),$$

the multiplying which on the right by $L_2(\varphi_t)$ results in the equality

$$\Omega_0^t(P) = L_2(\varphi_t) B_1(\varphi_t) \Omega_0^t(P_1) C_2(\varphi). \qquad (4.1.21)$$

In view of (4.1.19) the equality (4.1.21) becomes

$$\Omega_0^t(P) = L_2(\varphi_t) B_1(\varphi_t) \Omega_0^t(P_1) [L_2(\varphi) B_1(\varphi)]^{-1} = C_2(\varphi_t) \Omega_0^t(P_1) C_2^{-1}(\varphi),$$

and proves the $C^{r-1}(\mathcal{T}_m)$-equivalence of the matriciants $\Omega_0^t(P)$ and $\Omega_0^t(P_1)$.

The matriciant $\Omega_0^t \left(\frac{\partial a}{\partial \varphi} \right)$ specifies the distance between the solutions of the unperturbed system of equations starting at close points of M.

In fact, if $\psi - \varphi = \Theta$, $\|\Theta\| < \delta$, then for the distance between $x(t, f(\psi)) = f(\varphi_t(\psi))$ and $x(t, f(\varphi)) = f(\varphi_t(\varphi))$ the following estimate is valid

$$\rho(x(t, f(\psi)), x(t, f(\varphi))) = \|x(t, f(\psi)) - x(t, f(\varphi))\|$$

$$= \|x(t, f(\varphi + \Theta)) - x(t, f(\varphi))\|$$

$$\leq \left\| \frac{\partial x(t, f(\psi))}{\partial f} F(\varphi) \Theta \right\| + O(\|\Theta\|^2)$$

$$\leq \left\| \Omega_0^t \left(\frac{\partial X(f)}{\partial x} \right) F(\varphi) \Theta \right\| + O(\|\Theta\|^2) \qquad (4.1.22)$$

$$= \left\| F(\varphi_t(\psi)) \Omega_0^t \left(\frac{\partial a}{\partial \varphi} \right) \Theta \right\| + O(\|\Theta\|^2)$$

$$\leq K \left\| \Omega_0^t \left(\frac{\partial a}{\partial \varphi} \right) \Theta \right\| + O(\|\Theta\|^2), \quad K = \text{const},$$

which proves the desired statement.

The matriciant of the equation

$$\frac{d\Theta}{dt} = \frac{\partial a(\varphi_t)}{\partial \varphi} \Theta, \tag{4.1.23}$$

specifies stability on of the solutions of the unperturbed system of equations. It seems natural to refer to the equation (4.1.23) as a variational equation of the solutions of the unperturbed system of equations on M.

Equality (4.1.14) unambiguously relates $\Omega_0^t\left(\frac{\partial a}{\partial \varphi}\right)$ and $\Omega_0^t\left(\frac{\partial X(f)}{\partial x}\right)$. It implies, in particular, that $F(\varphi_t)\Omega_0^t\left(\frac{\partial a}{\partial \varphi}\right)$ is a solution of the variational equation (4.1.11) and for it the inequality

$$\frac{1}{c_1}\left\|\Omega_0^t\left(\frac{\partial X(f)}{\partial x}\right)F(\varphi)\right\| \leq \left\|\Omega_0^t\left(\frac{\partial a}{\partial \varphi}\right)\right\| \leq c_1\left\|\Omega_0^t\left(\frac{\partial X(f)}{\partial x}\right)F(\varphi)\right\| \tag{4.1.24}$$

is satisfied, where $c_1 = \text{const} \geq 1$.

The matriciant $\Omega_0^t(P)$ specifies the distance between M and the solutions of unperturbed system of equations starting near M. Obviously, if $y \notin M$, $\rho(M,y) < \delta_1$, then the expansion

$$y = f(\varphi) + B(\varphi)g, \quad \|g\| < \delta, \tag{4.1.25}$$

is valid, and therefore

$$\begin{aligned} x(t,y) &= x(t, f(\varphi) + B(\varphi)g) = f(\varphi(t)) + B(\varphi(t))h(t) \\ &= f(\varphi(t)) + B(\varphi(t))\partial h + O(\|g\|^2) \\ &= f(\varphi(t)) + B(\varphi(t))\Omega_0^t(P)g + O(\|g\|^2), \end{aligned} \tag{4.1.26}$$

where $\varphi(t) = \varphi(t,\varphi,g)$, $h(t) = h(t,\varphi,g)$. From (4.1.26) we deduce the estimate

$$\begin{aligned} \rho(M, x(t,y)) &\leq \|B(\varphi(t))\|\,\|\Omega_0^t(P)\|\,\|g\| + O(\|g\|^2) \\ &\leq K\|\Omega_0^t(P)\|\,\|g\| + O(\|g\|^2), \quad K = \text{const}, \end{aligned}$$

which proves the desired assumption.

The matriciant $\Omega_0^t(P)$ characterizes the stability of the manifold M. It is natural to refer to the equation

$$\frac{dg}{dt} = P(\varphi_t)g \tag{4.1.27}$$

as the variation equation of the invariant manifold M.

The equality (4.1.14) shows that $B(\varphi(t))\Omega_0^t(P)$ is not the solution of the variational equation (4.1.11) for $R_t \neq 0$ and leads to the estimation of the matriciant $\Omega_0^t(P)$ in the form

$$\|\Omega_0^t(P)\| \leq c_2 \left\|\Omega_0^t \left(\frac{\partial X(f)}{\partial x}\right) B(\varphi)\right\| \tag{4.1.28}$$

where $c_2 = \text{const} \geq 1$.

Then, we shall elucidate the part of the matrix R_t in the determination of solutions of the unperturbed system of equations starting near M.

For y defined by (4.1.25), $f(y)$ is called a projection of y on M.

As

$$x(t, f(\varphi) + B(\varphi)g) = f(\varphi(t)) + B(\varphi(t))h(t), \tag{4.1.29}$$

then $f(\varphi(t))$ is the projection of the solution $x(t, y)$ on M. The matrix R_t specifies the distance between the projection of the solution $x(t, y)$ on M and the solution $x(t, f(\varphi)) = f(\varphi_t)$ on M starting with the projection y on M. Evidently, since

$$\varphi(t) - \varphi_t = \varphi(t, \varphi, g) - \varphi(t, \varphi, 0) = \left.\frac{\partial \varphi(t, \varphi, g)}{\partial g}\right|_{g=0} g + O(\|g\|^2)$$

$$= \Theta(t, 0, g) + O(\|g\|^2) = R_t g + O(\|g\|^2),$$

where $\Theta(t, 0, g)$, $g(t, g)$ is a solution of the system (4.1.11) taking the value of Θ, g, for $t = 0$ then

$$\|f(\varphi(t)) - f(\varphi_t)\| \leq \|f(\varphi_t + R_t g + O(\|g\|^2)) - f(\varphi_t)\|$$

$$\leq \|F(\varphi_t) R_t g\| + O(\|g\|^2) \leq K\|R_t\|\|g\| + O(\|g\|^2),$$

where $K = \text{const}$.

The matrix R_t is a characteristic of stability of the solutions starting on M by way of the magnitude of the distance between these solutions and the projections of the solutions starting near M.

The choice of the matrix B in the change (4.1.4) essentially influences the properties of the matrix R_t. Let us demonstrate this. By (4.1.13) we have

$$R_t = \Phi_t \Omega_0^t(P), \quad \Phi_t = \int_0^t \Omega_s^t \left(\frac{\partial a}{\partial \varphi}\right) Q(\varphi_s) \Omega_t^s(P) \, ds.$$

Hence Φ_t is a solution of the equation

$$\frac{d\Phi}{dt} + \Phi P(\varphi_t) = \left(\frac{\partial a(\varphi_t)}{\partial \varphi}\right) \Phi + Q(\varphi_t). \tag{4.1.30}$$

Under certain assumptions on $\Omega_0^t\left(\frac{\partial a}{\partial \varphi}\right)$ and $\Omega_0^t(P)$ the equation (4.1.30) has a solution in $C^l(\mathcal{T}_m)$

$$\Phi = \Phi(\varphi), \quad \Phi \in C^l(\mathcal{T}_m), \quad r-1 \geq l \geq 0. \tag{4.1.31}$$

In this case, in view of the initial value $R_0 = 0$ we get the expression for R_t via $\Phi(\varphi)$

$$R_t = \Phi(\varphi_t)\Omega_0^t(P) - \Omega_0^t\left(\frac{\partial a}{\partial \varphi}\right)\Phi(\varphi). \tag{4.1.32}$$

Further (4.1.14) and (4.1.32) imply

$$\Omega_0^t\left(\frac{\partial X(f)}{\partial x}\right) B(\varphi) = F(\varphi_t)\Phi(\varphi_t)\Omega_0^t(P) - F(\varphi_t)\Omega_0^t\left(\frac{\partial a}{\partial \varphi}\right)\Phi(\varphi) + B(\varphi_t)\Omega_0^t(P)$$

$$= -\Omega_0^t\left(\frac{\partial X(f)}{\partial x}\right) F(\varphi)B(\varphi) + [B(\varphi_t) + F(\varphi_t)\Phi(\varphi_t)]\Omega_0^t(P),$$

and hence,

$$\Omega_0^t\left(\frac{\partial X(f)}{\partial x}\right) B_1(\varphi) = B_1(\varphi_t)\Omega_0^t(P), \tag{4.1.33}$$

where

$$B_1(\varphi) = F(\varphi)\Phi(\varphi) + B(\varphi). \tag{4.1.34}$$

Thus, the matrix $R_t = R_t(B)$ has the form (4.1.32) for B and for B_1, $R_t = R_t(B_1)$, that makes the properties of $R_t(B)$ and $R_t(B_1)$ essentially different. It is clear that $R_t \equiv 0$ if and only if

$$Q(\varphi) = 0, \quad \forall \varphi \in \mathcal{T}_m. \tag{4.1.35}$$

Therefore, there exists, under the condition of solvability in $C^l(\mathcal{T}_m)$ of equation (4.1.30), the matrix $B_1 \in C^l(\mathcal{T}_m)$, for which $Q = Q(B_1)$ satisfies condition (4.1.35). The matrix $B_1(\varphi_t)\Omega_0^t(P)$ is then a solution of the variational equation (4.1.11) and for $\Omega_0^t(P)$ the inequality of the form (4.1.24) is valid

$$\frac{1}{c_2}\left\|\Omega_0^t\left(\frac{\partial X(f)}{\partial x}\right) B_1(\varphi)\right\| \leq \|\Omega_0^t(P)\| \leq c_2 \left\|\Omega_0^t\left(\frac{\partial X(f)}{\partial x}\right) B_1(\varphi)\right\|. \tag{4.1.36}$$

It should be noted that the solvability in $C^l(\mathcal{T}_m)$ of equation (4.1.30) is equivalent to the existence in the system of variational equations (4.1.12) of the invariant manifold

$$\Theta = \Phi(\varphi_t)g, \tag{4.1.37}$$

and to the reduction of this system by the change of variables

$$\Theta = \Theta_1 + \Phi(\varphi_t)g, \tag{4.1.38}$$

to the block-diagonal form with the matrix coefficients

$$\text{diag}\left\{\frac{\partial a(\varphi_t)}{\partial \varphi}, P(\varphi_t)\right\}. \tag{4.1.39}$$

The conditions of the solvability of equation (4.1.30) in $C^l(\mathcal{T}_m)$ are discussed in Section 4.5 of the present Chapter.

4.2 Exponential Stability and Dichotomy Conditions for Linear Extensions of Dynamical Systems on a Torus

In the perturbation theory of invariant tori the properties of the operator L are of major importance, $L: C^{l+1}(\mathcal{T}_m) \to C^l(\mathcal{T}_m)$:

$$L = \frac{\partial}{\partial \varphi} a(\varphi) - P(\varphi), \tag{4.2.1}$$

where $(a, P) \in C^l(\mathcal{T}_m)$, $l \geq 0$. The system of equations corresponding to (4.2.1) for the characteristics

$$\frac{d\varphi}{dt} = a(\varphi), \quad \frac{dh}{dt} = P(\varphi)h, \tag{4.2.2}$$

is usually referred to as the linear extension of the dynamical system on a torus. It is assumed in the theory that the operator SL is at least coercive, where $S = S(\varphi)$ is any non-degenerate symmetric matrix from $C^l(\mathcal{T}_m)$: $(SLu, u) \geq \gamma\|u\|_0^2$ $\forall u \in C^l(\mathcal{T}_m)$, where (u, v) is a scalar product in $L^2(\mathcal{T}_m)$.

In the well known theory a weak regularity of the operator L is assumed, i.e. the existence of an operator pseudoinverse to L in the form of the integral operator determined by the Green function of the system (4.2.2). In the first case it necessarily follows that Ker $L = \{0\}$, in the second case Ker L can have a zero part Ker$_b$ L consisting of the function u from the space $C(\mathcal{T}_m; a)$ such that

$$u \not\equiv 0, \quad \lim_{|t|\to\infty} u(\varphi_t) = 0, \quad \forall \varphi \in \mathcal{T}_m.$$

In addition, $C(\mathcal{T}_m; a)$ is a subspace of $C(\mathcal{T}_m)$ consisting of the functions u, for which $u(\varphi_t)$ possesses a continuous derivative in t such that

$$\frac{du(\varphi_t)}{dt} = \dot{u}(\varphi_t), \quad \forall (t, \varphi) \in R \times \mathcal{T}_m, \tag{4.2.3}$$

where $\dot{u}(\varphi_t) \in C(\mathcal{T}_m)$.

In the coincident parts of both theories the exponential dichotomy of the system of equations (4.2.2) is required from L. The conditions ensuring the dichotomy are discussed below. It should be noted that these conditions which are established in terms of the properties of Lyapunov forms quadratic in h were first set out by Kulik [10]. The conditions below are expressed via the properties of solutions to the system (4.2.2).

We agree to consider h belonging to R^n and to designate the matricant $\Omega_0^t(P)$ by $\Omega_0^t(\varphi)$, stressing its dependence on φ,

$$\Omega_0^t(P) = \Omega_0^t(\varphi). \tag{4.2.4}$$

The exponential dichotomy of the system (4.2.2) is understood in the following sense.

For all $\varphi \in \mathcal{T}_m$, the space R^n is representable as the direct sum of the subspaces $E^+ = E^+(\varphi)$ and $E^- = E^-(\varphi)$ of auxiliary dimensions n_1 and $n - n_1 = n_2$ so that for $h \in E^+$ the solution $\varphi_\tau h_t = h_t(\varphi, h) = \Omega_0^t(\varphi)h$ of the system (4.2.2) satisfies the inequality

$$\|h_t(\varphi, h)\| \leq K \exp\{-\gamma(t - \tau)\} \|h_\tau(\varphi, h)\|, \quad \forall t \geq \tau, \tag{4.2.5}$$

and for $h \in E^-$ the inequality

$$\|h_t(\varphi, h)\| \leq K \exp\{\gamma(t - \tau)\} \|h_\tau(\varphi, h)\|, \quad \forall t \leq \tau, \tag{4.2.6}$$

for arbitrary $\tau \in R$ and $\varphi \in \mathcal{T}_m$ and positive $K \geq 1$ and γ independent of φ, h, t and τ.

The limiting case of exponential dichotomy, when $n_1 = n$ is called exponential stability of the system (4.2.2). In this case the inequality

$$\|\Omega_0^t(\varphi)\| \leq K \exp\{-\gamma t\}, \quad \forall t \in R^+ = [0, \infty), \tag{4.2.7}$$

is satisfied.

We consider the exponentially stable system (4.2.2). In this case (4.2.7) implies

$$\|\Omega_0^T(\varphi)\| \leq d = \text{const} < 1, \tag{4.2.8}$$

for $T \geq (1/\gamma) \ln(K/d) > 0$. Conversely, for some constants $T > 0$ and $d < 1$ let the inequality (4.2.8) be satisfied. Since for $a \in C_{\text{Lip}}(\mathcal{T}_m)$ and $P \in C(\mathcal{T}_m)$ the functions φ_t, $h_t = \Omega_0^t(\varphi)h$ form a dynamical system in $\mathcal{T}_m \times R^n$, then $\varphi_{t+\tau} = \varphi_t(\varphi_\tau)$ and $h_t = h_t(\varphi_\tau, h_\tau)$ and therefore

$$\Omega_0^{t+\tau}(\varphi) = \Omega_0^t(\varphi_\tau)\Omega_0^\tau(\varphi), \quad \forall (t, \tau, \varphi) \in R \times R \times \mathcal{T}_m. \tag{4.2.9}$$

Furthermore, (4.2.9) implies the equality

$$\begin{aligned}\Omega_0^t(\varphi) &= \Omega_0^{t-[t/T]T}(\varphi_{[t/T]T})\Omega_0^{[t/T]T}(\varphi) \\ &= \Omega_0^{t-[t/T]T}(\varphi_{[t/T]T})\Omega_0^T(\varphi_{([t/T]-1)T}) \ldots \Omega_0^T(\varphi),\end{aligned} \tag{4.2.10}$$

where $[t/T]$ denotes the integer part of the number t/T. Then (4.2.10) and (4.2.8) yield the estimate

$$\|\Omega_0^t(\varphi)\| \leq \|\Omega_0^{t-[t/T]T}(\varphi_{[t/T]T})\| d^{[t/T]}$$

$$= \|\Omega_0^{t-[t/T]T}(\varphi_{[t/T]T})\| \exp\left\{\left(\ln\frac{d}{T}\right)\left[\frac{t}{T}\right]T\right\}$$

$$= \left\|\Omega_0^{t-[t/T]T}(\varphi_{[t/T]T}) \exp\left\{-\left(\ln\frac{d}{T}\right)\left(t-\left[\frac{t}{T}\right]T\right)\right\}\right\| \qquad (4.2.11)$$

$$\times \exp\left\{\left(\ln\frac{d}{T}\right)t\right\} \leq \max\left\|\Omega_0^\tau(\varphi)\exp\left\{-\left(\ln\frac{d}{T}\right)\tau\right\}\right\|\exp\{-\gamma t\}$$

$$= K\exp\{-\gamma t\}, \quad \forall t \in R^+,$$

where max is taken with respect to $(\tau,\varphi) \in [0,T] \times \mathcal{T}_m$,

$$\gamma = \left|\ln\frac{d}{T}\right| > 0, \quad K = \max\left\|\Omega_0^\tau(\varphi)\exp\left\{-\left(\ln\frac{d}{T}\right)\tau\right\}\right\|.$$

The inequality (4.2.11) means the exponential stability of the system (4.2.2).

Thus, the following result is valid.

Theorem 4.2.1 *For the system (4.2.2) be for $a \in C_{\text{Lip}}(\mathcal{T}_m)$ and $P \in C(\mathcal{T}_m)$ exponentially stable, it is necessary and sufficient that positive constants T and $d < 1$ exist, such that*

$$\|\Omega_0^T(\varphi)\| \leq d, \quad \forall \varphi \in \mathcal{T}_m. \qquad (4.2.12)$$

Now consider the exponentially dichotomous system (4.2.2). For such a system there exists a projector in $C(\mathcal{T}_m)$ when $a \in C_{\text{Lip}}(\mathcal{T}_m)$ and $P \in C_{\text{Lip}}(\mathcal{T}_m)$

$$C \in C(\mathcal{T}_m), \quad C^2(\varphi) = C(\varphi), \qquad (4.2.13)$$

such that

$$\|\Omega_0^T(\varphi)\,C(\varphi)\| \leq K\exp\{-\gamma t\}, \quad \forall t \in R^+, \qquad (4.2.14)$$

$$\|\Omega_0^T(\varphi)\,C_1(\varphi)\| \leq K\exp\{\gamma t\}, \quad \forall t \in R^-, \qquad (4.2.15)$$

where $C_1(\varphi) = E - C(\varphi)$, $K = \text{const} \geq 1$, $\gamma = \text{const} > 0$, $R^- \in (-\infty, 0]$, φ is an arbitrary value in $\mathcal{T}_m$, E is an n-dimensional identity matrix. Moreover, the matrix $C(\varphi)$ satisfies the condition

$$\Omega_0^t(\varphi)\,C(\varphi) = C(\varphi_t)\Omega_0^t(\varphi), \quad \forall (t,\varphi) \in R \times \mathcal{T}_m. \qquad (4.2.16)$$

Inequalities (4.2.14) and (4.2.15) provide the existence of the positive constants T and $d > 1$ such that

$$\|\Omega_0^T(\varphi)C(\varphi)\| \leq d, \quad \|\Omega_0^{-T}(\varphi)C_1(\varphi)\| \leq d, \quad \forall \varphi \in \mathcal{T}_m. \tag{4.2.17}$$

Conversely, let there exist a matrix C satisfying the conditions (4.2.13) and (4.2.16) and positive constants T and $d > 1$ satisfying the inequalities (4.2.17). Then for $t \in R^+$ we have the estimate

$$\|\Omega_0^T(\varphi)C(\varphi)\| \leq \|\Omega_0^{t-[t/T]T}(\varphi_{[t/T]T})\| \|\Omega_0^{[t/T]T}(\varphi) C^2(\varphi)\|$$
$$= \|\Omega_0^{t-[t/T]T}(\varphi_{[t/T]T}) \Omega_0^T(\varphi_{([t/T]-1)T}) \ldots \Omega_0^T(\varphi_T) \Omega_0^T(\varphi) C^2(\varphi)\|$$
$$= \|\Omega_0^{t-[t/T]T}(\varphi_{[t/T]T}) \Omega_0^T(\varphi_{([t/T]-1)T}) \ldots \Omega_0^T(\varphi_T) C(\varphi_T) \Omega_0^T(\varphi) C(\varphi)\|$$
$$\leq \|\Omega_0^{t-[t/T]T}(\varphi_{[t/T]T}) \Omega_0^T(\varphi_{([t/T]-1)T}) \ldots \Omega_0^T(\varphi_T) C^2(\varphi_T)\| d \tag{4.2.18}$$
$$\leq \cdots \leq \|\Omega_0^{t-[t/T]T}(\varphi_{[t/T]T})\| d^{[t/T]}$$
$$= \|\Omega_0^{t-[t/T]T}(\varphi_{[t/T]T})\| \exp\left\{\left(\ln \frac{d}{T}\right)\left[\frac{t}{T}\right]T\right\}$$
$$= \left\|\Omega_0^{t-[t/T]T}(\varphi_{[t/T]T}) \exp\left\{-\left(\ln \frac{d}{T}\right)\left(t - \left[\frac{t}{T}\right]T\right)\right\}\right\| \exp\left\{\left(\ln \frac{d}{T}\right)t\right\}$$
$$= \max \left\|\Omega_0^\tau(\varphi) \exp\left\{-\left(\ln \frac{d}{T}\right)\tau\right\}\right\| \exp\{-\gamma t\} \leq K \exp\{-\gamma t\},$$

where max is taken with respect to $(\tau, \varphi) \in [0, T] \times \mathcal{T}_m$,

$$\gamma = \left|\ln \frac{d}{T}\right| > 0, \quad K = \max \left\|\Omega_0^\tau(\varphi) \exp\left\{-\left(\ln \frac{d}{T}\right)\tau\right\}\right\|.$$

A similar estimate is valid for $t \in R^-$:

$$\|\Omega_0^t(\varphi)C_1(\varphi)\| \leq K \exp\{\gamma t\},$$
$$\gamma = \left|\ln \frac{d}{T}\right| > 0, \quad K = \max \left\|\Omega_0^\tau(\varphi) \exp\left\{\left(\ln \frac{d}{T}\right)\tau\right\}\right\|, \tag{4.2.19}$$

where max is taken with respect to $(\tau, \varphi) \in [-T, 0] \times \mathcal{T}_m$.

Together with the estimates (4.2.18) and (4.2.19), conditions (4.2.13) and (4.2.16) denote exponential dichotomy of the system (4.2.2).

We weaken the condition (4.2.16). Let us show that for the condition to be satisfied it suffices that

$$\Omega_0^t(\varphi)C(\varphi) = C(\varphi_t(\varphi))\Omega_0^t(\varphi), \quad \forall (\tau, \varphi) \in I_\delta \times \mathcal{T}_m, \tag{4.2.20}$$

where $I_\delta = [-\delta, \delta]$, δ is an arbitrary small positive number.

In fact, when (4.2.20) is satisfied, for arbitrary $t \in I_\delta$ and $\tau \in I_\delta$ the equality which is obtained from (4.2.20) by the replacement of φ by $\varphi_\tau(\varphi)$ is valid:

$$\Omega_0^t(\varphi_\tau(\varphi))C(\varphi_\tau(\varphi)) = C(\varphi_{\tau+1}(\varphi))\Omega_0^t(\varphi_\tau(\varphi)).$$

Since
$$\Omega_\tau^t(\varphi_\Theta(\varphi)) = \Omega_{\tau+\Theta}^{t+\Theta}(\varphi), \quad \forall (t, \tau, \Theta, \varphi) \in R \times R \times R \times \mathcal{T}_m, \qquad (4.2.21)$$

then (4.2.21) is reduced to the form

$$\Omega_0^{t+\tau}(\varphi)\Omega_\tau^0(\varphi)C(\varphi_\tau(\varphi)) = C(\varphi_{t+\tau}(\varphi))\Omega_0^{t+\tau}(\varphi)\Omega_\tau^0(\varphi). \qquad (4.2.22)$$

When $t \in I_\delta$, (4.2.21) implies $C(\varphi_\tau(\varphi)) = \Omega_0^\tau(\varphi)C(\varphi)\Omega_\tau^0(\varphi)$ and therefore (4.2.22) becomes

$$\Omega_0^{t+\tau}(\varphi)C(\varphi)\Omega_\tau^0(\varphi) = C(\varphi_{t+\tau}(\varphi))\Omega_0^{t+\tau}(\varphi)\Omega_\tau^0(\varphi),$$

and proves that
$$\Omega_0^{t+\tau}(\varphi)C(\varphi) = C(\varphi_{t+\tau}(\varphi))\Omega_0^{t+\tau}(\varphi). \qquad (4.2.23)$$

The relation (4.2.23) shows the validity of (4.2.20) on the set $(t, \varphi) \in I_{2\delta} \times \mathcal{T}_m$. Hence it the validity of (4.2.16) obviously follows.

Thus, the following result holds true.

Theorem 4.2.2 *For the system of the equations (4.2.2) to be exponentially dichotomous for $a \in C_{\text{Lip}}(\mathcal{T}_m)$ and $P \in C_{\text{Lip}}(\mathcal{T}_m)$, it is necessary and sufficient that there exist the matrix C which satisfies the conditions (4.2.13), (4.2.20) and the positive constants T and $d < 1$, for which the inequalities (4.2.17) are satisfied.*

The advantages of the above results in application are obvious.

4.3 Roughness Conditions for the Green Function of the Linear Extension of a Dynamical System on a Torus with the Index of Smoothness l

Let the Green function be determined by

$$G_0(\tau, \varphi) = \begin{cases} \Omega_\tau^0(\varphi)C(\varphi_\tau(\varphi)), & \tau \leq 0, \\ -\Omega_\tau^0(\varphi)C_1(\varphi_\tau(\varphi)), & \tau > 0, \end{cases} \qquad (4.3.1)$$

$$C \in C(\mathcal{T}_m), \quad C_1 = E - C, \quad \int_{-\infty}^{+\infty} \|G_0(\tau, \varphi)\| d\tau \leq K, \qquad (4.3.2)$$

where $K = \text{const} > 0$.

Definition 4.3.1 The function $G_0(\tau, \varphi)$ is referred to as *rough* for $a \in C^l(\mathcal{T}_m)$, $P \in C^l(\mathcal{T}_m)$, $l \geq 1$, *with the index of smoothness* l, provided that a constant $\delta > 0$ is found such that the system of equations

$$\frac{d\varphi}{dt} = a(\varphi) + a_1(\varphi), \quad \frac{dh}{dt} = P(\varphi)h, \tag{4.3.3}$$

where $a_1 \in C^l(\mathcal{T}_m)$,

$$|a_1|_l \leq \delta, \tag{4.3.4}$$

has the Green function $\bar{G}_0(\tau, \varphi)$ satisfying the condition

$$|\bar{G}_0(\tau, \varphi) f(\varphi(\tau))|_l \leq K \exp\{-\gamma |\tau|\} |f|_l, \tag{4.3.5}$$

where f is an arbitrary function from $C^l(\mathcal{T}_m)$, $K = \text{const} > 1$, $\gamma = \text{const} > 0$, $|f|_l$ is the differential norm of the function $f \in C^l(\mathcal{T}_m)$

$$|f|_l = \max_{0 \leq |P| \leq l} |D^P f|_0 = \max \left| \frac{\partial^P f}{\partial \varphi_1^{P_1} \ldots \partial \varphi_1^{P_m}} \right|_0,$$

$$|f|_0 = \max_{\varphi \in \mathcal{T}_m} \|f(\varphi)\|,$$

$\varphi(t) = \varphi(t, \varphi)$ is a solution of the equations of the angular variables of the system (4.3.3)

$$|P| = \sum_{\nu=1}^{m} P_\nu.$$

The main result on roughness result for the Green function of the exponentially dichotomous system with the index of smoothness l is as follows.

Theorem 4.3.1 *Let* $a \in C^l(\mathcal{T}_m)$, $P \in C^l(\mathcal{T}_m)$, $l \geq 1$, *and the system (4.2.2) be exponentially dichotomous. We assume that the Green function* $G_0(\tau, \varphi)$ *of this system satisfies the inequality*

$$\|G_0(\tau, \varphi)\| \leq K \exp\{-\gamma |\tau|\}, \quad \forall \tau \in R, \tag{4.3.6}$$

and that the matriciant $\Omega_0^\tau(\partial a/\partial \varphi)$ *of the system (4.1.23) satisfies the inequality*

$$\left\| \Omega_0^\tau \left(\frac{\partial a}{\partial \varphi} \right) \right\| = K_1 \exp\{\alpha |\tau|\}, \quad \forall \tau \in R, \tag{4.3.7}$$

where $K = \text{const} \geq 1$, $K_1 = \text{const} \geq 1$, $\gamma = \text{const} \geq 0$, $\alpha = \text{const} \geq 0$. *Then, provided that*

$$\gamma > l\alpha, \tag{4.3.8}$$

then the Green function $G_0(\tau,\varphi)$ is rough with the index of smoothness l.

Passing to the *proof* of the theorem, we construct a form which is quadratic in h, by means of which inequality (4.3.6) is reconstructed with the index of exponential damping which differs arbitrarily little from γ. To this end we fix an arbitrary small $\varepsilon > 0$ and, taking λ on the interval $\mathcal{I} = [0, \gamma - \varepsilon]$, consider the matrix

$$S_\lambda(\varphi) = S_1(\varphi,\lambda) - S_2(\varphi,\lambda), \qquad (4.3.9)$$

for

$$S_1(\varphi,\lambda) = \int_0^{+\infty} C^*(\varphi)(\Omega_0^\tau(\varphi))^* \Omega_0^\tau(\varphi) C(\varphi)\, e^{2\lambda\tau}\, d\tau, \qquad (4.3.10)$$

$$S_2(\varphi,\lambda) = \int_0^{+\infty} C_1^*(\varphi)(\Omega_0^\tau(\varphi))^* \Omega_0^\tau(\varphi) C_1(\varphi)\, e^{2\lambda\tau}\, d\tau, \qquad (4.3.11)$$

where $C(\varphi) = G_0(0,\varphi)$, $\Omega_0^\tau(\varphi) = \Omega_0^\tau(P)$.

The estimates (4.3.6) imply the convergence of the integrals (4.3.10) and (4.3.11) uniformly on $\varphi \in \mathcal{T}_m$. Moreover, for these integrals the inequalities

$$S_1(\varphi_t, \lambda) = \int_0^{+\infty} C^*(\varphi_t)(\Omega_0^\tau(\varphi_t))^* \Omega_0^\tau(\varphi_t) C(\varphi_t)\, e^{2\lambda\tau}\, d\tau,$$

$$= \int_t^{+\infty} (\Omega_t^0(\varphi))^* C^*(\varphi)(\Omega_0^\tau(\varphi))^* \Omega_0^\tau(\varphi) C(\varphi) \Omega_t^0(\varphi)\, e^{2\lambda\tau - t}\, d\tau, \qquad (4.3.12)$$

$$S_2(\varphi_t, \lambda) = \int_{-\infty}^t (\Omega_t^0(\varphi))^* C^*(\varphi)(\Omega_0^\tau(\varphi))^* \Omega_0^\tau(\varphi) C(\varphi) \Omega_t^0(\varphi)\, e^{2\lambda\tau - t}\, d\tau,$$

are valid, deduced in a standard manner from (4.2.16) and (4.2.21). Then

$$\frac{dS_1(\varphi_t)}{dt} = -C^*(\varphi_t)C(\varphi_t) - P^*(\varphi_t)S_1(\varphi_t) - S_1(\varphi_t)P(\varphi_t) - 2\lambda S_1(\varphi_t),$$

$$\frac{dS_2(\varphi_t)}{dt} = C_1^*(\varphi_t)C_1(\varphi_t) - P^*(\varphi_t)S_2(\varphi_t) - S_2(\varphi_t)P(\varphi_t) - 2\lambda S_2(\varphi_t),$$

which yields the equality

$$\frac{dS_\lambda(\varphi_t)}{dt} = -[C^*(\varphi_t)C(\varphi_t) + C_1^*(\varphi_t)C_1(\varphi_t)] - P^*(\varphi_t)S_\lambda(\varphi_t) \qquad (4.3.13)$$
$$- S_\lambda(\varphi_t)P(\varphi_t) - 2\lambda S_\lambda(\varphi_t) = \dot{S}_\lambda(\varphi_t).$$

Hence it follows that the quadratic form

$$V_\lambda(\varphi_t, h_t) = (S_\lambda(\varphi_t)h_t, h_t), \tag{4.3.14}$$

where $h_t = \Omega_0^t(P)h_0$, satisfies the inequality

$$\frac{dV_\lambda(\varphi_t, h_t)}{dt} = -[(C^*(\varphi_t)h_t, C(\varphi_t)h_t) + (C_1^*(\varphi_t)h_t, C_1(\varphi_t)h_t)] \\ - 2\lambda V_\lambda(\varphi_t, h_t). \tag{4.3.15}$$

For the system of equations

$$\frac{d\varphi}{dt} = a(\varphi), \quad \frac{dh}{dt} = (P(\varphi) + \lambda E)h, \tag{4.3.16}$$

the function $G_0(t, \varphi, \lambda) = \exp\{-\lambda\tau\} G_0(\tau, \varphi)$ is the Green function which satisfies the uniqueness conditions

$$G_0(0, \varphi, \lambda) = C(\varphi) = C^2(\varphi), \quad \Omega_0^t(P + \lambda E)C(\varphi) = C(\varphi_t)\Omega_0^t(P + \lambda E).$$

This suffices for the system (4.3.16) to be exponentially dichotomous and the separatrix manifolds $E^+(\varphi)$ and $E^-(\varphi)$ of the systems (4.2.2) and (4.3.16) coincide.

The matrix $S_\lambda(\varphi)$ is taken so as to coincide with the matrix constructed for the system (4.3.16) according to the scheme of proof of the Theorem 4.3.1 on necessary conditions for the invariant torus dichotomy. Therefore, the matrix $S_\lambda(\varphi)$ satisfies the conditions of this theorem and hence, satisfies the correlations

$$S_\lambda \in C'(\mathcal{T}_m, a), \quad \det S_\lambda(\varphi) \neq 0, \quad \forall(\varphi, \lambda) \in \mathcal{T}_m \times \mathcal{I}, \tag{4.3.17}$$

$$\hat{S}_\lambda(\varphi) = \dot{S}_\lambda(\varphi) + P^*(\varphi)S_\lambda(\varphi) + S_\lambda(\varphi)P(\varphi) + 2\lambda S_\lambda(\varphi) \in \mathfrak{N}^-, \tag{4.3.18}$$

where $\mathfrak{N}^-$ is a set of negative definite symmetric matrices.

The separatrix manifold $E^+(\varphi_t)$ of the system (4.3.16) belongs, as has been proved when establishing sufficient conditions for the exponential dichotomy of the invariant torus, to the cone

$$(S_\lambda(\varphi_t)h_t^\lambda, h_t^\lambda) \geq 0, \quad \forall t \in R^+, \tag{4.3.19}$$

where φ_t, $h_t^\lambda = h_t^\lambda(\varphi, h) = h_t e^{\lambda t}$ are the solutions of the system (4.3.16). The cone (4.3.19) coincides with the cone

$$(S_\lambda(\varphi_t)h_t, h_t) \geq 0, \quad \forall t \in R^+, \tag{4.3.20}$$

and so the separatrix manifold $E^+(\varphi_t)$ belongs to the cone (4.3.20). For sufficiently small ε we have the estimate

$$\frac{d[V_\lambda(\varphi_t, h_t) + \varepsilon(h_t, h_t)]}{dt} = -2\lambda[V_\lambda(\varphi_t, h_t) + \varepsilon(h_t, h_t)]$$
$$+ 2\lambda\varepsilon(h_t, h_t) - 2[\|C(\varphi_t)h_t\|^2 + \|C_1(\varphi_t)h_t\|^2] + 2\varepsilon(P(\varphi_t)h_t, h_t)$$
$$\leq -2\lambda[V_\lambda(\varphi_t, h_t) + \varepsilon(h_t, h_t)] - \left(\frac{1}{2} - 2\lambda\varepsilon - 2M\varepsilon\right)\|h_t\|^2 \quad (4.3.21)$$
$$\leq -2\lambda[V_\lambda(\varphi_t, h_t) + \varepsilon(h_t, h_t)],$$

where $M = \max_{\varphi \in \mathcal{T}_m} \|P(\varphi)\|$.

The inequalities (4.3.20) and (4.3.21) imply the estimate

$$V_\lambda(\varphi_t, h_t^+) + \varepsilon\|h_t^+\|^2 \leq \exp\{-2\lambda t\}\left[\|V_\lambda(\varphi, h^+)\|^2 + \varepsilon\|h^+\|^2\right], \quad \forall t \in R^+,$$

where $h_t^+ = H_t(\varphi, h^+)$ is a solution of the system (4.2.2) starting on $E^+(\varphi)$. From the latter inequality we get

$$\|h_t^+\|^2 \leq K_2 \exp\{-\lambda t\}\|h^+\|^2, \quad \forall t \in R^+, \quad (4.3.22)$$

where K_2 is taken from the condition $K_1/(\sqrt{2}\varepsilon) + 1 \leq K_2$.

Similarly we find the estimate

$$\|h_t^-\|^2 \leq K_2 \exp\{\lambda t\}\|h^-\|^2, \quad \forall t \in R^- \quad (4.3.23)$$

for the solutions $h_t^- = H_t(\varphi, h^-)$ of the system (4.2.2) starting on the separatrix manifold $E^-(\varphi)$.

Estimates (4.3.22) and (4.3.23) and the properties of the Green function for the exponentially dichotomous system give the estimate

$$\|G_0(\tau, \varphi)\| \leq K_2 \exp\{-\lambda|\tau|\}, \quad \forall \tau \in R, \quad (4.3.24)$$

which yields the required reconstruction of the inequality (4.3.6) with respect to $V_\lambda(\varphi, h)$ for $\lambda = \gamma - \varepsilon$.

Let us demonstrate that the exponential dichotomy of the systems (4.3.16) for all $\lambda \in \mathcal{J}$ implies the generality of the separatrix manifolds $E^+(\varphi)$ and $E^-(\varphi)$ of these systems.

Actually, let $G_0(\tau, \varphi, \lambda)$ designates the Green function of the system (4.3.16). Then $G_0(\tau, \varphi, \lambda) = \exp\{-\lambda\tau\}G_0(\tau, \varphi, 0)$ for $\lambda \in [0, \lambda_1]$, where $\lambda_1 > 0$. Let λ_1 be the largest of the values of $\mathcal{J}$ for which $G_0(\tau, \varphi, \lambda) = \exp\{-\lambda\tau\}G_0(\tau, \varphi, 0)$. This means that

$$G_0(0, \varphi, \lambda) = G_0(0, \varphi, 0) = C(\varphi), \quad \forall \lambda \in [0, \lambda_1],$$
$$G_0(0, \varphi, \lambda_1) \not\equiv C(\varphi). \quad (4.3.25)$$

Replacing λ by $\lambda_1 + \mu$ in the system (4.3.16) one gets

$$G_0(\tau, \varphi, \lambda_1 + \mu) = \exp\{-\mu\tau\} G_0(\tau, \varphi, \lambda_1), \quad \text{for all} \quad |\mu| \leq \delta,$$

for sufficiently small $\delta > 0$. However we then have the equality

$$C(\varphi) = G_0(0, \varphi, \lambda_1 - \delta) = G_0(0, \varphi, \lambda_1),$$

which contradicts inequality (4.3.25). The contradiction obtained proves that

$$G_0(\tau, \varphi, \lambda) = \exp\{-\lambda\tau\} G_0(\tau, \varphi, 0),$$

for all $\lambda \in \mathcal{J}$. This shows that under the assumption made systems (4.3.16) have the same separatrix manifolds $E^+(\varphi)$ and $E^-(\varphi)$.

Now we consider the system of equations

$$\frac{d\varphi}{dt} = a(\varphi) + a_1(\varphi), \quad \frac{dh}{dt} = (P(\varphi) + \lambda E)h, \quad (4.3.26)$$

where $a \in C^l(\mathcal{T}_m)$ and satisfies the condition (4.3.3) and $\lambda \in \mathcal{J}$.

The matrix $S_\lambda(\varphi)$ can be smoothed for $l > 1$ so that the smoothed matrix $S_\lambda^n(\varphi) \in C^1(\mathcal{T}_m)$ and preserves the properties of the matrix $S_\lambda(\varphi)$

$$S_\lambda^n(\varphi) = (S_\lambda^n(\varphi))^*,$$

$$\lim_{n \to \infty} \left[\|S_\lambda^n(\varphi) - S_\lambda(\varphi)\| + \left\|\frac{\partial S_\lambda^n(\varphi)}{\partial \varphi} a(\varphi) - \dot{S}_\lambda(\varphi)\right\| \right] = 0, \quad (4.3.27)$$

uniformly in $(\varphi, \lambda) \in \mathcal{T}_m \times \mathcal{J}$. For sufficiently large n the relations (4.3.17) and (4.3.27) yield

$$\det S_\lambda^n(\varphi) \neq 0, \quad \forall (\varphi, \lambda) \in \mathcal{T}_m \times \mathcal{J}. \quad (4.3.28)$$

We designate the solutions of the system (4.3.26) by $\varphi(t)$, $h^\lambda(t)$ such that $\varphi(0) = 0$ and $h^\lambda(0) = h$. We set

$$V_\lambda^n(\varphi, \lambda) = (S_\lambda^n(\varphi)h, h), \quad (4.3.29)$$

and consider $dV_\lambda^n(\varphi(t), h^\lambda(t)) \, dt$. Then we have

$$\frac{dV_\lambda^n(\varphi(t), h^\lambda(t))}{dt} = \left(\left[\frac{\partial S_\lambda^n(\varphi)}{\partial \varphi} a(\varphi(t)) + \frac{\partial S_\lambda^n(\varphi)}{\partial \varphi} a_1(\varphi(t))\right.\right.$$

$$\left.\left. + S_\lambda^n(\varphi(t))P(\varphi(t)) + P^*(\varphi(t))S_\lambda^n(\varphi(t)) + 2\lambda S_\lambda^n(\varphi(t))\right] h^\lambda(t), h^\lambda(t)\right)$$

$$\leq \left([\dot{S}_\lambda(\varphi(t)) + S_\lambda(\varphi(t))P(\varphi(t)) + P^*(\varphi(t))S_\lambda(\varphi(t))\right. \quad (4.3.30)$$

$$\left. + 2\lambda S_\lambda(\varphi(t))] h^\lambda(t), h^\lambda(t)\right) + \left(\left\|\frac{\partial S_\lambda^n(\varphi(t))}{\partial \varphi} a(\varphi(t))\right\| + \varepsilon_n\right) \|h^\lambda(t)\|^2$$

$$= (\hat{S}_\lambda(\varphi)h^\lambda(t), h^\lambda(t)) + (\varepsilon_n + K_n\delta)\|h^\lambda(t)\|^2,$$

where the constants ε and K_n are determined by the conditions

$$\left\|\frac{\partial S_\lambda^n(\varphi)}{\partial \varphi} a(\varphi) - \dot{S}_\lambda(\varphi)\right\| + \left\|(S_\lambda^n(\varphi) - S_\lambda(\varphi))P(\varphi)\right\|$$
$$+ \left\|P^*(\varphi)(S_\lambda^n(\varphi) - S_\lambda(\varphi))\right\| + 2\lambda \left\|S_\lambda^n(\varphi) - S_\lambda(\varphi)\right\| \leq \varepsilon_n,$$
$$\left\|\frac{\partial S_\lambda^n(\varphi)}{\partial \varphi} a_1(\varphi)\right\| \leq K_n \|a_1(\varphi)\| \leq K_n \delta.$$

For sufficiently small ε and δ the relation (4.3.30) and the equalities (4.3.13) and (4.3.18) imply

$$\frac{dV_\lambda^n(\varphi(t), h^\lambda(t))}{dt} \leq -[\|C^*(\varphi(t))h^\lambda(t)\|^2 + \|C_1(\varphi(t))h^\lambda(t)\|^2]$$
$$- 2\varepsilon_n \|h^\lambda(t)\|^2 \leq \left(\frac{1}{2} - 2\varepsilon_n\right)\|h^\lambda(t)\|^2 \leq -\frac{1}{4}\|h^\lambda(t)\|^2. \tag{4.3.31}$$

The inequality (4.3.31) demonstrates negative definiteness of the matrix $\hat{S}_\lambda^n(\varphi)$ of coefficients of the quadratic form $dV_\lambda^n(\varphi(t), h^\lambda(t))/dt$

$$\hat{S}_\lambda^n(\varphi) = \frac{dS_\lambda^n(\varphi)}{d\varphi}(a(\varphi) + a_1(\varphi)) + S_\lambda^n(\varphi)P(\varphi) + P^*(\varphi)S_\lambda^n(\varphi) + 2\lambda S_\lambda^n(\varphi). \tag{4.3.32}$$

Alongside (4.3.28) this is sufficient to claim exponential dichotomy of the system (4.3.26) for all $\lambda \in \mathcal{J}$. As proved above, for the exponential dichotomy of the systems (4.3.26) the separatrix manifolds $E^+(\varphi)$ and $E^-(\varphi)$ of these systems are general when all $\lambda \in \mathcal{J}$. Moreover, the manifold $E^+(\varphi)$ belongs to the cone

$$(S_\lambda^n(\varphi(t))h(t), h(t)) \geq 0, \quad \forall t \in R^+, \tag{4.3.33}$$

where $\varphi(t), h(t)$ is a solution of the system of equations (4.3.2), $\varphi(0) = \varphi$, $h(0) = h$.

Consider the function $V_\lambda^n(\varphi(t), h^\lambda(t))$. We have

$$\frac{dV_\lambda^n(\varphi(t), h^\lambda(t))}{dt} = \left(\left[\frac{\partial S_\lambda^n(\varphi(t))}{\partial \varphi}a(\varphi(t)) + \frac{\partial S_\lambda^n(\varphi(t))}{\partial \varphi}a_1(\varphi(t))\right.\right.$$
$$\left.\left. + S_\lambda^n(\varphi(t))P(\varphi(t)) + P^*(\varphi(t))S_\lambda^n(\varphi(t))\right]h(t), h(t)\right)$$
$$\leq ([\dot{S}_\lambda(\varphi(t)) + S_\lambda(\varphi(t))P(\varphi(t)) + P^*(\varphi(t))S_\lambda(\varphi(t))]h(t), h(t))$$
$$+ 2\varepsilon_n \|h(t)\|^2 = -[\|C(\varphi(t))h(t)\|^2 + \|C_1(\varphi(t))h(t)\|^2] \tag{4.3.34}$$
$$- 2\lambda(S_\lambda(\varphi(t))h(t), h(t)) + 2\varepsilon_n \|h(t)\|^2$$
$$\leq -\left(\frac{1}{2} - 2\varepsilon_n\right)\|h(t)\|^2 - 2\lambda V_\lambda^n(\varphi(t), h(t))$$
$$\leq -2\lambda V_\lambda^n(\varphi(t), h(t)).$$

In the same way as the inequality (4.3.22) was obtained from (4.3.15) and (4.3.20), (4.3.33) and (4.3.34) lead to the inequality

$$\|h^+(t)\|^2 \le K_3 \exp\{-\lambda t\}\|h^+\|^2, \quad \text{for all} \quad t \in R^+, \tag{4.3.35}$$

where $h^+(t) = h^+(t, \varphi, h^+)$ is a solution of the system (4.3.3) starting on $E^+(\varphi)$ with $K_3 = \text{const} \ge 1$.

Similarly the inequality

$$\|h^-(t)\|^2 \le K_3 \exp\{\lambda t\}\|h^-\|^2, \quad \text{for all} \quad t \in R^-, \tag{4.3.36}$$

is derived, where $h^-(t) = h^-(t, \varphi, h^-)$ is a solution of the system (4.3.3) starting on $E^-(\varphi)$. Correlations (4.3.35) and (4.3.36) yield an estimate of the Green function $\bar{G}_0(\tau, \varphi)$ of the system (4.3.3) of the type

$$\|\bar{G}_0(\tau, \varphi)\|^2 \le K_3 \exp\{-\lambda|\tau|\}, \quad \text{for all} \quad \tau \in R. \tag{4.3.37}$$

Setting $\lambda = \gamma - \varepsilon$ in (4.3.37) we arrive at

$$\|\bar{G}_0(\tau, \varphi)\|^2 \le K_3 \exp\{-(\gamma - \varepsilon)|\tau|\}, \quad \text{for all} \quad \tau \in R. \tag{4.3.38}$$

Consider the matriciant $\Omega_0^\tau \left(\frac{\partial a}{\partial \varphi} + \frac{\partial a_1}{\partial \varphi}\right)$ of the system of the equations

$$\frac{d\Theta}{dt} = \left[\frac{\partial a(\varphi(t))}{\partial \varphi} + \frac{\partial a_1(\varphi(t))}{\partial \varphi}\right]\Theta. \tag{4.3.39}$$

We get an estimate for this matriciant from the inequalities (4.3.4), (4.3.7):

$$\left\|\Omega_0^\tau\left(\frac{\partial a}{\partial \varphi} + \frac{\partial a_1}{\partial \varphi}\right)\right\| \le K_1 \exp\{(\alpha + K_1\delta)|\tau|\}, \quad \text{for all} \quad \tau \in R.$$

For $K_1\delta < \varepsilon$ this estimate gives

$$\left\|\Omega_0^\tau\left(\frac{\partial a}{\partial \varphi} + \frac{\partial a_1}{\partial \varphi}\right)\right\| \le K_1 \exp\{(\alpha + \varepsilon)|\tau|\}, \quad \text{for all} \quad \tau \in R.$$

Hence, in view of the inequality (4.3.8) and the theorem on the smoothness of the Green function it follows that

$$|\bar{G}_0(\tau, \varphi)f(\varphi(\tau))|_l \le K_4 \exp\{-\gamma_1|\tau|\}|f|_l, \quad \text{for all} \quad \tau \in R, \tag{4.3.40}$$

where f is an arbitrary function from $C^l(\mathcal{T}_m)$ and $K_4 = \text{const} \ge 1$,

$$\gamma_1 = \gamma - l\alpha - (l+1)\varepsilon > 0, \quad 0 < \varepsilon < \frac{\gamma - l\alpha}{l+1}. \tag{4.3.41}$$

Inequality (4.3.40) indicates the means roughness of the Green function $G_0(\tau,\varphi)$ with the index of smoothness $l \geq 1$.

the exponential dichotomy of the system (4.2.2) exhausts the possibility of the existence of a unique Green function of the system (4.2.2). There is only the case left when the system (4.2.2) has a non-unique Green function $G_0(\tau,\varphi)$. Naturally, the exponential damping indices may prove to be special for each of these functions, and more subject to abrupt changes under small perturbations of the system (4.2.2). Thus, a general index of exponential damping should be derived for all Green functions of the system (4.2.2). The possibility of doing this is provided by the investigations of non-unique Green function. In the framework of these investigations, alongside the system (4.2.2) we consider the "extended" system of equations of the type

$$\frac{d\varphi}{dt} = a(\varphi), \quad \frac{dh}{dt} = P(\varphi)h, \quad \frac{dq}{dt} = Q(\varphi)q - P^*(\varphi)h, \tag{4.3.42}$$

the matrix $Q(\varphi)$ of which belongs to $C^l(\mathcal{T}_m)$ and is taken so that the system (4.3.42) is exponentially dichotomous. In the partial case, $Q(\varphi) = -E$ can be such a matrix. The Green function $G_0^1(\tau,\varphi)$ of the system (4.3.42) is

$$\begin{pmatrix} G_0(\tau,\varphi) & \Omega_0^\tau(\varphi)C_{12}(\varphi_\tau) \\ R_\tau G_0(\tau,\varphi) + (\Omega_0^\tau(\varphi))^*C_{21}(\varphi_\tau) & R_\tau \Omega_\tau^0(\varphi)C_{12}(\varphi_\tau) + G_1(\tau,\varphi) \end{pmatrix}, \tag{4.3.43}$$

and has the estimate

$$\|G_0^1(\tau,\varphi)\| \leq K \exp\{-\gamma|\tau|\}, \quad \text{for all} \quad \tau \in R, \tag{4.3.44}$$

where $K = \text{const} \geq 1$ and $\gamma = \text{const} > 0$. The estimate (4.3.44) implies that the estimates

$$\|G_0(\tau,\varphi)\| \leq K \exp\{-\gamma|\tau|\}, \quad \|\Omega_0^\tau(\varphi)C_{12}(\varphi_\tau)\| \leq K \exp\{-\gamma|\tau|\}, \tag{4.3.45}$$

are valid for all $\tau \in R$. In addition, there exists a one-parameter family of the Green functions of the system (4.2.2)

$$G_0(\tau,\varphi,\mu) = G_0(\tau,\varphi) + \mu\Omega_\tau^0(\varphi)C_{12}(\varphi_\tau), \tag{4.3.46}$$

where μ is an arbitrary value of R. The index γ specified by the condition (4.3.44) is referred to as *the general damping index* of the family of *Green functions* of the system (4.2.2).

From the result proved above the following theorem is easily derived on the roughness of a non-unique Green function of the system (4.2.2) with the index of smoothness l.

Theorem 4.3.2 *Let $a \in C^l(\mathcal{T}_m)$, $P \in C^l(\mathcal{T}_m)$, $l \geq 1$. The system (4.2.2) is assumed to have the Green function $G_0(\tau,\varphi)$ which satisfies the inequality (4.3.6), and the matrix $\Omega_0^\tau\left(\frac{\partial a}{\partial \varphi}\right)$ of the equations (4.1.17) satisfies the inequality (4.3.7). Then, provided inequality (4.3.8) is satisfied and γ is the general damping index of the family of Green functions of the system (4.2.2), the Green function $G_0(\tau,\varphi)$ is rough with the index of smoothness l.*

4.4 A Theorem of Perturbation Theory of an Invariant Torus of a Dynamical System

We return to the system of equations (4.1.1) and assume that the unperturbed system of equations has an invariant torus (4.1.2.) satisfying the conditions (4.1.3) with a well-organized neighbourhood. The change of variables (4.1.6) allows the system (4.1.1) to be reduced to the system (4.1.7)

$$\frac{d\varphi}{dt} = a(\varphi, h, \varepsilon), \quad \frac{dh}{dt} = P(\varphi, h, \varepsilon)h + f(\varphi, \varepsilon), \tag{4.4.1}$$

where a, P, f are functions of the space $C^l(\mathcal{T}_m \times K_\delta \times I)$, $I = [0, \varepsilon_0]$, ε_0 is a sufficiently small positive number, $l = r - 1 \geq 1$, and

$$f(\varphi, 0) = f(\varphi), \quad a(\varphi, 0, 0) = a(\varphi), \quad P(\varphi, 0, 0) = P(\varphi). \tag{4.4.2}$$

For the function $u \in C^S_{\text{Lip}}(\mathcal{T}_m)$ we designate the value $|u|_S + K$ by $|u|_{S,\text{Lip}}$, where K is a Lipschitz constant of the s-th derivatives of the function u.

We write down the variational equation for the invariant torus of the unperturbed system

$$\frac{d\varphi}{dt} = a(\varphi), \quad \frac{dh}{dt} = P(\varphi)h, \tag{4.4.3}$$

and the variational equation for the solutions on the invariant torus

$$\frac{d\varphi}{dt} = a(\varphi), \quad \frac{d\Theta}{dt} = \frac{\partial a(\varphi)}{\partial \varphi}\Theta. \tag{4.4.4}$$

On the basis of the results of the previous section the following theorem is deduced.

Theorem 4.4.1 *Let the right-hand side of the system of equations satisfy the above-cited conditions. We assume that the system (4.4.3) possesses the Green function $G_0(\tau,\varphi)$ which satisfies the inequality*

$$\|G_0(\tau,\varphi)\| \leq K_1 \exp\{-\gamma|\tau|\}, \quad \text{for all} \quad \tau \in R, \tag{4.4.5}$$

while the fundamental matrix of the solutions $\Omega_0^t\left(\frac{\partial a}{\partial \varphi}\right)$ of the system (4.4.4) satisfies the inequality

$$\left\|\Omega_0^t\left(\frac{\partial a}{\partial \varphi}\right)\right\| \leq K_2 \exp\{\alpha|\tau|\}, \quad \text{for all} \quad \tau \in R, \tag{4.4.6}$$

where γ is the general damping index for the family of Green functions of the system (4.4.3), $K_\nu = \text{const} \geq 1$, $\nu = 1, 2$, $\alpha = \text{const} \geq 0$. Then, if

$$\gamma > l\alpha, \tag{4.4.7}$$

a sufficiently small number $\varepsilon_0 > 0$ can be indicated, such that for any $\varepsilon \in I$ the system (4.4.1) has the invariant torus

$$M(\varepsilon): h = u(\varphi, \varepsilon), \tag{4.4.8}$$

where

$$u \in C_{\text{Lip}}^{l-1}(\mathcal{T}_m), \quad \forall \varepsilon \in I, \quad |u|_{l-1,\text{Lip}} \leq K_3|f|_l, \tag{4.4.9}$$

and $K_3 = \text{const} \geq 1$ does not depend on ε.

In fact, the conditions of Theorem 4.4.1 ensure the validity of Theorem 4.3.2, and hence the roughness of the Green function $G_0(\tau, \varphi)$ with the index of smoothness l. This suffices for all the conditions and all the assertions of Theorem 4.4.1 to be satisfied.

It should be noted that in case of uniqueness of the Green function $G_0(\tau, \varphi)$ Theorem 4.4.1 remains valid with the correction that γ is the damping index for the function $G_0(\tau, \varphi)$ determined by the inequality (4.4.5).

Provided that the Green function $G_0(\tau, \varphi)$ is not unique, the conditions of Theorem 4.4.1 are satisfied for every function $G_0(\tau, \varphi, \mu)$ of the family (4.3.46). Then the system of equations (4.4.1) has the invariant torus

$$M(\varepsilon, \mu): h = u(\varphi, \varepsilon, \mu), \tag{4.4.10}$$

possessing all the properties of the torus $M(\varepsilon)$. In this set up there should exist a family of invariant tori of the system (4.4.1) for all $\varepsilon \in I$ in the neighbourhood M.

4.5 Green Function for a Linear Matrix Equation

Let the system of equations

$$\frac{d\varphi}{dt} = a(\varphi), \quad \frac{dX}{dt} + XB(\varphi) = A(\varphi)X + Q(\varphi), \tag{4.5.1}$$

be given, where A is an n-dimensional and B is p-dimensional square matrix, X and Q are $(n \times p)$-dimensional rectangular matrices, a, A, B, Q belong to the space $C^l(\mathcal{T}_m)$. The solutions of the system (4.5.1) are to be found which belong to $C'(\mathcal{T}_m, a)$. This problem arises, in particular, when the problem of block diagonalization of the block-triangular system of the variational equations (4.1.5) is treated.

We write down the homogeneous system of equations corresponding to (4.5.1)

$$\frac{d\varphi}{dt} = a(\varphi), \quad \frac{dX}{dt} + XB(\varphi) = A(\varphi)X. \tag{4.5.2}$$

Let

$$H_{\nu j} = [0, \ldots, e_\nu, \ldots, 0], \quad \nu = 1, \ldots, n; \quad j = 1, \ldots, p, \tag{4.5.3}$$

denote an $(n \times p)$-dimensional matrix, the j-th column of which equals the ν-th unit vector e_ν of the space R^n.

We designate by $C^{\nu j}(\varphi)$, $C_1^{\nu j}(\varphi)$ a pair of $(n \times p)$-dimensional matrices from $C(\mathcal{T}_m)$ which satisfy the condition

$$C^{\nu j}(\varphi) + C_1^{\nu j}(\varphi) = H_{\nu j}, \quad \nu = 1, \ldots, n; \quad j = 1, \ldots, p. \tag{4.5.4}$$

Set

$$G_0^{\nu j}(\tau, \varphi) = \begin{cases} \Omega_\tau^0(A) C^{\nu j}(\varphi_\tau) \Omega_0^\tau(B), & \tau \leq 0 \\ -\Omega_\tau^0(A) C_1^{\nu j}(\varphi_\tau) \Omega_0^\tau(B), & \tau > 0 \end{cases} \tag{4.5.5}$$

and using $G_0^{\nu j}(\tau, \varphi)$ as blocks form the matrix

$$G_0(\tau, \varphi) = \{G_0^{\nu j}(\tau, \varphi)\}, \quad \nu = 1, \ldots, n; \quad j = 1, \ldots, p. \tag{4.5.6}$$

The matrix $G_0(\tau, \varphi)$ is referred to as the Green function of the system (4.5.2) whenever

$$\int_{-\infty}^{\infty} \|G_0^{\nu j}(\tau, \varphi)\| \, d\tau \leq K, \tag{4.5.7}$$

for some $\nu = 1, \ldots, n$, $j = 1, \ldots, p$, and any $K = \text{const}$.

We represent the matrix $Q(\varphi)$ via its elements $\{q_{\nu j}(\varphi)\}$, $\nu = 1, \ldots, n$, $j = 1, \ldots, p$, in the form

$$Q(\varphi) = \sum_{\nu, j}^{n, p} q_{\nu j}(\varphi) H_{\nu j}, \tag{4.5.8}$$

where the summation is made with respect to ν from 1 to n, and with respect to j from 1 to p. Let us show that the existence of the Green function $G_0(\tau, \varphi)$ of the system (4.5.2) is sufficient for the solution to equation (4.5.1) to exist in $C'(\mathcal{T}_m, a)$ and be represented as

$$U(\varphi) = \sum_{\nu,j}^{n,p} \int_{-\infty}^{\infty} G_0^{\nu j}(\tau, \varphi) q_{\nu j}(\varphi) \, d\tau. \qquad (4.5.9)$$

The existence of the function $U(\varphi)$ and its belonging to the space $C'(\mathcal{T}_m, a)$ is proved by the application of the inequality (4.5.7), proceeding in the same way as in the proof of the basic relations for the Green function $G_0(\tau, \varphi)$. It only remains to show that

$$U(\varphi_t) = \sum_{\nu,j}^{n,p} \int_{-\infty}^{\infty} G_0^{\nu j}(\tau, \varphi_t) q_{\nu j}(\varphi_t(\varphi_t)) \, d\tau \qquad (4.5.10)$$

together with φ_t satisfies the system (4.5.1). We have

$$U^{\nu j}(\varphi_t) = \int_{-\infty}^{0} \Omega_\tau^0(\varphi_t; A) C^{\nu j}(\varphi_\tau(\varphi_t)) \Omega_0^\tau(\varphi_t; B) q_{\nu j}(\varphi_\tau(\varphi_t)) \, d\tau$$

$$- \int_{0}^{\infty} \Omega_\tau^0(\varphi_t; A) C_1^{\nu j}(\varphi_\tau(\varphi_t)) \Omega_0^\tau(\varphi_t; B) q_{\nu j}(\varphi_\tau(\varphi_t)) \, d\tau \qquad (4.5.11)$$

$$= \int_{-\infty}^{t} \Omega_\tau^t(A) C^{\nu j}(\varphi_\tau) \Omega_t^\tau(B) q_{\nu j}(\varphi_\tau) \, d\tau$$

$$- \int_{t}^{\infty} \Omega_\tau^0(A) C_1^{\nu j}(\varphi_\tau) \Omega_t^\tau(B) q_{\nu j}(\varphi_\tau) \, d\tau.$$

Then differentiating (4.5.11) in t yields

$$\frac{dU^{\nu j}(\varphi_t)}{dt} = [C^{\nu j}(\varphi_t) + C_1^{\nu j}(\varphi_t)] q_{\nu j}(\varphi_t) + A(\varphi_t) U^{\nu j}(\varphi_t)$$
$$- U^{\nu j}(\varphi_t) B(\varphi_t) = A(\varphi_t) U^{\nu j}(\varphi_t) - U^{\nu j}(\varphi_t) B(\varphi_t) + H_{\nu j} q_{\nu j}(\varphi_t). \qquad (4.5.12)$$

Summing the equalities (4.5.12) completes the proof.

Let us show that the Green function introduced which corresponds to (4.5.2) is a linear extension of the dynamical system on a torus. Let

$$X = [x_1, \ldots, x_p], \quad Q = [Q_1, \ldots, Q_p], \qquad (4.5.13)$$

where x_ν and Q_ν are the columns of the matrices X and Q respectively, $\nu = 1, \ldots, p$. From them we form the column vectors

$$\hat{X} =: [x_1, \ldots, x_p] \quad \text{and} \quad \hat{Q} =: [Q_1, \ldots, Q_p],$$

where colon means that the vectors $x_1, \ldots, x_p$ and $Q_1, \ldots, Q_p$ are placed one under another in the order of succession.

The system of equations (4.5.1) is rewritten in the form of a linear extension of the dynamical system on a torus

$$\frac{d\varphi}{dt} = a(\varphi), \quad \frac{d\hat{x}}{dt} = P(\varphi)\hat{x} + \hat{Q}(\varphi), \qquad (4.5.14)$$

where $P(\varphi)$ is an np-dimensional square matrix determined relative to A and B. The fundamental matrix of solutions of the homogeneous system corresponding to (4.5.14) is designated by $\Omega_0^t(A, B)$ and its Green function by $G_0(\tau, \varphi; A, B)$. The vectors

$$H_{\nu j}, \quad \nu = 1, \ldots, n; \quad j = 1, \ldots, p \qquad (4.5.15)$$

formed from the matrix (4.5.3) specify a complete set of the unique dynamical vectors of the space R^{np}. Therefore, the fundamental matrix of the solutions $\Omega_0^t(A, B)$ corresponds to the system of the solutions

$$\Omega_0^t(A) H_{\nu j} \Omega_t^0(B), \quad \nu = 1, \ldots, n; \quad j = 1, \ldots, p, \qquad (4.5.16)$$

of the system (4.5.2). Also, the correspondence lies in the fact that (4.5.16) is $\Omega_0^t(A, B)\hat{H}_{\nu j}$ which is represented in the form of the matrix (4.5.16)

$$\Omega_0^t(A) H_{\nu j} \Omega_t^0(B) = \left[\left(\Omega_0^t(A, B)\hat{H}_{\nu j}\right)_1, \ldots, \left(\Omega_0^t(A, B)\hat{H}_{\nu j}\right)_p \right],$$

where $\left(\Omega_0^t(A, B)\hat{H}_{\nu j}\right)_1, \ldots, \left(\Omega_0^t(A, B)\hat{H}_{\nu j}\right)_p$ are n-dimensional blocks of the solution $\Omega_0^t(A, B)\hat{H}_{\nu j}$.

Let

$$C(\varphi) = G_0(0, \varphi; A, B). \qquad (4.5.17)$$

We determine the system of matrices with respect to $C(\varphi)$:

$$\begin{aligned} C^{\nu j}(\varphi) &= \left[\left(C(\varphi)\hat{H}_{\nu j}\right)_1, \ldots, \left(C(\varphi)\hat{H}_{\nu j}\right)_p \right], \\ C_1^{\nu j}(\varphi) &= \left[\left(C_1(\varphi)\hat{H}_{\nu j}\right)_1, \ldots, \left(C_1(\varphi)\hat{H}_{\nu j}\right)_p \right], \end{aligned} \qquad (4.5.18)$$

for each $\nu = 1, \ldots, n$, $j = 1, \ldots, p$. The fact that

$$C(\varphi)\hat{H}_{\nu j} + C_1(\varphi)\hat{H}_{\nu j} = \hat{H}_{\nu j}, \qquad (4.5.19)$$

implies the relation for the matrices (4.5.18):

$$C^{\nu j}(\varphi) + C_1^{\nu j}(\varphi) = H_{\nu j}, \quad \nu = 1, \ldots, n; \quad j = 1, \ldots, p. \qquad (4.5.20)$$

The inequality for the Green function $G_0(\tau, \varphi; A, B)$

$$\int_{-\infty}^{\infty} \|G_0(\tau, \varphi; A, B)\| \, d\tau \leq K, \qquad (4.5.21)$$

yields similar inequalities for $G_0(\tau, \varphi; A, B)\hat{H}_{\nu j}$.

The form of the function

$$G_t(\tau, \varphi; A, B)\, \hat{H}_{\nu j} = \begin{cases} \Omega_\tau^t(A, B) C(\varphi_\tau) \hat{H}_{\nu j}, & \text{for } t \geq \tau, \\ -\Omega_\tau^t(A, B) C_1(\varphi_\tau) \hat{H}_{\nu j}, & \text{for } t < \tau, \end{cases}$$

which is a solution of the homogeneous system (4.5.14) implies that the matrix

$$G_t^{\nu j}(\tau, \varphi) = \begin{cases} \Omega_\tau^t(A) C^{\nu j}(\varphi_\tau) \Omega_t^\tau(B), & \text{for } t \geq \tau, \\ -\Omega_\tau^t(A) C_1^{\nu j}(\varphi_\tau) \Omega_t^\tau(B), & \text{for } t < \tau, \end{cases}$$

coincides with the matrix

$$\left[(G_t(\tau, \varphi; A, B)\hat{H}_{\nu j})_1, \ldots, (G_t(\tau, \varphi; A, B)\hat{H}_{\nu j})_p \right].$$

Therefore,

$$G_0^{\nu j}(\tau, \varphi) = \left[(G_0(\tau, \varphi; A, B)\hat{H}_{\nu j})_1, \ldots, (G_0(\tau, \varphi; A, B)\hat{H}_{\nu j})_p \right],$$

which shows that $G_0^{\nu j}(\tau, \varphi)$ satisfies the inequalities (4.5.7).

According to the above arguments, the existence of the Green function $G_0(\tau, \varphi; A, B)$ implies the existence of the Green function $G_0(\tau, \varphi)$ of the system (4.5.2). The converse is obvious. Hence the one-to-one correspondence between the Green function determined above for the system (4.5.2) and the Green function of the linear extension of the dynamical system on a torus corresponding to (4.5.2) follows.

It is also obvious that the existence of the Green function $G_0(\tau, \varphi)$ of the system (4.5.2) with the index of smoothness l ensures that the solution $U(\varphi)$ of the system (4.5.1) belongs to the space $C^1(\mathcal{T}_m)$ for $l \geq 1$. This provides the solution of the problem of block diagonalization of the variational equation (4.1.12) of the system (4.1.5) cited below.

Theorem 4.5.1 *Let the system of equations*

$$\frac{d\varphi}{dt} = a(\varphi), \qquad \frac{d\Phi}{dt} + \Phi P(\varphi) = \frac{\partial a}{\partial \varphi} \Phi, \qquad (4.5.22)$$

possess the Green function with the index of smoothness l. Then the system of variational equations (4.1.12) and the block-diagonal system of the equations

$$\frac{d\Theta}{dt} = \frac{\partial a(\varphi_t)}{\partial \varphi}\Theta, \quad \frac{dq}{dt} = P(\varphi_t)g, \qquad (4.5.23)$$

are $C'(\mathcal{T}_m, a)$-equivalent, $r - 2 \geq l \geq 0$.

Since $\partial a/\partial \varphi \in C^{r-2}(\mathcal{T}_m)$, then the existence of the Green function of the system (4.5.23) is possible virtually, only with the index of smoothness $r - 2 \geq l$. Provided that such a function exists, then there exists a solution to equation (4.1.38) in $C'(\mathcal{T}_m)$ and the change (4.1.38) determined by this equation realizes the $C'(\mathcal{T}_m)$-equivalence of the systems (4.1.11) and (4.5.23).

4.6 On the Problem of Structure of Some Regular Linear Extensions of Dynamical Systems on a Torus

Consider the system of differential equations

$$\frac{d\varphi}{dt} = a(\varphi), \quad \frac{dh}{dt} = H(\varphi)h, \qquad (4.6.1)$$

where $\varphi \in \mathcal{T}_m$, $h \in R^n$, $t \in R$, $a(\varphi) \in C_{\text{Lip}}(\mathcal{T}_m)$ and $H(\varphi) \in C^0(\mathcal{T}_m)$, which is weakly regular, i.e. it has at least one Green function of the problem on a bounded invariant manifold $G_0(\tau, \varphi)$.

As has been established earlier, if $z \in R^n$ and H^* is a transposed matrix, then the system

$$\frac{d\varphi}{dt} = a(\varphi), \quad \frac{dh}{dt} = H(\varphi)h, \quad \frac{dz}{dt} = h - H^*(\varphi)z,$$

is regular, i.e. it has a unique Green function.

In case of two weakly regular systems

$$\frac{d\varphi}{dt} = a(\varphi), \quad \frac{dh}{dt} = H_i(\varphi)h, \quad i = 1, 2.$$

where $\varphi \in \mathcal{T}_m$, $h \in R^n$, $t \in R$, $a(\varphi) \in C_{\text{Lip}}(\mathcal{T}_m)$ and $H_i(\varphi) \in C^0(\mathcal{T}_m)$, a matrix can be constructed

$$P(\varphi) = \begin{pmatrix} \left[H_1 + \frac{1}{2}(H_2 + H_2^*) - I_n\right] & [H_1^* + H_2] & 0 \\ \left[-H_1 + \frac{1}{2}(H_2 - H_2^*) + I_n\right] & -H_1^* & 0 \\ \left[H_1 + \frac{1}{2}(H_2^* - H_2) + I_n\right] & -[H_2 + H_1^*] & -H_2^* \end{pmatrix},$$

such that the system

$$\frac{d\varphi}{dt} = a(\varphi), \quad \frac{dz}{dt} = P(\varphi)z, \quad z \in R^{3n},$$

is regular.

We will generalize this problem. Consider k systems of differential equations

$$\frac{d\varphi}{dt} = a(\varphi), \quad \frac{dh}{dt} = H_i(\varphi)h, \quad i = 1, \ldots, k, \tag{4.6.2}$$

where $\varphi \in \mathcal{T}_m$, $h \in R^n$, $t \in R$, $a(\varphi) \in C_{\text{Lip}}(\mathcal{T}_m)$, $H_i(\varphi) \in C^0(\mathcal{T}_m)$, every of which is weakly regular, i.e. possesses at least one Green function of the problem on the invariant torus $G_0(\tau, \varphi)$. There arises a problem of finding a matrix for $k \geq 3$

$$P(\varphi) = \bar{P}(H_1, \ldots, H_k) \tag{4.6.3}$$

with the dimensions $(k+1)n \times (k+1)n$, such that the system

$$\frac{d\varphi}{dt} = a(\varphi), \quad \frac{dz}{dt} = P(\varphi)z, \quad z \in R^{(k+1)n}, \tag{4.6.4}$$

will be regular, i.e. have a unique Green function.

Consider first the case when $k = 3$. We write down the system (4.6.4) as

$$\frac{d\varphi}{dt} = a(\varphi),$$

$$\frac{dz_1}{dt} = P_{11}(\varphi)z_1 + \cdots + P_{14}(\varphi)z_4,$$

$$\cdots\cdots\cdots\cdots\cdots\cdots\cdots\cdots\cdots\cdots \tag{4.6.5}$$

$$\frac{dz_4}{dt} = P_{41}(\varphi)z_1 + \cdots + P_{44}(\varphi)z_4,$$

$$z_i \in R^n,$$

and compute the derivative of the quadratic form

$$v(z) = 2[\langle z_1, z_2\rangle + \langle z_1, z_3\rangle + \langle z_1, z_4\rangle + \langle z_2, z_3\rangle + \langle z_2, z_4\rangle + \langle z_3, z_4\rangle],$$

along the solutions of the system (4.6.5). We get

$$\dot{v}(z) = 2[\langle \dot{z}_1, z_2\rangle + \langle z_1, \dot{z}_2\rangle + \langle \dot{z}_1, z_3\rangle + \langle z_1, \dot{z}_3\rangle$$
$$+ \langle \dot{z}_1, z_4\rangle + \langle z_1, \dot{z}_4\rangle + \langle \dot{z}_2, z_3\rangle + \langle z_2, \dot{z}_3\rangle \tag{4.6.6}$$
$$+ \langle \dot{z}_2, z_4\rangle + \langle z_2, \dot{z}_4\rangle + \langle \dot{z}_3, z_4\rangle + \langle z_3, \dot{z}_4\rangle]$$

$$= 2\big[\langle P_{11}(\varphi)z_1 + \cdots + P_{14}(\varphi)z_4, z_2\rangle + \langle z_1, P_{21}(\varphi)z_1 + \cdots + P_{24}(\varphi)z_4\rangle$$
$$+ \langle P_{11}(\varphi)z_1 + \cdots + P_{14}(\varphi)z_4, z_3\rangle + \langle z_1, P_{31}(\varphi)z_1 + \cdots + P_{34}(\varphi)z_4\rangle$$
$$+ \langle P_{11}(\varphi)z_1 + \cdots + P_{14}(\varphi)z_4, z_4\rangle + \langle z_1, P_{41}(\varphi)z_1 + \cdots + P_{44}(\varphi)z_4\rangle$$
$$+ \langle P_{21}(\varphi)z_1 + \cdots + P_{24}(\varphi)z_4, z_3\rangle + \langle z_2, P_{31}(\varphi)z_1 + \cdots + P_{34}(\varphi)z_4\rangle$$
$$+ \langle P_{21}(\varphi)z_1 + \cdots + P_{24}(\varphi)z_4, z_4\rangle + \langle z_2, P_{41}(\varphi)z_1 + \cdots + P_{44}(\varphi)z_4\rangle$$
$$+ \langle P_{31}(\varphi)z_1 + \cdots + P_{34}(\varphi)z_4, z_4\rangle + \langle z_3, P_{41}(\varphi)z_1 + \cdots + P_{44}(\varphi)z_4\rangle\big]$$
$$= \langle (SP(\varphi) + P^*(\varphi)S)z, z\rangle$$

where the following designation is used

$$S = \begin{pmatrix} 0 & I & I & I \\ I & 0 & I & I \\ I & I & 0 & I \\ I & I & I & 0 \end{pmatrix}. \tag{4.6.7}$$

Let the quadratic form (4.6.6) satisfy the inequality

$$\dot{v}(z) \geq \|z_1\|^2. \tag{4.6.8}$$

For this it is sufficient to assume that

$$SP(\varphi) = \operatorname{diag}\{B_0(\varphi), 0, 0, 0\} + M(\varphi), \tag{4.6.9}$$

where the matrix functions $M(\varphi)$ and $B_0(\varphi)$ have the following properties

$$M^*(\varphi) \equiv -M(\varphi), \quad \langle B_0(\varphi)z_1, z_1\rangle \geq \beta_0\|z_1\|^2, \quad \beta_0 > 0. \tag{4.6.10}$$

Consequently, in view of the equality (4.6.9) the matrix $P(\varphi)$ can be represented as

$$P(\varphi) = S^{-1}\big[\operatorname{diag}\{B_0(\varphi), 0, 0, 0\} + M(\varphi)\big]$$
$$= \frac{1}{3}\begin{pmatrix} -2I & I & I & I \\ I & -2I & I & I \\ I & I & -2I & I \\ I & I & I & -2I \end{pmatrix}$$
$$\times \begin{pmatrix} B_0(\varphi) & M_{12}(\varphi) & M_{13}(\varphi) & M_{14}(\varphi) \\ -M_{12}^*(\varphi) & 0 & M_{23}(\varphi) & M_{24}(\varphi) \\ -M_{13}^*(\varphi) & -M_{23}^*(\varphi) & 0 & M_{34}(\varphi) \\ -M_{14}^*(\varphi) & -M_{24}^*(\varphi) & M_{34}^*(\varphi) & 0 \end{pmatrix} \tag{4.6.11}$$
$$= \{P_{ij}(\varphi)\}_{i,j=1}^{4},$$

where

$$P_{11}(\varphi) = \frac{1}{3}\left(-2B_0(\varphi) - \sum_{i=2}^{4} M_{1i}^*(\varphi)\right),$$

$$P_{21}(\varphi) = \frac{1}{3}(B_0(\varphi) + 2M_{12}^*(\varphi) - M_{13}^*(\varphi) - M_{14}^*(\varphi)),$$

$$P_{22}(\varphi) = \frac{1}{3}(M_{12}(\varphi) - M_{23}^*(\varphi) - M_{24}^*(\varphi)), \qquad (4.6.12)$$

$$P_{33}(\varphi) = \frac{1}{3}(M_{13}(\varphi) + M_{23}(\varphi) - M_{34}^*(\varphi)),$$

$$P_{44}(\varphi) = \frac{1}{3}(M_{14}(\varphi) + M_{24}(\varphi) + M_{34}(\varphi)).$$

We introduce the designations

$$\bar{P}(\varphi) = \{P_{ij}(\varphi)\}_{i,j=2}^{4}, \quad \bar{z} = \{z_2, z_3, z_4\}, \qquad (4.6.13)$$

and assume the existence of the quadratic form

$$\bar{v}(\varphi, \bar{z}) = \langle S(\varphi)\bar{z}, \bar{z}\rangle, \qquad (4.6.14)$$

which has a positive definite derivative along the solutions of the system of equations

$$\frac{d\varphi}{dt} = a(\varphi), \quad \frac{d\bar{z}}{dt} = \bar{P}(\varphi)\bar{z}, \qquad (4.6.15)$$

i.e.

$$\dot{\bar{v}}(\varphi, \bar{z}) = \langle [\dot{S}(\varphi) + S(\varphi)\bar{P}(\varphi) + \bar{P}^*(\varphi)S(\varphi)]\bar{z}, \bar{z}\rangle \geq \|\bar{z}\|^2. \qquad (4.6.16)$$

Under these assumptions, if the parameter $p > 0$ takes sufficiently large values, then the derivative due to the system (4.6.4) of the quadratic form

$$pv(z) + \bar{v}(\varphi, \bar{z}) \qquad (4.6.17)$$

is positive definite. Since the quadratic form (4.6.17) is non-degenerate for large values of the parameter $p > 0$, the system (4.6.5) is regular.

The matrix $\bar{P}(\varphi)$ in the system (4.6.15) is of a specific block form (4.6.11). It is then natural to assume the satisfying of the following equalities:

$$\begin{aligned} P_{22}(\varphi) &= -H_1^*(\varphi), & P_{23}(\varphi) &= 0, \\ P_{33}(\varphi) &= -H_2^*(\varphi), & P_{24}(\varphi) &= 0, \\ P_{44}(\varphi) &= -H_3^*(\varphi), & P_{34}(\varphi) &= 0. \end{aligned} \qquad (4.6.18)$$

From here the matrices $M_{ij}(\varphi)$, $i = 1, 2, 3$, $j = 2, 3, 4$ can be determined unambiguously. In view of the equalities (4.6.12) equations (4.6.18) become

(I) $\quad M_{12}(\varphi) - M_{23}^*(\varphi) - M_{24}^*(\varphi) = -3H_1^*(\varphi),$
$\quad M_{13}(\varphi) + M_{23}(\varphi) - M_{34}^*(\varphi) = -3H_2^*(\varphi),$
$\quad M_{14}(\varphi) + M_{24}(\varphi) - M_{34}(\varphi) = -3H_3^*(\varphi),$

(II) $\quad M_{13}(\varphi) - 2M_{23}(\varphi) - M_{34}^*(\varphi) = 0,$
$\quad M_{14}(\varphi) - 2M_{24}(\varphi) + M_{34}^*(\varphi) = 0,$
$\quad M_{14}(\varphi) + 2M_{24}(\varphi) - 2M_{34}(\varphi) = 0.$

(4.6.18a)

We extract the first and second equations of the system (II) from the second and third equations of the system (I) respectively. Then we have

$$M_{23}(\varphi) = -H_2^*(\varphi), \quad M_{24}(\varphi) = M_{34}(\varphi) = -H_3^*(\varphi),$$
$$M_{12}(\varphi) = -3H_1^*(\varphi) - H_2(\varphi) - H_3(\varphi),$$
$$M_{13}(\varphi) = -2H_2^*(\varphi) - H_3(\varphi), \quad M_{14}(\varphi) = -H_3^*(\varphi).$$

Thus, if three systems of the equations (4.6.2) ($k = 3$) are weakly regular, then the system of equations (4.6.4) is regular provided that

$$P(\varphi) = \frac{1}{3} \begin{pmatrix} -2I & I & I & I \\ I & -2I & I & I \\ I & I & -2I & I \\ I & I & I & -2I \end{pmatrix}$$

$$\times \begin{pmatrix} B_0(\varphi) & (-3H_1^* - H_2 - H_3) & (-2H_2^* - H_3) & -H_3^* \\ (3H_1 + H_2^* + H_3^*) & 0 & -H_2^* & -H_3^* \\ (2H_2 + H_3) & H_2 & 0 & -H_3^* \\ H_3 & H_3 & H_3 & 0 \end{pmatrix},$$

where $B_0(\varphi) \in C^0(\mathcal{T}_m)$ is an arbitrary matrix of definite sign

$$|\langle B_0(\varphi)x, x \rangle| \geq \beta_0 \|x\|^2, \quad \beta_0 = \text{const} > 0. \tag{4.6.19}$$

Now we assume $k > 3$. Let the matrix S be

$$S = \begin{pmatrix} 0 & I & I & \cdots & I \\ I & 0 & I & \cdots & I \\ I & I & 0 & \cdots & I \\ \vdots & \vdots & \vdots & \ddots & \vdots \\ I & I & I & \cdots & 0 \end{pmatrix}$$

Then its inverse is

$$\begin{pmatrix} -(k-1)I & I & I & \cdots & I \\ I & -(k-1)I & I & \cdots & I \\ I & I & -(k-1)I & \cdots & I \\ \vdots & \vdots & \vdots & \ddots & \vdots \\ I & I & I & \cdots & -(k-1)I \end{pmatrix}$$

Similarly to (4.6.11) consider the matrix $P(\varphi)$

$$P(\varphi) = \frac{1}{k} \begin{pmatrix} -(k-1)I & I & I & \cdots & I \\ I & -(k-1)I & I & \cdots & I \\ I & I & -(k-1)I & \cdots & I \\ \vdots & \vdots & \vdots & \ddots & \vdots \\ I & I & I & \cdots & -(k-1)I \end{pmatrix}$$

(4.6.20)

$$\times \begin{pmatrix} B_0(\varphi) & M_{12}(\varphi) & M_{13}(\varphi) & \cdots & M_{1(k+1)}(\varphi) \\ -M_{12}^*(\varphi) & 0 & M_{23}(\varphi) & \cdots & M_{2(k+1)}(\varphi) \\ -M_{13}^*(\varphi) & -M_{23}^*(\varphi) & 0 & \cdots & M_{3(k+1)}(\varphi) \\ \vdots & \vdots & \vdots & \ddots & \vdots \\ -M_{1(k+1)}^*(\varphi) & -M_{2(k+1)}^*(\varphi) & -M_{3(k+1)}^*(\varphi) & \cdots & 0 \end{pmatrix}$$

We compose the systems of equations as in (4.6.18a)

$$\begin{aligned} M_{12}(\varphi) - M_{23}^*(\varphi) - M_{24}^*(\varphi) - \cdots - M_{2(k+1)}^*(\varphi) &= -kH_1^*(\varphi), \\ M_{13}(\varphi) + M_{23}(\varphi) - M_{34}^*(\varphi) - \cdots - M_{3(k+1)}^*(\varphi) &= -kH_2^*(\varphi), \\ M_{14}(\varphi) + M_{24}(\varphi) - M_{34}(\varphi) - \cdots - M_{4(k+1)}^*(\varphi) &= -kH_3^*(\varphi), \end{aligned}$$ (4.6.20a)

$$\cdots\cdots\cdots\cdots\cdots\cdots\cdots\cdots\cdots\cdots\cdots\cdots\cdots\cdots\cdots\cdots$$

$$M_{1(k+1)}(\varphi) + M_{2(k+1)}(\varphi) + M_{3(k+1)}(\varphi) + \cdots + M_{k(k+1)}(\varphi) = -kH_k^*(\varphi),$$

$$\begin{aligned} M_{13}(\varphi) - (k-1)M_{23}(\varphi) - M_{34}^*(\varphi) - \cdots + M_{3(k+1)}^*(\varphi) &= 0, \\ M_{14}(\varphi) - (k-1)M_{24}(\varphi) - M_{34}(\varphi) - \cdots - M_{4(k+1)}^*(\varphi) &= 0, \\ M_{15}(\varphi) - (k-1)M_{25}(\varphi) + M_{35}(\varphi) + \cdots - M_{5(k+1)}^*(\varphi) &= 0, \end{aligned}$$ (4.6.20b)

$$\cdots\cdots\cdots\cdots\cdots\cdots\cdots\cdots\cdots\cdots\cdots\cdots\cdots\cdots\cdots\cdots$$

$$M_{1(k+1)}(\varphi) - (k-1)M_{2(k+1)}(\varphi) + M_{3(k+1)}(\varphi) + \cdots + M_{k(k+1)}(\varphi) = 0,$$

and

$$\begin{aligned} M_{14}(\varphi) + M_{24}(\varphi) - (k-1)M_{34}(\varphi) - M_{45}^*(\varphi) - \cdots - M_{4(k+1)}^*(\varphi) &= 0, \\ M_{15}(\varphi) + M_{25}(\varphi) - (k-1)M_{35}(\varphi) + M_{45}(\varphi) - \cdots + M_{5(k+1)}^*(\varphi) &= 0, \end{aligned}$$

$$\cdots\cdots\cdots\cdots\cdots\cdots\cdots\cdots\cdots\cdots\cdots\cdots\cdots\cdots\cdots\cdots$$

$$M_{1(k+1)}(\varphi) + M_{2(k+1)}(\varphi) - (k-1)M_{3(k+1)}(\varphi) + \cdots + M_{k(k+1)}(\varphi) = 0.$$

We extract the first equation of the system (4.6.20b) from the second equation of the system (4.6.20a). We get
$$kM_{23}(\varphi) = -kH_2^*(\varphi).$$
Similar operations result in
$$\begin{aligned}
M_{23}(\varphi) &= -H_2^*(\varphi), \\
M_{24}(\varphi) &= M_{34}(\varphi) = -H_3^*(\varphi), \\
M_{25}(\varphi) &= M_{35}(\varphi) = M_{45}(\varphi) = -H_4^*(\varphi), \\
&\cdots\cdots\cdots\cdots\cdots\cdots\cdots\cdots\cdots\cdots\cdots\cdots\cdots \\
M_{2(k+1)}(\varphi) &= M_{3(k+1)}(\varphi) = M_{4(k+1)}(\varphi) = \cdots = M_{k(k+1)}(\varphi) = -H_k^*(\varphi).
\end{aligned} \quad (4.6.21)$$

Further we find that
$$\begin{aligned}
M_{12}(\varphi) &= -kH_1^*(\varphi) - \sum_{i=2}^{k} H_i(\varphi), \\
M_{13}(\varphi) &= -(k-1)H_2^*(\varphi) - \sum_{i=3}^{k} H_i(\varphi), \\
&\cdots\cdots\cdots\cdots\cdots\cdots\cdots\cdots\cdots\cdots \\
M_{1p}(\varphi) &= -(k-p+2)H_2^*(\varphi) - \sum_{i=p}^{k} H_i(\varphi), \\
M_{1(k+1)}(\varphi) &= -H_k^*(\varphi).
\end{aligned} \quad (4.6.22)$$

Thus, the following result is proved.

Theorem 4.6.1 *Let k systems of differential equations (4.6.2) ($k \geq 1$) be weakly regular, then the system of equations (4.6.4) is regular provided that the matrix $P(\varphi)$ has the form of (4.6.20)*

$$P(\varphi) = \frac{1}{k} \begin{pmatrix} -(k-1)I & I & I & \cdots & I \\ I & -(k-1)I & I & \cdots & I \\ I & I & -(k-1)I & \cdots & I \\ \vdots & \vdots & \vdots & \ddots & \vdots \\ I & I & I & \cdots & -(k-1)I \end{pmatrix}$$

$$\times \begin{pmatrix} B_0(\varphi) & M_{12}(\varphi) & M_{13}(\varphi) & \cdots & M_{1(k+1)}(\varphi) \\ -M_{12}^*(\varphi) & 0 & M_{23}(\varphi) & \cdots & M_{2(k+1)}(\varphi) \\ -M_{13}^*(\varphi) & -M_{23}^*(\varphi) & 0 & \cdots & M_{3(k+1)}(\varphi) \\ \vdots & \vdots & \vdots & \ddots & \vdots \\ -M_{1(k+1)}^*(\varphi) & -M_{2(k+1)}^*(\varphi) & -M_{3(k+1)}^*(\varphi) & \cdots & 0 \end{pmatrix}$$

where the $n \times n$-dimensional matrices M_{ij} are determined by equalities (4.6.21) and (4.6.22), and the matrix $B_0(\varphi)$ satisfies the condition (4.6.19).

Remark 4.6.1 In the case when $k = 1$ and $B_0(\varphi) = I$ the matrix (4.6.20) becomes

$$P(\varphi) = \begin{pmatrix} H_1(\varphi) & 0 \\ I & -H_1^*(\varphi) \end{pmatrix}.$$

If we admit $k = 2$ and $B_0(\varphi) = 2I$ in (4.6.20), then we get the matrix

$$P(\varphi) = \begin{pmatrix} \left[H_1 + \frac{1}{2}(H_2 + H_2^*) - I\right] & [H_1^* + H_2] & 0 \\ \left[-H_1 + \frac{1}{2}(H_2 - H_2^*) + I\right] & -H_1^* & 0 \\ \left[H_1 + \frac{1}{2}(H_2^* - H_2) + I\right] & -[H_2 + H_1^*] & -H_2^* \end{pmatrix},$$

where $H_1 = H_1(\varphi)$, $H_1^* = H_1^*(\varphi)$, $H_2 = H_2(\varphi)$, $H_2^* = H_2^*(\varphi)$.

Corollary 4.6.1 Let all k systems (4.6.2) be similar, i.e.

$$H_1(\varphi) \equiv H_2(\varphi) \equiv \ldots \equiv H_k(\varphi), \quad B_0(\varphi) = I.$$

Then the condition of weak regularity of the system

$$\frac{d\varphi}{dt} = a(\varphi), \quad \frac{dh}{dt} = H(\varphi)h,$$

implies, for any value of $k \geq 1$, the regularity of the system

$$\frac{d\varphi}{dt} = a(\varphi),$$

$$\frac{dh_1}{dt} = \left[\frac{k+1}{2}H(\varphi) + \frac{k-1}{2}H^*(\varphi) - \frac{k-1}{2}I\right]h_1$$

$$+ \sum_{i=1}^{k}(k-i)(H(\varphi) + H^*(\varphi))h_{i+1},$$

$$\frac{dh_2}{dt} = \left[-\frac{k-1}{2}(H(\varphi) + H^*(\varphi)) + \frac{1}{k}I\right]h_1 - H^*(\varphi)h_2,$$

$$\frac{dh_3}{dt} = \left[-\frac{k-3}{2}(H(\varphi) + H^*(\varphi)) + \frac{1}{k}I\right]h_1$$

$$- (H(\varphi) + H^*(\varphi))h_2 - H^*(\varphi)h_3,$$

$$\ldots \ldots \ldots \ldots \ldots \ldots \ldots \ldots \ldots \ldots$$

$$\frac{dh_{k+1}}{dt} = \left[-\frac{k-1}{2}(H(\varphi) + H^*(\varphi)) + \frac{1}{k}I\right]h_1$$

$$- (H(\varphi) + H^*(\varphi))(h_2 + h_3 + \cdots + h_k) - H^*(\varphi)h_{k+1}, \quad H_i \in R^n.$$

4.7 Invariant Manifolds of Autonomous Differential Equations and Lyapunov Functions with Alternating Signs

We will consider the system of differential equations

$$\frac{d\varphi}{dt} = a(\varphi), \quad \frac{dx}{dt} = A(\varphi)x, \qquad (4.7.1)$$

where $\varphi \in \mathcal{T}_m$, $x \in R^n$, $t \in R$. It is known that the system (4.7.1) is regular, if and only if there exists a non-degenerate quadratic form

$$V(\varphi, x) = \langle S(\varphi)x, x \rangle, \quad S^*(\varphi) \equiv S(\varphi) \in C'(\mathcal{T}_m, a), \quad \det S(\varphi) \neq 0,$$

possessing a positive definite derivative due to the system (4.7.1)

$$\dot{V}(\varphi, x) = \langle [\dot{S}(\varphi) + S(\varphi)A(\varphi) + A^*(\varphi)S(\varphi)]x, x \rangle \geq \|x\|^2, \qquad (4.7.2)$$

and which is weakly regular, if there exists a quadratic form

$$V(\varphi, y) = \langle \bar{S}(\varphi)y, y \rangle,$$

which is possibly already degenerate at some points $\varphi \in \mathcal{T}_m$, with the matrix of coefficients $\bar{S}(\varphi) \in C'(\mathcal{T}_m, a)$ possessing a positive definite derivative due to the system

$$\frac{d\varphi}{dt} = a(\varphi), \quad \frac{dy}{dt} = -A^*(\varphi)y, \qquad (4.7.1a)$$

i.e. the inequality

$$\dot{V}(\varphi, y) = \langle [\dot{\bar{S}}(\varphi) - \bar{S}(\varphi)A^*(\varphi) - A(\varphi)\bar{S}(\varphi)]y, y \rangle \geq \|y\|^2, \qquad (4.7.3)$$

is valid.

In this regard an important problem arises of finding such quadratic forms $V(\varphi, x)$, $V(\varphi, y)$ or at least indicating a coefficient criterion for its existence.

We note that if the degenerate matrix $\bar{S}(\varphi)$ exists which satisfies condition (4.7.3), then the extended system of equations

$$\frac{d\varphi}{dt} = a(\varphi), \quad \frac{dx}{dt} = A(\varphi)x, \quad \frac{dy}{dt} = x - A^*(\varphi)y, \qquad (4.7.4)$$

is regular. Moreover, the non-degenerate quadratic form $V(\varphi, x, y)$ which possesses a positive definite derivative due to the system (4.7.4) can be taken in the form

$$V(\varphi, x, y) = \lambda \langle x, y \rangle + \langle \bar{S}(\varphi)y, y \rangle, \qquad (4.7.5)$$

where λ is a sufficiently large fixed parameter.

We represent the system (4.7.1) as

$$\frac{d\varphi}{dt} = a(\varphi),$$

$$\frac{dx_1}{dt} = A_{11}(\varphi)x_1 + A_{12}(\varphi)x_2, \qquad (4.7.6)$$

$$\frac{dx_2}{dt} = A_{21}(\varphi)x_1 + A_{22}(\varphi)x_2,$$

where $x_1 \in R^{n_1}$, $x_2 \in R^{n_2}$, and assume the existence of some quadratic form $v_1(\varphi, x_1, x_2) = \langle S_1(\varphi)x, x \rangle$, the derivative of which due to the system (4.7.6) satisfies the inequality

$$\dot{v}_1(\varphi, x_1, x_2) \geq \|x_1\|^2. \qquad (4.7.7)$$

Moreover, we assume for the appropriate part of the system of equations (4.7.6)

$$\frac{d\varphi}{dt} = a(\varphi), \quad \frac{dx_2}{dt} = A_{22}(\varphi)x_2, \qquad (4.7.8)$$

the existence of the quadratic form $v_2(\varphi, x_2) = \langle S_2(\varphi)x_2, x_2 \rangle$ which has a positive definite derivative due to the system (4.7.8)

$$\dot{v}_2(\varphi, x_2) \geq \|x_2\|^2. \qquad (4.7.9)$$

Now similarly to the quadratic form (4.7.5), consider the following quadratic form with the parameter

$$V(\varphi, x) = \lambda v_1(\varphi, x_1, x_2) + v_2(\varphi, x_2), \qquad (4.7.10)$$

and estimate its derivative due to the system (4.7.6). We have

$$\dot{V}(\varphi, x) = \lambda \|x_1\|^2 + \langle \dot{S}_2(\varphi)x_2, x_2 \rangle + 2\langle S_2(\varphi)x_2, A_{21}(\varphi)x_1 + A_{22}(\varphi)x_2 \rangle$$

$$\geq \lambda \|x_1\|^2 + \|x_2\|^2 - 2\|S_2 A_{21}\|_0 \|x_1\| \|x_2\|.$$

It is clear then, that for sufficiently large values of the parameter λ the derivative $\dot{V}(\varphi, x)$ is positive definite.

We note further that the inequalities (4.7.7) and (4.7.9) can be represented as

$$\langle [\dot{S}_1(\varphi) + S_1(\varphi)A(\varphi) + A^*(\varphi)S_1(\varphi)]x, x \rangle \geq \|P_1 x\|^2, \qquad (4.7.11)$$

$$\langle [\dot{\hat{S}}(\varphi) + \hat{S}(\varphi)A(\varphi) + A^*(\varphi)\hat{S}(\varphi)]P_2 x, P_2 x \rangle \geq \|P_2 x\|^2, \qquad (4.7.12)$$

where $P_1 = \text{diag}\{I_{n_1}, 0\}$, $P_2 = I_n - P_1$, $\hat{S}(\varphi) = \text{diag}\{0, S_2(\varphi)\}$. Assuming that the matrices P_1 and P_2 are not, generally speaking, projecting in the inequalities (4.7.11) and (4.7.12), we prove the following result.

Theorem 4.7.1 *Let there exist two $(n \times n)$-dimensional matrices $S_i(\varphi) \in C'(\mathcal{T}_m, a)$, $i = 1, 2$, which satisfy the inequalities*

$$\langle [\dot{S}_1(\varphi) + S_1(\varphi)A(\varphi) + A^*(\varphi)S_1(\varphi)]x, x \rangle \geq \|M_1(\varphi)x\|^2, \tag{4.7.13}$$

$$\langle [\dot{S}_2(\varphi) + S_2(\varphi)A(\varphi) + A^*(\varphi)S_2(\varphi)]M_2(\varphi)x,\, M_2(\varphi)x \rangle \geq \|M_2(\varphi)x\|^2, \tag{4.7.14}$$

where $M_i(\varphi) \in C^0(\mathcal{T}_m)$ are some matrices with the properties

$$\det[M_1(\varphi) + M_2(\varphi)] \neq 0, \quad \forall \varphi \in \mathcal{T}_m, \tag{4.7.15}$$

$$\|[M_1(\varphi) + M_2(\varphi)]^{-1}\| \leq K, \quad K = \mathrm{const} < \infty, \tag{4.7.16}$$

$$M_1(\varphi)M_2(\varphi) \equiv M_2(\varphi)M_1(\varphi), \quad \forall \varphi \in \mathcal{T}_m. \tag{4.7.17}$$

Then the derivative of the quadratic form

$$V(\varphi, x) = \lambda \langle S_1(\varphi)x, x \rangle + \langle S_2(\varphi)x, x \rangle, \tag{4.7.18}$$

due to system (4.7.1) is positive definite for sufficiently large values of the parameter λ.

Proof First we consider the case when

$$M_1(\varphi) + M_2(\varphi) \equiv I_n, \quad \forall \varphi \in \mathcal{T}_m, \tag{4.7.19}$$

where I_n is an n-dimensional identity matrix. It is clear that the conditions (4.7.15)–(4.7.17) are satisfied in this case.

Further we will use the designation

$$W_i(\varphi) = \dot{S}_i(\varphi) + S_i(\varphi)A(\varphi) + A^*(\varphi)S_i(\varphi), \tag{4.7.20}$$

where $S_i(\varphi) \in C'(\mathcal{T}_m, a)$, $i = 1, 2, \ldots$, are some $(n \times n)$-dimensional matrices.

Putting down the derivative of the quadratic form (4.7.18) due to the system (4.7.1) and incorporating the inequality (4.7.13) we get

$$\dot{V}(\varphi, x) = \lambda \langle S_1(\varphi)x, x \rangle^{\cdot} + \langle S_2(\varphi)x, x \rangle^{\cdot} \geq \lambda \|M_1(\varphi)x\|^2 + \langle W_2(\varphi)x, x \rangle. \tag{4.7.21}$$

We add and extract the expression

$$\langle W_2(\varphi)M_2(\varphi)x, M_2(\varphi)x \rangle$$

from the right-hand side of the inequality (4.7.21). In view of the inequality (4.7.14) we have

$$\dot{V}(\varphi, x) \geq \lambda \|M_1(\varphi)x\|^2 + \|M_2(\varphi)x\|^2 \\ + \langle [W_2(\varphi) - M_2^*(\varphi)W_2(\varphi)M_2(\varphi)]x, x \rangle. \tag{4.7.22}$$

Applying identity (4.7.19) we estimate the third term in the right-hand side of the inequality (4.7.22) as

$$|\langle [W_2(\varphi) - M_2^*(\varphi)W_2(\varphi)M_2(\varphi)]x, x\rangle|$$
$$= |\langle [M_1^*(\varphi) + M_2^*(\varphi)W_2(\varphi)(M_1(\varphi) + M_2(\varphi)) - M_2^*(\varphi)W_2(\varphi)M_2(\varphi)]x, x\rangle|$$
$$\leq |\langle [M_1^*(\varphi)W_2(\varphi)M_1(\varphi)]x, x\rangle| + |\langle [M_1^*(\varphi)W_2(\varphi)M_2(\varphi)]x, x\rangle|$$
$$+ |\langle [M_2^*(\varphi)W_2(\varphi)M_1(\varphi)]x, x\rangle| \leq \|W_2\|_0 \|M_1 x\|^2 + 2\|W_2\|_0 \|M_1 x\| \|M_2 x\|.$$

The obtained estimate enables us to continue the inequality (4.7.22)

$$\dot{V}(\varphi, x) \geq (\lambda - \|W_2(\varphi)\|_0)\|M_1(\varphi)x\|^2 + \|M_2(\varphi)x\|^2$$
$$- 2\|W_2\|_0 \|M_1 x\| \|M_2 x\|. \tag{4.7.22a}$$

The right-hand side of the inequality (4.7.22a) is treated as the quadratic form of two variables σ_1, σ_2

$$\varphi(\sigma_1, \sigma_2) = (\lambda - \|W_2\|_0)\sigma_1^2 - 2\|W_2\|_0 \sigma_1 \sigma_2 + \sigma_2^2,$$

and its least value is found on the unit circle $\sigma_1 = \cos t$, $\sigma_2 = \sin t$. As a result of appropriate transformations we get

$$\varphi(\cos t, \sin t) = \frac{\lambda - \|W_2\|_0 - 1}{2} \cos 2t - \|W_2\|_0 \sin 2t + \frac{\lambda - \|W_2\|_0 + 1}{2}$$
$$\geq -\left[\left(\frac{\lambda - \|W_2\|_0 - 1}{2}\right)^2 + \|W_2\|_0^2\right]^{1/2} + \frac{\lambda - \|W_2\|_0 + 1}{2}$$
$$\geq \frac{\lambda - \|W_2\|_0 - \|W_2\|_0^2}{\lambda - \|W_2\|_0 + 1} = \gamma_2(\lambda).$$

It is clear, that $\gamma_2(\lambda)$ is positive, provided that

$$\lambda > \|W_2\|_0 + \|W_2\|_0^2. \tag{4.7.23}$$

In addition, inequality (4.7.22a) can be complemented as follows

$$\dot{V}(\varphi, x) \geq \gamma_2(\lambda)(\|M_1(\varphi)x\|^2 + \|M_2(\varphi)x\|^2)$$
$$\geq \frac{\gamma_2(\lambda)}{2}(\|M_1(\varphi)x\|^2 + \|M_2(\varphi)x\|^2) = \frac{\gamma_2(\lambda)}{2}\|x\|^2.$$

Thus, for the fixed values of the parameter λ which satisfy inequality (4.7.23) the derivative of the quadratic form (4.7.18) due to the system (4.7.1) is positive definite.

Now instead of identity (4.7.19) we assume that the conditions (4.7.15)–(4.7.19) are satisfied. Designate
$$M(\varphi) = M_1(\varphi) + M_2(\varphi), \quad \bar{M}_i(\varphi) = M^{-1}(\varphi)M_i(\varphi).$$
Clearly, for the matrices $\bar{M}_i(\varphi)$ the identity
$$\bar{M}_1(\varphi) + \bar{M}_2(\varphi) \equiv I_n$$
is already satisfied. It remains to be verified that the inequalities (4.7.13), (4.7.14) are satisfied with the matrices $\bar{M}_i(\varphi)$, $i = 1, 2$. For inequality (4.7.13) we have
$$\langle W(\varphi)x, x\rangle \geq \|M_1(\varphi)x\|^2 = \|M(\varphi)\bar{M}_1(\varphi)x\|^2 \geq K^{-2}\|\bar{M}_1(\varphi)x\|^2.$$
Note that the satisfaction of (4.7.17) yields $\bar{M}_2(\varphi) = M^{-1}(\varphi)M_2(\varphi) \equiv M_2(\varphi) \times M^{-1}(\varphi)$. Therefore after the change of variables $x = M^{-1}(\varphi)z$ we have in inequality (4.7.14)
$$\langle W_2(\varphi)\bar{M}_2(\varphi)z, \bar{M}_2(\varphi)z\rangle \geq \|\bar{M}_2(\varphi)z\|^2.$$
This completes the proof of Theorem 4.7.1.

Remark 4.7.1 The conditions of Theorem 4.7.1 ensure weak regularity of the system of equations (4.7.1a). If it is additionally assumed that
$$|\det S_1(\varphi)| \geq \delta = \text{const} > 0, \quad \forall \varphi \in \mathcal{T}_m,$$
then both systems (4.7.1) and (4.7.1a) are regular.

Remark 4.7.2 If the satisfaction of the identity (4.7.17) is not assumed in the conditions of Theorem 4.7.1, then the assertion of Theorem 4.7.1 becomes false.

This can be demonstrated by a simple example of the system of two equations
$$\frac{dx_i}{dt} = -x_1 + x_2, \quad i = 1, 2.$$
In this case inequality (4.7.13) is satisfied for the constant matrices
$$S_1 = \begin{pmatrix} 0 & -1/2 \\ -1/2 & 1 \end{pmatrix}, \quad M_1 = \begin{pmatrix} 1 & -1 \\ 0 & 0 \end{pmatrix},$$
and the inequality (4.7.14) is valid for
$$S_2 = \text{diag}\{0, 1\} = M_2.$$
Thus, all conditions of Theorem 4.7.1 are satisfied, except for the identity (4.7.17). However, there exists no quadratic form has a sign-definite derivative due to the initial system, since its solutions are of the form
$$x_1(t) = c_1 t + c_2, \quad x_2(t) = c_1 t + c_2 + c_1, \quad c_i = \text{const}.$$
Let us prove the following result employing designation (4.7.20)

Theorem 4.7.2 *Let there exist two symmetric matrices $S_i(\varphi) \in C'(\mathcal{T}_m, a)$, for which*

$$\langle W_i(\varphi)x, x \rangle \geq \|M_i(\varphi)x\|^2, \quad i = 1, 2, \tag{4.7.24}$$

where the matrices $M_i(\varphi) \in C^0(\mathcal{T}_m)$ satisfy the conditions (4.7.15), (4.7.16). Then the derivative of the quadratic form (4.7.18) due to (4.7.1) is positive definite for all values of $\lambda > 0$. Moreover, if one of the matrices $S_i(\varphi)$ satisfies the inequality

$$|\det S_i(\varphi)| \geq \delta = \text{const} > 0, \tag{4.7.25}$$

the sum $\lambda S_1(\varphi) + S_2(\varphi)$ is non-degenerate for all $\lambda > 0$, $\varphi \in \mathcal{T}_m$, and hence, both of the systems (4.7.1), (4.7.1a) are regular.

Proof In view of the inequalities (4.7.24) the derivative due to the system (4.7.1) of the quadratic form (4.7.18) is estimated as

$$\dot{V}(\varphi, x) \geq \lambda \|M_1(\varphi)x\|^2 + \|M_2(\varphi)x\|^2 \geq \lambda(\lambda^2 + 1)^{-1}(\|M_1(\varphi)x\|^2 + \|M_2(\varphi)x\|^2)$$

$$\geq \frac{(\lambda^2 + 1)^{-1}\lambda}{2}(\|M_1(\varphi)x\|^2 + \|M_2(\varphi)x\|^2)$$

$$\geq \frac{(\lambda^2 + 1)^{-1}\lambda}{2}\|(M_1(\varphi)x + M_2(\varphi))x\|^2 \geq K^{-2}\frac{(\lambda^2 + 1)^{-1}\lambda}{2}\|x\|^2.$$

Let the inequality (4.7.25) be satisfied for $i = 2$. Then, evidently the inequality

$$\det(\lambda S_1(\varphi) + S_2(\varphi)) \neq 0$$

is satisfied for sufficiently small values of $\lambda > 0$. Consequently, for sufficiently small $\lambda > 0$ the quadratic form (4.7.1) is positive definite. Therefore, all quadratic forms having a sign-definite derivative due to the system (4.7.1) are non-degenerate, and particularly the form (4.7.18) also, for all $\lambda > 0$.

The case when $|\det S_2(\varphi)| \geq \delta = \text{const} > 0$, is treated in the same way.

This completes the proof of Theorem 4.7.2.

Remark 4.7.3 If the inequality (4.7.25) is not required to be satisfied, but it is assumed only that

$$\det S_i(\varphi) \neq 0, \quad \forall \varphi \in \mathcal{T}_m,$$

then this is not sufficient for

$$\det(\lambda S_1(\varphi) + S_2(\varphi)) \neq 0, \quad \forall \varphi \in \mathcal{T}_m.$$

We represent the system (4.7.1) as

$$\frac{d\varphi}{dt} = a(\varphi), \quad \frac{dx_i}{dt} = \sum_{j=1}^{k} A_{ij}(\varphi)x_j, \quad i = 1, \ldots, k, \tag{4.7.26}$$

where $x_i \in R^{n_i}$, $n_1 + \cdots + n_k = n$, and designate

$$\hat{A}_k(\varphi) = A_{kk}(\varphi), \quad \hat{A}_{k-1}(\varphi) = \begin{pmatrix} A_{k-1\,k-1}(\varphi) & A_{k-1\,k}(\varphi) \\ A_{k\,k-1}(\varphi) & A_{k\,k}(\varphi) \end{pmatrix}, \quad \ldots, \quad \hat{A}_1(\varphi) = A(\varphi),$$

$$\hat{x}_k = x_k, \quad \hat{x}_{k-1} = \begin{pmatrix} x_{k-1} \\ x_k \end{pmatrix}, \quad \ldots, \quad \hat{x}_1 = x.$$

Now, for every system of equations

$$\frac{d\varphi}{dt} = a(\varphi), \quad \frac{d\hat{x}_i}{dt} = \hat{A}_i(\varphi)\hat{x}_i, \quad i = 1, \ldots, k, \tag{4.7.27}$$

we assume that there exists a matrix $\hat{S}_i(\varphi) \in C'(\mathcal{T}_m, a)$ which satisfies the inequality

$$\langle [\dot{\hat{S}}_i(\varphi) + \hat{S}_i(\varphi)\hat{A}_i(\varphi) + \hat{A}_i^*(\varphi)\hat{S}_i(\varphi)]x_i, x_i \rangle \geq \|x\|^2,$$
$$i = 1, \ldots, k. \tag{4.7.28}$$

There arises a question: whether a quadratic form exists under these conditions which has a sign-definite derivative due to the system (4.7.1)?

Incorporating the designation (4.7.20) for the matrices $W_i(\varphi)$ and designating in addition, $S_1(\varphi) = \hat{S}_1(\varphi)$, $S_i(\varphi) = \text{diag}\{0, \hat{S}_i(\varphi)\}$, $i = 2, \ldots, k$, we rewrite the inequalities (4.7.28) as

$$\langle W_1(\varphi)x, x \rangle \geq \|P_1 x\|^2,$$

$$\left\langle W_2(\varphi)\left(\sum_{i=2}^k P_i\right)x, \left(\sum_{i=2}^k P_i\right)x \right\rangle \geq \|P_2 x\|^2,$$

$$\left\langle W_3(\varphi)\left(\sum_{i=3}^k P_i\right)x, \left(\sum_{i=3}^k P_i\right)x \right\rangle \geq \|P_3 x\|^2, \tag{4.7.28a}$$

$$\cdots\cdots\cdots\cdots\cdots\cdots\cdots\cdots\cdots\cdots\cdots\cdots\cdots\cdots$$

$$\langle W_{k-1}(\varphi)(P_{k-1} + P_k)x, (P_{k-1} + P_k)x \rangle \geq \|P_{k-1}x\|^2,$$

$$\langle W_k(\varphi)P_k x, P_k x \rangle \geq \|P_k x\|^2,$$

where

$$P_1 = \text{diag}\{I_{n_1}, 0, \ldots, 0\}, \quad P_2 = \text{diag}\{0, I_{n_2}, \ldots, 0\}, \quad \ldots, \quad P_k = \text{diag}\{0, 0, \ldots, I_{n_k}\}.$$

Generalizing the inequalities (4.7.28a) to the case when the matrices P_i are not, generally speaking, projecting, we will prove the following result.

Theorem 4.7.3 *Let there exist k $(n \times n)$-dimensional symmetric matrices $S_i \in C'(\mathcal{T}_m, a)$ which satisfy the inequalities*

$$\langle W_1(\varphi)x, x \rangle \geq \|M_1(\varphi)x\|^2,$$

$$\left\langle W_2(\varphi)\left[\sum_{i=2}^{k} M_i(\varphi)\right]x, \left[\sum_{i=2}^{k} M_i(\varphi)\right]x \right\rangle \geq \|M_2(\varphi)x\|^2, \quad (4.7.29)$$

$$\cdots\cdots\cdots\cdots\cdots\cdots\cdots\cdots\cdots\cdots\cdots$$

$$\langle W_{k-1}(\varphi)[M_{k-1}(\varphi) + M_k(\varphi)]x, [M_{k-1}(\varphi) + M_k(\varphi)]x \rangle \geq \|M_{k-1}(\varphi)x\|^2,$$

$$\langle W_k(\varphi)M_k(\varphi)x, M_k(\varphi)x \rangle \geq \|M_k(\varphi)x\|^2,$$

where $(n \times n)$-dimensional matrices $M_i(\varphi) \in C^0(\mathcal{T}_m)$, $i = 1, \ldots,$ are such that its sum

$$M(\varphi) = \sum_{i=1}^{k} M_i(\varphi), \quad (4.7.30)$$

has the properties

$$\det M(\varphi) \neq 0, \quad \|M^{-1}(\varphi)\| \leq K, \quad K = \text{const} < \infty,$$
$$M(\varphi)M_1(\varphi) \equiv M_1(\varphi)M(\varphi), \quad \forall \varphi \in \mathcal{T}_m. \quad (4.7.31)$$

Then the derivative due to the system (4.7.1) of the quadratic form

$$V(\varphi, x) = \lambda^{3^{k-2}} \langle S_1(\varphi)x, x \rangle + \lambda^{3^{k-3}} \langle S_2(\varphi)x, x \rangle$$
$$+ \cdots + \lambda \langle S_{k-1}(\varphi)x, x \rangle + \langle S_k(\varphi)x, x \rangle \quad (4.7.32)$$

is positive definite for sufficiently large values of the parameter $\lambda > 0$.

Proof Considering the matrix $\bar{S}_{k-1}(\varphi) = \lambda S_{k-1}(\varphi) + S_k(\varphi)$ and taking into account the inequalities (4.7.29) we verify the inequalities

$$\langle \bar{W}_{k-1}(\varphi)[M_{k-1}(\varphi) + M_k(\varphi)]x, [M_{k-1}(\varphi) + M_k(\varphi)]x \rangle$$
$$\geq \gamma_k(\lambda) \|[M_{k-1}(\varphi)x + M_k(\varphi)x]x\|^2, \quad (4.7.33)$$

where

$$\gamma_k(\lambda) = \frac{\lambda - \|W_k\|_0 - \|W_k\|_0^2}{2(\lambda - \|W_k\|_0 + 1)}.$$

Also, the parameter $\lambda > 0$ is taken from the condition

$$\lambda - \|W_k\|_0 - \|W_k\|_0^2 > 0.$$

Displaying the left-hand side of the inequality (4.7.33) in detail and estimating it by means of the last two inequalities (4.7.29) we obtain

$$\lambda \langle W_{k-1}[M_{k-1} + M_k]x, [M_{k-1} + M_k]x \rangle + \langle W_k[M_{k-1} + M_k]x, [M_{k-1} + M_k]x \rangle$$

$$\geq \lambda \|M_{k-1}x\|^2 + \|M_k x\|^2 - \|W_k\|_0 \|M_{k-1}x\|^2 - 2\|W_k\|_0 \|M_{k-1}x\| \|M_k x\|$$

$$\geq 2\gamma_k(\lambda)(\|M_{k-1}x\|^2 + \|M_k x\|^2) \geq \gamma_k(\lambda)\|(M_{k-1} + M_k)x\|^2.$$

Consider further the matrix

$$\bar{S}_{k-2}(\varphi) = \sigma S_{k-2}(\varphi) + \bar{S}_{k-1}(\varphi) = \sigma S_{k-2}(\varphi) + \lambda S_{k-1}(\varphi) + S_k(\varphi),$$

with two so far independent parameters σ and λ. We estimate the following expression

$$\left\langle \bar{W}_{k-2}(\varphi)\left[\sum_{i=k-2}^{k} M_i(\varphi)\right]x, \left[\sum_{i=k-2}^{k} M_i(\varphi)\right]x \right\rangle$$

$$= \sigma \left\langle W_{k-2}(\varphi)\left[\sum_{i=k-2}^{k} M_i(\varphi)\right]x, \left[\sum_{i=k-2}^{k} M_i(\varphi)\right]x \right\rangle$$

$$+ \left\langle \bar{W}_{k-1}(\varphi)\left[\sum_{i=k-2}^{k} M_i(\varphi)\right]x, \left[\sum_{i=k-2}^{k} M_i(\varphi)\right]x \right\rangle$$

$$\geq \sigma \|M_{k-2}(\varphi)x\|^2 + \gamma_k(\lambda)\|[M_{k-1}(\varphi)x + M_k(\varphi)]x\|^2$$

$$- \|\bar{W}_{k-1}\|_0 \|M_{k-2}(\varphi)x\|^2 - 2\|\bar{W}_{k-1}\|_0 \|[M_{k-1}(\varphi)x + M_k(\varphi)]x\| \|M_{k-2}(\varphi)x\|$$

$$\geq \bar{\gamma}_{k-1}(\lambda, \sigma)\left\|\left[\sum_{i=k-2}^{k} M_i(\varphi)\right]x\right\|^2,$$

where

$$\bar{\gamma}_{k-1}(\lambda, \sigma) = [\gamma_k(\lambda)(\sigma - \|\bar{W}_{k-1}\|_0) - \|\bar{W}_{k-1}\|_0^2]2^{-1}(\sigma - \|\bar{W}_{k-1}\|_0 + \gamma_k(\lambda))^{-1}.$$

Since $\gamma_k(\lambda) \to 1/2$, $\|\bar{W}_{k-1}\|^2 \sim \lambda^2$ as $\lambda \to \infty$, then it is sufficient to set $\sigma = \lambda^3$ for the inequality $\bar{\gamma}_{k-1}(\lambda, \sigma) > 0$ to be satisfied. We designate

$$\gamma_{k-1}(\lambda) = \bar{\gamma}_{k-1}(\lambda, \lambda^3).$$

The next step is performed for the matrix

$$\bar{S}_{k-3}(\varphi) = \sigma S_{k-3}(\varphi) + \lambda^3 S_{k-2}(\varphi) + \lambda S_{k-1}(\varphi) + S_k(\varphi).$$

We have

$$\left\langle \bar{W}_{k-3}(\varphi)\left[\sum_{i=k-3}^{k} M_i(\varphi)\right]x, \left[\sum_{i=k-3}^{k} M_i(\varphi)\right]x \right\rangle \geq \bar{\gamma}_{k-2}(\lambda,\sigma)\left\|\left[\sum_{i=k-3}^{k} M_i(\varphi)\right]x\right\|^2,$$

where

$$\bar{\gamma}_{k-2}(\lambda,\sigma) = [\gamma_{k-1}(\lambda)(\sigma - \|\bar{W}_{k-2}\|_0) - \|\bar{W}_{k-2}\|_0^2]2^{-1}$$
$$\times (\sigma - \|\bar{W}_{k-2}\|_0 + \gamma_{k-1}(\lambda))^{-1}.$$

As $\gamma_{k-1}(\lambda) \to 1/4$ and $\|\bar{W}_{k-2}\|_0^2 \sim \lambda^6$ as $\lambda \to \infty$, then for the inequality $\gamma_{k-2}(\lambda,\sigma) > 0$ to be satisfied for sufficiently large values of λ and σ it is sufficient to let $\sigma = \lambda^{3^2}$. Also, we designate

$$\gamma_{k-2}(\lambda) = \bar{\gamma}_{k-2}(\lambda,\lambda^{3^2}).$$

Proceeding in a similar way with the consideration of the sum of the matrices $\sigma S_{k-2} + \bar{S}_{k-3}$ etc. we finally establish the following: For the matrix

$$\bar{S}_2(\varphi) = \lambda^{3^{k-3}} S_2(\varphi) + \lambda^{3^{k-4}} S_3(\varphi) + \cdots + \lambda S_{k-1}(\varphi) + S_k(\varphi),$$

the inequality

$$\langle \bar{W}_2(\varphi)[M(\varphi) - M_1(\varphi)]x, [M(\varphi) - M_1(\varphi)]x \rangle \geq \gamma_1 \|[M(\varphi) - M_1(\varphi)]x\|^2 \quad (4.7.34)$$

is valid, where the matrix $M(\varphi)$ is specified by (4.7.30).

Since the matrices $M(\varphi)$ and $M_1(\varphi)$ have the properties (4.7.31), it follows from the first inequality (4.7.29) that

$$\langle W_1(\varphi)M(\varphi)x, M(\varphi)x \rangle \geq \gamma_0 \|M_1(\varphi)x\|^2, \quad (4.7.35)$$

where $\gamma_0 = \text{const} > 0$. The last inequalities (4.7.34) and (4.7.35) allow us to conclude that the derivative of the quadratic form (4.7.32) due to the system (4.7.1) is positive definite for sufficiently large fixed values of the parameter $\lambda > 0$.

In conclusion we suggest some problems to the reader which illustrate the theory presented.

(1) Make sure that the system

$$\frac{d\varphi}{dt} = \sin\varphi, \quad \frac{dx}{dt} = (\cos\varphi)x, \quad \frac{dy}{dt} = (\sin\varphi)x - (\cos\varphi)y,$$

has a unique Green function $G_0(\tau,\varphi)$.

(2) Determine the quadratic form V, the derivative of which due to the system
$$\dot{x}_1 = (\tanh t)x_1 + 13x_3,$$
$$\dot{x}_2 = (\sin t)x_2 + x_3,$$
$$\dot{x}_3 = x_2 - (\sin t)x_3$$
is positive definite.

(3) Prove that the system
$$\dot{x}_1 = (\sin t)x_1 + 3x_2,$$
$$\dot{x}_2 = x_1 - (\sin t)x_2 + 13x_3,$$
$$\dot{x}_3 = -(\tanh t)x_3,$$
has an infinite set of the Green functions $G(t,\tau)$.

(4) Prove that the system
$$\dot{x}_1 = \left[\arctan\left(\ln\left(1+t^2\right)\right)\right]x_1 + (2+\sin t)x_2,$$
$$\dot{x}_2 = \left[1+\ln\left(1+t^2\right)\right]x_1 - \left[\arctan\left(\ln\left(1+t^2\right)\right)\right]x_2,$$
is regular on R.

(5) Verify that the system of equations
$$\frac{d\varphi}{dt} = 1,$$
$$\frac{dx_1}{dt} = (\sin\varphi)x_1 + 5(\cos^2\varphi)x_2,$$
$$\frac{dx_2}{dt} = x_1 - (\sin\varphi)x_2,$$
has a unique Green function $G_0(\tau,\varphi)$.

4.8 Comments and References

Section 4.0 The investigations made in Chapter 4 are similar to those by Friedrichs [1], Hale [1], Kulik [13, 15], Palmer [1], Saker [1] and Samoilenko and Kulik [10 – 13].

Sections 4.1 – 4.4 Theorems 4.1.1, 4.2.1, 4.2.2, 4.3.1, 4.3.2, 4.4.1 are due to Samoilenko [6].

Section 4.5 The results of this section are new.

Section 4.6 The basic idea of the proof of Theorem 4.6.1 is due to Kulik [14], A.Kulik and V.Kulik [2].

Sections 4.7 Theorems 4.7.1, 4.7.2 are due to Mitropolsky and Kulik [1].

References

Akhiezer, N. I.
[1] *Lectures on Approximation Theory.* Moscow: Nauka, 1965. [Russian]

Andronov, A. A., Vitt, A. A., and Khaikin, S. E.
[1] *Theory of Oscillations.* Reading, Mass.: Addison-Wesley, 1966.

Anosov, D. V.
[1] Geodesic currents on Rimannian manifolds of negative curvature. *Coll. Math. Inst. Acad. Nauk SSSR* **90** (1967) 1–210. [Russian]

Arnol'd V. I.
[1] *Mathematical Methods of Classical Mechanics.* (from 1974 corrected reprint of the second (1989) edition). New York: Springer-Verlag, 1995.

Bakaj, A. S. and Stepanovsky, Yu. P.
[1] *Adiabatic Invariants.* Kiev: Naukova Dumka, 1981. [Russian]

Barbashin, E. A.
[1] *Introduction to Stability Theory.* Moscow: Nauka, 1968. [Russian]
[2] *Liapunov Functions.* Moscow: Nauka, 1970. [Russian]

Barbashin, E. A. and Krasovsky, N. N.
[1] On stability motion in the whole. *Dokl. Akad. Nauk SSSR* **86** (1952) 453–456. [Russian]

Baris, Ya. S. and Fodchuk, V. I.
[1] Investigation of bounded solutions of nonlinear nonregularly perturbed systems by integral manifold method. *Ukr. Mat. Zh.* **22** (1970) 3–11. [Russian]

Bibikov, Yu. N.
[1] On existence of invariant tori in equilibrium state neighbourhood of system of differential equations. *Dokl. Akad. Nauk SSSR* **158** (1969) 9–13. [Russian]

Bibikov, Yu. N., and Pliss, V. A.
[1] On existence of invariant tori in neighbourhood of zero solution of system of ordinary differential equations. *Diff. Uravn.* **3** (1967) 1864–1881. [Russian]

Birkhoff G.
[1] *Dynamic Systems.* Providence, R.I., American Mathematical Society, 1927.

Blinov, I. N.
[1] On a problem of reducibility of systems of linear differential equations with quasi-periodic coefficients. *Izv. Akad. Nauk SSSR, Ser. Mat.* **31** (1967) 349–354. [Russian]

Bogoliubov, N. N.
[1] *On Some Statistical Methods in Mathematical Physics.* Kiev: Izdat. Akad. Nauk Ukr.SSR, 1945. [Russian]
[2] *On Quasi-Periodic Solutions in the Problems of Nonlinear Mechanics. Memoirs of the First Summer Mat. School*, Part 1. Kiev: Naukova Dumka, 1964. [Russian]
[3] *Selected Works. Part 1: Dynamical Theory.* New York: Gordon and Breach Science Publishers, 1990.

Bogoliubov, N. N. and Mitropolsky, Yu. A.
[1] The method of integral manifolds in nonlinear mechanics. *Contributions to Differential Equations*, **2** (1963) 123–196.
[2] *Asymptotic Methods in the Theory of Non-Linear Oscillations.* New York: Gordon and Breach Science Publishers, 1961.

Bogoliubov, N. N., Mitropolsky, Yu. A. and Samoilenko, A. M.
[1] *The Method of Accelerated Convergence in Non-Linear Mechanics.* Kiev: Naukova Dumka, 1969. [Russian]

Bohl, P. G.
[1] *Selected Papers.* Riga: Izdat. Akad. Nauk Latv.SSR, 1961. [Russian]
[2] Uber Differentialgleichungen. *J. Reine Angew. Math.* **144** (1914) 284–313.

Bohr, H.
[1] *Collected Mathematical Works. Almost Periodic Functions.* **2**, Kobenhavn: Dansk mathematisk rorening, 1952.

Bronstein, I. U.
[1] Linear extensions and Lyapunov functions. *Izv. Akad. Nauk MSSR, Ser. Phys.-Tekhn. Mat. Nauk* No 3 (1983) 16–20. [Russian]

Bronstein, I. U. and Cherniy, V. F.
[1] Linear extensions satisfying condition of exponential dichotomy. *Izv. Akad. Nauk MSSR. Ser. Phys.-Tekhn. Mat. Nauk* No 3 (1976) 12–16. [Russian]

Bylov, B. F.
[1] On reducibility of a system of linear equations to diagonal form. *Mat. Sb.* No 3 (1965) 338–344. [Russian]

Bylov, B. F., Vinograd, R. E., Lin, V. Ya. and Lokutsievsky, O. V.
[1] On topological causes of anomalous behaviour of some almost periodic systems. *Problems of Asymptotic Theory of Nonlinear Oscillations.* (Eds.:...), Kiev: Naukova Dumka, (1977) pp. 54–61. [Russian]

Cesari, L.
[1] *Asymptotic Behaviour and Stability Problems in Ordinary Differential Equations.* Berlin, etc.: Springer-Verlag, 1971.

Cheban, D. M.
[1] Non-autonomous dissipative dynamic systems. *Dokl. Akad. Nauk SSSR* **286** (1968) 824–827. [Russian]

Cheresiz, V. M.
[1] Stable and conditionally stable almost periodic solutions of Y-monotone systems. *Sib. Mat. Zh.* **15** (1974) 162–176. [Russian]

Coppel, W. A.
[1] *Dichotomies in Stability Theory.* Berlin: Springer-Verlag, 1978.
[2] Dichotomies and Lyapunov Functions. *J. Differential Eqns* **52** (1984) 58–65.
[3] Dichotomies and reducibility. *J. Differential Eqns* **3** (1967) 500–521; II, ibid **4** (1968) 386–398.

Daletsky, Yu. L. and Krein, M. G.
[1] *Stability of Solutions of Differential Equations in Banach Space.* Moscow: Nauka, 1970. [Russian]

Demidovich, B. P.
[1] *Lectures on Mathematical Theory of Stability.* Moscow: Nauka, 1967. [Russian]

Erugin, N. P.
[1] A. M. Lyapunov methods and problems of stability in the whole. *Prikl. Mat. Mekh.* **17** (1953) 389–400. [Russian]
[2] A. M. Lyapunov's first method. Mech. *Mechanics in SSSR with 50 Years.* (Ed. L. I .Sedov). Moscow: Nauka, **1** (1967) pp, 67–86. [Russian]

Feshchenko, S. F., Shkil', N. I. and Nikolenko, L. D.
[1] *Asymptotic Methods in the Theory of Linear Differential Equations.* Kiev: Naukova Dumka, 1966. [Russian]

Filatov, A. N. and Sharova, L. V.
[1] *Integral Inequalities in the Theory of Nonlinear Oscillations.* Moscow: Nauka, 1976. [Russian]

Friedrichs, K. O.
[1] Symmetric hyperbolic linear differential equations. *Com. Pure Appl. Math.* **7** (1954) 345–392.

Gantmacher, F. R.
[1] *The Theory of Matrices.* **1, 2**. New York: Chelsea Publ. Co.,1974.

Golets, B. I. and Kulik, V. L.
[1] To the problem on dichotomy for solutions of systems of linear differential equations. *Ukr. Mat. Zh.* **24** (1972) 528–531. [Russian]

Grebenikov, E. A. and Riabov, Yu. A.
[1] *New Qualitative Methods in Celestial Mechanics.* Moscow: Nauka, 1971. [Russian]

Griebienikow, J. A. and Riabov, J. A.
[1] *Metoda Usrednienia w Mechanice Nieliniiowej.* Warsawa: PWN, 1982. [Polska]

Grujić, Lj. T., Martynyuk, A. A. and Ribbens-Pavella, M.
[1] *Large Scale Systems Stability under Structural and Singular Perturbations.* Kiev: Naukova Dumka, 1984. [Russian] English Translation, New York: Springer-Verlag, 1987.

Hale, J. K.
[1] Integral manifolds of perturbed differential systems. *Annals of Math.* **73** (1961) 496–531.
[2] *Oscillations in Nonlinear Systems.* New York: McGraw-Hill Company, Inc., 1963.

Hartman, P.
[1] *Differential Equations.* New York, etc.: John Wiley and Sons, 1964.

Hayashi, C.
[1] *Nonlinear Oscillations in Physical Systems.* New York, etc.: McGraw-Hill Company, Inc., 1964.

Izobov, N. A.
[1] Low estimate for minimal index of linear systems. *Diff. Uravn.* **14** (1978) 1576–1588. [Russian]
[2] To the theory of Lyapunov characteristic indices of linear and quasi-linear differential systems. *Mat. Sb.* **28** (1980) 459–476. [Russian]

Kolmogorov, A. N. and Fomin, S. V.
[1] *Elements of Theory of Functions and Functional Analysis.* Moscow: Nauka, 1968. [Russian]

Krasnosel'sky, M. A., Burd, V. Sh. and Kolesov, Yu. S.
[1] *Nonlinear Almost Periodic Oscillations.* Moscow: Nauka, 1970. [Russian]

Krein, M. G.
[1] *Lectures on Stability Theory of Solutions of Differential Equations in Banach Space.* Kiev: Izdat. Inst. Mat. Akad. nauk Ukr.SSR, 1964. [Russian]

Kryloff, N. and Bogoliuboff N.
[1] *New Methods of Nonlinear Mechanics.* Moscow-Leningrad: ONTI, 1934. [Russian]
[2] *Introduction to Non-Linear Mechanics.* Princeton University Press, 1947.

Kulik, V. L.
[1] Three-block decomposibility of linear extensions of dynamic systems on torus. *Coll.: Some Problems of the Theory of Asymptotic Methods of Nonlinear Mechanics.* (Ed. Mitropolsky, Yu. A.). Kiev: Izdat. Inst. Mat. Akad. Nauk Ukr. SSR (1976) 124–130. [Russian]
[2] Dependence of Green function of the problem on invariant torus on a parameter. *Ukr. Mat. Zh.* **30** (1978) 545–551. [Russian]
[3] Linear algebraic systems of equations with smooth coefficients given on m-dimensional torus. *Coll.: Analytical Methods in Nonlinear Mechanics.* (Ed. Mitropolsky, Yu. A.). Kiev: Izdat. Inst. Mat. Akad. Nauk Ukr.SSR (1981) 54–62.[Russian]
[4] Quadratic forms and dichotomy of solutions of systems of linear differential equations. *Ukr. Mat. Zh.* **34** (1982) 43–49. [Russian]
[5] Reversibility of a theorem on decomposibility of linear extensions of dynamic systems on torus. *Ukr. Mat. Zh.* **35** (1983) 67–72. [Russian]
[6] On solvability of linear algebraic systems with variable coefficients given on torus. *Coll.: Approximate Methods for Investigating Nonlinear Oscillations.* (Ed.Mitropolsky, Yu. A.). Kiev: Izdat. Inst. Mat. Akad. Nauk Ukr.SSR (1983) 112–120. [Russian]
[7] Green function of a problem on bounded solutions for linear systems of differential equations. *Diff. Uravn.* **20** (1984) 570–577. [Russian]
[8] On a connection between quadratic forms and Green function of linear extension of dynamic systems on torus. *Ukr. Mat. Zh.* **36** (1984) 258–262. [Russian]
[9] Weakly regular linear systems of differential equations. *Ukr. Mat. Zh.* **37** (1985) 501–506. [Russian]
[10] Regularity of systems of linear differential equations of block-triangular form. *Mat. Zametki* **40** (1986) 484–491. [Russian]
[11] Sign variable Lyapunov functions and preservation of invariant tori under perturbations. *Ukr. Mat. Zh.* **39** (1987) 45–52. [Russian]
[12] Bounded solutions of systems of linear differential equations. *Ukr. Mat. Zh.* **39** (1987) 727–732. [Russian]
[13] Of the existence of Lyapunov functions with variable sign for linear expansions of dynamical systems on torus. *Ukr. Mat. Zh.* **42** (1990) 1747–1751.

[Russian]
[14] Regular in R linear systems of differential equations. *Diff. Uravn.* **29** (1993) 1699–1704. [Russian]
[15] On some classes of regular linear expansions of dynamical systems. *Ukr. Mat. Zh.* **46** (1994) 1479–1485. [Russian]

Kulik, V. L. and Eremenko, V. A.
[1] On quasi-periodic solutions of linear systems of differential equations with a degenerating matrix at a derivative. *Ukr. Mat. Zh.* **32** (1980) 746–753. [Russian]

Kulik, A. N. and Kulik, V. L.
[1] Lyapunov functions and dichotomy on semi-axes of linear systems. *Diff. Uravn.* **20** (1984) 233–241. [Russian]
[2] Regular linear expansions of dynamical systems on torus. *Ukr. Mat. Zh.* **45** (1993) 1170–1173. [Russian]

Kurosh, A. G.
[1] *A Course of Higher Algebra.* Moscow: Nauka, 1986. [Russian]

Kurzweil, J.
[1] Invariant manifolds in the theory of functional differential equations. *Int. Conf. Nonlinear Osc.* E. Berlin, September 1975.

Lappo-Danilevsky, I. A.
[1] *Applications of Functions of Matrices to the Theory of Linear Systems of Ordinary Differential Equations.* Moscow: Gostekhizdat, 1957. [Russian]

LaSalle, J. P. and Lefschetz, S.
[1] *Stability by Lyapunov's Direct Method with Applications.* New York: Academic Press, !966.

Letov, A. M.
[1] *Dynamics of a Flight and Control.* Moscow: Nauka, 1969. [Russian]

Levitan, B. M.
[1] *Almost-Periodic Functions.* Moscow: Gostekhizdat, 1953. [Russian]

Lomov, S. A.
[1] *Introduction in the General Theory of Singular Perturbations.* Providence, RI.: American Mathematical Society, 1992.

Lyapunov, A. M.
[1] *The General Problem of the Stability of Motion.* Kharkov: Kharkov Mathematical Society, 1892. [Russian] English Translation by Tylor and Francis, London, 1992.

Lyashchenko, N. Ya.
[1] On asymptotic stability of solutions of systems of differential equations. *Dokl. Akad. Nauk SSSR* **96** (1954) 237–239. [Russian]

Majzel, A. D.
[1] On stability of solutions of systems of differential equations. *Trudy Ural. Politekh. In-ta, Ser. Mat.* **51** (1954) 20–50. [Russian]

Malkin, I. G.
[1] *Some Problems of the Theory of Nonlinear Oscillations.* Moscow: Gostekhizdat, 1956. [Russian]
[2] Theory of stability of motion. Moscow: Nauka, 1966. [Russian]

Mandel'shtam, L. L.
[1] *Lectures on the Theory of Oscillations.* Moscow: Nauka, 1972. [Russian]

Martynyuk, A. A.
[1] *Stability Analysis: Nonlinear Mechanics Equations.* (Stability and Control: Theory, Methods and Applications Series, ISSN 1023–6155; Vol.2), New-York, Gordon and Breach Publishers, 1995.

Martynyuk, A. A. and Gutowski, R.
[1] *Integral Inequalities and Stability of Motion.* Kiev: Naukova Dumka, 1979. [Russian]

Martynyuk, A. A. and Obolensky, A. Yu.
[1] On stability of Wazewsky autonomous systems. *Diff. Uravn.* **16** (1980) 1392–1407. [Russian]

Martynyuk, D. I. and Tsyganovsky, N. S.
[1] Invariant manifolds of systems with delay under impulse impact. *Diff. Uravn.* **15** (1979) 1783–1795. [Russian]

Massera, J. L. and Schaeffer, J. J.
[1] *Linear Differential Equations and Function Space.* New York, etc.: Academic Press, 1966.

Mel'nikov, G. I.
[1] Some problems of Lyapunov direct method. *Dokl. Akad. Nauk SSSR* **110** (1956) 326–330. [Russian]

Millionshchikov, V. M.
[1] Proof of attainability central indices of linear systems. *Sib. Mat. Zh.* **10** (1969) 99–104. [Russian]

Mitropolsky, Yu. A.
[1] *Problems of the Asymptotic theory of Nonstationary Vibrations.* Jerusalem: Israel Program for Scientific Translations, 1965.
[2] On a construction of general solution of nonlinear differential equations by the method providing "accelerated" convergence. *Ukr. Mat. Zh.* **16** (1964) 475–501. [Russian]
[3] *Averaging Method in Nonlinear Mechanics.* Kiev: Naukova Dumka, 1971. [Russian]

[4]   The method of integral manifolds in the theory of nonlinear oscillations. *Proc. of Inter. Simpos. on Nonlinear Differential Equations and Nonlinear Mechanics.* (Eds.: LaSalle, P. and Lefschetz, S.). New York, London: Academic Press. XIV (1963) 1–15.

Mitropolsky, Yu. A. and Kulik, V. L.
[1]   Invariant manifolds of autonomous differential equations and Lyapunov functions with variable sign. *Diff. Uravn.* **28** (1992) 398–405. [Russian]
[2]   Lyapunov functions and bounded solutions of linear systems of differential equations. *Ukr. Mat. Zh.* **38** (1985) 726–787. [Russian]
[3]   On application of quadratic forms in the theory of invariant manifolds. *Ukr. Mat. Zh.* **37** (1985) 306–316. [Russian]
[4]   Differentiable invariant manifolds of dynamical systems. *Diff. Uravn.* **22** (1986) 1523–1532. [Russian]

Mitrpolsky, Yu. A. and Lykova, O. B.
[1]   *Integral Manifolds in Nonlinear Mechanics.* Moscow: Nauka, 1973. [Russian]

Mitropolsky, Yu. A. and Samoilenko, A. M.
[1]   *Some Problems of the Theory of Multifrequency Oscillations.* Preprint Akad. Nauk Ukr.SSR, Inst. Mat. 77.14 (1977). [Russian]

Mitropolsky, Yu. A., Samoilenko, A. M. and Kulik, V. L.
[1]   *Investigations of Linear Systems of Differential Equations with the Help of Quadratic Forms.* Preprint Akad. Nauk Ukr.SSR, Inst. Mat. 82.10 (1982). [Russian]
[2]   Application of quadratic forms to investigation of systems of linear differential equations. *Diff. Uravn.* **21** (1985) 726–787. [Russian]

Mitropolsky, Yu. A., Samoilenko, A. M. and Martynyuk, D. I.
[1]   *System of Evolutionary Equations with Periodic and Conditionally Periodic Coefficients.* Kiev: Naukova Dumka, 1984. [Russian]

Mishchenko, E. F. and Rozov, N. Kh.
[1]   *Differential Equations with a Small Parameter and Relaxational Oscillations.* Moscow: Nauka, 1975. [Russian]

Mozer, J.
[1]   Fast converging of iteration method and nonlinear differential equations. *Usp. Mat. Nauk* **23** (1968) 179–238. [Russian]

Molchanov, A. M.
[1]   Uniform asymptotic of linear systems. *Dokl. Akad. Nauk SSSR* **173** (1967) 519–522. [Russian]

Mukhamadiev, E. O.
[1]   On construction of functions with sign variable derivative along trajectories of autonomous systems. *Dokl. Akad. Nauk SSSR* **26** (1983) 333–336. [Russian]

Myshkis, A. D.
[1] On one geometric lemma and its application to Lyapunov stability theory. *Dokl. Akad. Nauk SSSR* **55** (1947) 299–302. [Russian]

Neimark, Yu. I.
[1] *Method of Point Mappings in the Theory of Nonlinear Oscillations.* Moscow: Nauka, 1972. [Russian]

Nemytsky, V. V.
[1] Oscillations in autonomous systems. *Coll.: The V-th Summer Mat. School.* Kiev: Naukova Dumka, (1968) 436–472. [Russian]

Nemytsky, V. V. and Stepanov, V. V.
[1] *Qualitative Theory of Differential Equations.* Moscow-Leningrad: Gostekhizdat, 1947. [Russian]

Ordynskaya, Z. P. and Kulik, V. L.
[1] On exponential dichotomy of systems of differential equations. *Sib. Mat. Zh.* **24** (1983) 128–135. [Russian]

Osipenko, G. S.
[1] Perturbations of dynamic systems in a neighbourhood of invariant manifolds. *Diff. Uravn.* **16** (1980) 620–628. [Russian]

Palmer, K. J.
[1] Exponential dichotomy and Fredholm operators. *Proc. Amer. Math. Soc.* **104** (1988) 149–156.
[2] A diagonal dominance criterion for exponential dichotomy. *Bull. Austral. Math. Soc.* (to appear).

Perestyuk, N. A.
[1] Invariant sets of one class of discontinuous dynamic systems. *Ukr. Mat. Zh.* **36** (1984) 63–69. [Russian]

Perov, A. I.
[1] *Variational Methods in the Theory of Nonlinear Oscillations.* Voronezh: Izdat. Voronezh. Un-ta, 1981. [Russian]

Perov, A. I. and Trubnikov, Yu. V.
[1] Monotone differential equations, IV. *Diff. Uravn.* **14** (1978) 1190–1202. [Russian]

Persidsky, K. P.
[1] *Stability Theory of Solutions of Differential Equations. Probabilities Theory, Collected Works.* Alma-Ata: Nauka, 1976. [Russian]

Persidsky, S. K.
[1] On some theorems of Lyapunov's second method. *Diff. Uravn.* **5** (1969) 678–687. [Russian]

Pliss, V. A.
[1] *Non-Local Problems of Oscillation Theory.* Moscow: Nauka, 1964. [Russian]
[2] Bounded solutions of nonhomogeneous linear systems of differential equations. *Coll.: Problems of Asymptotic Theory of Nonlinear Oscillations.* (Ed. ...). Kiev: Naukova Dumka, 1977, pp. 168–173. [Russian]

Povzner, Ya.
[1] Existence theory in the whole for nonlinear system and index of error of nonlinear operator. *Sib. Mat. Zh.* **5** (1964) 377–386. [Russian]

Poincaré, H.
[1] Sur les courbes définies par une équation différentielle. *J. de Mathématiques, série 3,* **7** (1881–1882) 375–422; *série 8* (1882) 251–296.
[2] *Selected Papers.* Moscow: Nauka, 1971. [Russian]

Rozov, N. Kh.
[1] Asymptotic calculation of close to discontinuous periodic solutions of systems of the second order differential equations. *Dokl. Akad. Nauk SSSR* **145** (1962) 38–40. [Russian]

Rejzin', L. E.
[1] *Local Equivalence of Differential Equations.* Riga: Zinatne, 1971. [Russian]

Rouche, N., Habets, P. and Laloy, M.
[1] *Stability Theory by Liapunov's Direct Method.* New York, etc.: Springer-Verlag, 1977.

Rozenvasser, E. N.
[1] *Lyapunov Indices in the Theory of Linear Systems of Control.* Moscow: Nauka, 1977. [Russian]

Rumjantsev, V. V.
[1] Method of Lyapunov functions in the theory of stability of motion. *Coll.: Mechanics in USSR for 50 Years.* (Ed. L. I. Sedov). Moscow: Nauka, (1968) 7–66. [Russian]

Saker, R. J. and Sell, G. R.
[1] A spectral theory for linear differential systems. *J. Diff. Eqns* **27** (1978) 320–358.

Samoilenko, A. M.
[1] To perturbation theory of invariant manifolds of dynamic systems. *Coll.: Proc. of the V-th Intern. Conference on Nonlinear Oscillations.* Kiev: Izdat. Inst. Mat. Akad. Nauk Ukr.SSR 1 (1970) 495–499. [Russian]
[2] On preservation of invariant torus under perturbations. *Izv. Akad. Nauk USSR. Ser. Mat.* **34** (1970) 1219–1240. [Russian]
[3] Quasi-periodic solutions of systems of linear algebraic equations with quasi-periodic coefficients. *Coll.: Analytical Methods of Investigating the Solutions of Nonlinear Differential Equations.* (Ed. Mitropolsky, Yu. A.). Kiev: Izdat.

Inst. Mat. Akad. Nauk Ukr.SSR (1975) 5–26. [Russian]
[4] Invariant toroidal manifolds of systems with slowly changing variables. *Coll.: Problems of Asymptotic Theory of Nonlinear Oscillations.* (Ed. Mitropolsky, Yu. A.). Kiev: Naukova Dumka, (1977) 181–191. [Russian]
[5] *Elements of Mathematical Theory of Multifrequency Oscillations.* Moscow: Nauka, 1987. [Russian]
[6] Linear expansions of dynamical systems on torus and Green functions. *Ukr. Mat. Zh.* **42** (1990) 1219–1224. [Russian]
[7] On some problems of perturbation theory of smooth invariant tora of dynamical systems. *Ukr. Mat. Zh.* **46** (1994) 1665–1669. [Russian]

Samoilenko, A. M. and Kulik, V. L.
[1] On Green function existence of the problem on invariant torus. *Ukr. Mat. Zh.* **27** (1975) 348–359. [Russian]
[2] On e-dichotomy and decomposibility of linear systems of equations. *Diff. Uravn.* **15** (1979) 755–756. [Russian]
[3] Exponential dichotomy of invariant torus of dynamic systems. *Diff. Uravn.* **15** (1979) 1434–1444. [Russian]
[4] On decomposibility of linearized systems of differential equations. *Ukr. Mat. Zh.* **34** (1982) 587–597. [Russian]
[5] On continuity of Green function problem on invariant torus. *Ukr. Mat. Zh.* **30** (1978) 779–788. [Russian]
[6] Decomposibility of linear extensions of dynamic systems on torus. *Dokl. Akad. Nauk Ukr.SSR* No 12 (1984) 23–27. [Russian]
[7] Sets of invariant tori of dynamic systems linear extensions. *Dokl. Akad. Nauk Ukr.SSR* No 8 (1985) 15–19. [Russian]
[8] Decomposibility problems of linear extensions of dynamic systems on torus. *Coll.: Approximate Methods for Analysis of Nonlinear Oscillations.* (Ed. Mitropolsky, Yu. A.). Kiev: Izdat. Inst. Mat. Akad. nauk Ukr.SSR (1984) 103–112. [Russian]
[9] Preservation conditions of invariant tori under perturbations. *Diff. Uravn.* **22** (1986) 1095–1096. [Russian]
[10] On regular over the whole axis linear systems of differential equations. *Diff. Uravn.* **26** (1990) 1338–1343. [Russian]
[11] Linear expansions of dynamical systems on torus and Green functions. *Ukr. Mat. Zh.* **42** (1990) 1219–1224. [Russian]
[12] To the problem on existence of Lyapunov Functions with variable sign for linear systems of differential equations. *Ukr. Mat. Zh.* **44** (1992) 114–120. [Russian]
[13] On regularity of differential equations linearized with respect to a part of variables. *Diff. Uravn.* **31** (1995) 773–777. [Russian]

Sobolev, S. L.
[1]   *Mathematical Physics Equations.* Moscow: Nauka, 1966. [Russian]

Starzhinsky, V. M. and Yakubovich, V. A.
[1]   *Linear Differential Equations with Periodic Coefficients.* Moscow: Nauka, 1972. [Russian]

Sternberg, S.
[1]   *Lectures on Differential geometry.* New York: Bhelsea, 1983.

Stepanov, V. V.
[1]   *The Course of Differential Equations.* Moscow: Fizmatgiz, 1959. [Russian]

Shkil', N. I.
[1]   *Asymptotic Methods in Differential Equations.* Kiev: Vyshcha shkola, 1971. [Russian]

Tikhonov. A. N.
[1]   On systems of differential equations containing parameters. *Mat. Sb.* **27** (1950) 147–156. [Russian]

Tkachenko, V. I.
[1]   On block diagonalization of almost-periodic systems. *Dokl. Akad. Nauk Ukr.SSR, Ser. A* No 6 (1983) 18–20. [Russian]

Trofimchuk, S. I.
[1]   Necessary conditions for existence of invariant manifold of dynamic systems linear extension on compact manifolds. *Ukr. Mat. Zh.* **36** (1984) 390–393. [Russian]

Vasil'eva, A .B. and Butuzov, V. F.
[1]   *Asymptotic Expansions of Solutions of Singularly Perturbed Equations.* Moscow: Nauka, 1973. [Russian]

Vishik, M. I. and Lyusternik, L. A.
[1]   Regularly degeneration and boundary layer for linear differential equations with a small parameter. *Usp. Mat. Nauk* **12** (1957) 3–122. [Russian]

Zadiraka, K. V.
[1]   On integral manifolds of a system of differential equations containing a small parameter. *Dokl. Akad. Nauk SSSR* **115** (1957) 646–649. [Russian]

Zhautykov, O. A.
[1]   On application of averaging method to a solution of one partial differential equations in oscillation theory. *Coll.: Approximate Methods of Solution of Differential Equations.* Kiev: Naukova Dumka, (1964) 52–61. [Russian]

Zhikov, V. V. and Levitan, B. M.
[1]   Favard theory. *Usp. Mat. Nauk* **32** (1977) 123–171. [Russian]

Zubov, V. I.
[1]   *Theory of Oscillations.* Moscow: Vyschaja shkola, 1979. [Russian]

Index

A

Arbitrary matrix of the orthogonal projection 63

B

Block diagonal form 28, 276
— diagonal system 29, 276
— triangular form 29, 55
Bohl's theorem 100
Bounded-invariant manifold 235
Boundedness of the matrix function 20

C

C'-block-split 270
C'-block-splittability 270
Canonical form 6
Cauchy problem 86, 138, 232
Closed sphere 8
Conic set 6

D

Decomposed system 34, 47
Degenerate invariant torus 116, 125
Derivative of definite sign 23, 76, 78
— to be negative definite 26, 76, 78
Derivatives of the Green function in the variables φ 218
Diagonal matrix functions 29

E

Eigenvalues of the matrix 36, 67
Exponential dichotomy (e-dichotomy) 1, 2, 29, 51, 78, 253, 317
— — of system 24, 133
— — of the invariant torus 125
— — of the trivial torus 125, 132, 137
— behaviour of solutions 42
Exponentially damping 74, 82
— dichotomous of systems 22, 23, 42, 77, 78, 81, 88
— dichotomous on the semiaxes 3, 22, 27, 29, 37
— dichotomous on the whole R axis 3, 29, 36, 50, 51, 57, 81, 88, 132, 141
— increasing 74
— stable invariant torus 112, 114, 137

F

Floquet-Lyapunov theory 117

G

General damping index 328
Green function
— — of the problem on solutions bounded on R 15
— — of the problem on invariant tori 142, 167, 171
— — of the problem on bounded invariant manifolds 232

— — non-unique   94, 154, 328, 330
— — of the system   134, 324, 328
— — derivatives in variables φ 218
— — derivatives in the parametr μ 211
— — of the exponentially dichotomous system   321
— — unique   163

H
Heine-Borel lemma   171
Hyperplane   6

I
Inhomogeneous equation   50
— system   3, 14, 22, 34, 49, 55
Invariant manifold   233, 235
— tori of dynamical systems   93
— torus   95, 123, 160, 161, 175
— — unique   169

J
Jordan form   126, 137, 291

L
Linear independent solutions   18
Linear extension of dynamical system   99
Lyapunov functions   1, 125, 138, 169
— matrix   12, 21, 23
— transformation   22, 29
— variables   12, 24, 28, 35, 40, 47, 49, 53

M
m-dimensional torus   93, 94
Matriciant   12, 72
Matrix of the orthogonal projection   21
— functions   15, 18
— non-degenerate   57
— — symmetric   316
— quadratic   13

N
Non-degenerate invariant torus   116
Non-trivial solutions   24
— torus   159, 161
Norm of the matrix functions   26

O
Orthogonal change of variables   6
Osgood theorem   138
Ostrogradsky–Liouville relation   98

P
Perron theorem   72
Perturbed system   83
Projection matrix   16, 21, 56
Problem on invariant tori   142

Q
Quadratic form   1, 2, 6, 23, 27
— — non-degenerate for all t 51

R
Rank of matrix   31
Riccati type equation   215

S
Schmidt orthogonal process   19
Separatrix invariant manifold   301
Solutions damps exponentially on $\pm\infty$ 18
Split form   262
Steklov functions   183
Strictly monotonous decreasing   31
Strict monotonicity of the functions   31
Structure of the Green function   168
Subspace   15
System of equations be exponentially stable   114
— — exponentially dichotomous   49

T
Theory of multifrequence oscillations   93
Toroidal manifold   215
Triangular form   28, 72
Trivial Torus   137

V
Vector-function   15

W
Weakly regular on R 37, 41, 335, 339
— — on the whole R axis   37

Other titles in the Stability and Control: Theory, Methods and Applications series.

Volume 13
Advances in Stability Theory at the End of the 20th Century
A. A. Martynyuk

Volume 14
Dichotomies and Stability in Nonautonomous Linear Systems
Yu. A. Mitropolsky, A. M. Samoilenko and V. L. Kulik

Volume 15
Almost Periodic Solutions of Differential Equations in Banach Spaces
Y. Hino, T. Naito, Nguyen Van Minh and Jong Son Shin

Volume 16
Functional Equations with Causal Operators
C. Corduneanu

Volume 17
Optimal Control of Growth of Wealth of Nations
E. N. Chukwu

Volume 18
Stability and Stabilization of Nonlinear Systems with Random Structure
I. Ia. Kats & A. A. Martynyuk

Volume 19
Lyapunov Functions in Differential Games
V. I. Zhukovskiy

Volume 20
Stability of Differential Equations with Aftereffect
N. V. Azbelev & P. M. Simonov